Surveys in Differential Geometry

Vol. 1: Lectures given in 1990
edited by S.-T. Yau and H. Blaine Lawson

Vol. 2: Lectures given in 1993
edited by C.C. Hsiung and S.-T. Yau

Vol. 3: Lectures given in 1996
edited by C.C. Hsiung and S.-T. Yau

Vol. 4: Integrable systems
edited by Chuu Lian Terng and Karen Uhlenbeck

Vol. 5: Differential geometry inspired by string theory
edited by S.-T. Yau

Vol. 6: Essay on Einstein manifolds
edited by Claude LeBrun and McKenzie Wang

Vol. 7: Papers dedicated to Atiyah, Bott, Hirzebruch, and Singer
edited by S.-T. Yau

Vol. 8: Papers in honor of Calabi, Lawson, Siu, and Uhlenbeck
edited by S.-T. Yau

Vol. 9: Eigenvalues of Laplacians and other geometric operators
edited by A. Grigor'yan and S-T. Yau

Vol. 10: Essays in geometry in memory of S.-S. Chern
edited by S.-T. Yau

Vol. 11: Metric and comparison geometry
edited by Jeffrey Cheeger and Karsten Grove

Vol. 12: Geometric flows
edited by Huai-Dong Cao and S.-T. Yau

Vol. 13: Geometry, analysis, and algebraic geometry
edited by Huai-Dong Cao and S.-T.Yau

Vol. 14: Geometry of Riemann surfaces and their moduli spaces
edited by Lizhen Ji, Scott A. Wolpert, and S.-T. Yau

Volume VIII

Surveys in Differential Geometry

Lectures on geometry and topology held in honor of Calabi, Lawson, Siu, and Uhlenbeck

edited by

Shing-Tung Yau

International Press
www.intlpress.com

Surveys in Differential Geometry, Vol. 8

Editor:
Shing-Tung Yau, Harvard University

ISBN 978-1-57146-179-7

Paperback reissue 2010. Previously published in 2003 under ISBN 1-57146-114-0 (clothbound).

Typeset using the LaTeX system.

Contents

Projective Planes, Severi Varieties and Spheres

Michael Atiyah and Jürgen Berndt

ABSTRACT. A classical result asserts that the complex projective plane modulo complex conjugation is the 4-dimensional sphere. We generalize this result in two directions by considering the projective planes over the normed real division algebras and by considering the complexifications of these four projective planes.

1. Introduction

There is an elementary but very striking result which asserts that the quotient of the complex projective plane CP^2 by complex conjugation is the 4-dimensional sphere. This result has attracted the attention of many geometers over the years, rediscovered afresh each time and with a variety of proofs. For some historical remarks about the origins of this theorem we refer to Arnold [**3**], where it is put in a more general context.

The purpose of the present paper is to give two parallel generalizations of this theorem. In the first we view CP^2 as the second member of the family of projective planes over the four normed real division algebras. This is close to Arnold's treatment and completes it by dealing with the octonions. The second generalization views CP^2 as the complex algebraic variety obtained by complexifying the real projective plane RP^2 and hence as the first member of the four algebraic varieties got by complexifying the four projective planes. This family of algebraic varieties has appeared in several contexts. First, in relation to Lie groups and the "magic square" of Freudenthal [**13**], somewhat independently in the characterization by Zak [**26**] of "Severi varieties", and also in the classification by Nakagawa and Takagi [**19**] of Kähler submanifolds with parallel second fundamental form in complex projective spaces. Since these varieties are not so widely known we shall give a brief account of them in an appendix. An extensive treatment based on the "magic square" can be found in [**15**].

Denoting by CP^2, HP^2 and OP^2 the projective planes over the complex numbers C, the quaternions H and the octonions O, and by S^d the sphere of dimension d, our first result may be formulated as follows.

There are natural diffeomorphisms

$$\begin{aligned} CP^2/O(1) &= S^4 , \\ HP^2/U(1) &= S^7 , \\ OP^2/Sp(1) &= S^{13} . \end{aligned} \tag{1.1}$$

REMARKS. 1) The first equality is the "folklore" theorem which provides our starting point. The second one is proved in [**3**] and independently in [**5**].

2) The maps from the projective planes to the spheres in (1.1) are fibrations outside the "branch locus" given by the preceding projective plane. The sense in which the equations in (1.1) are diffeomorphisms is explained in the next section.

3) The embeddings of the branch loci

$$RP^2 \subset S^4 \ , \ CP^2 \subset S^7 \ , \ HP^2 \subset S^{13} \tag{1.2}$$

are well-known and will be elaborated on in Section 3.

4) We shall later in Section 4 formulate (1.1) more precisely as Theorem A. This will include an explicit and simple construction of the maps from the projective planes to the spheres. These maps will also be compatible with the relevant symmetry group $SO(3)$, $SU(3)$ and $Sp(3)$.

Before formulating our second result we need to introduce the *complexifications* of the four projective planes. The first case is clear, it leads from RP^2 to CP^2. Note that the action of $SO(3)$ on RP^2 extends to a complex action of $SO(3,C)$ on CP^2 and this leaves invariant the complex curve $z_1^2 + z_2^2 + z_3^2 = 0$ which, for reasons that will be clear later, we denote by $CP^2(\infty)$. All this generalizes to the other projective planes. If we denote the projective planes by P_n $(n = 0,1,2,3)$, so that $\dim P_n = 2^{n+1}$, then their complexifications[1] $P_n(C)$ are complex algebraic varieties of complex dimension 2^{n+1}. The isometry group of P_n extends to an action of its complexification on $P_n(C)$ leaving invariant a complex hypersurface $P_n(\infty)$. Moreover $P_n(C)$ has a "real structure", i.e. a complex conjugation. The real points are just P_n, and $P_n(\infty)$ inherits a real structure with no real points. We can now state our second result.

For each $n = 0,1,2,3$ *we have a natural map*

$$\varphi_n : P_n(C) \to S^{d(n)} \ , \ d(n) = 3 \cdot 2^n + 1 \ , \tag{1.3}$$

which is a fibration outside the branch locus P_n *and the hypersurface* $P_n(\infty)$. *The fibres are the spheres* S^0, S^1, S^3, S^7.

REMARKS. 1) The case $n = 0$ of (1.3) is just the first case of (1.1).

2) The image of $P_n(\infty)$ under φ_n is the focal set of the branch locus P_n in $S^{d(n)}$ and turns out to be another embedding of P_n in $S^{d(n)}$: it is just the image under the antipodal map of the embedding $\varphi_n(P_n)$. For $n < 3$ this is just the embedding in (1.2), and for $n = 3$ it is the known fourth member of this sequence (as will be explained in Section 4).

3) The varieties $P_n(C)$ are in fact homogeneous spaces of a larger group as we shall explain later. For $n = 1$ the complexification of CP^2 is just the product $CP^2 \times CP^2$, while for $n = 2$ the complexification of HP^2 turns out to be the complex Grassmannian $Gr_2(C^6)$ of 2-planes in C^6, and for $n = 3$ the complexification of OP^2 is the exceptional Hermitian symmetric space $E_6/Spin(10)U(1)$.

---

[1] $P_n(C)$ should not be confused with the complex projective space of dimension n, for which we use the notation CP^n.

The paper is organized as follows. In Section 2 we review well-known elementary properties of the normed real division algebras and the associated geometries. In particular we explain the general notion of a branched fibration of which (1.3) is an illustration. In Section 3 we move on to projective planes and their automorphisms, where we pay special attention to the Cayley plane and its exceptional status. Section 4 examines the orbit structure for the projective planes and spheres in (1.1) with respect to the relevant symmetry groups. This enables us to establish (1.1). Section 5 is devoted to the formulation and proof of Theorem A, the more precise version of (1.1), while Section 6 deals similiarly with Theorem B, a more explicit version of (1.3). In Section 7 we establish projective versions of the two theorems, where the symmetry group is replaced by the larger group of projectivities of the relevant projective planes. In the appendix we elaborate more on the complexified projective planes $P_n(C)$.

We thank Friedrich Hirzebruch and Jean-Pierre Serre for drawing our attention to some of the relevant literature.

2. Projective Lines, Hopf Maps and Branching

We recall that there are four normed division algebras A_n over the reals, with $\dim A_n = 2^n$, namely

$n = 0$: R, the real field, of dimension 1;
$n = 1$: C, the complex field, of dimension 2;
$n = 2$: H, the non-commutative field of quaternions, of dimension 4;
$n = 3$: O, the non-associative algebra of octonions, of dimension 8.

To each of these we can associate the corresponding projective line A_nP^1 by adding a point ∞ to the algebra. This shows that the projective line is a sphere. The more classical definition of a projective line is to consider all lines (i.e. one-dimensional subspaces) in the 2-dimensional vector space A_n^2 over A_n. For $n = 3$ one has to be a little careful with the definition of a line because of the non-associativity of the octonions. A comprehensive introduction to the octonions can be found in the survey article [**7**].

Now consider the tautological line bundle over A_nP^1, where the fibre over each point in A_nP^1 is the corresponding line in A_n^2. It can also be obtained by cutting the sphere A_nP^1 into two closed discs and then gluing together the trivial line bundles over these two discs along the boundary by using the multiplication with elements of norm one in the normed division algebra A_n. We denote by $\mathfrak{L}_n$ the dual bundle of the tautological line bundle. The line bundle $\mathfrak{L}_n$ generates the reduced real K-theory of the sphere of dimension 2^n, and the octonionic line bundle $\mathfrak{L}_3$ induces a periodicity between the reduced real K-theories of higher-dimensional spheres, known as Bott periodicity.

Since each fibre of $\mathfrak{L}_n$ is equipped with a norm this line bundle naturally induces a sphere bundle $S(\mathfrak{L}_n)$ over A_nP^1. The manifolds $S(\mathfrak{L}_n)$ are again spheres. The fibrations $S(\mathfrak{L}_n) \to A_nP^1$ are usually called the Hopf fibrations (though for $n = 1$

it goes back originally to W.K. Clifford). In detail they are

$$\begin{aligned} S^1 &\xrightarrow{O(1)} RP^1 = S^1 , \\ S^3 &\xrightarrow{U(1)} CP^1 = S^2 , \\ S^7 &\xrightarrow{Sp(1)} HP^1 = S^4 , \\ S^{15} &\xrightarrow{S^7} OP^1 = S^8 . \end{aligned} \tag{2.1}$$

The first three fibrations are principal (i.e. group actions) while the last is not: S^7 is the set of elements of norm one in the octonions but is not a group since O is non-associative. The fact that every division algebra over R has dimension 2^n, $n = 0, 1, 2, 3$, can be proved topologically by showing that there are no further sphere fibrations beyond (2.1). More details about the construction of these fibrations can be found in §20 of [**22**]. The fibres in (2.1) will be denoted by Γ_n, so that

$$\begin{aligned} \Gamma_0 &= S^0 = O(1) , \\ \Gamma_1 &= S^1 = U(1) , \\ \Gamma_2 &= S^3 = Sp(1) , \\ \Gamma_3 &= S^7 . \end{aligned} \tag{2.2}$$

We will discuss the Hopf fibrations again in Section 3 in relation with group actions.

We now discuss the notion of "branching" as we shall encounter it in this paper. The classical situation occurs in complex variable theory where one Riemann surface can appear as the branched covering of another. We shall restrict ourselves to the simple case of double coverings, where the local model is the equation $w = z^2$. Although the group of order 2, given by $z \mapsto -z$, has a fixed point at $z = 0$, the quotient is still a smooth surface. The underlying topological reason is that, on the small circles $|z| = \epsilon$ surrounding the fixed point, we get the double covering in the first line of (2.1) so that the quotient is still a circle and hence is the boundary of a small disc.

If we take the product with R^{n-2} we get the more general situation where a group of order 2 acting on an n-dimensional manifold, with fixed-point components all of codimension 2, has a manifold as quotient. Again we refer to the fixed-point set as the branch locus of the double covering.

The purpose of this lengthy analysis of a familiar situation was to point out that each of the equations in (2.1) gives rise to a similar story, except that the finite group $O(1)$ is replaced by a higher-dimensional group (or sphere) Γ_n so that, outside the fixed-point set, we have a fibration. We shall refer to such fibrations as *branched fibrations.*

Consider the case of the second equation in (2.1) involving $U(1)$. The local model here is the action of $U(1)$ on $C^2 = R^4$ via complex scalars. The quotient is R^3 with S^2 being the boundary of the branch point. The geometry of the branched $U(1)$-fibration $R^4 \to R^3$ is fundamental in physics where it describes the geometry of a magnetic monopole. It was a major discovery by Dirac that the quantization of electric charge could be explained (in modern terms) by the $U(1)$-bundle above, over the complement of the point magnetic source at the origin. R^4 in this situation is now referred to as the Kaluza-Klein model of the Dirac monopole.

More generally, in current physical theories where space-time is viewed as having higher dimension than 4, a $U(1)$-action with a fixed manifold of codimension 4 is viewed as providing a charge on the branch locus (which has codimension 3 in the quotient). Examples of such situations were, for example, studied in detail in [**5**] and provided some of the early motivation for this paper.

In a similar way an $Sp(1)$-action with a fixed-point set of codimension 8 (and the standard action on $H^2 = R^8$) gives a branched fibration carrying an $Sp(1) = SU(2)$-"charge" on the branch locus, which has codimension 5 in the quotient.

Finally the last equation in (2.1) gives a similar story for branched fibrations with fibre S^7 and branch locus having codimension 9 in the quotient. Notice that, in this case, the fibration is not a group action.

In all these cases the quotient manifold has the induced topology but *not* the induced differentiable structure. In other words, it is not true that a differentiable function above, which is invariant under the group action, is a differentiable function below. For example for the double cover $w = z^2$ the function x^2, where $x = \mathrm{Re}(z)$, is not a differentiable function of $\mathrm{Re}(w), \mathrm{Im}(w)$. However, there *is* a natural differentiable structure on the quotient and we shall always use this and refer to it as the quotient structure. Note that, for holomorphic functions the invariants are indeed the functions of w and so the induced holomorphic structure on the quotient agrees with our differentiable quotient structure.

In the examples of branched fibrations which we shall study there will be a further group action in addition to the actions of the type in (2.1). These will be actions of "cohomogeneity one", i.e. an action of a connected Lie group G on a connected smooth manifold M whose generic orbit has codimension one. If G and M are compact such an action has a simple global structure: either there are no exceptional orbits and we have a fibration over the circle, or else there are just two exceptional orbits and the quotient is the closed unit interval, see e.g. [**17**]. In this second case the exceptional orbits have isotropy groups K_1 and K_2 and the generic orbit has isotropy group $K \subset K_1 \cap K_2$ when we consider the isotropy groups along a suitable path which connects the two exceptional orbits. Moreover, the homogeneous spaces

$$K_1/K = S^p \quad \text{and} \quad K_2/K = S^q$$

must both be spheres, where $p+1$ and $q+1$ are the codimensions of the two exceptional orbits. The normal sphere bundles of these two orbits are the maps

$$G/K \to G/K_1 \quad \text{and} \quad G/K \to G/K_2 \ .$$

Finally the manifold M, with its G-action, is entirely determined by the conjugacy class of the triple of subgroups K_1, K_2 and K.

The prototype example, which will be analyzed carefully in Section 4, is when $M = CP^2$ and $G = SO(3)$ acting with two exceptional orbits of codimension 2, namely RP^2 and S^2 (the conic $z_1^2 + z_2^2 + z_3^2 = 0$). Here $K_1 = S(O(1) \times O(2)) \cong O(2)$ and $K_2 = SO(2) \times SO(1) \cong SO(2)$ are embedded in $SO(3)$ so that $K = K_1 \cap K_2 = S(O(1) \times O(1)) \times SO(1) \cong Z_2$ and $p = q = 1$.

3. Projective Planes

In addition to the projective lines over the division algebras A_n we can consider higher-dimensional projective spaces. For $n = 0, 1, 2$, when A_n is associative, this gives us the classical projective spaces

$$RP^m \ , \ CP^m \ , \ HP^m \ (m \geq 2) \ .$$

For $n = 3$ however, $A_3 = O$ (the octonions) is not associative. In this case it is possible to construct a projective plane OP^2 (the Cayley plane), but not the projective spaces of higher dimension. In fact the non-associativity of O is related to the *non-Desarguesian* property of OP^2, and it is known (see e.g. [**24**]) that projective spaces of dimension ≥ 3 must be Desarguesian.

Just as with projective lines there are two ways of defining a projective plane over A_n. The first is to introduce the affine plane in the obvious way as pairs (x, y) of points x and y in A_n and then to compactify this by adding a projective line at infinity. This defines P_n as a manifold and the Hopf fibration appears naturally as the fibration of a spherical neighbourhood of the line at infinity. The fact that two lines meet in one point gets translated in this way to an assertion about the topology of the Hopf fibrations, namely that the Hopf invariant (or linking number of two fibres) is one. This is the fact which is used to show that the dimension of a division algebra over R must be 2^n, $n = 0, 1, 2, 3$: see [**2**] for a short proof.

This construction of P_n does not exhibit its homogeneity, and this is where an alternative construction is useful. For $n = 0, 1, 2$ the classical approach is to use the 3-dimensional vector space over A_n and to define P_n as the space of one-dimensional subspaces. This gives P_n as a homogeneous space of the relevant classical group

$$SL(3, R)\ ,\ SL(3, C)\ ,\ SL(3, H) \tag{3.1}$$

or of its compact form

$$SO(3)\ ,\ SU(3)\ ,\ Sp(3)\ . \tag{3.2}$$

The linear groups consist of projectivities, i.e. transformations preserving lines. The compact groups consist of isometries, where the projective plane is equipped with the Riemannian metric which is induced in the natural way from the Killing form of the group. For CP^2 the full isometry group is the extension of $SU(3)$ by Z_2 of complex conjugation. For the other two cases the isometry group is connected. In all cases the centre acts trivially so that it is really the adjoint group that acts effectively. The isotropy group for the actions of the three groups in (3.2) are

$$O(2) \cong S(O(1) \times O(2))\ ,\ U(2) \cong S(U(1) \times U(2))\ ,\ Sp(1) \times Sp(2)\ .$$

For $n = 3$ we cannot use this approach to construct the Cayley plane as there is no group $SL(3, O)$. However there is a substitute, both for the linear group and for its compact form, which plays the part of the fourth term of the sequences (3.1) and (3.2). For (3.2) we have the exceptional compact Lie group F_4 and for (3.1) we have the non-compact real form E_6^{-26} with character -26 of the exceptional complex Lie group $E_6(C)$. The group of projective transformations of the Cayley plane has been explicitly determined by Freudenthal in [**14**].

At this point it is easy to discuss the homogeneity of the Hopf fibrations in (2.1). If we fix a point o in P_n then the isotropy group at o of the connected isometry group of P_n acts transitively on the dual projective line o^* in P_n (the set of all antipodal points of o in P_n) and on the metric sphere bundle over o^* in P_n of sufficiently small radius. The projection of this sphere bundle from o onto o^* is just the Hopf fibration associated with A_n. The isotropy group of this action at a point in o^* acts transitively on the corresponding fibre.

The best way to unify all the projective planes P_n, and the associated groups of symmetries, is to introduce Jordan algebras. For a quite self-contained treatment in the octonionic case we refer to [**14**] and [**18**], the other cases work analogously and are easier to deal with. We summarize here the basic facts.

For each division algebra A_n we consider the real vector space H_n of 3×3 Hermitian matrices over A_n. Recall that in A_n we have a notion of conjugate $x \mapsto \bar{x}$ which fixes the "real" part and changes the sign of the "imaginary" part. Note that conjugation is an anti-involution of the algebra. Then, as usual, a matrix is Hermitian if its conjugate is equal to its transpose,

$$\bar{x}_{ij} = x_{ji} \quad (i, j = 1, 2, 3) .$$

We make H_n into a commutative (but non-associative) algebra by defining a multiplication

$$X \circ Y = \frac{1}{2}(XY + YX) , \tag{3.3}$$

where XY and YX denote usual matrix multiplication. Together with this multiplication H_n becomes a real Jordan algebra which we denote by J_n. The unit matrix I acts as an identity.

For $n = 0, 1, 2$ the groups in (3.2) act on H_n by

$$X \mapsto AXA^* \quad (A^* = \bar{A}^t)$$

and preserve the multiplication (3.3). Hence they are automorphisms of the Jordan algebra J_n. It can be shown that modulo their centres they are the full group $\mathrm{Aut}(J_n)$ of automorphisms of J_n, except for $n = 1$ when we get the identity component. For $n = 3$ the automorphism group $\mathrm{Aut}(J_3)$ provides an explicit model of the exceptional compact Lie group F_4.

For $n = 0, 1, 2$ we have a natural embedding of the projective plane P_n in H_n. We just associate to a one-dimensional subspace of the 3-dimensional vector space A_n^3 over A_n the Hermitian 3×3 matrix which represents orthogonal projection onto it. In terms of homogeneous coordinates (x_1, x_2, x_3), normalized so that $\|x_1\|^2 + \|x_2\|^2 + \|x_3\|^2 = 1$, this is given by

$$x \mapsto X = (x_i \bar{x}_j)_{i,j=1,2,3} .$$

Note that x and $x\lambda$ with λ in A_n, $\|\lambda\| = 1$, give the same matrix, and that X satisfies[2]

$$\begin{aligned} \operatorname{tr} X &= 1 , \\ \|X\|^2 &= 1 , \\ \det X &= 0 . \end{aligned} \tag{3.4}$$

Clearly the image of P_n in H_n is just the orbit of the diagonal matrix $\mathrm{Diag}(1, 0, 0)$ under the isometry group[3].

For $n = 3$ it can be shown that the orbit of $\mathrm{Diag}(1, 0, 0) \in J_3$ under the action of $F_4 = \mathrm{Aut}(J_3)$ provides a model for the Cayley plane P_3. The isotropy group is isomorphic to $Spin(9)$ and hence $OP^2 = P_3 = F_4/Spin(9)$ as a homogeneous space [**8**]. If we consider, as discussed above, the Hopf fibration $S^{15} \to S^8$ as a projection of a metric sphere bundle over o^* from a point o in $P_3 = OP^2$ onto the dual projective line o^*, and if $Spin(9)$ denotes the isotropy group of F_4 at o, then $Spin(9)$ acts transitively on S^{15} with isotropy group $Spin(7)$ and transitively

[2]The definition of the determinant for $n = 2$ (the quaternionic case) requires a little care, see Section 7.

[3]This still works for $P_1 = CP^2$, where the isometry group is disconnected.

on S^8 with isotropy group $Spin(8)$. Moreover, $Spin(8)$ acts transitively on the corresponding fibre S^7.

In all four cases J_n has three invariant polynomials of degrees $1, 2, 3$ as in (3.4), and we shall use the same notation. This follows from the fact that every element in J_n can be reduced by $\mathrm{Aut}(J_n)$ to real diagonal form. If the diagonal entries (the "eigenvalues") are $\lambda_1, \lambda_2, \lambda_3$ then the three invariant polynomials are

$$\begin{aligned} \mathrm{tr} &= \lambda_1 + \lambda_2 + \lambda_3 , \\ \| \ \|^2 &= \lambda_1^2 + \lambda_2^2 + \lambda_3^2 , \\ \det &= \lambda_1 \lambda_2 \lambda_3 . \end{aligned} \tag{3.5}$$

In all cases only the cubic polynomial det is invariant under the group of projectivities of the projective plane P_n. The fact that it is actually given by a polynomial needs to be proved (see Section 7).

4. Orbit Structures

As mentioned at the end of Section 2 the manifolds we are interested in have cohomogeneity one group actions compatible with the maps we need to construct to prove the identifications (1.1). We proceed to spell these out in detail beginning with the basic example of CP^2, which will be a model for the others.

We consider the action of $SO(3)$ on CP^2 via the natural embedding $SO(3) \subset SU(3)$. There are two special orbits, namely RP^2 with isotropy group $K_1 = S(O(1) \times O(2)) \cong O(2)$, and a 2-sphere S^2 (the conic $z_1^2 + z_2^2 + z_3^2 = 0$) with isotropy group $K_2 = SO(2) \times 1 \cong SO(2)$. The intersection

$$K = K_1 \cap K_2 = S(O(1) \times O(1)) \times 1 \cong Z_2 \tag{4.1}$$

consists of the two diagonal matrices $\mathrm{Diag}(\lambda, \lambda, 1)$ with $\lambda = \pm 1$. Each of the homogeneous spaces K_1/K and K_2/K is a circle. The generic orbit $SO(3)/K$ is 3-dimensional and it fibres over each of the two special orbits RP^2 and S^2 with S^1 as fibre.

One way to establish the identity

$$CP^2/O(1) = S^4$$

of (1.1) is to analyze the $SO(3)$-orbit structure of S^4 and compare it with that of CP^2. Here we can view S^4 as the unit sphere in the vector space of symmetric 3×3 real matrices of trace zero equipped with its usual norm. The orbits are determined by the three real eigenvalues $\lambda_1 \leq \lambda_2 \leq \lambda_3$ with $\lambda_1 + \lambda_2 + \lambda_3 = 0$. There are two special orbits $\lambda_1 = \lambda_2$ and $\lambda_2 = \lambda_3$ each of which is an RP^2, while the generic orbit is the real flag manifold of all full flags in R^3. Thus the three isotropy groups are

$$K_1' = S(O(1) \times O(2)) \ , \ K_2' = S(O(2) \times O(1)) \ , \ K' = K_1' \cap K_2' = S(O(1)^3) \ . \tag{4.2}$$

Since the 4-manifold with its $SO(3)$-orbit structure is determined by the conjugacy class of the triples of isotropy groups, comparison of (4.1) and (4.2) shows that there is a natural map

$$CP^2 \to S^4$$

compatible with the two $SO(3)$-actions. Moreover this map identifies the special orbit RP^2 in CP^2 with one of the two RP^2 in S^4, while outside this we have a double covering, given by the action of $O(1)$ on CP^2 as complex conjugation.

The orbit structures of HP^2 and OP^2 are quite similar as are those of the corresponding spheres. We consider first the action of $SU(3)$ on HP^2 via the embedding $SU(3) \subset Sp(3)$, and on S^7 as the unit sphere in the vector space of Hermitian 3×3 complex matrices of trace zero equipped with its usual norm. For S^7 we find two copies of CP^2 as special orbits and the complex flag manifold of all full flags in C^3 as generic orbit, so that the isotropy groups are

$$K_1' = S(U(1) \times U(2)) \ , \ K_2' = S(U(2) \times U(1)) \ , \ K' = K_1' \cap K_2' = S(U(1)^3) \ . \tag{4.3}$$

For HP^2 the two special orbits are CP^2 and a circle bundle over the dual CP^2 which is a 5-dimensional sphere S^5 (as one sees by using all quaternion lines, i.e. 4-spheres, determined by complex lines). The isotropy groups are

$$K_1 = S(U(1) \times U(2)) \ , \ K_2 = SU(2) \times 1 \ , \ K = K_1 \cap K_2 = S(U(1)^2) \times 1 \ .$$

Comparison with (4.3) shows the existence of a map[4]

$$HP^2 \to S^7$$

compatible with the two $SU(3)$-actions. It identifies the CP^2 in HP^2 with one of the two CP^2 in S^7, and on the complement $HP^2 \setminus CP^2$ it is an S^1-bundle over the complement $S^7 \setminus CP^2$. The $SU(3)$ determines a maximal subgroup $U(3)$ of $Sp(3)$, and the central $U(1)$ in this $U(3)$ acts trivially on the CP^2 and gives the fibres on $HP^2 \setminus CP^2$.

Finally consider the action of $Sp(3)$ on the Cayley plane OP^2 and on S^{13}, the unit sphere in the space of Hermitian 3×3 quaternion matrices with trace zero equipped with its usual norm. Again we have two special orbits in S^{13}, both copies of HP^2, and the generic orbit is the quaternionic flag manifold of all full flags in H^3, so that the isotropy groups are

$$K_1' = Sp(1) \times Sp(2) \ , \ K_2' = Sp(2) \times Sp(1) \ , \ K' = K_1' \cap K_2' = Sp(1)^3 \ . \tag{4.4}$$

For the Cayley plane, HP^2 is clearly a special orbit. The generic orbit is just the normal sphere bundle of HP^2 in OP^2 with fibre S^7. This is the unit sphere in the normal R^8 which is a representation of the isotropy group $K_1 = Sp(1) \times Sp(2)$. Note that this representation is *not* the standard action on H^2 since $Sp(1)$ acts trivially and only $Sp(2)$ acts in the standard manner, so that the generic isotropy group is $K = Sp(1) \times Sp(1) \times 1$. By considering all the Cayley lines (8-spheres) determined by quaternion lines we see that the other special orbit is fibred over the dual HP^2 with 3-sphere fibres and therefore is an 11-dimensional sphere S^{11}. Hence the isotropy groups are

$$K_1 = Sp(1) \times Sp(2) \ , \ K_2 = Sp(2) \times 1 \ , \ K = K_1 \cap K_2 = Sp(1)^2 \times 1 \ .$$

Comparison with (4.4) then shows the existence of a map

$$OP^2 \to S^{13}$$

compatible with the action of $Sp(3)$. HP^2 is the branch locus and outside this we have a 3-sphere fibration. The $Sp(3)$ is contained in a maximal subgroup $Sp(3)Sp(1)$ in F_4, where the $Sp(1)$ centralizes the $Sp(3)$ in F_4, see e.g. [**9**]. This $Sp(1)$ fixes the HP^2 in OP^2 pointwise, and the orbits through the other points in OP^2 are just the 3-spheres of the fibration. This establishes the identity $OP^2/Sp(1) = S^{13}$ in (1.1).

[4]The proof outlined here is essentially that of [**3**], [**5**].

We have thus established (1.1) from a complete description of the relevant orbit structures. In the next section we shall formulate and prove Theorem A, a more explicit version of (1.1) which does not rely on such a detailed knowledge of the orbit structure, and which provides such an explicit map.

The orbit structure of these group actions on the projective planes have also been studied in **[20]** in relation to homotopy theory.

5. An Explicit Map

In Section 3 we saw that in all four cases the projective plane P_n has a natural embedding in H_n, the vector space of 3×3 Hermitian matrices over the division algebra A_n. Moreover the formulae (3.4) show that P_n lies in the hyperplane $H_n(1)$ given by $\operatorname{tr} X = 1$ and on the sphere $\|X\|^2 = 1$. Note that the intersection of the hyperplane and the sphere is the sphere of one lower dimension with centre $I/3$ and radius ρ where I is the unit matrix $\operatorname{Diag}(1,1,1)$ and $\rho^2 = 2/3$. Since

$$\dim H_n = 3(2^n + 1)$$

we thus have an embedding

$$P_n \subset S^{d(n)} \; , \; d(n) = 3 \cdot 2^n + 1 \; . \tag{5.1}$$

For $n = 0, 1, 2$ these are the classical embeddings referred to in (1.2) and for $n = 3$ we have the corresponding one for the Cayley plane.

For $n \geq 1$ we have a natural inclusion of algebras

$$A_{n-1} \subset A_n$$

and hence, using the Euclidean metric given by the norm, an orthogonal projection $A_n \to A_{n-1}$ which induces a projection

$$\pi_n : H_n \to H_{n-1} \; .$$

Note that π_n commutes with the trace and hence it maps P_n into the affine hyperplane $H_{n-1}(1)$ in H_{n-1}. In this hyperplane we will choose the point $I/3$ as centre and use the shifted coordinate

$$\tilde{X} = X - I/3$$

for $H_{n-1}(0)$, the linear subspace of H_{n-1} given by $\operatorname{tr} X = 0$. Restricting π_n to $P_n \subset H_n$ we get a map

$$\pi_n : P_n \to H_{n-1}(1)$$

and a shifted map $\tilde{\pi}_n : P_n \to H_{n-1}(0)$. The following lemma will be crucial for our construction.

LEMMA 1. *The matrix $I/3$ does not lie in the image $\pi_n(P_n)$, so that $\tilde{\pi}_n(X) \neq 0$ for all $X \in P_n$.*

We postpone the proof till later. Assuming that this lemma is true we can rescale the maps $\tilde{\pi}_n$ to define a map

$$f_n : P_n \to S^{d(n-1)} \tag{5.2}$$

given by

$$f_n(X) = \frac{1}{3} I + \rho \frac{\tilde{\pi}_n(X)}{\|\tilde{\pi}_n(X)\|} \; . \tag{5.3}$$

Thus $f_n(P_n)$ lies in the sphere $S^{d(n-1)}$ of radius ρ in $H_{n-1}(1)$ centred at $I/3$.

The map (5.2) will be the map inducing the diffeomorphism of (1.1), but before formulating Theorem A we shall need a few further properties of f_n. First we note that, when $X \in P_{n-1} \subset P_n$, $\pi_n(X) = X$ and $\|X\| = 1$, so that $\|\tilde{\pi}_n(X)\| = \rho$ and $f_n(X) = X$. Hence, when restricted to P_{n-1}, the map f_n is just the standard embedding (5.1) for $n-1$.

Let us denote by G_n the groups of isometries of P_n which we already discussed above, namely

$$\begin{aligned} G_0 &= SO(3)\ , \\ G_1 &= SU(3)\ , \\ G_2 &= Sp(3)\ , \\ G_3 &= F_4\ . \end{aligned} \tag{5.4}$$

The space H_n of Hermitian matrices is a representation of G_n, which splits off a trivial factor (corresponding to the trace). When restricted to G_{n-1} ($n = 1, 2, 3$) it decomposes as

$$H_n = \dot{H}_{n-1} \oplus H_{n-1}^{\perp}\ , \tag{5.5}$$

so that the projection $\pi_n : H_n \to H_{n-1}$ is compatible with the action of G_{n-1}. Thus, assuming Lemma 1, the map f_n of (5.2) also commutes with the action of G_{n-1}.

In addition we have an action of the group Γ_{n-1} of elements of norm one of the division algebra A_{n-1} on H_n which commutes with the action of G_{n-1} and preserves the fibres of π_n. Recall that G_n is, modulo its centre, the automorphism group $\mathrm{Aut}(J_n)$ of the Jordan algebra J_n except for $n = 1$ when $\mathrm{Aut}(J_1)$ has another connected component induced by complex conjugation. Note that this complex conjugation does not extend to an automorphism of J_2 but to an anti-automorphism. Then Γ_{n-1} can be viewed as the centralizer of the connected component of $\mathrm{Aut}(J_{n-1})$ in $\mathrm{Aut}(J_n)$. Explicitly we have

$n = 1$: $\Gamma_0 = O(1) \cong Z_2$ acts by complex conjugation on H_1;

$n = 2$: $\Gamma_1 = U(1)$ acts by conjugation on H_2 with respect to the elements of norm one of $C \subset H$;

$n = 3$: $\Gamma_2 = Sp(1)$. The action of $Sp(1)$ on H_3 cannot be described in a similar fashion because of the non-associativity of the octonions, but the construction of $Sp(1)$ as a subgroup of $\mathrm{Aut}(J_3) = F_4$ is quite simple. Consider a root space decomposition of the Lie algebra of F_4 such that the Lie algebra of $G_2 = Sp(3)$ is determined by the two short simple roots and the adjacent long simple root. Then the maximal root determines the Lie algebra of $\Gamma_2 = Sp(1)$.

Note that Γ_{n-1} acts trivially on the summand H_{n-1} in the decomposition (5.5) and hence trivially on the projective plane $P_{n-1} \subset H_{n-1}$. The action of $\Gamma_0 = O(1)$ on $H_0^{\perp} \cong R^3$ is just by scalar multiplication by ± 1, the one of $\Gamma_1 = U(1)$ on $H_1^{\perp} \cong C^3$ is the scalar action $(\lambda, z) \mapsto \lambda^2 z$ ($\lambda \in U(1) \cong S^1 \subset C$, $z \in C^3$), and the one of $\Gamma_2 = Sp(1)$ on $H_2^{\perp} \cong H^3$ is by right multiplication. The action of Γ_{n-1} on the normal bundle of P_{n-1} in P_n is obtained from the one on $H_{n-1}^{\perp}$ by suitable restriction.

Finally we are in a position to formulate our promised refinement of (1.1).

THEOREM A. *For $n = 1, 2, 3$ the map*

$$f_n : P_n \to S^{d(n-1)}$$

defined by (5.3) induces a diffeomorphism

$$P_n/\Gamma_{n-1} \approx S^{d(n-1)} .$$

Moreover, this diffeomorphism is compatible with the natural action on P_n/Γ_{n-1} *and* $S^{d(n-1)}$ *of the group* G_{n-1} *of (5.4).*

REMARK. We have already observed that the action of Γ_{n-1} on the normal bundle of P_{n-1} in P_n is just the scalar action of the relevant field (for $n = 2$ it is the square of the scalar action), so that the quotient P_n/Γ_{n-1} is indeed a manifold, branched along P_{n-1} in the sense of Section 2. Outside the branch locus the action of Γ_{n-1} is free (except for $n = 2$ where each orbit is a double covering of the circle, in which case we may consider $\Gamma_1/Z_2 \cong \Gamma_1$ to get a free action).

The case $n = 1$ of Theorem A gives the known diffeomorphism of $P_1 = CP^2$ modulo complex conjugation with S^4. Since this case is basic to the others we shall prove it first.

Consider, as a preliminary, the complex projective line CP^1 viewed as embedded by idempotents in the affine space R^3 of Hermitian 2×2 complex matrices of trace one. The projection onto the real symmetric matrices of trace one is easily seen to be the standard projection of S^2 onto the disc D^2 (with centre $I/2$ and radius σ with $\sigma^2 = 1/2$), which identifies conjugate points. Note that the $O(2)$-orbits on S^2, the "circles of latitude", go into the concentric circles in the disc.

We are now ready to look at CP^2 and the projection π_1. For every RP^1 in RP^2 its complexification is a CP^1 in CP^2 and these fill out CP^2, intersecting only at points of RP^2. Thus $CP^2 \setminus RP^2$ is fibred over the dual RP^2 with fibre $CP^1 \setminus RP^1 = S^2 \setminus S^1$ (two open discs). Alternatively it is fibred over S^2 with fibre an open disc. Moreover, $SO(3)$ acts transitively on RP^2 (or S^2), the base of this fibration, and the isotropy group $S(O(1) \times O(2)) \cong O(2)$ acts on each fibre. Because the projection π_1 is compatible with this action of $SO(3)$ it is entirely determined by its restriction to a single fibre. But such a fibre can be taken to be given by the equation $z_3 = 0$, and so we are reduced to studying the projection of CP^1 which we have just done above. The first implication of this is to establish Lemma 1 for $n = 1$, because the scalar 3×3 matrix $I/3$ cannot lie in the subspace of 2×2 matrices (with zeroes in the third row and column). Moreover, the orbit analysis of the 2-dimensional case shows that the map f_1 sends the $SO(3)$-orbits of CP^2 (modulo conjugation) diffeomorphically onto the $SO(3)$-orbits of S^4, thus proving Theorem A for $n = 1$.

To deal with the cases $n = 2, 3$ of Theorem A, i.e. with HP^2 and OP^2, we shall choose appropriate embeddings of CP^2 in the higher projective planes and then use Theorem A for CP^2. The embeddings we want are not the standard ones given by the original inclusions $C \subset H \subset O$, but are suitable conjugates of these. Thus, for H, we choose the embedding $C \to H$ which takes $i \in C$ into $j \in H$, which gives a second embedding of CP^2 into HP^2, which we will denote by $CP^2(j)$ to distinguish it from the original CP^2. Note that

$$CP^2(j) \cap CP^2 = RP^2 . \tag{5.6}$$

Similarly we will choose a third embedding of CP^2 into OP^2 (not coming from HP^2). Consider the sphere S^6 of imaginary elements of norm one in $O = R^8$. This contains the S^2 of imaginary quaternions of norm one. Choose an element $e \in S^6$ of norm one which lies in R^4 orthogonal to $H \subset O$, and embed C in O by sending i to e. Since the exceptional compact Lie group[5] G_2 of automorphisms of the octonions acts transitively on S^6 we can find $g \in G_2$ which takes i into e. Since G_2 acts naturally on the exceptional Jordan algebra J_3 (so that $G_2 \subset F_4$) we get a copy $g(CP^2) \subset OP^2$. Note that

$$g(CP^2) \cap CP^2 = RP^2 . \tag{5.7}$$

Because our elements j and e were chosen orthogonal to i it follows that the projections π of HP^2 and OP^2 restrict to the standard projection of $CP^2(j)$ and $g(CP^2)$. Because of (5.6) and (5.7) these copies of CP^2 are transversal to the orbits of the relevant groups $SU(3)$ and $Sp(3)$. More precisely, each orbit of the larger group intersects our CP^2 in just one $SO(3)$-orbit. This follows by examining the corresponding groups. Consider first the embedding $CP^2(j) \subset HP^2$. This is an orbit of a copy of $SU(3)$ which we may denote by $SU(3)_j$. Clearly this intersects the original $SU(3) \subset Sp(3)$ precisely in $SO(3)$. Similarly $g(CP^2) \subset OP^2$ is an orbit of $g(SU(3))$ and we need to check that

$$SU(3) \cap g(SU(3)) = SO(3) .$$

But this is clear because the intersection must preserve

$$CP^2 \cap g(CP^2) = RP^2 .$$

This correspondence between the $SO(3)$-orbits on CP^2 and the orbits of $SU(3)$ on HP^2 and of $Sp(3)$ on OP^2 shows that Lemma 1 for $n = 2, 3$ follows from the case $n = 1$, which we have already proved. The correspondence also shows that the map f_n induces a diffeomorphism on the one-dimensional space (interval) of orbits. But each orbit in P_n is known and outside the branch locus it is just fibred over the corresponding fibre in $S^{d(n-1)}$. Together with the local behaviour near the branch locus this completes the proof of Theorem A.

6. The Complexified Projective Planes

As mentioned in Section 1 the four projective planes P_n have natural complexifications $P_n(C)$ as complex projective algebraic varieties. These have a "real structure", i.e. an anti-holomorphic involution τ, which has P_n as the real part (fixed by τ). We shall now examine these varieties in greater detail and study their symmetries.

Recall that $P_n \subset H_n$, the space of Hermitian 3×3 matrices over the division algebra A_n. It is the orbit of the diagonal matrix $\mathrm{Diag}(1, 0, 0)$ under the compact Lie group G_n (listed in (5.4)). Note that $P_n \subset H_n(1)$, the affine subspace of matrices of trace one. If we denote by $\mathfrak{P}_n$ the real projective space of the vector space H_n, so that $\dim \mathfrak{P}_n = 3 \cdot 2^n + 2$, then we can identify $\mathfrak{P}_n$ with the projective completion of the real affine space $H_n(1)$ and we have $P_n \subset \mathfrak{P}_n$.

The group G_n of isometries of P_n acts on the affine space $H_n(1)$. This action extends to an action of a larger group $\mathfrak{G}_n$ on the projective space $\mathfrak{P}_n$ which preserves P_n. This induces on P_n its group of projectivities. For $n = 0, 1, 2$ we have

$$\mathfrak{G}_n = SL(3, A_n)$$

as noted in (3.1), while $\mathfrak{G}_3$ is the non-compact real form E_6^{-26} of the exceptional complex Lie group $E_6(C)$. In all cases $\mathfrak{G}_n$ has a natural irreducible representation

[5]Note that this is not to be confused with the group $G_2 = Sp(3)$ of the sequence in (5.4).

on the real vector space H_n, which splits off a trivial factor (corresponding to the trace) when restricted to the compact subgroup G_n.

For $n = 0$ the representation of $\mathfrak{G}_0 = SL(3, R)$ in $H_0 = R^6$ is via the symmetric square $S^2(R^3)$, and the embedding $P_0 \subset \mathfrak{P}_0$ is the embedding

$$RP^2 \subset RP^5$$

induced by the diagonal (squaring) map $R^3 \to S^2(R^3)$ given by $x \mapsto x^2$. It is the (real) Veronese embedding, and we shall use the same term for all n.

Note that, if $V = R^3$, then $SL(V)$ has two inequivalent irreducible representations of dimension 6, namely $S^2(V)$ and $S^2(V^*)$, where V^* denotes the dual vector space of V. These become equivalent when restricted to $SO(3)$. A similar story holds for all n, so that these Veronese embeddings come in dual pairs

$$P_n \subset \mathfrak{P}_n \ , \ P_n^* \subset \mathfrak{P}_n^* \ ,$$

where P_n^* is the dual projective plane of P_n, representing its projective lines. Reduction to the compact group gives an identification

$$P_n \to P_n^*$$

by associating to each point p of the plane P_n the "opposite" projective line, which may be defined as the set of all antipodal points q on a closed geodesic through p, the "polar" of p in P_n.

We are now in a position to complexify everything. We get a complex Lie group $\mathfrak{G}_n(C)$ acting on the complex projective space $\mathfrak{P}_n(C)$. Since P_n is an orbit of $\mathfrak{G}_n$ we get as complexification an orbit $P_n(C)$ of $\mathfrak{G}_n(C)$, which defines the complexified projective plane. It is a complex projective manifold with

$$\dim_C P_n(C) = \dim_R P_n = 2^n \ .$$

Since $\mathfrak{P}_n(C)$ has a natural Hermitian metric we can define the maximal compact subgroup $\hat{G}_n$ of $\mathfrak{G}_n(C)$. This preserves the induced Kähler metric on $P_n(C)$ and it clearly contains G_n. Note that $P_n(C)$ equipped with this induced Kähler metric is a Hermitian symmetric space.

Explicitly, the groups $\mathfrak{G}_n(C)$ and $\hat{G}_n$ are given by the following table, where for greater clarity we also list the groups G_n and $\mathfrak{G}_n$:

n	0	1	2	3
$\mathfrak{G}_n(C)$	$SL(3,C)$	$SL(3,C) \times SL(3,C)$	$SL(6,C)$	$E_6(C)$
$\mathfrak{G}_n$	$SL(3,R)$	$SL(3,C)$	$SL(3,H)$	E_6^{-26}
$\hat{G}_n$	$SU(3)$	$SU(3) \times SU(3)$	$SU(6)$	E_6
G_n	$SO(3)$	$SU(3)$	$Sp(3)$	F_4

. (6.1)

The compact complex manifolds $P_n(C)$ are necessarily homogeneous spaces also of the maximal compact subgroup $\hat{G}_n$ of the complex Lie group $\mathfrak{G}_n(C)$. Explicitly we have

$$\begin{aligned}
P_0(C) &= SU(3)/S(U(1) \times U(2)) &&= CP^2 \ , \\
P_1(C) &= SU(3)^2/S(U(1) \times U(2))^2 &&= CP^2 \times CP^2 \ , \\
P_2(C) &= SU(6)/S(U(2) \times U(4)) &&= Gr_2(C^6) \ , \\
P_3(C) &= E_6/Spin(10)U(1) \ .
\end{aligned}$$

Here $Gr_2(C^6)$ denotes the complex Grassmannian of 2-planes in C^6. The identification of the isotropy groups is easy in the classical cases ($n = 0, 1, 2$) and follows from the representation theory of E_6 and F_4 for the last case [**1**]. The embeddings $P_n(C) \to \mathfrak{P}_n(C)$ are well-known classical embeddings for $n = 0, 1, 2$. For $n = 0$ it is the (complex) Veronese embedding $CP^2 \to CP^5$, for $n = 1$ it is the Segre embedding $CP^2 \times CP^2 \to CP^8$, and for $n = 2$ it is the Plücker embedding $Gr_2(C^6) \to CP^{14}$.

The homogeneous space P_n and $P_n(C)$, together with their isometry groups, are essentially given by the first two rows of Freudenthal's magic square

$\mathfrak{so}(3)$	$\mathfrak{su}(3)$	$\mathfrak{sp}(3)$	$\mathfrak{f}_4$
$\mathfrak{su}(3)$	$\mathfrak{su}(3) \oplus \mathfrak{su}(3)$	$\mathfrak{su}(6)$	$\mathfrak{e}_6$
$\mathfrak{sp}(3)$	$\mathfrak{su}(6)$	$\mathfrak{so}(12)$	$\mathfrak{e}_7$
$\mathfrak{f}_4$	$\mathfrak{e}_6$	$\mathfrak{e}_7$	$\mathfrak{e}_8$

, (6.2)

see for example [**13**]. According to Freudenthal, the entries in the first row describe 2-dimensional elliptic geometries, in the second row 2-dimensional projective geometries, in the third row 5-dimensional symplectic geometries, and in the last row metasymplectic geometries.

We now want to look at the action of G_n on $P_n(C)$ and study its orbit structure. We already know that P_n is one orbit, say $P_n = G_n/M_n$ where M_n is given by

$$\begin{array}{lcll} M_0 & = & S(O(1) \times O(2)) & \cong O(2)\ , \\ M_1 & = & S(U(1) \times U(2)) & \cong U(2)\ , \\ M_2 & = & Sp(1) \times Sp(2)\ , & \\ M_3 & = & Spin(9)\ . & \end{array}$$

Since $P_n(C)$ is the complexification of P_n the normal bundle N_n is isomorphic to the tangent bundle and hence the action of M_n on N_n is just the representation of M_n on the quotient of the Lie algebras

$$L(G_n)/L(M_n)\ . \qquad (6.3)$$

But for all these representations M_n is transitive on the unit sphere. It follows that the generic orbit of the action of G_n on $P_n(C)$ has codimension one. Moreover the generic isotropy groups are just the isotropy groups of M_n on the unit sphere in (6.3). Hence the generic orbits are

$$\begin{array}{lcl} n = 0 & : & SO(3)/O(1)\ , \\ n = 1 & : & SU(3)/U(1)\ , \\ n = 2 & : & Sp(3)/Sp(1) \times Sp(1)\ , \\ n = 3 & : & F_4/Spin(7)\ . \end{array}$$

For $n = 3$ this follows from the well-known fact that $S^{15} = Spin(9)/Spin(7)$, where the action of $Spin(9)$ on $S^{15} \subset R^{16}$ is via its irreducible spin representation [**8**].

In addition to the special orbit P_n in $P_n(C)$ there must be another special orbit. Since G_n acts on the affine space $H_n(1)$ it leaves invariant the hyperplane section of $P_n(C)$ at infinity, which we already denoted in Section 1 by $P_n(\infty)$.

After these preliminaries on the spaces $P_n(C)$ and the groups acting on them we now want to explicitly construct the maps referred to in (1.3). The method is very similar to that we used to construct the maps f_n of Theorem A, the essential point being a projection onto a linear space which misses the origin (Lemma 1) and hence can be normalized to map to the sphere. For P_n our projection was the orthogonal projection

$$\pi_n : H_n \to H_{n-1}$$

restricted to $P_n \subset H_n$. For such a method we need now to replace the projective embedding $P_n(C) \to \mathfrak{P}_n(C)$ by an embedding in a Euclidean space. Fortunately such an embedding exists for every complex projective algebraic variety which is a homogeneous space of a compact Lie group. We just have to pick an appropriate orbit in the Lie algebra. Thus we can find embeddings

$$P_n(C) \to L(\hat{G}_n)$$

compatible with the $\hat{G}_n$-action, where $\hat{G}_n$ is the group of isometries of $P_n(C)$ given by table (6.1).

We can be more precise. Note that $G_n \subset \hat{G}_n$ and that $\hat{G}_n/G_n$ is the compact dual of the non-compact symmetric space $\mathfrak{G}_n/G_n$, since $\hat{G}_n$ and $\mathfrak{G}_n$ are both real forms of the same complex Lie group $\mathfrak{G}_n(C)$. The compact symmetric space $\hat{G}_n/G_n$ is just the space that parametrizes all real P_n in $P_n(C)$, whereas the non-compact symmetric space $\mathfrak{G}_n/G_n$ is the space that parametrizes the standard metrics on P_n (up to overall scale) and so can be identified with the open set

$$H_n^+(1) \subset H_n(1)$$

consisting of matrices which are positive-definite (i.e. all X in $H_n(1)$ for which the eigenvalues $\lambda_1, \lambda_2, \lambda_3$ as in (3.5) are positive). Here we choose the diagonal matrix $I/3$ to define our base metric (base point of $\mathfrak{G}_n/G_n$). It follows that there is a natural isomorphism between the tangent spaces

$$L(\hat{G}_n)/L(G_n) \cong L(\mathfrak{G}_n)/L(G_n) \cong H_n(0) , \tag{6.4}$$

where $H_n(0)$ is the vector space obtained from the affine space $H_n(1)$ with $I/3$ as its origin.

P_n is the G_n-orbit of the matrix $\mathrm{Diag}(1,0,0)$ in $H_n(1)$ or equivalently the G_n-orbit of $\mathrm{Diag}(2/3,-1/3,-1/3)$ in $H_n(0)$. From (6.4) we get an orthogonal decomposition

$$L(\hat{G}_n) = L(G_n) \oplus H_n(0) . \tag{6.5}$$

The G_n-orbit P_n in $H_n(0)$ then generates the $\hat{G}_n$-orbit $P_n(C)$ in $L(\hat{G}_n)$. The projection

$$\sigma_n : L(\hat{G}_n) \to H_n(0)$$

defined by (6.5) restricts to give a map

$$\sigma_n : P_n(C) \to H_n(0) . \tag{6.6}$$

Parallel to Lemma 1 we now have

LEMMA 2. *The image of the map σ_n in (6.6) does not contain* 0.

Assuming for the moment the truth of Lemma 2 we can now define a map

$$\varphi_n : P_n(C) \to S^{d(n)} , \ d(n) = 3 \cdot 2^n + 1 ,$$

by the normalization

$$\varphi_n(Z) = \rho \frac{\sigma_n(Z)}{\|\sigma_n(Z)\|} , \tag{6.7}$$

where ρ with $\rho^2 = 2/3$ is inserted to ensure that, on restriction to $P_n \subset P_n(C)$, the map φ_n coincides with the standard inclusion $P_n \to H_n(0)$.

We are now in a position to formulate the main result of this section, refining (1.3),

THEOREM B. *The map $\varphi_n : P_n(C) \to S^{d(n)}$ defined by (6.7) for $n = 0, 1, 2, 3$ is a fibration outside the branch locus P_n and the hypersurface $P_n(\infty)$ which gets mapped to the antipodal P_n. The fibres are the norm one elements Γ_n of A_n (namely the spheres S^0, S^1, S^3, S^7) and φ_n commutes with the action of G_n.*

REMARK. The case $n = 0$ of Theorem B coincides with the case $n = 1$ of Theorem A, since $P_0(C) = P_1 = CP^2$ is the complexification of RP^2 and

$$L(\hat{G}_1) \cong \mathfrak{su}(3) \cong iH_1(0) .$$

Note that Lemma 2 for $n = 0$ reduces to Lemma 1 for $n = 1$. We can therefore use the case $n = 0$ of Lemma 2 and Theorem B for the other three cases. We will simply use the natural inclusions

$$P_0(C) \subset P_1(C) \subset P_2(C) \subset P_3(C) \tag{6.8}$$

and the corresponding inclusions

$$H_0(0) \subset H_1(0) \subset H_2(0) \subset H_3(0) .$$

The maps σ_n and φ_n are compatible with these inclusions. Moreover, for $n = 1, 2, 3$, the G_n-orbits in $P_n(C)$ intersect $P_{n-1}(C)$ in the G_{n-1}-orbits. Thus each inclusion in (6.8) induces naturally a diffeomorphism from the space of G_n-orbits onto the space of G_{n-1}-orbits (which are both closed intervals). Since the property in Lemma 2 of not containing 0 is G_n-invariant it is a property of orbits. Lemma 2 for $n \geq 1$ therefore follows from the case $n = 0$ (since the σ_n-image of any G_n-orbit of $P_n(C)$ contains the corresponding σ_0-image). Moreover, for the same reasons we see that φ_n maps the generic G_n-orbit in $P_n(C)$ onto a generic G_n-orbit in $S^{d(n)}$ in a smooth manner (i.e. by a diffeomorphism of the orbit parameter). The identification of the fibres as Γ_n follows from their explicit description. Finally, the branch locus P_n being preserved by φ_n, the other exceptional orbit $P_n(\infty)$ in $P_n(C)$ must go to the other (dual or antipodal) P_n in $S^{d(n)}$. This completes the proof of Theorem B.

In the appendix we will give more information about the map from $P_n(\infty)$ onto this dual P_n.

REMARK. As can be seen Theorems A and B are very similar. However, in one sense Theorem B was easier to prove by induction on n because we could use the natural inclusions (6.8), whereas the natural inclusions of the real projective plane were not transversal to the G_n-action and had to be replaced by a second set of embeddings. On the other hand, Theorem B has an exceptional fibre $P_n(\infty)$ for $n \geq 1$, whereas in Theorem A, f_n is a fibration outside the branch locus. Notice that, in each case, our branch locus P_n is embedded in a manifold of twice its dimension, namely

$$\begin{array}{ll} P_n \subset P_{n+1} & \text{(Theorem A)} , \\ P_n \subset P_n(C) & \text{(Theorem B)} . \end{array}$$

These coincide only for $n = 0$. For $n \geq 1$ even the normal bundles are different.

The four spheres $S^{d(n)}$ together with their cohomogeneity one actions by G_n appear in several contexts in differential geometry. We discuss briefly two of them.

a) *Isoparametric hypersurfaces.* A real-valued function f on a Riemannian manifold is isoparametric if the first and second differential parameter of f (i.e. $\|\text{grad}\, f\|^2$ and Δf) are constant along the level sets of f. The interest in such functions originated from geometrical optics. Any regular level set of an isoparametric function is called an isoparametric hypersurface. E. Cartan proved that a hypersurface in a space of constant curvature is isoparametric if and only if it has constant principal curvatures. The number of distinct principal curvatures of an isoparametric hypersurface in a sphere is $1, 2, 3, 4$ or 6. It is easy to show that the isoparametric hypersurfaces in spheres with 1 or 2 distinct principal curvatures are the distance spheres or Clifford tori, respectively. In **[12]** E. Cartan proved that the isoparametric hypersurfaces with 3 distinct principal curvatures exist only in $S^{d(n)}$, $n = 0, 1, 2, 3$, and moreover they are the regular orbits of the cohomogeneity one action of G_n on $S^{d(n)}$. For a survey about this topic see **[23]**.

b) *Positive curvature.* A classical problem is to classify all simply connected closed smooth manifolds which admit a Riemannian metric with positive sectional curvature. The standard examples of such manifolds are the spheres S^m and the projective spaces CP^m, HP^m and OP^2 ($m \geq 2$). Wallach **[25]** proved that the only simply connected closed smooth manifolds admitting a *homogeneous* Riemannian metric with positive sectional curvature are, apart from the even-dimensional spheres and the above projective spaces, precisely the regular orbits of the cohomogeneity one actions of G_n on $S^{d(n)}$, $n = 1, 2, 3$. Explicitly these are the flag manifolds $SU(3)/U(1)^2$, $Sp(3)/Sp(1)^3$ and $F_4/Spin(8)$ of all full flags in CP^2, HP^2 and OP^2. However, we should point out that the positive curvature metrics on these flag manifolds are not the induced metrics from the spheres with their standard metrics.

7. The Projective Version

So far we have concentrated on constructing maps compatible with the compact isometry groups G_n of the projective planes P_n. Now we shall extend these results appropriately to the non-compact groups $\mathfrak{G}_n$ of projectivities of P_n.

In Section 6 we saw that the embeddings $P_n \subset H_n(1)$, which are compatible with the action of G_n, extend naturally to embeddings

$$P_n \subset \mathfrak{P}_n$$

compatible with the action of $\mathfrak{G}_n$. Here $\mathfrak{P}_n$ is the real projective space associated with H_n, and so is the projective completion of the affine space $H_n(1)$. The vector space H_n is an irreducible representation of $\mathfrak{G}_n$ which splits off a one-dimensional trivial factor (corresponding to the trace) on restriction to G_n, so that $\mathfrak{G}_n$ acts on $\mathfrak{P}_n$ with G_n preserving the hyperplane at infinity and the central point given by the scalar multiples of $I/3$.

The projection $\pi_n : H_n \to H_{n-1}$ induces a corresponding projection

$$\pi_n : \mathfrak{P}_n \setminus \mathfrak{P}(H_{n-1}^{\perp}) \to \mathfrak{P}_{n-1} \tag{7.1}$$

which is defined in the complement of the "axis" of projection arising from $H_{n-1}^{\perp}$. The map (7.1) is compatible with the action of $\mathfrak{G}_{n-1}$. Note that P_n is contained in $H_n(1)$ and so does not intersect the axis of the projection (7.1). Hence we get a well-defined map

$$\pi_n : P_n \to \mathfrak{P}_{n-1} \tag{7.2}$$

compatible with the action of $\mathfrak{G}_{n-1}$. We propose to examine (7.2) with respect to the orbits of $\mathfrak{G}_{n-1}$.

We begin by looking at the action of $\mathfrak{G}_n$ on $\mathfrak{P}_n$ (and we shall then replace n by $n-1$). We recall that we have a cubic polynomial det in H_n whose vanishing defines a hypersurface $Z_n \subset \mathfrak{P}_n$. In the appendix which follows, the complexification $Z_n(C) \subset \mathfrak{P}_n(C)$ is discussed in detail. The group $\mathfrak{G}_n$ leaves Z_n invariant: this is clear for $n = 0, 1$, requires a little verification for $n = 2$, and is a classical result of E. Cartan [**11**] when $n = 3$ and $\mathfrak{G}_3 = E_6^{-26}$ is a real form of $E_6(C)$. Moreover Z_n contains P_n as a $\mathfrak{G}_n$-orbit and P_n is the singular locus of Z_n.

The group G_n acts on the affine part $H_n(1)$ of $\mathfrak{P}_n$ with $I/3$ as fixed point and its orbits are parametrized by three real eigenvalues

$$\lambda_1 \leq \lambda_2 \leq \lambda_3 \ , \quad \lambda_1 + \lambda_2 + \lambda_3 = 1 \ .$$

Z_n is given by $\lambda_1\lambda_2\lambda_3 = 0$, while P_n is given by $\lambda_1 = \lambda_2 = 0$ and $\lambda_3 = 1$. We can indicate the affine part of Z_n schematically by the following picture (which is actually what a 2-dimensional (affine) slice would look like: a real cubic curve):

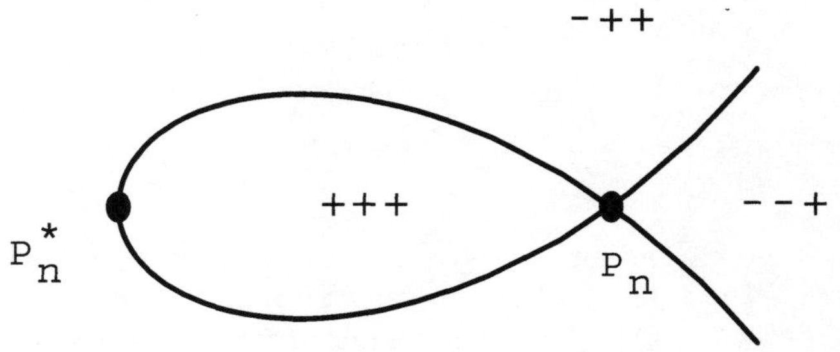

where the complement of Z_n in $H_n(1)$ is divided into regions, depending on the signs of the three eigenvalues as indicated. We shall focus attention on the bounded region of positive-definite matrices, which we denote by Λ_n^+, and its boundary which will be denoted by Σ_n. Notice that Σ_n is a semi-algebraic set (given by polynomial equations and inequalities), since it is only part of the real algebraic variety Z_n.

Note that Σ_n contains not only P_n, as a singular locus, but also another copy P_n^* given by

$$\lambda_1 = 0 \ , \ \lambda_2 = \lambda_3 = \frac{1}{2} \ .$$

Unlike P_n the copy P_n^* consists of smooth points of Z_n.

Our first key lemma is

LEMMA 3. *The set Λ_n^+ is convex and its boundary Σ_n is homeomorphic to a sphere of dimension $d(n) = 3 \cdot 2^n + 1$.*

The convexity follows from the fact that Z_n is a *cubic* hypersurface so that any line meets it in at most three points. Thus a chord of Σ_n cannot exit and then reenter Λ_n^+. Projection from any interior point, say $I/3$, then gives the required homeomorphism with the sphere.

REMARK. For the classical cases $n = 0, 1, 2$ this lemma is directly evident from the properties of eigenvalues of Hermitian matrices: in fact, as we see, it also holds for the octonionic case $n = 3$. The classical cases were used by Arnold [**3**] to establish Theorem A for $n = 1, 2$. He also had results for larger matrices. By contrast we stick to 3×3 matrices but handle also the Cayley case.

Our next lemma describes the $\mathfrak{G}_n$-orbit structure of $\mathfrak{P}_n$.

LEMMA 4. *The $\mathfrak{G}_n$-orbits on $\mathfrak{P}_n$ are as follows:*

(1) *Two open orbits, given by matrices in Λ_n^+ and by matrices with $\lambda_1 < 0 < \lambda_2 \leq \lambda_3$;*
(2) *Two orbits of codimension one, namely $\Sigma_n \setminus P_n$ (which is given by $\lambda_1 = 0 < \lambda_2 \leq \lambda_3$) and $Z_n \setminus \Sigma_n$ (which is given by $\lambda_1 < 0 = \lambda_2 < \lambda_3$);*
(3) *One orbit of codimension $2^n + 2$, namely P_n, given by $\lambda_1 = \lambda_2 = 0$ and $\lambda_3 = 1$.*

Proof. This follows from the fact that the $\mathfrak{G}_n$-orbits on H_n are characterized by the rank of the matrix and the sign of the eigenvalues (in the projective space $\mathfrak{P}_n$ a matrix X is also equivalent to $-X$).

Finally we shall need to know the $\mathfrak{G}_{n-1}$-orbits on P_n:

LEMMA 5. *The action of $\mathfrak{G}_{n-1}$ on P_n has two orbits, namely P_{n-1} and its complement.*

Proof. First we prove this for $n = 1$. We have to show that $SL(3, R)$ acts transitively on $CP^2 \setminus RP^2$. Let ξ be in this open set. Then $\xi \neq \bar{\xi}$, so we have a unique CP^1 joining ξ and $\bar{\xi}$. This CP^1 is the complexification of a suitable line RP^1 in RP^2. The subgroup of $SL(3, R)$ that leaves this RP^1 invariant induces on CP^1 an action of $GL(2, R)$. Now $SL(2, R)$ acting on $CP^1 = S^2$ acts transitively on each hemisphere (the model of the hyperbolic plane), and an element of determinant -1 switches the two hemispheres. Together with the fact that $SL(3, R)$ is transitive on lines in RP^2 this establishes the lemma for $n = 1$. To prove it for $n = 2, 3$ we use the second embeddings introduced in Section 5,

$$CP^2(j) \subset HP^2 \quad \text{and} \quad g(CP^2) \subset OP^2 \ ,$$

which according to (5.6) and (5.7) intersect the original CP^2 in RP^2. We recall that the orbits of the compact groups $Sp(3)$ and F_4 cut out on $CP^2(j)$ and $g(CP^2)$ the orbits of $SO(3)$. Since we have shown that $SL(3, R)$ acts transitively on $CP^2 \setminus RP^2$, and since

$$SL(3, R) = \mathfrak{G}_0 \subset \mathfrak{G}_1 \subset \mathfrak{G}_2 \ ,$$

Lemma 5 follows for $n = 2, 3$.

By definition the map $\pi_n : P_n \to \mathfrak{P}_{n-1}$ of (7.2) is the identity on P_{n-1}, and from Lemma 4 and Lemma 5 we see that π_n must map P_n into Σ_{n-1}, since this is the only compact union of two $\mathfrak{G}_{n-1}$-orbits which lies in the affine part $H_{n-1}(1)$ of $\mathfrak{P}_{n-1}$. It is easy to check that we cannot have $\pi_n(P_n) = P_{n-1}$: it is enough to check this for $n = 1$, when it cannot happen for dimension reasons ($\pi_1(P_1)$ being of dimension 4). Thus we deduce the following projective refinement of Theorem A:

THEOREM A′ *The map* $\pi_n : P_n \to \mathfrak{P}_{n-1}$ *has image* Σ_{n-1} *and* $P_n \setminus P_{n-1} \to \Sigma_{n-1} \setminus P_{n-1}$ *is a homogeneous fibration for the group* $\mathfrak{G}_{n-1}$.

REMARKS. 1) Note that Σ_{n-1} does not have a natural smooth structure compatible with the action of $\mathfrak{G}_{n-1}$, since P_{n-1} is a singular locus. However, if we restrict to the compact subgroup $G_{n-1} \subset \mathfrak{G}_{n-1}$, which fixes $I/3$, then projection from $I/3$ maps Σ_{n-1} to the sphere $S^{d(n-1)}$ (as shown by Lemma 3) and now the map

$$P_n \to S^{d(n-1)}$$

identifies the smooth structure of $S^{d(n-1)}$ with the quotient smooth structure of P_n as explained in Section 2. Thus Theorem A follows from Theorem A′ and Lemma 3. The smoothing of Σ_{n-1} by the radial projection effectively "rounds off the corners" as exemplified by the projection of a square from its centre onto a circle.

2) By considering the projective lines of P_n we get the following more precise picture of the geometry of Z_n in relation to P_n. Consider any line in P_n. This is a sphere of dimension 2^n embedded in the standard way as a real quadric in $\mathfrak{P}_n$ (lying in a linear subspace of dimension $2^n + 1$). Take its interior, an open ball, and its closure. This lies inside Σ_n, and Σ_n is filled up by the union of the closed balls. If we fix a standard metric on P_n, and hence a compact subgroup $G_n \subset \mathfrak{G}_n$, then the sphere becomes a round sphere in $H_n(1)$ and its centre lies on P_n^*, which is the locus of such centres under the action of G_n. This shows that P_n^* is the closest part of Σ_n to the centre $I/3$. For a fixed line in P_n the geometry of its interior is just hyperbolic geometry and the subgroup of $\mathfrak{G}_n$ preserving the line is the conformal group of the sphere or the isometry group of its interior. The corresponding complex picture is described in the appendix.

3) Since π_n is a linear projection it takes the interiors of real quadrics to the interiors of their projections. Remark 2 then shows that π_n maps $\Sigma_n \setminus P_n$ into the interior of Σ_{n-1}. Finally, since this latter is an open orbit of $\mathfrak{G}_{n-1}$, it follows that we get the whole of the interior.

We shall now consider to what extent there is a Theorem B′ analogous to Theorem A′. It is easy to see that there can be no version compatible with $\mathfrak{G}_{n-1}$, since this moves the projective plane P_{n-1}^* while, for Theorem B (for $n > 0$), P_{n-1}^* is a distinguished subspace of the sphere, being the image of the exceptional fibre. However the first part of Theorem A′ extends to give

THEOREM B′. *The image of the map* $\sigma_n : P_n(C) \to H_n(0)$ *is* Σ_n.

Proof. For $n = 0$ we have $\sigma_0 = \pi_1$ (of Theorem A′) and hence Theorem B′ for $n = 0$ follows from Theorem A′ for $n = 1$. For $n \geq 1$ we have a commutative diagram

$$\begin{array}{ccc} P_{n-1}(C) & \xrightarrow{\sigma_{n-1}} & H_{n-1}(0) \\ \downarrow & & \downarrow \\ P_n(C) & \xrightarrow[\sigma_n]{} & H_n(0) \end{array} .$$

Moreover the codimension one orbits of G_n in $P_n(C)$ cut out the codimension one orbits of G_{n-1} on $P_{n-1}(C)$. Hence the image of σ_n is the union of the G_n-orbits of the image of σ_{n-1}. But, from the characterization of the G_n-orbits in H_n by their eigenvalues, it then follows that

$$G_n(\Sigma_{n-1}) = \Sigma_n .$$

Thus Theorem B′ follows by induction on n.

8. Appendix

In Section 6 we studied the differential geometry of the complexified projective varieties $P_n(C)$. In this appendix we will review (without complete proofs) some of the algebraic geometry which is of independent interest and provides further background. All dimensions in this section are complex dimensions. For further details see [**15**].

The algebraic variety

$$P_n(C) \subset \mathfrak{P}_n(C) \quad (n = 0, 1, 2, 3)$$

of dimension 2^{n+1} has appeared as an orbit of the complex Lie group $\mathfrak{G}_n(C)$ in the complex projective space $\mathfrak{P}_n(C)$ of dimension $3 \cdot 2^n + 2$. The group $\mathfrak{G}_n(C)$ is the complexification of the compact Lie group $\hat{G}_n$ given in table (6.1). Explicitly we have the sequence of groups $\mathfrak{G}_n(C)$ $(n = 0, 1, 2, 3)$

$$SL(3, C) \ , \ SL(3, C) \times SL(3, C) \ , \ SL(6, C) \ , \ E_6(C) \ .$$

These have irreducible representations on the vector spaces $H_n(C) = H_n \otimes C$, which are the complexifications of the real vector spaces H_n of all 3×3 Hermitian matrices over the division algebra A_n. These representations have dimension $3(2^n + 1)$, explicitly (for $n = 0, 1, 2, 3$)

$$6 \ , \ 9 \ , \ 15 \ , \ 27 \ ,$$

and they projectivize to give the spaces $\mathfrak{P}_n(C)$.

We can also consider the dual representations on $H_n(C)^*$ giving dual projective spaces $\mathfrak{P}_n(C)^*$, and in these there is a unique compact orbit $P_n(C)^*$ of $\mathfrak{G}_n(C)$.

On $H_n(C)$ there is a unique (up to scalars) cubic polynomial invariant under $\mathfrak{G}_n(C)$ which we have denoted by det, for reasons explained in Section 3. For $n = 0, 1$ it is just the usual determinant. For $n = 2$ it is the $SL(6, C)$-invariant cubic on $\Lambda^2(C^6)$ given by the exterior cube into $\Lambda^6(C^6) \cong C$. For $n = 3$ with $\mathfrak{G}_3(C) = E_6(C) \subset SL(27, C)$ this invariant cubic was discovered by E. Cartan [**11**]. In all cases it defines a cubic hypersurface in $\mathfrak{P}_n(C)$ which we denote by $Z_n(C)$, and which contains $P_n(C)$.

The action of $\mathfrak{G}_n(C)$ on $\mathfrak{P}_n(C)$ has just three orbits, namely $P_n(C)$, $Z_n(C) \setminus P_n(C)$ and $\mathfrak{P}_n(C) \setminus Z_n(C)$. When $n = 0$ these just correspond to symmetric matrices of ranks $1, 2, 3$, and we could use the same terminology in the general case. In fact $Z_n(C) \setminus P_n(C)$ is the $\mathfrak{G}_n(C)$-orbit of the diagonal matrix $\mathrm{Diag}(1, 1, 0)$ of rank 2, and as we have already observed $P_n(C)$ is the $\mathfrak{G}_n(C)$-orbit of the diagonal matrix $\mathrm{Diag}(1, 0, 0)$. Points of rank 3 constitute the $\mathfrak{G}_n(C)$-orbit of the unit matrix. The complex Lie group $\mathfrak{G}_n(C)$ is just the identity component of the group of holomorphic transformations of $P_n(C)$.

The natural embedding of 2×2 matrices into 3×3 matrices by adding zeroes in the third row and column gives a linear subspace

$$\mathfrak{L}_n(C) \subset \mathfrak{P}_n(C) \ , \ \dim \mathfrak{L}_n(C) = 2^n + 1 \ ,$$

and its orbit under $\mathfrak{G}_n(C)$ fills out the whole of $Z_n(C)$. The intersection

$$\mathfrak{L}_n(C) \cap P_n(C) = L_n(C)$$

is a complex quadric of dimension 2^n. In fact, $L_n(C)$ inside $\mathfrak{L}_n(C)$ plays the same role (for 2×2 matrices) that $Z_n(C)$ does inside $\mathfrak{P}_n(C)$: it is the hypersurface given by the invariant *quadratic* $\det_2$ for 2×2 matrices.

The family of all transforms of $\mathfrak{L}_n(C)$ under $\mathfrak{G}_n(C)$ therefore cuts out on $P_n(C)$ a corresponding family of quadrics $L_n(C)$ with

$$\dim L_n(C) = \frac{1}{2} \dim P_n(C) .$$

The dual $\mathfrak{L}_n(C)^*$ of $\mathfrak{L}_n(C)$ is a linear subspace of $\mathfrak{P}_n(C)^*$ of dimension

$$(3 \cdot 2^n + 2) - (2^n + 1) - 1 = 2^{n+1} = \dim P_n(C)^* .$$

In fact, $\mathfrak{L}_n(C)^*$ is the tangent space to $P_n(C)^*$ at a point ℓ_n. The correspondence

$$L_n(C) \longleftrightarrow \ell_n$$

represents $P_n(C)^*$ as the parameter family of the quadrics $L_n(C)$ on $P_n(C)$, and the situation is symmetrical (or dual): points of $P_n(C)$ parametrize quadrics (of half the dimension) on $P_n(C)^*$.

Note that the quadrics $L_n(C)$ are all non-singular (since all points are of "rank 1"), whereas $Z_n(C)$ has $P_n(C)$ as a *singular locus*: a generic point of $Z_n(C)$ has rank 2, whereas points of $P_n(C)$ are of rank 1. Thus $Z_n(C)$ determines $P_n(C)$ (as its singular locus), and conversely $P_n(C)$ determines $Z_n(C)$, as the space generated by the linear spaces $\mathfrak{L}_n(C)$ spanned by the quadrics $L_n(C)$.

It may be helpful at this stage if we looked in detail at the special case $n = 0$, so that we are dealing with the classical embedding of CP^2 as the Veronese surface V in CP^5. The lines of CP^2 become conics on V and these lie in planes. The lines of CP^2 are parametrized by the dual CP^2 which can be identified with the dual Veronese surface V^* in $(CP^5)^*$. Thus the planes spanned by the conics in V form a 2-parameter family and they fill out a (cubic) hypersurface Z. Since every pair of distinct points on CP^2 lies on a unique line, every pair of distinct points of V lies on a unique conic. In particular it follows that the chordal variety of V (i.e. the closed subspace generated by all chords) is also the space generated by all the planes spanned by the conics and hence is the hypersurface Z. This is a very unusual situation for a surface in CP^5. On dimension grounds one could expect the chordal variety to be the whole ambient space. Equivalently, when the chordal variety is only a hypersurface, the projection from a generic point gives an *embedding*[6] (without singularities) in CP^4. In fact, it is a classical result of Severi **[21]** that the Veronese surface is the only surface (not contained in a hyperplane) in CP^5 with this property.

Zak **[26]** (see also **[16]**) has investigated this "Severi property" for higher dimensions when $V_d \subset CP^N$. The critical case is when $d = \frac{2}{3}(N-2)$. Zak proved the remarkable result that, firstly

$$d = 2^{n+1} \ , \ n = 0, 1, 2, 3 \ ,$$

and secondly that the only such varieties in these dimensions are the complexified projective planes $P_n(C)$ in their standard projective embeddings in $\mathfrak{P}_n(C)$. For this reason these varieties have been named Severi varieties **[15]**.

To see how this fits into the picture we have described we have to note that in all cases *the chordal variety of $P_n(C)$ is the cubic hypersurface $Z_n(C)$*. The proof

[6]Over the reals this gives an embedding $RP^2 \subset RP^4$. Since this is of degree 2 it lifts to the double cover S^4 giving the embedding $RP^2 \subset S^4$ of (1.2).

is very similar to the case $n = 0$, but with a caveat. Given any two distinct points x and y in $P_n(C)$ there are just two possibilities, either

(i) the projective line in $\mathfrak{P}_n(C)$ containing x and y lies entirely on $P_n(C)$, or
(ii) there is a unique quadric $L_n(C)$ on $P_n(C)$ containing both x and y.

For $n = 0$ case (i) never happens: the classical Veronese variety contains no lines. Clearly (ii) is the generic situation and as before this implies that the *hypersurface* $Z_n(C)$ *(generated by the planes* $\mathfrak{L}_n(C)$ *spanned by the* $L_n(C)$*) is precisely the chordal variety of* $P_n(C)$. This shows that $P_n(C)$ does indeed have the Severi property in $\mathfrak{P}_n(C)$. The power of Zak's Theorem is that these are the only ones.

There is a striking resemblance between Zak's theorem in complex algebraic geometry and the classical results about division algebras and projective planes. It would be interesting to see if a purely topological proof of Zak's theorem could be found. We recall that the use of Steenrod squares enables one to prove that a projective plane must have dimension a power of 2 (analogous to the first part of Zak's Theorem), while K-theory is needed for the final part [**2**]. One is therefore tempted to expect a K-theory proof of Zak's theorem, particularly in view of the role that K-theory, as developed by Grothendieck, plays in algebraic geometry.

It is at this point that we should perhaps pass from the purely complex approach to the varieties $P_n(C)$ and introduce real structures. Recall that $P_n(C)$ has a complex conjugation, preserved only by the real subgroup $\mathfrak{G}_n$ of $\mathfrak{G}_n(C)$, and that the set of real points just recover the original projective planes P_n. The complex quadrics $L_n(C)$ on $P_n(C)$, parametrized by $P_n(C)^*$, include those preserved by conjugation and parametrized by P_n^*. These are just the complexifications of the projective lines in P_n (i.e. spheres of dimension 2^n). The incidence properties of these projective lines in P_n imply generically the corresponding properties of the complex quadrics $L_n(C)$ in $P_n(C)$, ensuring (ii) above. However, for $n \geq 1$, the exceptional case (i) does occur. For example, when $n = 1$, we have $P_1(C) = CP^2 \times CP^2$, and two distinct point pairs whose first components agree give case (i).

The manifolds $P_n(C)$, with their family of submanifolds $L_n(C)$ of middle dimension, are examples of generalized projective planes in the sense of Atsuyama [**6**], who studied these from the point of view of differential geometry.

Recall that we have the following subgroups of $\mathfrak{G}_n(C)$:

$$\begin{array}{ccc} G_n & \longrightarrow & \mathfrak{G}_n \\ \downarrow & & \downarrow \\ \hat{G}_n & \longrightarrow & \mathfrak{G}_n(C) \end{array},$$

where the groups in the first column are the maximal compact subgroups of those in the second column and preserve the metrics on P_n and $P_n(C)$ respectively. We can also introduce the groups $G_n(C)$, the complexification of G_n. Since the representation $H_n(C)$ splits off a trivial factor of dimension one (given by the trace) when restricted to $G_n(C)$, it follows that there is a hyperplane section $P_n(\infty)$ of P_n invariant under $G_n(C)$. In fact, $G_n(C)$ acts on $P_n(C)$ with just two orbits, $P_n(\infty)$ and its complement. We propose to examine the geometry of $P_n(\infty)$ which is a homogeneous space of $G_n(C)$ and so also of G_n. In particular we will see that we can reconstruct $P_n(C)$ canonically from $P_n(\infty)$.

It is instructive to consider first the simple case $n = 0$. Then $P_0(\infty)$ is a rational normal quartic curve, the image under the Veronese embedding of a conic in CP^2. This is the conic $z_1^2 + z_2^2 + z_3^2 = 0$ invariant under $G_0(C) = SO(3, C)$. Consider now the conics on $V = P_0(C)$ (the Veronese surface) which are complexifications of projective lines in $P_0 = RP^2$. In the "abstract" CP^2 which maps to V these just correspond to projective lines with real equations. Each such line meets the conic $z_1^2 + z_2^2 + z_3^2 = 0$ in a pair of conjugate points. The quotient of the conic by this involution is naturally identified with the *dual* RP^2. This is part of the content of Theorem B for $n = 0$. If we now consider all conics on V, they come from all projective lines on CP^2, and they meet the conic $z_1^2 + z_2^2 + z_3^2 = 0$ in any pair of points. Thus if we start with our rational normal curve $P_0(\infty)$ on $P_0(C)$ and consider the variety of all (unordered) pairs of points on $P_0(\infty)$ (i.e. its symmetric square) we get the dual CP^2. From this we can (by duality) recover the original CP^2 and also its Veronese embedding. Thus $P_0(\infty)$ determines the whole picture.

We want to show that this is typical of the general case (i.e. for all n).

We consider the real family of complex quadrics $L_n(C)$ parametrized by points of $P_n^* \subset P_n(C)^*$. These are just the complexifications of the projective lines of the projective plane P_n. Since any two distinct points of P_n are joined by a unique line they are never in the special position of (i). Dually this means that no two quadrics of the real family of $L_n(C)$ meet in more than one point. Since they already have one common point on P_n they meet nowhere else. In particular the family of quadrics of one lower dimension cut out on $P_n(\infty)$,

$$L_n(\infty) = L_n(C) \cap P_n(\infty) ,$$

are all disjoint. A dimension count shows that they must fill out the whole of $P_n(\infty)$, and since they are by construction parametrized by the dual space P_n^* we get a fibration

$$P_n(\infty) \to P_n^* \tag{8.1}$$

with fibre $L_n(\infty)$. This gives a more explicit description of the behaviour at infinity of the map of Theorem B. Note that, for $n = 0$, $L_0(\infty)$ has dimension zero and is a point-pair as we have already seen.

Motivated by the case $n = 0$ we now consider the full complex family of quadrics $L_n(C)$. These intersect $P_n(\infty)$ in the full complex family of quadrics (of one lower dimension) whose real members are the fibres of (8.1). This full family is therefore parametrized by the complexification $P_n(C)^*$ of P_n^*. Since we have cut down the symmetry from $\mathfrak{G}_n(C)$ to $G_n(C)$ there is a distinguished subset $P_n(\infty)^*$. These quadrics in $P_n(\infty)$ are *singular*, they arise from quadrics $L_n(C)$ which touch $P_n(\infty)$: note that these are never the real members (the fibres of (8.1)) since P_n^* and $P_n(\infty)^*$ are disjoint. All of this checks with what we saw for $n = 0$. Thus $P_n(\infty)$ contains a family of complex quadrics $L_n(\infty)$, parametrized by $P_n(C)^*$. This enables us to recover $P_n(C)^*$ and hence $P_n(C)$ from $P_n(\infty)$.

The fibration (8.1) can be viewed as a *twistor fibration*. For example when $n = 1$, $P_1(C) = CP^2 \times (CP^2)^*$ and $P_2(\infty)$ is the incidence locus. It is therefore the flag manifold of $SU(3)$ and (8.1) is the twistor fibration for CP^2 regarded as a 4-manifold with self-dual metric [**4**]. For $n = 2, 3$ the fibration (8.1) is a partial twistor fibration in the sense of Bryant [**10**], and it is given as an interesting example of a general theory. Explicitly, the two fibrations are

$$P_2(\infty) = \frac{Sp(3)}{Sp(1)U(2)} \longrightarrow \frac{Sp(3)}{Sp(1)Sp(2)} = HP^2 \cong P_2^*$$

and

$$P_3(\infty) = \frac{F_4}{Spin(7)U(1)} \longrightarrow \frac{F_4}{Spin(9)} = OP^2 \cong P_3^* ,$$

where the fibres are $L_2(\infty) = Sp(2)/U(2)$ and $L_3(\infty) = Spin(9)/Spin(7)U(1)$ respectively. Note that $Sp(2)/U(2)$ is isomorphic to the 3-dimensional quadric $SO(5)/SO(3)SO(2)$ and that $Spin(9)/Spin(7)U(1)$ is isomorphic to the 7-dimensional quadric $SO(9)/SO(7)SO(2)$.

References

[1] J.F. Adams: *Lectures on exceptional Lie groups.* University of Chicago Press, Chicago, 1996.

[2] J.F. Adams, M.F. Atiyah: K-theory and the Hopf invariant. *Quart. J. Math. Oxford II. Ser.* **17** (1966), 31-38.

[3] V.I. Arnold: Relatives of the quotient of the complex projective plane by complex conjugation. *Proc. Steklov Inst. Math.* **224** (1999), 46-56.

[4] M.F. Atiyah, N.J. Hitchin, I.M. Singer: Self-duality in four-dimensional Riemannian geometry. *Proc. Royal Soc. London, Ser. A* **362** (1978), 425-461.

[5] M. Atiyah, E. Witten: M-theory dynamics on a manifold of G_2 holonomy. To appear in *J. Theor. Math. Phys.*, hep-th/0107177.

[6] K. Atsuyama: Projective spaces in a wider sense, I. *Kodai Math. J.* **15** (1992), 324-340.

[7] J.C. Baez: The octonions. *Bull. Am. Math. Soc.* **39** (2002), 145-205.

[8] A. Borel: Le plan projectif des octaves et les spheres comme espaces homogenes. *C.R. Acad. Sci., Paris* **230** (1950), 1378-1380.

[9] A. Borel, J. de Siebenthal: Les sous-groupes fermes de rang maximum des groupes de Lie clos. *Comment. Math. Helv.* **23** (1949), 200-221.

[10] R.L. Bryant: Lie groups and twistor spaces. *Duke Math. J.* **52** (1985), 223-261.

[11] E. Cartan: *Sur la structure des groupes de transformations finis et continus.* Ph.D. Thesis, Paris, 1894.

[12] E. Cartan: Sur des familles remarquables d'hypersurfaces isoparamétriques dans les espaces sphériques. *Math. Z.* **45** (1939), 335-367.

[13] H. Freudenthal: Lie groups in the foundations of geometry. *Adv. Math.* **1** (1964), 145-190.

[14] H. Freudenthal: Oktaven, Ausnahmegruppen und Oktavengeometrie. *Geom. Dedicata* **19** (1985), 1-63.

[15] J.M. Landsberg, L. Manivel: The projective geometry of Freudenthal's magic square. *J. Algebra* **239** (2001), 477-512.

[16] R. Lazarsfeld, A. Van de Ven: *Topics in the geometry of projective space.* Birkhäuser, Basel, 1984.

[17] P.S. Mostert: On a compact Lie group acting on a manifold. *Ann. Math.* **65** (1957), 447-455. Erratum: ibid. **66** (1957), 589.

[18] S. Murakami: Exceptional simple Lie groups and related topics in recent differential geometry. In: *Differential geometry and topology,* Proc. Spec. Year, Tianjin/PR China 1986-87, Lect. Notes Math. 1369 (1989), 183-221.

[19] H. Nakagawa, R. Takagi: On locally symmetric Kaehler submanifolds in a complex projective space. *J. Math. Soc. Japan* **28** (1976), 638-667.

[20] T. Püttmann, A. Rigas: Isometric actions on the projective planes and embedded generators of homotopy groups. Preprint, 2002.

[21] F. Severi: Intorno ai punti doppi impropri di una superficie generale dello spazio a quattro dimensioni, e a'suoi punti tripli apparenti. *Rend. Circ. Mat. Palermo* **15** (1901), 33-51.

[22] N. Steenrod: *The topology of fibre bundles.* Princeton University Press, Princeton, 1974.

[23] G. Thorbergsson: A survey on isoparametric hypersurfaces and their generalizations. In: *Handbook of differential geometry, Volume I,* North-Holland, Amsterdam, 2000, 963-995.

[24] O. Veblen, J. Young: *Projective geometry.* Blaisdell Publishing Company, New York Toronto London, 1965.

[25] N.R. Wallach: Compact homogeneous Riemannian manifolds with strictly positive curvature. *Ann. Math. (2)* **96** (1972), 277-295.
[26] F.L. Zak: Severi varieties. *Math. USSR, Sb.* **54** (1986), 113-127.

University of Edinburgh, Department of Mathematics and Statistics, Mayfield Road, Edinburgh EH9 3JZ, United Kingdom
E-mail address: `atiyah@maths.ed.ac.uk`

University of Hull, Department of Mathematics, Cottingham Road, Hull HU6 7RX, United Kingdom
E-mail address: `j.berndt@hull.ac.uk`

Degeneration of Einstein metrics and metrics with special holonomy

Jeff Cheeger

CONTENTS

0. Introduction

A a riemannian manifold, M^n, is called *Einstein* if for some constant, λ, the metric, g, and Ricci tensor, Ric_{M^n}, satisfy

$$\mathrm{Ric}_{M^n} = \lambda g\,.$$

In a local coordinate system in which the coordinate functions are harmonic, the Einstein condition becomes a quasi-linear elliptic equation on the metric; see (1.2).

This paper is not intended as a *general* survey of the subject of the Einstein metrics. A number of topics which are of central importance are not even mentioned e.g. [**Ca1**], [**Yau1**], [**Yau2**]. Rather, we try to give a detailed overview of one *specific* line of progress due to Colding, Tian and the author (in various combinations) which has shed light on the following question:

What sorts of objects arise as limits of sequences of badly behaved Einstein metrics?

It is natural to make the normalization,

$$|\mathrm{Ric}_{M^n}| \leq n-1\,, \tag{0.1}$$

The author was partially supported by NSF Grant DMS 9303999 and the second by NSF Grant DMS 9504994 .

or equivalently, $|\lambda| \le n-1$, since this can be achieved by rescaling the metric; n-dimensional spaces with constant sectional curvature, ± 1, have $\lambda = \pm(n-1)$.

For $n \le 3$, Einstein metrics have constant curvature; [**Be**]. Thus, we will assume $n \ge 4$.

In dimension 4, if collapsing does not take place and the topology remains bounded, the limiting objects are Einstein manifolds with orbifold singularities; [**An1**]–[**An3**], [**Nak1**], [**Nak2**], [**Ti1**], [**Ti2**].

Important features of the 4-dimensional case persist in higher dimensions. In many instances, the limiting objects are known to have singularities of codimension ≥ 4. Conjecturally, the codimension 4 piece of the singular set, consists of singular points of orbifold type. At the infinitesimal level, this is known hold. As in dimension 4, the L_2-norm of curvature plays a distinguished role. However, in higher dimensions, the L_2-norm fails to be scale invariant. This, together with the appearance of nonisolated singularities, causes substantial difficulties.

The approach we describe starts with the development of a structure theory for limits of sequences of manifolds satisfying the weaker assumption,

$$\mathrm{Ric}_{M^n} \ge -(n-1)\,, \tag{0.2}$$

where to be more precise, we should write $\mathrm{Ric}_{M^n} \ge -(n-1)g$. This theory, is outlined in Sections 1–4 below; for an exposition with proofs see [**Ch5**].

Later, additional conditions are added: an upper bound on the Ricci tensor, the assumption that the metric is Einstein, or of special holonomy (e.g. Kähler-Einstein) and finally, an L_p bound on the full curvature tensor. The case, $p = 2$, is particularly significant. For compact manifolds with special holonomy and Ricci tensor normalized as in (0.1), the L_2-norm of the curvature is majorized by a certain topological invariant $C(M^n)$; see (0.10).

Let $\{M_i^n\}$ denote a sequence of compact riemannian manifolds satisfying (0.2) for which the diameters are uniformly bounded, $\mathrm{diam}(M_i^n) \le d$ for all i. By a fundamental observation of Gromov, any such sequence has a subsequence, $\{M_j^n\}$, which converges in a certain weak sense — the Gromov-Hausdorff sense — to some compact metric space Y. We write $M_j^n \xrightarrow{d_{GH}} Y$. Gromov's compactness theorem is a consequence of relative volume comparison, the control which the lower bound on Ricci curvature, (0.2), exerts over the volumes of metric balls.

Gromov-Hausdorff convergence of a sequence of metric spaces is the notion of convegence associated to the Gromov-Hausdorff distance between (isometry classes of) compact metric spaces.

Let (W_1, ρ_1), (W_2, ρ_2), denote compact metric spaces. The Gromov-Hausdorff distance, $d_{GH}((W_1, \rho_1), (W_2, \rho_2))$, is the infimum of those $\epsilon > 0$, for which there exists a metric, ρ, on the disjoint union, $W_1 \amalg W_2$, such that:

i) $\rho \,|\, W_k = \rho_k$, $k = 1, 2$.

ii) W_k is ϵ-dense in $W_1 \amalg W_2$, $k = 1, 2$.

Often, one just writes $d_{GH}(W_2, W_2)$, supressing the metrics ρ_1, ρ_2.

The Gromov-Hausdorff distance is an intrinsic generalization of the classical Hausdorff distance, $d_H(W_1, W_2)$, between compact subsets, W_1, W_2, of a metric space, (W, ρ):

$$d_H(W_1, W_2) := \max\left(\max_{w_1 \in W_1} \rho(w_1, W_2),\ \max_{w_2 \in W_2} \rho(w_2, W_1)\right).$$

Intuitively, the Hausdorff distance between two subsets of a metric space, or Gromov-Hausdorff distance between two metric spaces, is very small if, to the naked eye, the objects appear to be identical.

There is an extension of Gromov-Hausdorff convergence to sequences of pointed metric spaces $\{(W_i, y_i, \rho_i)\}$. Namely, pointed Gromov-Hausdorff convergence means Gromov-Hausdorff convergence of the sequence of balls, $\{B_r(y_i)\}$, for all $r < \infty$. With this understanding, Gromov's compactness theorem has a natural generalization to the case in which the bound on the diameter is dropped. This is used in defining the concept of "tangent cone", which plays a key role in the sequel.

By starting with *any* sequence of manifolds satisfying (0.2), no matter how badly behaved, and passing to a suitable subsequence, we obtain a limit space, Y, whose regularity and singularity structure we can examine. Clearly, the properties of Y will closely reflect those of the manifolds in our sequence.

A length space is a metric space such that every pair of points is joined by a continuous curve whose length is equal to the distance between them. Here, the length of a continuous curve is defined using partitions as in elementary calculus. It is not difficult to see that the Gromov-Hausdorff limit of a sequence of length spaces is again a length space. This completely general result is one of the few easily established properties of Gromov-Hausdorff limit spaces satisfying

$$\mathrm{Ric}_{M_i^n} \geq -(n-1)\,. \tag{0.3}$$

Since the elliptic equation satisfied by Einstein metrics is *nonlinear*, it is quite possible for limit spaces to have singularities. Many such examples of singular limit spaces are known; see e.g. [**Joy3**] and the references therein. In actuality, since there is no fixed background metric, the Einstein equation is *more* nonlinear than the partially analogous equations for minimal surfaces, harmonic maps and Yang-Mills fields. In and of itself, this makes the Einstein case harder to handle.

EXAMPLE 0.4. (Eguchi-Hansen manifolds) The unit sphere bundle of TS^2, the tangent bundle of the 2-sphere, is diffeomorphic to the real projective space $\mathbf{RP}(3)$. Thus, the complement of the unit disc bundle of TS^2 is diffeomorphic to $(1,\infty) \times \mathbf{RP}(3)$, and hence to a neighborhood of infinity in $\mathbf{R}^4/\mathbf{Z}_2$, where the action of $\mathbf{Z}_2$ is generated by the antipodal map.

In fact, TS^2 has a Ricci flat (hyper-Kähler) metric, g, the Eguchi-Hansen metric, which at infinity, becomes rapidly asymptotic to the space, $(1,\infty) \times \mathbf{RP}(3) \subset \mathbf{R}^4/\mathbf{Z}^2$, with its canonical flat metric; see [**EgHan**].

When $\epsilon \to 0$, the 1-parameter family, $(TS^2, \epsilon^2 g)$, converges in the pointed Gromov-Hausdorff sense, to the flat singular cone $\mathbf{R}^4/\mathbf{Z}_2$. The zero section, a homologically nontrivial 2-sphere, shrinks in the limit to the vertex of the cone. Since the condition of being Ricci flat is invariant under scaling of the metric, the space, $\mathbf{R}^4/\mathbf{Z}_2$, is the Gromov-Hausdorff limit of a sequence of Ricci flat manifolds. The singular set of this limit space has codimension 4.

Rescaling, rigidity and almost rigidity.

Let $M_i^n \xrightarrow{d_{GH}} Y$ satisfy (0.3). Any *individual* manifold, M_i^n, looks locally like $\mathbf{R}^n$ on a sufficiently small scale. Thus, the formation of singularities in the limit reflects the absence of *uniform* control over the scale on which the metric becomes standard.

The theory we will describe gives restrictions on the *Hausdorff dimension* of the singular set of Y. From this it follows that uniform control of the metric does exist, except perhaps on a set which is very small in a definite sense. We will also give restrictions on the *structure* of the singular set. These imply that there is a weaker sort of uniform control at all points of all manifolds in the approximating sequence.

Imagine that a manifold with $\mathrm{Ric}_{M^n} \geq -(n-1)$ and $\mathrm{diam}(M^n) = d$, is observed under a powerful microscope, so that all distances appear to have been multiplied by a large factor ϵ^{-1}. A ball of radius ϵ will appear to have radius 1 and the Ricci curvature will appear to be bounded below by $-(n-1)\cdot\epsilon^2$. Thinking more globally, we can just rescale the distance function on the whole manifold by a factor ϵ^{-1}. The Ricci curvature of the rescaled metric is bounded below by $-(n-1)\cdot\epsilon^2$. The rescaled diameter is $\epsilon^{-1}\cdot d$, which for ϵ^{-1} sufficiently large, is effectively infinite.

The rescaling idea, which is pervasive in science, strongly suggests that after suitable rescaling, the small scale properties of manifolds whose Ricci curvature has a definite lower bound should resemble the fixed scale properties of complete noncompact manifolds whose Ricci curvature is nonnegative.

Since ϵ, while very small, is nonetheless strictly positive, implementation of this approach requires quantitative versions of theorems on nonnegative Ricci curvature, allowing for small errors in both the hypotheses and conclusions. The relevant theorems are *rigidity* theorems, for example, the splitting theorem; [**Top2**], [**ChGl1**]. Their corresponding quantitative versions are called *almost rigidity theorems*; [**ChCo2**].

The implications of rescaling can be expressed directly in terms of limit spaces:

At the infinitesimal level, limit spaces have nonnegative Ricci curvature in a generalized sense.

The noncollapsed case.

Much, but not all, of our discussion will be restricted to the *noncollapsed* case, in which it is assumed apriori, that there exists $v > 0$, such that for all $m_i \in M_i^n$,

$$\mathrm{Vol}(B_1(m_i)) \geq v\,. \tag{0.5}$$

There are many specific examples of noncollapsed limit spaces with interesting properties.

The *collapsed* case is also of great interest, for example in connection with mirror symmetry; compare [**StYauZa**], [**GsWil**], [**KonSoi**]. Examples show that the situation in the collapsed case is much less constrained, and by the same token, much less is understood. To date, the main structure theorem in the collapsed case, asserting the rectifiability of Y with respect to any so-called renormalized limit measure, has not been significantly strengthened if (0.2) is replaced by the Einstein condition; see [**ChCo5**].

Concentration of curvature and formation of singular sets.

Behavior like that in Example 0.4 could not occur for noncollapsed Gromov-Hausdorff limit spaces with a uniform bound on *sectional* curvature. Such limit spaces are actually smooth manifolds and the metric space structure arises from a riemannian metric which, in harmonic coordinate systems, is $C^{1,\alpha}$, for any $\alpha < 1$; [**Ch1**], [**GvLP**], [**An2**], [**JoKar**].

For the family of Eguchi-Hansen metrics described in Example 0.4, the singular set consists of a single point at which the curvature concentrates in the limit in the L_2 sense. It does not concentrate in the L_p sense for $1 \leq p < 2$. In considerable generality, for noncollapsed limit spaces with a uniform lower Ricci curvature bound, the failure of curvature to concentrate sufficiently in the L_p sense prevents the formation of a singular set with positive $(n-2p)$-dimensional Hausdorff measure. The proof of this fact relies on certain previously established constraints on the the structure of singular sets in dimension $n-2p$.

Special holonomy.

Important examples of Einstein manifolds are furnished by manifolds with special holonomy.

By definition, the holonomy group, H_p of a riemannian manifold, M^n, at $p \in M^n$, is the group of the orthogonal transformations of the tangent space, M^n_p, obtained by parallel translating around smooth closed loops based at p. The holonomy group is always a Lie group; [**Yam**]. A smooth curve from p to q induces an isomorphism of corresponding holonomy groups. Up to conjugacy in the orthogonal group, this isomorphism is independent of the chosen path. ¿From now on we drop the dependence on the base point and just write H for the holonomy group.

The *restricted holonomy group*, H_0, is the identity component of H. Equivalently, it is the subgroup generated by loops which are contractible. In particular, H_0 is a normal subgroup of H.

A riemannian manifold, M^n, is said to have *special restricted holonomy*, if it is locally irreducible, H_0 is a proper subgroup of $\mathrm{SO}(n)$ and there exists some (possibly distinct) riemannian manifold, with restricted holonomy group H_0, which is not a symmetric space.

The special restricted holonomy groups, H_0, are:

$$\mathrm{U}(\frac{n}{2}), \quad \mathrm{SU}(\frac{n}{2}), \quad \mathrm{Sp}(\frac{n}{4})\mathrm{Sp}(1) \quad (n > 4),$$

$$\mathrm{Sp}(\frac{n}{4}), \quad \mathrm{G}_2, \quad \mathrm{Spin}(7);$$

see [**Ber**]; see also [**Si**], [**Al**], [**BrGr**].

Each of these groups occurs as the holonomy group of a compact simply connected riemannian manifold; see [**Joy3**]. In all cases, the representation of the restricted holonomy group is unique up to conjugacy $\mathrm{O}(n)$, and the action of the holonomy group on the unit sphere is transitive; [**Si**].

We make the following convention:

A manifold (which might not be locally irreducible) has special holonomy if H is contained in one of the above groups.

If $H \subset \mathrm{U}(\frac{n}{2})$, the dimension, n, is even and there is an almost complex structure, J, which is parallel, and hence, integrable. Equivalently, the Kähler form, $\omega = g(J\,\cdot\,,\,\cdot\,)$, is closed. The first Chern class is represented by the 2-form,

$$c_1 = \frac{1}{2\pi}\mathrm{Ric}_{M^n}(J\,\cdot\,,\,\cdot\,), \tag{0.6}$$

which, up to normalization, can also be viewed as the curvature of the anticanonical line bundle. The Einstein condition is equivalent to

$$c_1 = \frac{1}{2\pi}\lambda\omega\,. \tag{0.7}$$

If $H_0 = \mathrm{U}(\frac{n}{2})$, the underlying manifold is never Ricci flat. If $H_0 \subset \mathrm{SU}(\frac{n}{2})$, M^n it is always Ricci flat.

If $H \subset \mathrm{Sp}(\frac{n}{4})\mathrm{Sp}(1)$ then n is a multiple of 4. In this case, there is a distinguished canonically oriented 3-dimensional *parallel sub-bundle*, E, of the endomorphism bundle of the tangent bundle, such that any oriented orthonormal basis of a fibre consists of a triple, I, J, K, of anticommuting almost complex structures satisfying $IJ = K$. The 4-form, $\omega_I^2 + \omega_J^2 + \omega_K^2$, the sum of the Kähler forms associated to I, J, K, is parallel and is independent of the specific choice of I, J, K.

If $H_0 \subset \mathrm{Sp}(\frac{n}{4})$, the bundle E is flat. Thus, manifolds of this type carry 3 locally defined parallel almost complex structures I, J, K. Hence, the individual Kähler forms, ω_I, ω_J, ω_K, are parallel in this case. More generally, an endomorphism of the form, $rI + sJ + tK$, with $r^2 + s^2 + t^2 = 1$, is a parallel almost complex structure. With respect to any such complex structure, $\mathrm{Sp}(\frac{n}{4}) \subset \mathrm{SU}(\frac{n}{2})$. Therefore, this case is Ricci flat.

If $H_0 = \mathrm{Sp}(\frac{n}{4})\mathrm{Sp}(1)$, the bundle, E, does not have parallel local sections and in general, a manifold with holonomy, $\mathrm{Sp}(\frac{n}{4})\mathrm{Sp}(1)$, need not admit any integrable almost complex structure

For $n > 4$, M^n is Einstein, but not Ricci flat. The curvature tensor of E can be expressed in terms of the Ricci tensor and a triple of almost complex structures, I, J, K; compare (0.7) and see (10.3), (10.6). The curvature tensor of E is parallel and its norm is determined by the dimension, n, and Einstein constant.

The unit sphere bundle, $S(E)$, is called the *twistor space* associated to M^n. The natural metric for which the projection to M^n is a riemannian submersion has the property that the Ricci tensor and all of its covariant derivatives are bounded. In addition, there is a canonical integrable almost complex structure on $S(E)$, all of whose covariant derivatives are bounded. These facts allow many questions concerning manifolds with $H \subset \mathrm{Sp}(\frac{n}{4})\mathrm{Sp}(1)$ to be reduced to the corresponding questions for a slight generalization of the $\mathrm{U}(\frac{n}{2})$ case.

For $n = 4$, we have the canonical isomorphism $\mathrm{Sp}(1)\mathrm{Sp}(1)=\mathrm{SO}(4)$. Hence, $\mathrm{Sp}(1)\mathrm{Sp}(1)$ is *not* a special restricted holonomy group and the metric on the underlying manifold is not constrained. Einstein 4-manifolds with anti-self-dual curvature, are the analogs of higher dimensional manifolds with $H = \mathrm{Sp}(\frac{n}{4})\mathrm{Sp}(1)$. For such 4-manifolds, the curvature properties of the bundle, E, are as in the higher dimensional case and the associated twistor space plays an analogous role.

We note the inclusions

$$\mathrm{Sp}(\frac{n}{4}) \subset \mathrm{SU}(\frac{n}{2}) \subset \mathrm{U}(\frac{n}{2})\,,$$

$$\mathrm{Sp}(\frac{n}{4}) \subset \mathrm{Sp}(\frac{n}{4})\mathrm{Sp}(1)\,.$$

The cases, $H_0 \subset \mathrm{SU}(\frac{n}{2})$, $H_0 \subset \mathrm{Sp}(\frac{n}{4})$, can be viewed as limits of the $\mathrm{U}(\frac{n}{2})$, $\mathrm{Sp}(\frac{n}{4})\mathrm{Sp}(1)$ cases, in which the norm of the Ricci tensor tends to zero.

Manifolds with the restricted holonomy groups G_2, $\mathrm{Spin}(7)$, have dimension 7, 8, respectively. There are natural inclusions,

$$\mathrm{SU}(3) \subset \mathrm{G}_2 \subset \mathrm{Spin}(7)\,.$$

Each of the subgroups, $\mathrm{SU}(3) \subset \mathrm{G}_2$, $\mathrm{G}_2 \subset \mathrm{Spin}(7)$, is realized as the isotropy group of a point on the unit sphere of its representation space.

Together with the twistor construction, the above relations indicate that for instance of special holonomy, there is a close connection with the $\mathrm{U}(\frac{n}{2})$ or $\mathrm{SU}(\frac{n}{2})$ case. This comes in to play in the proofs of the main results discussed in Sections 5, 6, 9.

The cases, $H \subset \mathrm{U}(\frac{n}{2})$, $\mathrm{SU}(\frac{n}{2})$, are called Kähler, respectively, special-Kähler In the $\mathrm{SU}(\frac{n}{2})$ case, the canonical bundle is flat. If the flat connection is actually globally trivial, the manifold is called Calabi-Yau.

The cases, $H \subset \mathrm{Sp}(\frac{n}{4})$, $\mathrm{Sp}(\frac{n}{4})\mathrm{Sp}(1)$, are called hyper-Kähler, respectively quaternion-Kähler.

The cases, G_2, $\mathrm{Spin}(7)$, are called exceptional.

Anti-self-duality of curvature.

Let V denote a real inner product space space and let $W \subset \bigwedge^2(V)$. The orthogonal projection, $\pi_W : \bigwedge^2(V) \to W$, can be viewed as an element of the symmetric tensor product $S^2(\bigwedge^2(V))$. Let $A : S^2(\bigwedge^2(V)) \to \bigwedge^4(V)$ denote the canonical map given by antisymmetrization. With respect to an orthonormal basis, $\{w_j\}$, of W, we have

$$A(\pi_W) = \sum_j w_j \wedge w_j \,.$$

An oriented manifold, M^n, with holonomy group H_p, at $p \in M^n$ and Lie algebra, $\mathfrak{h}_p$, carries a canonical parallel $(n-4)$-form whose value at p is $A(\pi_{\mathfrak{h}_p})$. If the restricted holonomy group is $\mathrm{SO}(n)$, then $A(\pi_{\mathfrak{h}_p}) \equiv 0$.

If the holonomy group, H, is special, we put

$$* \,\Omega(p) = c(H) \cdot A(\pi_{\mathfrak{h}_p}) \,. \tag{0.8}$$

where $c(H)$ is an appropriately chosen of the constant. A calculation shows that the (suitably defined) trace free part, R_0, of the curvature, R, satisfies the anti-self-duality relation,

$$* \, R_0 = -\Omega \wedge R_0 \,; \tag{0.9}$$

compare **[Sa1]**, **[Sa2]**, **[BaHiSing1]**, **[Ti3]**.

If $H_0 \subset \mathrm{U}(\frac{n}{2})$, the form, $*\Omega$, is a multiple of the square of the Kähler form. If $H_0 \subset \mathrm{Sp}(\frac{n}{4})\mathrm{Sp}(1)$, then $*\Omega$ is a multiple of the form $\omega_I^2 + \omega_J^2 + \omega_K^2$. The groups, G_2, $\mathrm{Spin}(7)$, can actually be characterized as the subgroups of all linear transformations which preserve the form $*\Omega$; see Section 10.

Let p_1 denote the first Pontrjagin class. Put

$$C(M^n) = -(p_1 \cup [\Omega])(M^n) \,. \tag{0.10}$$

A multiple, $c_0(n)$, of the topological invariant, $C(M^n)$, bounds the L_2-norm of R_0. This is a direct consequence of (0.9) and Chern-Weil theory; see Section 9. Given the normalization of the Ricci curvature in (0.1), it follows that a multiple, $c(n)$, of $C(M^n)$, bounds the L_2-norm of the full curvature tensor; compare Conjecture 0.15 and Theorem 9.4.

Degenerations.

M. Anderson has conjectured that if Y is the Gromov-Hausdorff limit of a noncollapsing sequence of Einstein manifolds, then Y is a smooth Einstein manifold off a closed singular set of Hausdorff dimension $\leq n-4$. At least for limits of manifolds with special holonomy, this holds.

THEOREM 0.11. ([**Ch3**], [**ChTi2**]) Let $M_i^n \xrightarrow{d_{GH}} Y$, where the manifolds, M_i, have special holonomy, H, or, in case $n=4$, are Einstein with anti-self-dual curvature. Assume

$$|\mathrm{Ric}_{M_i^n}| \leq n-1\,,$$
$$\mathrm{Vol}(M_i^n) \geq v > 0\,.$$

Then there is a closed set, $\mathcal{S}$, with dim $\mathcal{S} \leq n-4$, such that $Y \setminus \mathcal{S}$ is a smooth n-manifold with holonomy contained in H, respectively an Einstein 4-manifold with anti-self-dual curvature.

The Kähler case of Theorem 0.11 is due to Cheeger and to Tian. It was first written up in [**Ch3**]. Ideas in the proof are similar to ones used in [**ChCoTi2**]. For the remaining cases, see [**ChTi2**]. The proof of Theorem 0.11 is described in Section 5.

REMARK 0.12. In Theorem 0.11, anti-self-duality of curvature does not enter and no assumption concerning the boundedness of the sequence of characteristic numbers, $\{C(M_i^n)\}$, is required.

Before stating the following corollary of Theorem 0.11 it is convenient to make a definition. (For the notion of "tangent cone", which appears in Corollary 0.13, see Section 3.)

A metric space, X, is said to have a singularity at $x \in X$ of *orbifold type* if some neighborhood of x is isometric to U/Γ, where U is a smooth riemannian manifold and Γ is a finite group of isometries of U.

Let P_c denote parallel translation along the smooth curve c.

For H a compact Lie group, the singularity will be called of *H-orbifold type* if the following additional condition holds. For all $p \in U$, there is a subgroup, H_p, conjugate in $\mathrm{O}(n)$ to H, such that for all $k \in \Gamma$ and all curves, c, from p to $k(p)$, the transformation, $dk^{-1} \circ P_c$, lies in H_p.

COROLLARY 0.13. ([**Ch3**], [**ChTi2**]) Let the assumptions be as Theorem 0.11. There exists $\mathcal{S}_{n-6} \subset \mathcal{S}$, with dim $\mathcal{S}_{n-6} \leq n-6$, such that for all $y \in (\mathcal{S} \setminus \mathcal{S}_{n-6})$, there exists at least one tangent cone of H-orbifold type, $\mathbf{R}^{n-4} \times \mathbf{R}^4/\Gamma$, where $\Gamma \subset H \subset \mathrm{SO}(n)$ acts trivially on $\mathbf{R}^{n-4}$ and freely on $\mathbf{R}^4 \setminus \{0\}$.

REMARK 0.14. In Theorem 0.11 and Conjecture 0.15 below, if $H_0 \subset \mathrm{Sp}(\frac{n}{4})\mathrm{Sp}(1)$ then $\mathcal{S}_{n-6}$ can be replaced by $\mathcal{S}_{n-8}$ (with dim $\mathcal{S}_{n-8} \leq n-8$); see Theorem 6.1.

The following Conjecture 0.15, is suggested by additional facts which are known to hold in the 4-dimensional case. Its solution is one main goal of the program described in this paper.

CONJECTURE 0.15. ([**ChTi2**]) Let the sequence of manifolds, $M_i^n \xrightarrow{d_{GH}} Y$, with special holonomy group, H, satisfy

$$|\mathrm{Ric}_{M_i^n}| \leq n-1\,,$$
$$\mathrm{Vol}(M_i^n) \geq v > 0\,.$$

Assume in addition that the sequence, $\{C(M_i^n)\}$, is bounded above. Then there exists $\mathcal{S}_{n-6} \subset \mathcal{S}$, with dim $\mathcal{S}_{n-6} \leq n-6$, such that for all $y \in (\mathcal{S} \setminus \mathcal{S}_{n-6})$, the singularities are of H-orbifold type.

REMARK 0.16. If Conjecture 0.13 holds, then $\mathcal{S}_{n-4}$, the codimension 4 piece of the singular set, is *calibrated* by the $(n-4)$-form Ω; see [**HarLaw**].

For $H \subset \mathrm{U}(\frac{n}{2})$, Conjecture 0.15 is due to Cheeger-Colding-Tian, who also gave somewhat weaker and more technical versions for Einstein manifolds in general; compare Sections 7, 8.

In addition to the information contained in Theorem 0.11 and Corollary 0.13, the assertions of the conjecture are known to hold at the *infinitesimal* level, off a subset with $(n-4)$-dimensional Hausdorff measure 0.

Off such a subset, the tangent cone is unique and is of H-orbifold type, $\mathbf{R}^{n-4} \times \mathbf{R}^4/\Gamma$, where the finite group, $\Gamma \subset H \subset \mathrm{SO}(n)$, acts freely on $\mathbf{R}^4 \setminus \{0\}$ and trivially on $\mathbf{R}^{n-4}$; see Theorem 7.14.

Information concerning behavior on the *local* level is also available: Bounded subsets of $\mathcal{S}$, have finite $(n-4)$-dimensional Hausdorff measure and are actually $(n-4)$-rectifiable; see Theorems 8.1, 9.4.

These matters, including the relevant definitions, will be explained at greater length in the sequel. Although the above mentioned results represent strong progress, Conjecture 0.15 still seems quite difficult.

Organization.

In Sections 1–3, we assume only the lower bound on Ricci curvature, $\mathrm{Ric}_{M^n} \geq -(n-1)$.

In Sections 4–6, we assume the 2-sided bound, $|\mathrm{Ric}_{M^n}| \leq n-1$, or in addition, that M^n is Einstein. In Sections 5, 6, we also assume that M^n has special holonomy.

In Sections 7, 8, we assume $\mathrm{Ric}_{M^n} \geq -(n-1)$, and also, a bound on the L_p-norm of the full curvature tensor.

In Sections 9, all of our assumptions come into play, the 2-sided bound, $|\mathrm{Ric}_{M^n}| \leq n-1$, special holonomy, and a bound on the topological invariant $C(M^n)$.

In Section 10, the appendix, we review some standard facts concerning special holonomy, including the definitions of the special holonomy groups, the computation of the Ricci tensor and the relations between the cases $\mathrm{Sp}(\frac{n}{4})\mathrm{Sp}(1)$, G_2, Spin(7) and the Kähler case.

Apart from Theorem 0.11 and Corollary 0.13, the main results on degenerations of Einstein metrics and metrics with special holonomy are Theorems 3.4, 4.2, 6.1, 7.11, 7.14, 9.4.

1. Preliminaries

We collect here, some basic properties of manifolds whose Ricci curvature is bounded below, on which the proofs of results described in later sections depend. Thus, although we will not supply proofs of those results, we will be able to give an indication of what the proofs involve. An exposition with complete proofs, of the results of Sections 1–4, is given in [**Ch4**].

Bochner's formula.

Let X^* denote the 1-form dual to the vector field X and let θ^* denote the vector field dual to the 1-form θ. Let $\Delta = \operatorname{div}\nabla = -\delta d$ denote the Laplacian on functions, ∇, the riemannian covariant derivative and $d\delta+\delta d$, the Hodge Laplacian on 1-forms.

Bochner's formula asserts:

$$\begin{aligned}\frac{1}{2}\Delta|X|^2 &= \langle \Delta X, X\rangle + \langle \nabla X, \nabla X\rangle \\ &= |\nabla X|^2 - \langle X, ((d\delta+\delta d)X^*)^*\rangle + \operatorname{Ric}(X,X)\,.\end{aligned}$$

Since $d(d\delta+\delta d) = (d\delta+\delta d)d$ and $df^* = \nabla f$, it follows that if $X = \nabla f$ is the gradient of a function, f, then

$$\frac{1}{2}\Delta|\nabla f|^2 = |\mathrm{Hess}_f|^2 + \langle \nabla f, \nabla\Delta f\rangle + \operatorname{Ric}(\nabla f, \nabla f)\,,$$

where $\mathrm{Hess}_h = \nabla(\nabla h)$.

For h harmonic,

$$\frac{1}{2}\Delta|\nabla h|^2 = |\mathrm{Hess}_h|^2 + \operatorname{Ric}(\nabla h, \nabla h)\,. \tag{1.1}$$

Note that if $\operatorname{Ric}_{M^n} \geq 0$ (e.g. $M^n = \mathbf{R}^n$) and $|\nabla h| \equiv 1$, we get $\nabla(\nabla h) \equiv 0$. By the de Rham decomposition theorem, every $p \in M^n$, is contained in a neighborhood, U, which is isometric to a subset of a riemannian product, $\mathbf{R}\times\overline{M}^{n-1}$, by an isometry with carries ∇h to the unit tangent vector field to the $\mathbf{R}$ factor. We think of such functions as locally linear in a generalized sense.

Local harmonic coordinates.

In a neighborhood of any point, $m \in M^n$, there exist many local coordinate systems with the property that the coordinate functions are harmonic. It a simple consequence of Bochner's formula that in such a local coordinate system, the Ricci tensor is given by

$$\operatorname{Ric}\left(\frac{\partial}{\partial x_i}, \frac{\partial}{\partial x_j}\right) = -\frac{1}{2}\Delta(g_{i,j}) + Q_{i,j}\left(g^{s,t}, \frac{\partial g_{k,\ell}}{\partial x_m}\right), \tag{1.2}$$

where $Q_{i,j}$ is quadratic in both $\frac{\partial g_{k,l}}{\partial x_m}$ and $g^{s,t}$. In each product, $\frac{\partial g_{k,l}}{\partial x_m} \cdot \frac{\partial g_{k',l'}}{\partial x_{m'}}$, one of k, ℓ, m, equals i, and one of k', ℓ', m' equals j.

In view of (1.2), in harmonic coordinates, the Einstein condition becomes a quasi-linear elliptic equation on the metric.

Relative Volume comparison.

Let $\underline{M}^n_H$ denote the complete, simply connected, n-dimensional space of constant curvature $\equiv H$. Let $\underline{p} \in \underline{M}^n_H$,

THEOREM 1.3. ([**Bish**], [**GvLP**]) If $\operatorname{Ric}_{M^n} \geq (n-1)H$, then the following functions are nondecreasing:

$$\frac{\operatorname{Vol}(\partial B_r(p))}{\operatorname{Vol}(\partial B_r(\underline{p}))}\downarrow,$$

$$\frac{\mathrm{Vol}(B_r(p))}{\mathrm{Vol}(B_r(\underline{p}))} \downarrow .$$

Monotonicity inequalities play a key role in geometric analysis. They are assertions to the effect that behavior of a certain quantity on a fixed scale, controls its behavior on all smaller scales. The relations in Theorem 1.3 are the *only* monotonicity inequalities known to hold for manifolds with Ricci curvature bounded below. In particular:

Volume behavior for a given ball of a fixed size, controls from below, volume behavior for all smaller concentric balls.

Relative volume comparison is a consequence of the formula for the first variation of area, together with so-called *mean curvature* comparison off the set, $\{p\} \cup \mathcal{C}_p$, where $\mathcal{C}_p$ denotes the *cut locus* of p; see below.

Gromov's compactness theorem.

THEOREM 1.4. (**[GvLP]**) The collection of (isometry classes of) compact riemannian manifolds satisfying $\mathrm{Ric}_{M^n} \geq -(n-1)$, $\mathrm{diam}(M^n) \leq d$, is precompact with respect to the Gromov-Hausdorff distance.

By a standard argument based on the doubling property for the riemannian measure implied by Theorem 1.3, the manifolds in Theorem 1.4 form a uniformly totally bounded collection i.e. for all $\epsilon > 0$, there exists $N(\epsilon, d, n)$, such that every such M^n, contains an ϵ-dense set with at most $N(\epsilon, d, n)$ members. (Any maximal set for which each pair of distinct points has distance at least ϵ, is easily seen to have this property.)

A pigeon hole argument shows that any uniformly totally bounded collection of compact metric spaces is totally bounded with respect to the Gromov-Hausdorff distance.

Finally, from a construction like that which associates to a metric space its completion, it follows that the ideal points which must be added to complete the above collection with respect to the Gromov-Hausdorff distance, are actually realized by compact metric spaces.

Mean curvature comparison and Laplacian comparison.

Let $m(x)$ denote the mean curvature in the the direction of inward normal of distance sphere, $\partial B_r(p)$, at the point $x \in \partial B_r(p)$. Let $\underline{m}(r)$ denote the mean curvature of the distance sphere, $\partial B_r(\underline{p})$, where $\underline{p} \in \underline{M}^n_H$. Mean curvature comparison states that $\mathrm{Ric}_{M^n} \geq (n-1)H$ implies

$$m(x) \leq \underline{m}(r) .$$

Initially, mean curvature comparison only makes sense at points at which the distance function is smooth i.e. off the set $\{p\} \cup \mathcal{C}_p$. This version of mean curvature comparison is proved by elementary arguments from the theory of ordinary differential equations.

Let $\tilde{\Delta}$ denote the Laplacian on $\partial B_r(p)$, with respect to the induced metric. In view of the formula for the Laplacian in geodesic polar coordinates, at smooth points of $\partial B_r(p)$, we have

$$\Delta = \frac{\partial^2}{\partial r^2} + m\frac{\partial}{\partial r} + \tilde{\Delta} .$$

It follows that mean curvature comparison has an immediate generalization to Laplacian comparison:

$$\Delta f(r) \leq \underline{\Delta} f(r) \qquad (\text{if } f' \geq 0),$$
$$\Delta f(r) \geq \underline{\Delta} f(r) \qquad (\text{if } f' \leq 0).$$

Mean curvature comparison is the special case of Laplacian comparison in which $f(r) = r$.

It is of crucial importance that Laplacian comparison has a sense and is valid at *all* points of M^n, including the set $\{p\} \cup \mathcal{C}_p$. It holds in the distribution sense and in the sense of barriers, which concept is very closely related to (what subsequently became known as) *viscosity solutions.* Since distance functions have natural barriers and the strong maximum principle extends to the barrier setting, Laplacian comparison is a powerful tool in the study of Ricci curvature. These fundamental results were obtained by Calabi in 1957; [**Ca2**].

Let the metric on $\underline{M}^n_H$ be given in geodesic polar coordinates by

$$\underline{g} = dr^2 + \underline{k}^2 g^{S^{n-1}},$$

where

$$\underline{k} = \begin{cases} \sin \sqrt{H} r / \sqrt{H} & H > 0 \\ r & H = 0 \\ \sinh \sqrt{-H} r / \sqrt{-H} & H < 0 . \end{cases}$$

Denote the Laplacian in geodesic polar coordinates on $\underline{M}^n_H$ by

$$\underline{\Delta} = \frac{\partial^2}{\partial r^2} + \underline{m}\frac{\partial}{\partial r} + \tilde{\underline{\Delta}} .$$

The following functions on the spaces, $\underline{M}^n_H$, are particularly useful for applying Laplacian comparison.

For $n \geq 3$, the function,

$$\underline{G}(r) = \frac{1}{(n-2)\mathrm{Vol}(S^{n-1})} \int_r^\infty \underline{k}^{1-n}(s) ds ,$$

satisfies $\underline{\Delta}\, \underline{G} = 0$, $\underline{G}(0) = \infty$ and $\underline{G}' < 0$.

The function, $\underline{G}(\overline{x,p})$, is the Green's function with singularity at $\underline{p} \in \underline{M}^n_H$. (For $n = 2$, the definition must be slightly modified.)

If $H \equiv 0$, then $\underline{k} = r$ and

$$\underline{G} = \frac{1}{(n-2)\mathrm{Vol}(S^{n-1})} r^{2-n} .$$

The smooth function,

$$\underline{U} = \int_0^r \underline{k}^{-(n-1)}(s) \left(\int_0^s \underline{k}^{n-1}(u) du \right) ds ,$$

satisfies $\underline{\Delta}\, \underline{U} = 1$, $\underline{U}(0) = 0$, $\underline{U}' \geq 0$ and $|\nabla \underline{U}(r)| = \mathrm{Vol}(B_r(\underline{p}))/\mathrm{Vol}(\partial B_r(\underline{p}))$.

If $H \equiv 0$, then

$$\underline{U} = \frac{r^2}{2n} .$$

Given $R > 0$, put $\underline{G}_R = \underline{G} - \underline{G}(R)$ and $\underline{U}_R = \underline{U} - \underline{U}(R)$.

Set $c_1 = -\frac{\underline{U}'(R)}{\underline{G}'(R)} > 0$. We have $\underline{G}''_R \geq 0$, $\lim_{r\to 0} \underline{G}'_R = -\infty$, and $\underline{U}'' \geq 0$, $\underline{U}'(0) = 0$. Thus, if we let

$$\underline{L}_R = c_1 \, \underline{G}_R + \underline{U}_R \,,$$

then we have $\Delta \underline{L}_R = 1$, $\underline{L}'_R \leq 0$ on $(0, R]$, and $\underline{L}_R(R) = 0$.

The function, $\underline{L}_R$, plays a role in the Abresch-Gromoll inequality discussed below.

The segment inequality.

The segment inequality is a sharpened version of the Poincaré inequality. It is of particular importance in proving almost rigidity theorems in the collapsed case.

Given a nonnegative function, $g \geq 0$, put

$$\mathcal{F}_g(x_1, x_2) = \inf\nolimits_\gamma \int_0^\ell g(\gamma(s))ds \,,$$

where the inf is taken over all minimal geodesics, γ, from x_1 to x_2, and s denotes arclength.

THEOREM 1.5. (**[ChCo3]**) Let $\mathrm{Ric}_{M^n} \geq -(n-1)$ and let $A_1, A_2 \subset B_r(p)$, with $r \leq R$. Then

$$\int_{A_1 \times A_2} \mathcal{F}_g(x_1, x_2) \leq c(n,R) r \big(\mathrm{Vol}(A_1) + \mathrm{Vol}(A_2)\big) \times \int_{B_{2R}(p)} g \,.$$

One specific consequence of the segement inequality is a lower bound on the first eigenvalue of the Laplacian for Dirichlet problem on a metric ball $B_R(p)$; compare [**B**], [**LiSch**], [**Li**]. To see this, let $f : B_R(p) \to \mathbf{R}$ satisfy $f \,|\, B_R(p) \equiv 0$ and extend f to vanish identically outside $B_R(p)$. Take $g = |\nabla f|$, $A_1 = A_2 = B_R(p)$.

The proof of the segment inequality employs the techniques which lead to relative volume comparison.

The Cheng-Yau gradient estimate.

THEOREM 1.6. (**[CgYau]**) Let

$$\mathrm{Ric}_{M^n} \geq (n-1)H \qquad (\text{on } B_{R_2}(p)) \,,$$

and let $u : B_{R_2}(p) \to \mathbf{R}$ satisfy

$$u \geq 0 \,,$$
$$\Delta u = K(u) \,.$$

Then for $R_1 < \mathrm{R}_2$, on $B_{R_1}(p)$,

$$\frac{|\nabla u|^2}{u^2} \leq \max\big(2u^{-1}K(u), c(n, R_1, R_2, H) + 2u^{-1}K(u) - 2K'(u)\big) \,.$$

The Cheng-Yau gradient estimate, which we have not stated in utmost generality, can be further specialized to the particularly important case in which u satisfies an equation of the form $\Delta u = a \cdot u + b$. This class of functions includes harmonic functions and eigenfunctions of Δ.

Note that by slightly changing the form of the function, $K(u)$, the assumption, $u \geq 0$, can be achieved by adding a suitable constant to the function u. The resulting estimate then depends on the size of this constant. In particular, if u is harmonic, then so is $u+c$, for any constant c.

The Abresch-Gromoll inequality.

Let $\overline{A_{R_1,R_2}(p)}$ denote the metric annulus, $B_{R_2}(p)\setminus\overline{B_{R_1}(p)}$.

Let $\mathrm{Ric}_{M^n}\geq -(n-1)$ and let $f: B_{R_2}(p)\to\mathbf{R}$. The Abresch-Gromoll inequality provides a bound for $f(p)$ under the assumptions:

i) $f\,|\,\partial B_{R_2}(p)\geq 0$,

ii) Δf has a definite upper bound on $\overline{A_{R_1,R_2}(p)}$,

iii) $f(x)$ has a definite upper bound at some $x\in A_{R_1,R_2}$,

iv) the Lipschitz constant of f has a definite bound on $B_{R_1}(p)$.

At the heart of Abresch-Gromoll inequality is a beautiful quantitative extension of the maximum principle.

Let the functions, $\underline{L}_R$, $\underline{G}_R$ be defined as in the subsection on Laplacian comparison.

THEOREM 1.7. ([**AbGl**]) Let $\mathrm{Ric}_{M^n}\geq -(n-1)$ and let $\overline{B_{R_2}(p)}\subset\Omega$. For $\delta>0$, let $f:\overline{B_{R_2}(p)}\to\mathbf{R}$ satisfy $f\,|\,\partial B_{R_2}(p)\geq 0$ and

$$\Delta f\leq\delta \qquad (\text{on } \overline{A_{R_1,R_2}(p)})\,,$$

$$\mathrm{Lip}\, f\leq c \qquad (\text{on } B_{R_1}(p))\,.$$

If for some $x\in A_{R_1,R_2}(p)$ and $t>0$,

$$f(x)<(\delta\underline{L}_{R_2}+t\underline{G}_{R_2})(R) \qquad (R=\overline{x,p})\,,$$

then

$$f(p)<(\delta\underline{L}_{R_2}+t\underline{G}_{R_2})(R_1)+cR_1\,.$$

Since we assume $\mathrm{Lip}\, f\leq c$ on $B_{R_1}(p)$, to get the bound on $f(p)$, it suffices to show that somewhere on $\partial B_{R_1}(p)$, we have $f\leq(\delta\underline{L}_{R_2}+t\underline{G}_{R_2})(R_1)$. This in turn, is a quantitive extension of the maximum principle (on the annulus $A_{R_1,R_2}(p)$). It is proved with the aid of Laplacian comparison. To see the relevance of the maximum principle, consider the limiting case $\delta=0$.

Abresch and Gromoll used their inequality in proving a weakened quantitative extension of the splitting theorem. For this, the metric is rescaled so that the lower bound on Ricci curvature is close to 0. Later, this weakened version played a significant role in the proof of the full quantitative extension of the splitting theorem; see Theorem 2.3 of Section 2.

The cutoff function with bounded Laplacian.

The the proof of the following theorem, which gives the existence of a cutoff function with bounds on the Laplacian, is based on ideas which are related to those used in proving the Abresch-Gromoll inequality. It is required for the proof of the almost splitting theorem.

THEOREM 1.8. ([**ChCo3**]) Let $\mathrm{Ric}_{M^n}\geq -(n-1)$ and let $B_{R_2}(p)\subset M^n$. Then for all $R_1<R_2$, there exists $\phi: M^n\to[0,1]$, such that

$$\phi \,|\, B_{R_1}(p) \equiv 1\,,$$

$$\operatorname{supp}\phi \subset B_{R_2}(p),$$

$$|\nabla\phi| \leq c(n, R_1, R_2)\,,$$

$$|\Delta\phi| \leq c(n, R_1, R_2)\,.$$

2. Almost rigidity and rescaling

Typically, geometric properties of a riemannian manifold which are excluded if the curvature satisfies a certain strict inequality, can occur only under very restricted circumstances if the strict inequality is relaxed to a weak one. Theorems to this effect are known as *rigidity theorems*.

The splitting theorem.

Nonnegative Ricci curvature tends to oppose noncompactness. In particular, according to Myers' theorem, if the Ricci curvature has a positive lower bound, $\mathrm{Ric}_{M^n} \geq (n-1)H > 0$, then

$$\operatorname{diam}(M^n) \leq \frac{\pi}{\sqrt{H}}\,.$$

While manifolds with nonnegative Ricci curvature (e.g. 2-dimensional paraboloids) can certainly be noncompact, the situation becomes highly constrained if the manifold is noncompact in a sufficiently strong sense.

A geodesic, $\gamma : (-\infty, \infty) \to M^n$, is called a *line*, if each finite segment of γ is minimal.

THEOREM 2.1. ([**ChGl2**]) If M^n is complete, $\mathrm{Ric}_{M^n} \geq 0$, and M^n contains a line, then M^n splits as an isometric product, $M^n = \mathbf{R} \times \overline{M}^{n-1}$, for some $\overline{M}^{n-1}$.

In proving Theorem 2.1, one starts by considering a *ray*, $\gamma : [0, \infty) \to M^n$. By definition, each finite segment of γ is minimal. Associated to γ is its *Buseman function*, defined by

$$b_\gamma(x) = \lim_{s\to\infty} \overline{x, \gamma(s)} - s\,.$$

By using the triangle inequality, it is easily seen that the limit exists. The pointwise Lipschitz constant of b_γ is $\equiv 1$.

In the special case in which $\gamma : (-\infty, \infty) \to \mathbf{R}^n$ is the x-axis, b_γ is the linear harmonic function $-x$. In the general case, one shows by means of Laplacian comparison that b_γ is superharmonic.

For γ a line, let $b_\pm$ denote the Buseman functions associated to the rays $\gamma \,|\, [0, \infty)$, $-\gamma \,|\, [0, -\infty)$, which form the two halves of γ. The functions, b_+, b_-, are superharmonic and by the triangle inequality $b_+ + b_- \geq 0$. Since γ is a line, it also follows that $b_+ + b_- \,|\, \gamma \equiv 0$.

From the strong maximum principle, we now get $b_+ + b_- \equiv 0$. Hence, $b_+ = -b_-$ is both superharmonic and subharmonic. So b_+, b_-, are harmonic functions. In particular, they are smooth.

Clearly, $|\nabla b_+| \equiv 1$ and by Bochner's formula, $\mathrm{Hess}_{b_+} \equiv 0$ i.e. ∇b_+ is parallel. Now, the existence of an isometric splitting, given by the level surfaces of b_+ and the integral curves of ∇b_+, follows from the de Rham decomposition theorem.

REMARK 2.2. For the case in which the sectional curvature is nonnegative, the splitting theorem is due to Toponogov, whose proof relied on his comparison theorem for geodesic triangles; see [**Top1**], [**Top2**] and compare [**ChGl1**]; compare also [**EschHei**].

Making the discussion quantitative.

The quantitative generalization of the splitting theorem is most easily stated as an extension of the splitting theorem to pointed Gromov-Hausdorff limit spaces whose Ricci curvature is nonnegative in a generalized sense.

THEOREM 2.3. ([**ChCo2**]) Let the sequence of complete manifolds, M_i^n, satisfy, $(M_i^n, m_i) \xrightarrow{d_{GH}} (Y, y)$,

$$\mathrm{Ric}_{M_i^n} \geq -(n-1)\delta_i \qquad (\text{where } \delta_i \to 0)\,. \tag{2.4}$$

If Y contains a line, then Y splits as an isometric product, $Y = \mathbf{R} \times \overline{Y}$, for some length space $\overline{Y}$.

By arguing by contradiction, it is a simple exercise to translate Theorem 2.3 into an equivalent quantitative statement concerning manifolds whose Ricci curvature is bounded below by a small negative constant.

Note in this connection, that a line in Y need not be the limit of a sequence of lines in the manifolds, M_i^n, of the approximating sequence. However, a line in Y can always be viewed as the limit of a sequence of triangles in the manifolds, M_i^n, the excesses of which converge to zero. (The excess of a triangle is the minimum of the three numbers which are expressible as the sum of the lengths two sides minus the length of a third.)

To prove Theorem 2.3, one establishes the equivalent statement for smooth manifolds, in which the hypotheses of the splitting theorem are only satisfied up to small errors. The idea is to make quantitative, each of the steps in the argument which was outlined above. This turns out to be less straightforward than one might initially suppose.

Let $\mathrm{Ric}_{M^n} \geq -(n-1)\delta$, for some very small δ. Assume that $p, q_+, q_- \in M^n$ determine a triangle with small excess. Specifically let

$$\overline{p, q_-} + \overline{p, q_+} - \overline{q_-, q_+} \leq \delta\,,$$
$$\overline{p, q_\pm} \geq \delta^{-1}\,.$$

Put $b_+(x) = \overline{x, q_+} - \overline{p, q_+}$. Fix a ball, $B_R(p)$, where R can be taken arbitrarily large if δ is sufficiently small. Let $\mathbf{b}_+$ denote the harmonic function on $B_R(p)$ whose boundary values coincide with those of b_+. In the ideal case, that of Theorem 2.1, we would have $b_+ = \mathbf{b}_+$. So we would like to know that $b_+, \mathbf{b}_+$ are close in a suitably strong sense.

With the help of the Abresch-Gromoll inequality, it follows that $b_+ + b_- - \overline{q, q_+}$ is small on $B_R(p)$. By a straightforward argument based on Laplacian comparison and the maximum principle, we get that $b_+, \mathbf{b}_+$ are uniformly close on $B_R(p)$. From this, together with the lower bound for the smallest eigenvalue of the Dirichlet problem on $B_R(p)$ (see Theorem 1.5) we find that the gradients, $\nabla b_+, \mathbf{b}_+$ are close in the L_2 sense. (Here and below, the riemannian measure on $B_R(p)$ is normalized so as to make the volume of $B_R(p)$ equal to 1.)

In particular, $|\nabla \mathbf{b}_+|$ is close to the constant function, 1 in the L_2 sense. From Bochner's formula, used in concert with the cutoff function with bounded Laplacian (of Theorem 1.8), and the fact that $|\nabla \mathbf{b}_+|$ has small oscillation, it follows that $|\mathrm{Hess}_{\mathbf{b}_+}|$ is small in the the L_2 sense on $B_{\frac{R}{2}}(p)$; compare (1.1).

Finally, and this is the most significant step, by using the segment inequality, the gradient estimate and the information established above, one shows by an argument based on the first variation formula, that $B_{\frac{R}{2}}(p)$ is close in the Gromov-Hausdorff sense, to a ball in some product space, $\mathbf{R} \times \overline{Y}$, where b_+ corresponds to the $\mathbf{R}$ coordinate; see [**ChCo3**] and compare [**Co1**]–[**Co4**].

REMARK 2.5. A lower bound on volume is not required in Theorem 2.1, nor is the noncollapsing assumption is needed in Theorem 2.3.

Volume cones are metric cones.

Next, we consider the equality cases in the relative volume comparison theorem, Theorem 1.3. As before, let the metric on $\underline{M}^n_H$ be given in geodesic polar coordinates by $dr^2 + \underline{k}^2 g^{S^{n-1}}$.

THEOREM 2.6. Let $\mathrm{Ric}_{M^n} \geq (n-1)H$ on the annulus, $A_{r_1,r_2}(p)$, and assume that

$$\frac{\mathrm{Vol}(\partial B_{r_1}(p))}{\mathrm{Vol}(\partial B_{r_2}(p))} = \frac{\mathrm{Vol}(\partial B_{r_1}(\underline{p}))}{\mathrm{Vol}(\partial B_{r_2}(\underline{p}))} .$$

Then the metric on $A_{r_1,r_2}(p)$ is of the form,

$$dr^2 + \underline{k}^2 \tilde{g} ,$$

for some smooth riemannian metric, $\tilde{g}$, on $\partial B_{r_1}(p)$.

In essence, to prove Theorem 2.6, one follows the proof of relative volume comparison and notes that in the resulting string of inequalities, equality must hold at every stage. For instance, the Schwarz inequality applied to the second fundamental form of a distance sphere enters in the proof of mean curvature comparison. Since we must be in the equality case, we get the key fact that everywhere on such a distance sphere, the second fundamental form must be diagonal.

So far, we have only considered the equality case of the first inequality in Theorem 1.3. In actuality, given Theorem 2.6, it is easily seen that the equality case of the second inequality in Theorem 1.3 can only occur for annuli in the space of constant curvature $\underline{M}^n_H$ i.e. $\tilde{g} = g^{S^{n-1}}$. However, if as in Theorem 2.3, we include limit spaces, then new examples arise.

We will assume nonnegative Ricci curvature in the extended sense of (2.4), since this is the condition which figures in the applications. The generalization for arbitrary lower bounds on Ricci curvature is straightforward.

EXAMPLE 2.7. Let $Y^2 \subset \mathbf{R}^3$ denote some 2-dimensional cone. By rounding off the cone tip, one sees that Y^2 is the Gromov-Hausdorff limit of a sequence of 2-dimensional surfaces which are convex and hence of nonnegative curvature. Thus, Y^2 has nonnegative Ricci curvature in the sense of (2.4). The volume function of a ball, $B_r(y)$, centered at the cone tip, y, satisfies

$$\frac{\mathrm{Vol}(B_r(y))}{\mathrm{Vol}(B_r(\underline{p}))} \equiv \frac{\mathrm{Vol}(B_1(y))}{\pi}\,,$$

or equivalently,

$$\frac{\mathrm{Vol}(B_r(y))}{\mathrm{Vol}(B_1(y))} \equiv \frac{\mathrm{Vol}(B_1(\underline{p}))}{\mathrm{Vol}(B_1(\underline{p}))}\,.$$

Here, $\mathrm{Vol}(B_r(y))$ is 2-dimensional area, the model space, $\underline{M}_0^2$, is $\mathbf{R}^2$ with its usual flat metric and $\mathrm{Vol}(B_1(\underline{p})) = \pi$.

For Z a metric space, denote by $C(Z)$, the *metric cone*, on Z. By definition, $C(Z)$ is the completion of metric space, $(0,\infty)\times Z$, with metric,

$$\overline{(r_1,z_1),\,(r_2,z_2)} = r_1^2 + r_2^2 - 2r_1r_2\cdot\cos\overline{z_1,z_2} \qquad (\text{if } \overline{z_1,z_2}\le\pi)\,,$$

and

$$\overline{(r_1,z_1),\,(r_2,z_2)} = r_1 + r_2 \qquad (\text{if } \overline{z_1,z_2}>\pi)\,.$$

To see that the definition is reasonable, think of the case of 3-dimensional cones with arbitrary cross-section in Euclidean space $\mathbf{R}^3$.

Let z^* denote the vertex of $C(Z)$. Thus, z^* is the unique point which is added when $(0,\infty)\times Z$, with the cone metric, is completed to get $C(Z)$.

Note that if there exist z_1, z_2, with $\overline{z_1,z_2}\ge\pi$, then $(1,z_1)$, $(1,z_2)$ lie on a line which passes through z^*. If $\overline{z_1,z_2}>\pi$, this line does *not* split off as an isometric factor. Thus, by Theorem 2.3, cones for which there exist such pairs of points do not arise as limits of sequences of spaces satisfying (2.4).

EXAMPLE 2.8. For $n>1$, the n-fold ramified covering space of the plane, $\mathbf{R}^2$ (ramified at a single point) is a cone with cross-section the circle, $\mathbf{S}_{2n\pi}$, with intrinsic diameter $n\pi$ (i.e. the circle of circumference $2n\pi$). These cones does not arise as limit spaces satisfying (2.4).

The following theorem asserts that for limit spaces with nonnegative Ricci curvature in the sense of (2.4), volumes cones are metric cones.

THEOREM 2.9. (**[ChCo3]**) Let $M_i^n \xrightarrow{d_{GH}} Y$ satisfy (2.4). For $\underline{p}\in\mathbf{R}^n$ and all $s\in[0,r]$, assume

$$\frac{\mathrm{Vol}(B_s(p))}{\mathrm{Vol}(B_1(p))} = \frac{\mathrm{Vol}(B_s(\underline{p}))}{\mathrm{Vol}(B_1(\underline{p}))}\,.$$

Then for some length space Z, with

$$\mathrm{diam}(Z)\le\pi\,,$$

the ball, $B_r(p)$, is isometric to the ball, $B_r(z^*)$, in the cone $C(Z)$.

As in the case of the generalized splitting theorem, Theorem 2.3, an argument by contradiction allows Theorem 2.9 to be reformulated as an almost rigidity theorem for smooth manifolds whose Ricci curvature is bounded below by a small negative constant $-(n-1)\delta$. It is this equivalent formulation which is actually proved. The proof is similar in spirit to that of Theorem 2.3, but the technical details are somewhat more complicated.

Rescaling and almost splitting.

We return to the rescaling idea introduced in Section 0, which suggested a program for understanding the small scale structure of manifolds with $\mathrm{Ric}_{M^n} \geq -(n-1)$.

Let $\mathrm{Ric}_{M^n} \geq -(n-1)$ and $\mathrm{diam}(M^n) \geq 2$. Let $\gamma : [-L, L] \to$ be a geodesic segment of length $2L \geq 2$. If we rescale distance by a factor ϵ^{-1}, where $0 < \epsilon << 1$, and $1 << \epsilon^{-1}L$, then the ball, $B_\epsilon(\gamma(0))$, in the original metric, now looks to the naked eye, like a ball of radius 1, with center on in a line in a noncompact manifold with nonnegative Ricci curvature — or better — in a noncompact limit space, whose Ricci curvature is nonnegative in the sense of (2.4).

The generalized splitting theorem, Theorem 2.3, implies that this rescaled ball appears to be a ball in some isometric product space, $\mathbf{R} \times X$. Unless the space, X, is indistinguishable from a point, we can repeat the construction in some direction tangent to X, etc. In this sense, the geometry has a self-regulating feature:

The more directions, the more small scale approximate splitting.

Rescaling, almost volume cones and almost metric cones.

Theorem 2.9 above, is particularly useful in the noncollapsed case.

Let $\mathrm{Ric}_{M^n} \geq -(n-1)$ and let

$$\mathrm{Vol}(B_1(p)) \geq v\,.$$

The decreasing function, $\mathrm{Vol}(B_s(p))/\mathrm{Vol}(B_s(\underline{p}))$, satisfies

$$1 \geq \frac{\mathrm{Vol}(B_s(p))}{\mathrm{Vol}(B_s(\underline{p}))} \geq \frac{v}{\mathrm{Vol}(B_1(\underline{p}))}\,.$$

It follows that for $r \leq 1$ and all $\delta > 0$, the relation,

$$(1-\delta)\frac{\mathrm{Vol}(\partial B_{\frac{1}{2}r}(p))}{\mathrm{Vol}(\partial B_{\frac{1}{2}r}(\underline{p}))} \leq \frac{\mathrm{Vol}(\partial B_r(p))}{\mathrm{Vol}(\partial B_r(\underline{p}))}\,,$$

is violated for the end points of at most a definite number, $N(\delta)$, of disjoint intervals $[\delta r, r]$.

When combined with rescaling, this leads to the conclusion that tangent cones at points of noncollapsed limit spaces are metric cones.

3. The structure of limit spaces

Let

$$\mathrm{Ric}_{M_i^n} \geq -(n-1)\,, \tag{3.1}$$

and let

$$(M_i^n, \overline{m}_i) \xrightarrow{d_{GH}} (Y, \overline{y})\,.$$

Tangent cones.

The basic notion for studying the infinitesimal structure of such limit spaces is that of a *tangent cone*, Y_y, at $y \in Y$. Intuitively, tangent cones are obtained by blowing up the space at a point and taking a limit, possibly after passing to a subsequence. In case Y happens to be a smooth riemannian manifold, this proceedure recovers the usual tangent space with its canonical metric.

Let ρ denote the distance function (i.e. the metric) of Y. Gromov's compactness theorem implies that every pointed sequence, $\{(Y, y, r_i^{-1}\rho)\}$, has a subsequence which converges in the pointed Gromov-Hausdorff sense, to some space $(Y_y, y_\infty, \rho_\infty)$. The limit, Y_y, of such a subsequence is called a *tangent cone* at y.

For a given $y \in Y$, the tangent cone need not be unique.

The regular and singular sets.

The *regular set*, $\mathcal{R}$, is the set of those points, $y \in Y$, such that there exists $k \leq n$ for which every tangent cone is isometric to $\mathbf{R}^k$.

As expected, in the noncollapsed case, it turns out that only $k = n$ is possible. Even if Y is collapsed, almost all points are regular, provided "almost all" is defined in an appropriate measure theoretic sense.

By definition, the *singular set*, $\mathcal{S}$, is the complement of the regular set.

The regular set need *not* be open; equivalently, the singular set need not be closed. One can easily construct convex 2-dimensional surfaces, Y^2, which are limits of suitable sequences of polyhedral surfaces, for which $\mathcal{S}(Y^2)$ is a countable set, with limit points that are regular points. It follows by "rounding the corners", that such Y^2 are limit spaces with positive curvature in the sense of (2.4); compare Example 2.7.

Iterated tangent cones have nonnegative Ricci curvature.

Let ρ denote the distance function of the riemannian manifold, M^n. An obvious diagonal argument gives the key fact that every tangent cone is itself a pointed limit space,

$$(M^n_{k(j)}, p_{k(j)}, r^{-1}_{k(j)}\rho_{k(j)}) \xrightarrow{d_{GH}} (Y_y, y_\infty, \rho_\infty)\,,$$

where the rescaled manifolds, $(M^n_{k(j)}, r^{-1}_{k(j)}\rho_{k(j)})$, satisfy

$$\liminf_{k(j)\to\infty}\ \mathrm{Ric}_{M^n_{k(j)}} \geq 0\,;$$

compare (2.4). By a similar argument, it follows that any tangent cone, Y_{y_1}, at $y_1 \in Y_y$, is a limit space satisfying (2.4). More generally, if we pass to tangent cones of tangent cones an arbitrary number of times, then such *iterated* tangent cones have nonnegative Ricci curvature in the generalized sense.

From the generalized splitting theorem, Theorem 2.3, we get:

THEOREM 3.2. ([**ChCo3**]) If $M_i^n \xrightarrow{d_{GH}} Y$ satisfies (3.1), then any iterated tangent cone which contains a line, splits off this line as an isometric factor.

The noncollapsed case.

From now on we assume the noncollapsing condition

$$\mathrm{Vol}(B_1(m_i)) \geq v > 0\,. \tag{3.3}$$

The ϵ-regular set.

The ϵ-regular set, $\mathcal{R}_\epsilon \supset \mathcal{R}$, consists of those points, $y \in Y$, such that for every tangent cone, Y_y, we have

$$d_{GH}(B_1(y_\infty), B_1(0)) < \epsilon\,,$$

where $B_1(0)$ denotes the unit ball in $\mathbf{R}^n$.

There exists $\epsilon(n) > 0$, such that for $\epsilon \leq \epsilon(n)$, the interior of $\mathcal{R}_\epsilon$ is homeomorphic to a smooth connected topological manifold which contains the regular set $\mathcal{R}$. The proof depends on a fundamental theorem of Colding to the effect that if $\mathrm{Ric}_{M^n} \geq (n-1)H$, then a ball, $B_r(p) \subset M^n$, is Gromov-Hausdorff close to the ball, $B_r(\underline{p}) \subset \underline{M}^n$, if and only if $\mathrm{Vol}(B_r(p))$ is close to $\mathrm{Vol}(B_r(\underline{p}))$; see [**Co2**], [**Co3**].

The proof that the interior of $\mathcal{R}_\epsilon$ is manifold also uses an *intrinsic* version of a classical result of Reifenberg; see [**ChCo3**]. Reifenberg's theorem says that if $C \subset \mathbf{R}^N$ is a closed subset with the property that for all $p \in C$ and all $r \leq 1$, the ball, $B_r(p)$, is sufficiently Hausdorff close (on its own scale) to a ball in some affine subspace, $V^n_{p,r} \subset \mathbf{R}^N$, then C is a topological n-manifold; see [**Reif**].

In the Einstein case, we can assume that $\epsilon(n)$ has been chosen such that for $\epsilon \leq \epsilon(n)$, we have $\mathcal{R}_\epsilon = \mathcal{R}$. In this case, $\mathcal{R}$ is a smooth Einstein manifold; see Theorem 4.2.

Noncollapsed tangent cones are metric cones.

From Theorem 2.9 and the previous discussion (compare also the discussion at the end of Section 2) we obtain:

THEOREM 3.4. ([**ChCo3**]) Let the noncollapsed limit space, $M^n_i \xrightarrow{d_{GH}} Y^n$, satisfy (3.1), (3.3). Then every (possibly iterated) tangent cone is a metric cone, $C(Z)$, on some length space, X, with $\mathrm{diam}(X) \leq \pi$.

Blow up arguments.

Theorems 3.2, 3.4, provide two respects in which noncollapsed iterated tangent cones are better behaved than arbitrary noncollapsed limit spaces satisfying (0.3). This enables certain statements concerning arbitrary noncollapsed limit spaces with $\mathrm{Ric}_{M^n_i} \geq -(n-1)$, to be proved by "blow up" arguments; [**Fe**]. These are proofs by contradiction, in which the main step consists of showing that if a desired property were ever to fail, it would already fail for some tangent cone. After repeating this step sufficiently many times, with *suitably chosen* iterated tangent cones, one arrives at a situation in which the geometry of the iterated tangent cone has improved to such an extent, that the desired property actually holds. This contradiction establishes that the property holds in general.

Theorem 3.7 below is proved by a blow up argument.

Volume convergence.

We recall the definition of ℓ-dimensional Hausdorff measure.

Let W denote a metric space. For $A \subset W$, consider the collection of countable coverings, $\{B_{r_i}(w_i)\}$, of A, such that $\sup_i r_i \leq \eta$. Here we allow $\eta = \infty$. Put

$$\mathcal{H}^\ell_\eta(A) = \inf_{\{B_{r_i}(w_i)\}} \omega_\ell \sum_i r_i^\ell \,,$$

where $\omega_\ell > 0$ is a certain explicit constant, which for ℓ an integer, is equal to the volume of the unit ball in $\mathbf{R}^\ell$.

It is clear that $\mathcal{H}^\ell_\eta(A)$ is a nonincreasing function of η. Define the ℓ-dimensional spherical Hausdorff measure of A by

$$\mathcal{H}^\ell(A) = \lim_{\eta \to 0} \mathcal{H}^\ell_\eta(A) \,,$$

where the value, $\mathcal{H}^\ell(A) = \infty$, is permitted.

It follows easily that there is a unique value, dim A, the *Hausdorff dimension* of A, with $0 \leq \dim A \leq \infty$, such that, $\mathcal{H}^\ell(A) = \infty$ for $\ell < \dim A$, and $\mathcal{H}^\ell(A) = 0$ for $\dim A < \ell$.

THEOREM 3.5. ([**ChCo3**]) Let the noncollapsed limit spaces, Y_i^n, satisfy (3.1), (3.3), for all i. If $(Y_i^n, y_i) \xrightarrow{d_{GH}} (Y^n, y)$, then for all $r < \infty$,

$$\lim_{i\to\infty} \mathcal{H}^n(\overline{B_r(y_i)}) = \mathcal{H}^n(\overline{B_r(y)})\,.$$

In follows from Theorem 3.5 that a noncollapsed limit space, Y^n, has Hausdorff dimension, n and (by Theorem 2.1) that n-dimensional Hausdorff measure on Y^n satisfies relative volume comparison.

REMARK 3.6. The special case of Theorem 3.5 in which Y_i^n, Y^n are smooth riemannian manifolds was conjectured by Anderson-Cheeger and proved by Colding; [**Co4**]. This was very significant for the development of the theory.

The natural filtration on the singular set.

For $0 \leq k \leq n-1$, let $\mathcal{S}_k \subset \mathcal{S}$ consist of those points for which no tangent cone splits off a factor, $\mathbf{R}^{k+1}$, isometrically. (As motivation, think of the k-skeleton of a simplicial complex.)

We have $\mathcal{S}_{n-1} = \mathcal{S}$. Equivalently, if some tangent cone at y is isometric to $\mathbf{R}^n$, then so is every other tangent cone at y.

By Theorem 3.7 below, $\dim \mathcal{S}_k \leq k$. Hence, $\dim \mathcal{S} \leq n-1$. From this and (3.3), it follows in particular that the regular set, $\mathcal{R}$, has full measure with respect to $\mathcal{H}^n$. In fact, Theorem 3.8 below gives $\mathcal{S} = \mathcal{S}_{n-2}$. Together with Theorem 3.7, this implies that the singular set has Hausdorff codimension 2.

THEOREM 3.7. ([**ChCo3**]) Let $M_i^n \xrightarrow{d_{GH}} Y^n$ satisfy (3.1), (3.3). Then

$$\dim \mathcal{S}_k \leq k\,.$$

The blow up argument used in proving Theorem 3.7 depends on Theorems 3.2, 3.4. For a closely related result (with a very different proof) for subsets of $\mathbf{R}^n$, see [**Wh**], which also contains additional related references.

THEOREM 3.8. ([**ChCo3**]) Let Y^n satisfy (3.1), (3.3). Then

$$\mathcal{S} = \mathcal{S}_{n-2}\,,$$

and

$$\dim \mathcal{S} \leq n-2\,.$$

If $y \in \mathcal{S}_{n-1} \setminus \mathcal{S}_{n-2}$, then by definition, some tangent cone, Y_y, splits off a factor $\mathbf{R}^{n-1}$. Since $y \in \mathcal{S}$, it follows that Y_y is not isometric to $\mathbf{R}^n$. It is easy to see that the only remaining possibility is that Y_y is closed half space $\mathbf{R}^{n-1} \times \mathbf{R}_+$. So to prove Theorem 3.8, it suffices to show that this potential tangent cone does not actually occur. This is done by an argument which is essentially topological. The idea is that a limit of closed manifolds should not have a nonempty topological boundary.

The complex filtration in the Kähler case.

Here and in Sections 7, 8, results on Kähler manifolds which depend only on the lower bound $\mathrm{Ric}_{M_i^n} \geq -(n-1)$, will be included in our general discussion of that case. Thus, although $\mathrm{U}(\frac{n}{2})$ is a special holonomy group, only those theorems on Kähler manifolds which depend on the 2-sided bound, $|\mathrm{Ric}_{M_i^n}| \leq n-1$, will be bracketed with results on special holonomy.

By definition, if M^n is Kähler, there is a parallel almost complex structure J. Thus, $J^2 = -1$, $\nabla J = 0$.

Suppose M^n splits off a line isometrically, $M^n = \mathbf{R} \times \overline{M}^{n-1}$. If v denotes the parallel vector field tangent to the $\mathbf{R}$ factor of this splitting, then $\nabla v = 0$. Hence, $\nabla J(v) = 0$ as well.

By the de Rham decomposition theorem, it follows that the vector fields, $v, J(v)$, are tangent to the $\mathbf{R}^2$-factor of a *local* isometric splitting of the metric. In addition, the factor, $\mathbf{R}^2$, can be naturally identified with the complex plane $\mathbf{C}$. However, the local factor, $\mathbf{C}$, need *not* be global. To see this, consider the case of a 2-dimensional cylinder $\mathbf{R} \times \mathbf{S}^1 = \mathbf{C}^2/\mathbf{Z}$.

If we view the real line, $\mathbf{R}$, as the collapsed Gromov-Hausdorff limit of a sequence of cylinders for which the $\mathbf{S}^1$ factor shrinks to a point, we see that the local isometric factor corresponding to $J(v)$ (the $\mathbf{S}^1$ factor) can disappear in the limit.

It turns out that for noncollapsed limit spaces satisfying (2.4) which are metric cones, the above complications do not occur. Roughly speaking, an almost parallel vector field on a cone is always almost the gradient of the coordinate function corresponding to the factor, $\mathbf{R}$, of some (global) almost isometric splitting. The proof of this fact relies on a technical theorem from the general theory of metric measure spaces for which the measure satisfies a doubling condition and for which a (suitably formulated) Poincaré inequality holds; see Theorem 16.32 of [**Ch2**].

THEOREM 3.9. ([**ChCoTi2**]) Let $M_i^n \xrightarrow{d_{GH}} Y^n$ satisfy (3.1), (3.3) and assume that M_i^n is Kähler for all i. Then $\mathcal{S}_{2i+1} = \mathcal{S}_{2i}$ for all i. In particular, all strata have even codimension. Moreover, every tangent cone, Y_y, has a parallel almost complex structure. If $Y_y = \mathbf{R}^j \times C(X)$ denotes the isometric splitting for which j is maximal, then $\mathbf{R}^j = \mathbf{C}^{\frac{j}{2}}$.

In Theorem 3.9, we were intentionally vague about the sense in which there exists a parallel almost complex structure on the cone $C(X)$. In the Kähler-Einstein, in which the regular part is a smooth manifold, the exact meaning is clear.

We will give a version of Theorem 3.9 which covers all cases of special holonomy; see Theorem 6.1.

4. Einstein limit spaces

We now specialize the discussion to the Einstein case,

$$\mathrm{Ric}_{M_i^n} = \lambda g_i\,,$$

where we make the normalization

$$|\mathrm{Ric}_{M_i^n}| \leq n-1\,. \tag{4.1}$$

THEOREM 4.2. ([**ChCo3**]) There exists $\epsilon(n) > 0$, such that if the sequence of Einstein manifolds, $M_i^n \xrightarrow{d_{GH}} Y^n$, satisfies (3.3), (4.1), then for $\epsilon < \epsilon(n)$, the

ϵ-regular set, $\mathcal{R}_\epsilon$, is a smooth Einstein manifold. In particular, the singular set, $\mathcal{S} = \mathcal{S}_{n-2}$, is a closed set with dim $\mathcal{S} \leq n-2$.

Theorem 4.2 is a quite direct consequence of Theorem 3.7 and the following ϵ-regularity theorem of M. Anderson, which states that balls of almost maximal volume are standard.

THEOREM 4.3. ([**An2**]) There exists $\delta = \delta(n) > 0$ and constants, $K_k(n) < \infty$, such that if M^n is Einstein, with

$$|\mathrm{Ric}_{M^n}| \leq n-1\,,$$

and $B_r(p) \subset M^n$, satisfies

$$\mathrm{Vol}(B_r(p)) \geq (1-\delta)\mathrm{Vol}(B_r(\underline{p}))\,,$$

then $B_{\frac{r}{2}}(p)$ is the domain of a harmonic coordinate system in which

$$|g_{i,j}|_{C^k} \leq K_k r^{-k}\,,$$
$$\det(g_{i,j}) > K_0 > 0\,.$$

Anderson's theorem is proved by a blow up argument which depends on the fact that in a local harmonic coordinate system, the Einstein equation is a quasi-linear elliptic equation on the metric; see (1.2).

REMARK 4.4. Anderson also shows that if the Einstein condition is removed in Theorem 4.3, then there exists a harmonic coordinate system as above, in which the metric satisfies corresponding $C^{1,\alpha}$-bounds, for all $\alpha < 1$. Thus, if the Einstein condition is dropped in Theorem 4.2, the ϵ-regular set, $\mathcal{R}_\epsilon$, is a $C^{1,\alpha}$-riemannian manifold for all $\alpha < 1$. Essentially, this the only difference between the case in which both (4.1) *and* the Einstein condition are assumed, and that in which only (4.1) is assumed.

From Theorem 4.3 and the behavior of (4.1) under scaling we get:

THEOREM 4.5. ([**ChCo3**]) Let the sequence, $M_i^n \xrightarrow{d_{GH}} Y^n$, satisfy (3.3), (4.1). Then the regular part of any iterated tangent cone is a smooth Ricci flat Einstein manifold.

The elementary fact that in dimension 3, Einstein manifolds have constant curvature, immediately implies:

COROLLARY 4.6. Let the sequence, $M_i^n \xrightarrow{d_{GH}} Y$, satisfy (3.3), (4.1). If for $k \leq 3$, an interated tangent cone splits isometrically as $\mathbf{R}^{n-k} \times C(X)$, then the regular part, $\mathcal{R}(Y_y)$, is flat.

Consider an iterated tangent cone of the form, $\mathbf{R}^{n-3} \times C(X)$, for which. the cross-section, X, is smooth. From Corollary 4.6, it follows that, $\mathbf{R}^{n-3} \times \mathbf{R}^3/\mathbf{Z}_2 = \mathbf{R}^{n-3} \times C(\mathbf{RP}(2))$ is the only nontrivial possibility. However, a slicing argument (like the one explained in Section 5) shows that if such a tangent cone actually did exist, then the manifold, $\mathbf{RP}(2)$, would bound some (nonorientible) 3-manifold. Since, the Euler characteristic of $\mathbf{RP}(2)$ satisfies $\chi(\mathbf{RP}(2)) = 1$ and the Euler characteristic of an even dimensional manifold which bounds is even, this is impossible.

Thus, we get:

COROLLARY 4.7. If $M_i^n \xrightarrow{d_{GH}} Y$ satisfies (3.3), (4.1) and $\mathcal{S}_{n-2}(Y_y)\backslash\mathcal{S}_{n-3}(Y_y) = \emptyset$, for all iterated tangent cones, Y_y, then Y satisfies $\mathcal{S} = \mathcal{S}_{n-4}$.

Given our previous discussion, it is natural to sharpen the conjecture of Anderson which was mentioned in Section 0.

CONJECTURE 4.8. If $M_i^n \xrightarrow{d_{GH}} Y^n$ satisfies (3.3), (4.1), then $\mathcal{S} = \mathcal{S}_{n-4}$.

Proving Conjecture 4.8 amounts to ruling out the occurance of tangent cones of the form, $\mathbf{R}^{n-2} \times C(\mathbf{S}^1_{2d})$, where $\mathbf{S}^1_{2d}$ denotes the circle of circumference $2d < 2\pi$. At least for the case of special holonomy, this can be done; see Theorem 0.11, whose proof is discussed in Section 5.

CONJECTURE 4.9. If $M_i^n \xrightarrow{d_{GH}} Y^n$ satisfies (3.3), (4.1), then $\mathcal{R}$ has finite topological type.

The conclusion of Conjecture 4.9 fails if the 2-sided bound on Ricci curvature, (3.3), is replaced by the lower bound (3.1); see [**Men1**].

5. Special holonomy and codimension 4 singularities

In this section we indicate the proof of Theorem 0.11.

Theorem 0.11 states that for noncollapsed Gromov-Hausdorff limits of sequences of manifolds with special holonomy (or Einstein 4-manifolds with anti-self-dual curvature) satisfying

$$|\mathrm{Ric}_{M_i^n}| \le n-1\,,$$
$$\mathrm{Vol}(M_i^n) \ge v\,,$$

we have

$$\mathcal{S} = \mathcal{S}_{n-4}\,.$$

By Corollary 4.7, to prove Theorem 0.11, it suffices to rule out tangent cones of the form $\mathbf{R}^{n-2} \times C(\mathbf{S}^1)$, which are not isometric to $\mathbf{R}^n$.

Those cases in which H is contained in one of the groups, $\mathrm{SU}(\frac{n}{2})$, $\mathrm{Sp}(\frac{n}{4})$, G_2, Spin(7) are not difficult to handle. It suffices pass to the limit and examine the situation away from the singular set $\mathbf{R}^{n-2} \times x^* \subset \mathbf{R}^{n-2} \times C(\mathbf{S}^1)$. In the cases, $H \subset \mathrm{U}(\frac{n}{2})$, $H \subset \mathrm{Sp}(\frac{n}{2})\mathrm{Sp}(1)$, it is necessary to take into account what is happening near the singularity which is assumed to be developing as the limit is approached.

The easier cases; $\mathcal{S} = \mathcal{S}_{n-4}$.

We will explain the argument for the case $H \subset \mathrm{SU}(\frac{n}{2})$. The cases, G_2, Spin(7), can be treated by essentially the same argument; see [**ChTi2**] and compare Theorem 6.1.

Suppose Y_y is a tangent cone of the form $\mathbf{R}^{n-2} \times C(\mathbf{S}^1)$. Since $H \subset \mathrm{U}(\frac{n}{2})$, it follows easily from Anderson's ϵ-regularity theorem, Theorem 4.3, that there is a limiting parallel almost complex structure, J, on the flat open manifold, $\mathcal{R}(Y_y)$, the regular part of Y_y. The holonomy group, $H(\mathcal{R}(Y_y))$, satisfies $H(\mathcal{R}(Y_y)) \subset \mathrm{SU}(\frac{n}{2})$, where $\mathrm{SU}(\frac{n}{2})$ is defined with respect to this J. Given the riemannian product structure, it now follows that the isometric factor, $\mathbf{R}^{n-2}$, is actually a complex subspace, $\mathbf{C}^{\frac{n}{2}-1}$, and in addition, that $C(\mathbf{S}^1) = \mathbf{C}$. Otherwise, the complex determinant of a holonomy transformation would not be equal to 1.

The remaining cases; $\mathcal{S} = \mathcal{S}_{n-4}$.

By means of the twistor space construction, the case, $H \subset \mathrm{Sp}(\frac{n}{4})\mathrm{Sp}(1)$, can be reduced to a slight generalization of the case, $H \subset \mathrm{U}(\frac{n}{2})$, discussed below; see [**ChTi2**] for details.

In the $\mathrm{U}(\frac{n}{2})$ case ([**Ch3**]), Theorem 0.11 comes down to an ϵ-regularity theorem. The basic idea behind it and our subsequent ϵ-regularity theorems, is illustrated by the following elementary 2-dimensional example. Since $n = 2$, technical issues concerning slicing which are present in higher dimensions do not enter.

A model example.

Consider a conical piece of 2-dimensional surface, Y^2, in the shape of a paper cup (of the sort which is often dispensed at a water cooler). Assume that except at the cone tip, Y^2 is smooth and intrinsically flat. We ask if it is possible to replace Y^2 by a smooth manifold with boundary, which coincides with Y^2 except in a tiny neighborhood, U, of the cone tip, and for which the total amount of curvature contained in U is very small. Although U might be too small to observe directly, we can conclude that the proposed smoothing is impossible, by noting that the resulting metric would not obey the Gauss-Bonnet formula for manifolds with boundary.

Let M^2 denote the purported smooth manifold. The interior term in the Gauss-Bonnet formula is a very small number, ϵ, which is the sum contributions from an "observable region" and a "too small to be directly observable region". The boundary term, x, which is equal to the circumference of the bounding circle divided by 2π, satisfies $0 < x < 1$, and is a definite amount away from both 0 and 1 (since we imagine a cone in the shape of cup that could actually hold water). Thus, we get $0 < \epsilon + x < 1$. However, the Euler characteristic of M^2 is an integer; a contradiction.

REMARK 5.1. Note that in obtaining the above contradiction, it was *not* sufficient to assume that the amount of the curvature (measured in the L_1-norm) in a neighborhood of the rounded cone tip was small. In applying the Gauss-Bonnet formula, we also used the cone structure of the observable region.

Slicing.

For ϵ very small, let the metric on M^n satisfy $\mathrm{Ric}_{M^n} \geq -(n-1)\epsilon$. Assume that for some $\ell >> 1$, $\epsilon << 1$, the ball, $B_\ell(m)$, is ϵ-close in the Gromov-Hausdorff sense, to a ball, $B_\ell((\underline{0}, x^*))$, in $\mathbf{R}^{n-2} \times C(\mathbf{S}^1)$.

We want to "slice" M^n in such a way that the generic slice resembles the manifold M^2 in the model example. By bringing in the first Chern form, c_1, this slicing allows the $\mathrm{U}(\frac{n}{2})$ case of Theorem 0.11 to be treated by an argument which is very similar to the one used in the model example.

Since the same slicing strategy is also employed in our subsequent ϵ-regularity theorems, we begin more generally with a cone of the form $\mathbf{R}^k \times C(X)$, where $0 \leq k \leq n-2$.

From the proof of the generalized splitting theorem, Theorem 2.3, it follows that there exist harmonic functions, $\mathbf{b}_1, \ldots, \mathbf{b}_k$, on $B_1(m)$, which are very "close" to the coordinate functions of the factor, $\mathbf{R}^k$, on the ball, $B_3((\underline{0}, x^*))$. The notion of closeness is in the "Gromov-Hausdorff sense" which is meaningful since the functions involved are uniformly continuous.

Let u denote the distance function from the closed set $\mathbf{R}^k \times x^* \subset \mathbf{R}^k \times C(X)$. The function, u^2, satisfies the equation $\Delta u^2 = 2(n-k)$.

From the proof of Theorem 2.9, the "almost volume annulus implies almost metric annulus" theorem, there exists a function, $\mathbf{u} : B_3(m) \to \mathbf{R}_+$, which is very close to the function, u, in the Gromov-Hausdorff sense, such that $\Delta \mathbf{u}^2 = 2(n-k)$

We have maps, $\Phi : B_3(m) \to \mathbf{R}^k$, $\Lambda : B_3(m) \to \mathbf{R}^k \times \mathbf{R}_+$, given by $\Phi = (\mathbf{b}_1, \ldots, \mathbf{b}_k)$, $\Lambda = (\mathbf{b}_1, \ldots, \mathbf{b}_k, \mathbf{u})$.

We would like to know that when intersected with sublevel sets of the the function, $\mathbf{u}$, generic fibres of the map, Φ, resemble (k-dimensional versions of) the surface, M^2, in the model example.

The first step is to show that the notion of "generic" makes good sense i.e. that the images of the maps, Φ, Λ, have almost full measure. Here, the point of departure is the theorem on volume convergence; see Theorem 3.5.

The next step uses the coarea formula and the Cheng-Yau gradient estimate to establish that the volumes of almost all of the fibres are close to the volumes of the corresponding sets in the space $\mathbf{R}^k \times C(X)$.

In proving the above assertions (and additional ones which play a role in Section 7) the fact that the functions, $\mathbf{b}_j$, $\mathbf{u}$, satisfy suitable elliptic equations is crucial for obtaining the required estimates. Among other things, Bochner formulas are used; for details, see [**ChCoTi2**].

At this point, we specialize to the case, $k = n-2$, $X = \mathbf{S}^1$, $H \subset \mathrm{U}(\frac{n}{2})$. (For the ϵ-regularity theorems of Section 7, it will be necessary to continue the discussion of the general case.)

Let S_{2d} denote the circle of circumference $2d$. Assume that $H \subset \mathrm{U}(\frac{n}{2})$ and that there exists a tangent cone of the form, $\mathbf{R}^{n-2} \times C(\mathbf{S}^1_{2d})$, which is not isometric to $\mathbf{R}^n$ i.e. $2d < 2\pi$. Then we can find a Kähler manifold, M_i^n, for which a rescaled ball, $B_1(m_i)$, is as close as we like to the ball, $B_1((\underline{0}, x^*))$, in the Gromov-Hausdorff sense.

As above, the 2-dimensional area of a generic slice, Σ_i^2, of M_i^n, is close to the area of the ball $B_1(x^*) \subset C(\mathbf{S}^1)$. Since the metric on M_i^n has been rescaled, the pointwise norm of the first Chern form, c_1 is very small, and we have

$$\lim_{i\to\infty} \int_{\Sigma_i^2} c_1 = 0\,.$$

On the other hand, according the theory of differential characters, the above integral is equal mod $\mathbf{Z}$, to the associated differential character, $\widehat{c_1}$, evaluated on $\partial\Sigma_i^2$; [**ChSi**]. The crucial point is that the quantity, $\widehat{c_1}(\partial\Sigma_i^2) \in \mathbf{R}/\mathbf{Z}$, is determined entirely from the geometry of TM_i^n at $\partial\Sigma_i^2$. In fact, as $i \to \infty$, the secondary geometric invariant, $\widehat{c_1}(\partial\Sigma_i^2)$, converges to the mod $\mathbf{Z}$ reduction of the boundary term in Gauss-Bonnet formula for $C(S^1_{2d})$. As in the model example, this is nonzero mod $\mathbf{Z}$, unless $2d = 2\pi$ i.e. unless $C(S^1_{2d}) = \mathbf{R}^2$. Thus, we get a contradiction.

To see why the boundary term in the Gauss-Bonnet formula appears, note that near $\partial\Sigma_i^2$, as $i \to \infty$, the convergence of M_i^n to $\mathbf{R}^{n-2} \times C(\mathbf{S}^1_{2d})$, takes place in the C^∞ topology (since we are away from the singularity which is purported to be developing). Thus, the tangent bundle, TM_i^n, converges to the Whitney sum of a flat trivial bundle of dimension $n-2$ and the tangent bundle to $C(\mathbf{S}^1)$.

By the Whitney sum formula in Chern Weil theory, the Chern form, $c_1(TM_i^n)$, converges, to the Chern form $c_1(\Sigma_i^2)$. Since $\dim \Sigma_i^2 = 2$, the first Chern form is the Euler form i.e. the Gauss-Bonnet form of $C(\mathbf{S}^1_{2d})$. According to [**ChSi**], these relations continue to hold at the more refined level of differential characters. Moreover, when evaluated on $\partial\Sigma_i^2$, the Euler character is just the boundary term in the Gauss-Bonnet formula.

REMARK 5.2. In the above argument, we used the first Chern form. This required a Kähler structure and a 2-sided bound on Ricci curvature. Had we instead attempted to use the Gauss-Bonnet formula for the induced metric on a slice, then given our assumptions, it would not have been possible to show that the interior term approached zero as $i \to \infty$; compare the discussion in Section 7.

6. Special holonomy and tangent cones

In this section we indicate the proof of Corollary 0.13 and give some additional results on the classification of tangent cones for the case of special holonomy; for details see **[Ch3]**, **[ChTi2]**.

Corollary 0.13 states that if $M_i^n \xrightarrow{d_{GH}} Y$, where the manifolds, M_i^n, have special holonomy and satisfy the 2-sided bound, $|\mathrm{Ric}_{M_i^n}| \le n-1$, then for all $y \in \mathcal{S}_{n-4} \setminus \mathcal{S}_{n-6}$, there exists at least one tangent cone, Y_y, of H-orbifold type,

$$Y_y = \mathbf{R}^{n-4} \times \mathbf{R}^4/\Gamma\,.$$

By definition, if $y \in \mathcal{S}_{n-4} \setminus \mathcal{S}_{n-5}$, then there exists some tangent cone of the form, $\mathbf{R}^{n-4} \times C(X^3)$. By Corollary 4.6, the regular part of this cone is Ricci flat. From Theorem 0.11, it follows that X^3 is smooth. It follows that $Y_y = \mathbf{R}^{n-4} \times \mathbf{R}^4/\Gamma$ as above.

It remains only to show

$$\mathcal{S}_{n-5} \setminus \mathcal{S}_{n-6} = \emptyset\,,$$

and if $H \subset \mathrm{Sp}(\frac{n}{4})\mathrm{Sp}(1)$, then

$$\mathcal{S}_{n-5} \setminus \mathcal{S}_{n-8} = \emptyset\,.$$

This is a consequence of the following Theorem 6.1.

Recall the inclusions $\mathrm{SU}(\frac{n}{2}) \subset \mathrm{U}(\frac{n}{2})$, $\mathrm{Sp}(\frac{n}{4}) \subset \mathrm{Sp}(\frac{n}{4})\mathrm{Sp}(1)$, and $\mathrm{SU}(3) \subset \mathrm{G}_2 \subset \mathrm{Spin}(7)$.

THEOREM 6.1. (**[ChTi2]**) Let the sequence, $M_i^n \xrightarrow{d_{GH}} Y^n$, of manifolds with special holonomy, H, satisfy (3.3), (4.1).

i) If $H \subset \mathrm{U}(\frac{n}{2})$, then

$$\mathcal{S}_{n-i} \setminus \mathcal{S}_{n-i-1} = \emptyset \qquad (i \neq 2k, \text{ for some } k \in \mathbf{Z})\,.$$

ii) If $H \subset \mathrm{Sp}(\frac{n}{4})\mathrm{Sp}(1)$, then

$$\mathcal{S}_{n-i} \setminus \mathcal{S}_{n-i-1} = \emptyset \qquad (i \neq 4k, \text{ for some } k \in \mathbf{Z})\,.$$

iii) If $H \subset \mathrm{Spin}(7)$, then

$$\mathcal{S}_{n-i} \setminus \mathcal{S}_{n-i-1} = \emptyset \qquad (i \neq 4,\, 6,\, 7,\, 8).$$

7. Integral bounds on curvature

In this section (and the next) we describe the effect of adding an integral bound on curvature to the hypotheses (3.1), (3.3). Recall that (3.1) specifies only a lower bound on Ricci curvature and not a 2-sided bound as in Sections 5, 6. However, the results apply in particular to Einstein manifolds.

The integral bound is an extra apriori assumption. But in the case of special holonomy, an L_2 bound on curvature is implied by a bound on the topological invariant, $C(M^n)$, of (0.10); see Section 9.

Let $1 \leq p \leq \frac{n}{2}$. In the presence of an L_p bound on curvature, there are two basic statements which hold without further qualification:

i) The singular set has Hausdorff dimension $\leq (n-2p)$. In fact, $\mathcal{H}^{n-2p}(\mathcal{S}) = 0$ unless p is an integer.

ii) If p is an integer, the tangent cone is unique for $\mathcal{H}^{n-2p}$-a.e. $y \in \mathcal{S}$ and is of orbifold type $Y_y = \mathbf{R}^{n-2p} \times \mathbf{R}^{2p}/\Gamma$.

At least in the Kähler case and for manifolds with special holonomy, an additional fact holds:

iii) For p an integer, bounded subsets of the singular set have have finite $(n-2p)$-dimensional Hausdorff measure and are actually $(n-2p)$-rectifiable.

If the Kähler or special holonomy assumption is omitted in iii), rectifiability of bounded subsets is known for that part, $\mathcal{N}$, of $\mathcal{S}$, where "nonexceptional" tangent cones are present. This awkward point might just be a reflection of the present state of our technology. If $p = 1$ or $p = \frac{n}{2}$, then $\mathcal{N} = \mathcal{S}$.

There are three main steps in the proofs of i) and the statement in iii) concerning the finiteness of Hausdorff measure. The first of these is also the starting point for proving ii).

a) Show that if $k \leq 2p$, then for $\mathcal{H}^{n-2p}$-a.e. $y \in \mathcal{S}_{n-k}$, there exists at least one tangent cone of type $\mathbf{R}^{n-k} \times C(\mathbf{S}^{k-1}/\Gamma)$, $(k > 2)$, or $\mathbf{R}^{n-2} \times C(S^1_{2d})$, $(k = 2)$.

b) Prove an ϵ-regularity theorem to the effect that if a sequence satisfying (3.1), (3.3), converges to a tangent cone, Y_y, as in a), then when measured in the L_k sense, a definite amount of curvature must concentrate in the limit at y.

c) Show that for bounded subsets of $\mathcal{S}$, the curvature concentration required in b) can only occur on a subset with finite $(n-2p)$-dimensional Hausdorff measure, whose intersection with $\mathcal{S} \setminus \mathcal{S}_{n-2p}$ has vanishing $(n-2p)$-dimensional Hausdorff measure.

Step c) is proved by a standard covering argument. The assertion in step b) should be compared to that of Corollary 0.13.

L_1 curvature bounds.

Step a) is trivial for $p = 1$.

For $p = 1$, the ϵ-regularity theorem, step b), incorporates a direct generalization of the argument in the model case which was described in Section 5. For simplicity, we consider balls of radius 1; the case of arbitrary radius follows by a simple scaling argument.

THEOREM 7.1. ([**ChCoTi2**]) For all $v, \eta > 0$, there exists $\epsilon = \epsilon(\eta, v, n) > 0$, $\ell = \ell(\eta, v, n) < \infty$, $\delta = \delta(\eta, v, n) > 0$, such that if M^n satisfies

$$\mathrm{Ric}_{M^n} \geq -(n-1)\epsilon\,, \tag{7.2}$$

$$\mathrm{Vol}(B_1(m)) \geq v\,, \tag{7.3}$$

$$⨍_{B_{3\epsilon}(m)} |\mathrm{R}| \leq \delta\,, \tag{7.4}$$

and for some $0 < d \leq \pi$ and $(\underline{0}, x^*) \in \mathbf{R}^{n-2} \times C(\mathbf{S}^1_{2d})$,

$$d_{GH}(B_\ell(m), B_\ell((\underline{0}, x^*))) \leq \ell^{-1}\,, \tag{7.5}$$

then for $B_\epsilon(0) \subset \mathbf{R}^n$, we have

$$d_{GH}(B_1(m), B_1(0)) \leq \eta\,. \tag{7.6}$$

As in the proof of Theorem 0.11, the idea is to reduce to the 2-dimensional case by employing a slicing argument. Since we don't have a first Chern form at our disposal, we want to use the intrinsic Gauss-Bonnet formula on a generic slice. So we must continue the discussion of Section 5.

The next step in that discussion is to show that the second fundamental form of a generic level set of the map Φ is small in the L_2-sense. This, together with (7.4) and the Gauss curvature equation, gives that the L_1-norm of the intrinsic curvature on a generic slice is small. Similar properties hold for the map Λ.

At this point, the proof can be completed by employing on a generic slice, essentially the same argument as in the model example explained in Section 5.

Suppose we add to assumptions, (3.1), (3.3), the integral bound,

$$⨍_{B_1(m_i)} |\mathrm{R}| \leq c\,. \tag{7.7}$$

Since c) above follows by a standard covering argument, we get:

THEOREM 7.8. ([**ChCoTi2**]) Let $M_i^n \xrightarrow{d_{GH}} Y$ satisfy (3.1), (3.3), (7.7). Then bounded subsets of the singular set, $\mathcal{S}$, have finite $(n-2)$-dimensional Hausdorff measure.

L_p curvature bounds; $p > 1$.

Assume

$$⨍_{B_1(m_i)} |\mathrm{R}|^{2p} \leq c\,. \tag{7.9}$$

Flatness of $\mathcal{R}(Y_y)$.

The proof of step a) begins by establishing a weaker statement. To avoid inessential technical difficulties, we explain the argument in the Einstein case where, by Theorem 4.2, the regular part, $\mathcal{R}$, is a smooth Einstein manifold.

Relation (7.9), together with a covering argument, easily implies that for $\mathcal{H}^{n-2p}$-a.e. $y \in Y$, if we rescale the ball, $B_r(y)$, to a ball of radius 1, and let $r \to 0$, then

$$⨍_{B_1(y)\cap\mathcal{R}} |\mathrm{R}|^p \to 0 \qquad (\text{as } r \to 0)\,. \tag{7.10}$$

To see this, let $\{U_i\}$ denote a sequence of neighborhoods whose intersection is some bounded subset of $\mathcal{S}$. Then

$$\lim_{i\to\infty}\int_{U_i\cap\mathcal{R}}|R|^p = 0\,,$$

and since we let $r \to 0$, in the covering argument, we need only consider balls which are contained in the U_i.

From (7.10) it follows that there exists $A \subset \mathcal{S}$, with $\mathcal{H}^{n-2p}(\mathcal{S}\setminus A) = 0$, such that $\mathcal{R}(Y_y)$ is flat, for all $y \in A$.

Existence of tangent cones of orbifold type.

If $y \in (\mathcal{S}_{n-k}\setminus\mathcal{S}_{n-k-1})$, there exists some tangent cone of the form $\mathbf{R}^k \times C(X)$. By induction on p, the cross-section, X can be assumed to be smooth. Otherwise, we would have $\dim\mathcal{S} \geq n-k+1 > n-2p$. (Here, we use $\dim\mathcal{S}(C(X)) = \dim\mathcal{S}(X)+1$.) It follows that X is of the form $\mathbf{S}^{k-1}/\Gamma$ which gives step a).

The ϵ-regularity theorem.

For p a half integer, an argument which closely resembles the one given at the end of Section 4 can be applied away from the singularity.

For n even and $p = \frac{n}{2}$ there is no problem in generalizing Theorem 7.1, since there is no second fundamental form to contend with.

As of this moment, for p an integer, with $1 < p < \frac{n}{2}$, the proof of Theorem 7.1 has not been generalized to the case in which an L_p curvature bound is assumed and the cone, $\mathbf{R}^{n-2}\times C(\mathbf{S}^1_{2d})$, is replaced by a cone $\mathbf{R}^{n-2p}\times C(X)$, where $X = \mathbf{S}^{n-2p-1}/\Gamma$. A technical difficulty arises when one attempts to estimate L_{2p}-norm of the second fundamental form of a generic slice.

The difficulty can be illustrated by considering the case, $\mathbf{R}\times C(X)$, in which there is just a single harmonic function $\mathbf{b}$. By a standard formula, the second fundamental form of a level surface of $\mathbf{b}$ satisfies

$$|II|^2 = \frac{|\mathrm{Hess}_{\mathbf{b}}|^2 - |\mathrm{Hess}_{\mathbf{b}}(\,\cdot\,,N)|^2}{|\nabla\mathbf{b}|^2}\,,$$

where $N = \nabla\mathbf{b}/|\nabla\mathbf{b}|$. So to estimate the L_{2p}-norm of the second fundamental form, one must bound an expression with $|\mathrm{Hess}_{\mathbf{b}}|^{2p}$ in the numerator and $|\nabla\mathbf{b}|^{2p}$ in the denominator.

The L_p bound on curvature, together with a higher order Bochner formula, yields an L_{2p} bound on $|\mathrm{Hess}_{\mathbf{b}}|$. This bound, which is also crucial for the results on rectifiability of singular sets, is explained at greater length in Section 8.

To get the required bound on the denominator, it would suffice show that at *all* points of a generic slice, $|\nabla b|$ is close to 1. Thinking (a bit loosely) in terms of the Sobolev imbedding theorem, we note that the L_{2p} bound on $|\mathrm{Hess}_{\mathbf{b}}|$ gives an L_{2p} bound on the norm of $\nabla|\nabla\mathbf{b}|$. However, since the slices have dimension $2p$, it would seem that what is actually needed is a bound on the $L_{2p'}$-norm of $\nabla|\nabla\mathbf{b}|$, for some $p' > p$. Such a bound is lacking; compare the discussion of Section 8.

If on the other hand, we are given an $L_{p'}$ bound on curvature, with $p' > p$, then we do get that $|\nabla\mathbf{b}|$ is close to 1, everywhere on a generic slice.

By a covering argument like that which plays a role in Theorem 7.8, we obtain:

THEOREM 7.11. ([**ChCoTi2**], [**Ch3**]) Let $M_i^n \xrightarrow{d_{GH}} Y$ satisfy (3.1), (3.3), (7.9). Then

$$\dim \mathcal{S} \leq n - 2p\,,$$
$$\mathcal{H}^{n-2p}(\mathcal{S} \setminus \mathcal{S}_{n-k}) = 0 \qquad (k < 2p)\,,$$
$$\mathcal{H}^{n-2p}(\mathcal{S}) = 0 \qquad (p \text{ not an integer})\,.$$

Finiteness of Hausdorff measure.

Because of the difficulty in generalizing Theorem 7.1 to the case, $p > 1$, at present there is only a partial analog of Theorem 7.11 in the real case. In the complex case, one has an analog of Theorem 7.1 in which higher Chern forms, c_p, are used in the same way as c_1 is used in proving Theorem 0.11; compare Sections 5, 9. This leads to:

THEOREM 7.12. ([**ChCoTi2**], [**Ch3**]) Suppose that the sequence of Kähler manifolds, $M_i^n \xrightarrow{d_{GH}} Y$, satisfy (3.3), (4.1), (7.9). Then bounded subsets of the singular set, $\mathcal{S}$, have finite $(n-2p)$-dimensional Hausdorff measure.

The real case.

In the real case, the best we can do is to use Pontrjagin forms in place of Chern forms. This requires our assuming that $p = 2k$ is an *even* integer. However, the ϵ-regularity theorem still does not work perfectly. The reason is that there are certain tangent cones, $\mathbf{R}^{n-4k} \times C(\mathbf{S}^{4k-1}/\Gamma)$, for which all secondary geometric invariants associated to Pontrjagin polynomials of degree k vanish; see [**ChSi**]. This means that we can only deal with the subset, $\mathcal{N} \subset \mathcal{S}$, for which there exists a cone, $\mathbf{R}^{n-4k} \times C(\mathbf{S}^{4k-1}/\Gamma)$, which is *not* of this type.

If, for instance, $k = 1$, then the tangent cones which must be excluded are those of the form $\mathbf{R}^{n-4} \times C(\mathbf{S}^3/\Gamma)$, where $\mathbf{S}^3/\Gamma$ is a lens space, $L_{p,q}$, with $q^2 \equiv -1 \bmod p$.

As in Theorem 7.12, we get:

THEOREM 7.13. ([**ChCoTi2**], [**Ch3**]) Let $M_i^n \xrightarrow{d_{GH}} Y$, satisfy (3.3), (4.1), (7.9), for $p = 2k$. Then bounded subsets of the subset, $\mathcal{N} \subset \mathcal{S}$, have finite $(n-4k)$-dimensional Hausdorff measure.

Uniqueness of tangent cones.

The statement concerning the existence of tangent cones of orbifold type can be strengthened.

THEOREM 7.14. ([**ChCoTi2**], [**Ch3**]) Let $M_i^n \xrightarrow{d_{GH}} Y$ satisfy (3.3), (4.1), (7.9). Then there exists $A \subset \mathcal{S}$, with $\mathcal{H}^{n-2p}(\mathcal{S} \setminus A) = 0$, such that for all $y \in A$, the tangent cone is unique and of orbifold type $\mathbf{R}^{n-2p} \times \mathbf{R}^{2p}/\Gamma$.

For $p > 1$, Theorem 7.14 is proved by a deformation argument.

It is not difficult to see that (in all situations in geometric analysis in which tangent cones arise) the space of all tangent cones is connected in a suitable topology. So in the context of Theorem 7.14, one can start with a tangent cone of the form $\mathbf{R}^{n-2p} \times C(\mathbf{S}^{2p-1}/\Gamma)$ and try to show that it cannot be deformed within the space of tangent cones.

If we knew that *every* tangent cone split off an isometric factor, $\mathbf{R}^{n-2p}$, it would follow as in the proof of existence of tangent cones of orbifold type, that every tangent cone is of the form $\mathbf{R}^{n-2p}\times C(\mathbf{S}^{2p-1}/\Gamma)$. Although apriori, Γ might depend on the particular tangent cone, since spherical space forms cannot be deformed, we would be done. (Here, the assumption, $p > 1$, enters.)

A volume comparison argument.

Since we *can't* assume that every tangent cone splits off $\mathbf{R}^{n-2p}$ isometrically, we must bring in some additional information.

A general consequence of relative volume comparison is that, for all tangent cones at a given point, the $(n-1)$-dimensional Hausdorff measure of the cross-section is the same. Using this and a volume comparison argument which depends on knowing that for every tangent cone, Y_y, the regular part, $\mathcal{R}(Y_y)$, is flat, the proof of Theorem 7.11 can be completed.

Specifically, one shows that among all tangent cones which are sufficiently close to one of the form $\mathbf{R}^{n-2p}\times C(\mathbf{S}^{2p-1}/\Gamma)$, the cone, $\mathbf{R}^{n-2p}\times C(\mathbf{S}^{2p-1}/\Gamma)$, is the *unique* one for which the $(n-1)$-dimensional Hausdorff measure, of the cross-section is maximal.

REMARK 7.15. The volume comparison argument discussed above also uses the rigidity of spherical space forms (for $p > 1$), to locate inside of Y_y, a copy of $\mathbf{S}^{2p-1}/\Gamma$. This being done, one views Y_y as a tube around $\mathbf{S}^{2p-1}/\Gamma$ and uses the flatness of $\mathcal{R}(Y_y)$.

REMARK 7.16. If the conclusion of the volume comparison argument could be shown to hold without without appealing to the flatness of the regular part, $\mathcal{R}(Y_y)$, it would follow that the tangent cone is unique at *all* points of $\mathcal{S}_{n-4}\setminus\mathcal{S}_{n-5}$. Alternatively, this would follow if the flatness of $\mathcal{R}(Y_y)$ could be established without using the integral bound on curvature (7.9). Knowing uniqueness of the tangent cone at all points of $\mathcal{S}_{n-4}\setminus\mathcal{S}_{n-6}$ would be helpful in proving Conjecture 0.15.

REMARK 7.17. For the case $p = 1$, there is a completely different proof of uniqueness of tangent cones, which is closely related to the proof of rectifiablity of singular sets discussed in Section 8. This argument is valid for all p. However, it does not work in the absence of the integral bound (7.9).

8. Rectifiability of singular sets

A metric space, W, is called d-rectifiable if $0 < \mathcal{H}^d(W) < \infty$ and there exists a countable collection of subsets, C_j, with $\mathcal{H}^d(W\setminus\bigcup_j C_j) = 0$, such that each C_j is bi-Lipschitz to a subset of $\mathbf{R}^d$.

Rectifiability can be thought of as a weak measure theoretic generalization of the property of being a Lipschitz manifold. In particular, first order calculus makes sense on rectifiable spaces.

Proving that a space is rectifiable can be a step in the process of establishing additional regularity properties; compare [**HarShiff**], [**Ti3**]. It is hoped that the rectifiability theorems of this section and Section 9 will be useful in proving Conjecture 0.15.

While rectifiability plays a central role in geometric measure theory, most of the classical theory is concerned with subsets of $\mathbf{R}^N$; compare [**Fe**], [**Ma**], [**Sim1**]–[**Sim3**]. As a consequence, the standard criteria for establishing that a set is rectifiable do not apply in our (intrinsic) context.

Our main results are strengthenings of Theorems 7.8, 7.12, 7.13.

THEOREM 8.1. ([**Ch3**]) Let $M_i^n \xrightarrow{d_{GH}} Y$ satisfy (3.3), (4.1), (7.9). Then bounded subsets of the singular set, $\mathcal{S}$, are $(n-2)$-rectifiable.

THEOREM 8.2. ([**Ch3**]) Let the sequence of Kähler manifolds, $M_i^n \xrightarrow{d_{GH}} Y$, satisfy (3.3), (4.1), (7.9). Then bounded subsets of the singular set, $\mathcal{S}$, are $(n-2p)$-rectifiable.

THEOREM 8.3. ([**Ch3**]) Let $M_i^n \xrightarrow{d_{GH}} Y$, satisfy (3.3), (4.1), (7.9), for $p = 2k$. Then bounded subsets of the subset, $\mathcal{N} \subset \mathcal{S}$, are $(n-4k)$-rectifiable.

Theorems 8.1–8.3 are proved by directly constructing the sets, C_j, and bi-Lipschitz maps from the C_j to a Euclidean space of the appropriate dimension. Apart from the differences in the hypotheses of the underlying ϵ-regularity theorems, the proofs in all three cases are essentially the same.

To fix notation, we suppose that our integral bound is on the L_p-norm of the curvature and hence, that $\dim \mathcal{S} = n - 2p$.

To grasp the main ideas, there is no harm in thinking of the Einstein case, in which we have a closed singular set and a smooth regular part.

Tangent cones and locally defined Lipschitz maps.

Consider $y \in A$, where $A \subset \mathcal{S}$ is as in Theorem 7.14. As in the discussion of slicing in Section 5, a tangent cone of the form $\mathbf{R}^{n-2p} \times C(\mathbf{S}^{2p-1}/\Gamma)$, gives rise to a locally defined map $\Phi_i = (\mathbf{b}_{i,1}, \ldots, \mathbf{b}_{i,n-2p})$, from each approximating manifold, M_i^n, to $\mathbf{R}^{n-2p}$, where the $\mathbf{b}_{i,j}$ are harmonic functions.

By the Cheng-Yau gradient estimate, the maps, Φ_i, have a uniform gradient bound. Thus we can pass to the limit and get a locally defined Lipschitz map, $\Phi = (\mathbf{b}_1, \ldots, \mathbf{b}_{n-2p})$, from a neighborhood of $y \in Y$, to $\mathbf{R}^{n-2p}$.

If we can show that when restricted to $\mathcal{S}$, the map, Φ, is *bi-Lipschitz* on some subset, $C \subset \mathcal{S}$, with $\mathcal{H}^{n-2p}(C) > 0$, then by standard soft argument, the proof can be completed.

Intuitively, we would like to find $C \subset \mathcal{S}$, with $\mathcal{H}^{n-2p}(C) > 0$, such that $\Phi \,|\, C$ is nondegenerate in the sense that the "gradients" of our Lipschitz functions, $\mathbf{b}_j$, are linearly independent. In the simplest case, $n - 2p = 1$, we want the "gradient" of $\mathbf{b} \,|\, C$ to be nonvanishing. (We use quotes because the sense in which $\mathbf{b}_j \,|\, \mathcal{S}$ has a gradient is not immediately clear.)

Volume nondegeneracy of Φ.

The first step is to show that $\Phi \,|\, \mathcal{S}$, is nondegenerate in a weaker sense. Namely, $(n-2p)$-dimensional measure of the range of $\Phi \,|\, \mathcal{S}$ has a definite lower bound. This statement entails kind of "gap" phenomenon: If A is nonempty, then $\mathcal{H}^{n-2p}(A) > 0$.

In fact, the gap theorem is just a quantitative version of the ϵ-regularity theorem. If near $y \in A$, most slices (i.e. fibres of Φ) did not intersect the singular set, we could apply the ϵ-regularity theorem directly on the limit space (to the union of such fibres) and conclude that $y \notin \mathcal{S}$.

Reduction to the existence of $\nabla\Phi$ on $\mathcal{S}$.

We observed above that $\Phi \,|\, \mathcal{S}$ *does not* totally compress volume. To complete the argument, it would suffice to know that $\Phi \,|\, \mathcal{S}$ has the additional property that if for $\mathcal{H}^{n-2p}$-a.e. $y \in A$, it compresses distance in some direction (i.e. if Φ is nowhere bi-Lipschitz) then it *does* compress volume.

Nice maps such as linear maps, or more generally, differentiable maps, do have this property. For instance, if a linear map defined on $\mathbf{R}^n$ has a nontrivial kernel, then its image is lower dimensional and hence, has vanishing n-dimensional volume. So we would like to know that near generic points of A, the map, Φ looks asymptotically (generalized) linear i.e. differentiable, in some generalized sense; compare (1.1).

In essence, we have reduced the problem of showing that the "gradient" of Φ does not vanish on a subset of A with positive measure, to showing that $\nabla\Phi$ exists on such a subset!

It is not difficult to see that rather than working directly on the singular set, it suffices to show that for $y \in A$, the restriction of Φ to the *regular part*, $\mathcal{R}$, looks asymptotically generalized linear as we approach y.

A higher order Bochner formula.

In our situation a function is (approximately) generalized linear if its Hessian is *small* in the integral sense; compare the discussion of Bochner's formula in Section 1 and the proofs of the splitting theorem and in its generalization, described in Section 2.

Let $f_{m,1}$ denote the average of f over the ball of radius 1.

Let $h : B_1(m) \to \mathbf{R}$ be harmonic and put

$$\sup_{B_{\frac{3}{4}}(m)} |\nabla h| = V\,.$$

By using in tandem the Bochner formulas for $\Delta|\nabla h|^2$, $\Delta|\mathrm{Hess}_h|^2$, one can derive an estimate for the quantities,

$$⨍_{B_{\frac{1}{2}}(m)} |\mathrm{Hess}_h|^{2p}\,, \qquad ⨍_{B_{\frac{1}{2}}(m)} |\nabla \mathrm{Hess}_h|^2 \cdot |\mathrm{Hess}_h|^{2p-4}\,, \tag{8.4}$$

in terms of

$$V, \quad ⨍_{B_1(m)} \big||\nabla h|^2 - (|\nabla h|^2)_{m,1}\big|\,. \quad ⨍_{B_1(m)} |R|^p\,; \tag{8.5}$$

see **[Ch3]**,

For fixed V, the quantities in (8.4) will be as small as one likes, if the second two quantities in (8.5) are sufficiently small.

If we apply the above estimate to the function, $\mathbf{b}_{i,j}$, then the first term in (8.5) is bounded and the second vanishes in the limit as $i \to \infty$. However, since as $i \to \infty$, we are converging to a (rescaled) neighborhood of a *singular* point, $y \in A$, by the ϵ-regularity theorem, the curvature must be concentrating. Thus, while bounded, the third term in (8.5) is definitely *not* going to zero as $i \to \infty$.

A removability property of the singular set.

The above estimate can be passed to the limit space, Y, and applied to the functions $\mathbf{b}_j$ on the regular part $\mathcal{R}$. However, since apparently, a contribution from the curvature which has concentrated on the singular set must be included, what this accomplishes is not immediately clear.

The same considerations which led in Section 7, to the conclusion that tangent cones are flat on their regular parts, show that near $y \in A$, after suitable rescaling, the integral of $|R|^p$ over the *regular set*, $\mathcal{R}$, is small.

Thus, it would suffice to know that if near $y \in A$, we want to control the L_{2p}-norm of $\mathrm{Hess}_{\mathbf{b}_j}$ on $\mathcal{R}$, then only the curvature on $\mathcal{R}$ contributes to the estimate i.e. the curvature which concentrates in the limit on $\mathcal{S}$ can be ignored.

This assertion, which says that in a certain sense, the singular set is "removable", turns out to hold. To prove it, we attempt to repeat the derivation of (8.4), (8.5), while working on the smooth incomplete manifold $\mathcal{R}$.

Apart from a several integrations by parts, the argument is purely local. The whole point is to show is that the integrations by parts don't lead to a "residue term" attached to the singular set.

Given that bounded subsets of $\mathcal{S}$ have finite $(n-2p)$-dimensional Hausdorff measure, the estimates which were gotten by passing the estimates in the smooth case to the limit, provide enough control on the functions, $\mathbf{b}_j$, to prove the absence of a residue attached to the singular set. Here, although our ultimate interest is in $|\mathrm{Hess}_{\mathbf{b}_j}|_{L_{2p}}$, the estimate on the term involving $|\nabla \mathrm{Hess}_{\mathbf{b}_j}|$ is needed as well.

9. Anti-self-duality of curvature and the structure of singular sets

In dimension 4, the notion of anti-self-dual (or self-dual) curvature tensor, one which satisfies $*F = -F$ (or $*F = F$) is of crucial importance in gauge theory and of significance in riemannian geometry. The key point is the identity

$$\mp p_1(M^4) = \frac{1}{8\pi^2}\int_{M^4} |F|^2\,, \tag{9.1}$$

where p_1 denotes the first Pontrjagin class. This relation, which follows directly from Chern-Weil theory, implies that the underlying connection is an absolute minimum for the Yang-Mills functional.

The class of riemannian manifolds with special holonomy provides a framework for extending anti-self-duality to higher dimensions. For such manifolds, there is a parallel $(n-4)$-form Ω; compare [**Joy3**], [**Sa1**], [**Sa2**] and see (0.8). The curvature tensor of a connection on a bundle whose base space, M^n, has such a parallel $(n-4)$-form, is called Ω-*anti-self-dual* if

$$*\,F = -F \wedge \Omega\,; \tag{9.2}$$

compare [**Ti3**], [**BaHiSing1**], [**BaHiSing2**]. From now on, we will supress the dependence on Ω and say F satisfying (9.2) is anti-self-dual.

In [**Ti3**], a program was initiated, whose aim is to extend anti-self-duality and its consequences to higher dimensional gauge theory. Strong information was obtained on the degeneration problem, and a compactification of the moduli space was introduced, for the case in which the anti-self-dual connection on an auxilliary bundle varies, while the riemannian connection stays fixed. In the present paper, by contrast, we are concerned with the situation in which the anti-self-dual connection is the riemannian connection itself.

As noted in [**Ti3**], for certain purposes, it is useful to require only that the suitably defined trace free part, F_0, satisfies (9.2), and sometimes, that the trace of F is harmonic.

If M^n has special holonomy, then the trace free part, R_0, of its riemann curvature tensor, R, satisfies

$$* R_0 = -R_0 \wedge \Omega\,; \tag{9.3}$$

compare [**Joy3**], [**Sa1**], [**Sa2**].

With the exception of the cases, $H_0 = \mathrm{U}(\frac{n}{2})$, $H_0 = \mathrm{Sp}(\frac{n}{4})\mathrm{Sp}(1)$, we have $R_0 = R$.

In the $\mathrm{U}(\frac{n}{2})$ case, the part, $R - R_0$, is determined by the Ricci tensor and the almost complex structure.

In the $\mathrm{Sp}(\frac{n}{4})\mathrm{Sp}(1)$ case, $(n > 4)$, $R - R_0$ is determined locally by the anti-commuting almost complex structures on the fibre (generated by I, J, K) and the Ricci tensor $\mathrm{Ric}_{M^n} = \lambda g$.

As in the case of 4-manifolds with anti-self-dual curvature, for manifolds with special holonomy, the L_2-norm of the full curvature tensor can be bounded in terms of a topological invariant, $C(M^n)$, and a bound on the norm of $F - F_0$, where in the riemannian case, the norm of $R - R_0$ can be bounded in terms of the Ricci tensor. The characteristic number, $C(M^n)$, is defined by

$$C(M^n) = (-p_1 \cup [\Omega])(M^n)\,.$$

Just as in Section 8 (compare also Section 7) we have:

THEOREM 9.4. ([**ChTi2**]) Let the sequence, $M_i^n \xrightarrow{d_{GH}} Y$, of manifolds with special holonomy and $\{C(M_i^n)\}$ bounded above, satisfy (3.1), (3.3). Then bounded subsets of the singular set, $\mathcal{S}$, are $(n-4)$-rectifiable.

REMARK 9.5. In light of Theorems 0.11, 6.1 and Corollary 0.13, the $(n-4)$-rectifiablity of $\mathcal{S}$ can be thought of as asserting that in a weak sense, $\mathcal{S}_{n-4}$ is calibrated by the $(n-4)$-form Ω; see [**HarLaw**].

REMARK 9.6. The statements of Conjecture 0.15 and Theorem 9.4 can be generalized by replacing the sequence of manifolds, $\{M_i^n\}$, by a sequence of H-orbifolds with codimension 4 singularities. Part of the interest of this degree of generality pertains to the quaternion-Kähler case with *positive* scalar curvature. As Blaine Lawson pointed out to us, there are many examples of orbifolds with this property, while conjecturally, in higher dimensions, the only smooth examples are symmetric spaces; see[**GaLaw**], [**HerHer**].

10. Appendix; Review of special holonomy

In this appendix, we recall some material concerning aspects of special holonomy which enter in the body of the paper. Specifically we discuss the Einstein condition and the role played by complex structures (or partial complex structures) in the various cases.

The cases: $\mathbf{U}(\frac{n}{2})\,,\mathbf{SU}(\frac{n}{2})$.

Let M^n be Kähler, with Hermitian almost complex structure, I, satisfying $\nabla I = 0$. Let $e_1, I(e_1), \ldots, e_{\frac{n}{2}}, I(e_{\frac{n}{2}})$ denote an orthonormal basis for the tangent space, M_p^n, at $p \in M^n$. The Kähler form, ω, is given by

$$\omega = e_1^* \wedge I(e_1^*) + \cdots + e_{\frac{n}{2}}^* \wedge I(e_{\frac{n}{2}})\,.$$

By the holonomy theorem of Ambrose-Singer (or by direct calculation) the curvature transformations are contained in $\mathfrak{u}(\frac{n}{2})$ and the riemann curvature tensor can be regarded as an element of $\bigwedge^2 \otimes \mathfrak{u}(\frac{n}{2})$. With this understanding, the relation $R(x,y,u,v) = R(x,y,I(u),I(v))$ and the Jacobi identity give the well known formula

$$\begin{aligned}
\langle \omega, R\rangle(x,y) &= \sum_{i=1}^{\frac{n}{2}} R(e_i, I(e_i), x, y) \\
&= -\sum_{i=1}^{\frac{n}{2}} R(x, e_i, I(e_i), y) - \sum_{i=1}^{\frac{n}{2}} R(I(e_i), x, e_i, y) \\
&= +\sum_{i=1}^{\frac{n}{2}} R(x, e_i, e_i, I(y)) + \sum_{i=1}^{\frac{n}{2}} R(x, I(e_i), I(e_i), y) \\
&= \operatorname{Ric}(x, I(y))\,.
\end{aligned} \tag{10.1}$$

Thus, if we regard $R \in S^2(\mathfrak{u}(\frac{n}{2}))$, then M^n is Kähler-Einstein if and only if relative to the decomposition of $S^2(\mathfrak{u}(\frac{n}{2}))$ induced by the decomposition, $\mathfrak{u}(\frac{n}{2}) = \mathfrak{su}(\frac{n}{2}) \oplus \mathbf{R}\cdot\omega$, the mixed components of R vanish. Similarly, M^n is Ricci flat if and only if $R \in S^2(\mathfrak{su}(\frac{n}{2}))$.

The cases: $\mathbf{Sp}(\frac{n}{4})\mathbf{Sp}(1)$, $\mathbf{Sp}(\frac{n}{2})$.

Let M^n be hyper-Kähler, with parallel anti-commuting parallel almost complex structures I, J, K, satisfying $IJ = K$. In this case, n is a multiple of 4. The group, $\mathrm{Sp}(\frac{n}{4})$ consists of those orthogonal transformations which commute with I, J, K.

Let $e_1, I(e_1), J(e_1), K(e_1), e_2, \ldots, K(e_{\frac{n}{4}})$ denote an orthonormal basis for the tangent space, M^n_p, at $p \in M^n$.

Let ω_I, ω_J, ω_K denote the the Kähler forms of I, J, K, respectively.

Since, M^n is Kähler with respect to I, by (10.1), we have $\langle \omega_I, R\rangle(x,y) = \operatorname{Ric}(x, I(y))$. Since $IJ = -JI$ and M^n is Kähler with respect to J,

$$\begin{aligned}
\langle \omega, R\rangle(x,y) &= \sum_{i=1}^{\frac{n}{4}} R(e_i, I(e_i), x, y) + \sum_{i=1}^{\frac{n}{4}} R(J((e_i), IJ(e_i)), x, y) \\
&= \sum_{i=1}^{\frac{n}{4}} R(e_i, I(e_i), x, y) - \sum_{i=1}^{\frac{n}{4}} R(J((e_i), JI(e_i)), x, y) \\
&= \sum_{i=1}^{\frac{n}{4}} R(e_i, I(e_i), x, y) - \sum_{i=1}^{\frac{n}{4}} R(((e_i), I(e_i)), x, y) \\
&= 0,.
\end{aligned} \tag{10.2}$$

Thus, $\operatorname{Ric}_{M^n} \equiv 0$ in the hyper-Kähler case.

Let $\mathrm{Id} \in \mathrm{SO}(n)$ denote the identity transformation. Linear transformations of the form $r\cdot I + s\cdot J + t\cdot K + u\cdot \mathrm{Id}$, with $r^2+s^2+t^2+u^2 = 1$, form a subgroup of $\mathrm{SO}(n)$. This subgroup, which is isometric to $\mathrm{Sp}(1)$, commutes with the action of $\mathrm{Sp}(\frac{n}{4})$ and intersects $\mathrm{Sp}(\frac{n}{4})$ in the element, $-\mathrm{Id}$. The group generated by $\mathrm{Sp}(\frac{n}{4})$ and $\mathrm{Sp}(1)$ is denoted $\mathrm{Sp}(\frac{n}{4})\mathrm{Sp}(1)$.

The Lie algebra of $\mathrm{Sp}(\frac{n}{4})\mathrm{Sp}(1)$ has the orthogonal orthogonal direct sum decomposition $\mathfrak{sp}(\frac{n}{4}) \oplus \mathfrak{sp}(1)$, where $\mathfrak{sp}(1) = \mathbf{R}\cdot\omega_I \oplus \mathbf{R}\cdot\omega_J \oplus \mathbf{R}\cdot\omega_K \oplus \mathbf{R}$.

We have $\langle \omega_I, \omega_J \rangle = \langle \omega_I, \omega_K \rangle = \langle \omega_J, \omega_K \rangle = 0$. Thus, if the group, $\mathrm{U}(\frac{n}{2})$, is defined with respect to any one of the almost complex structures, I, J, K, then the remaining two almost complex structures are contained in $\mathfrak{su}(\frac{n}{2})$.

If M^n is a riemannian manifold with $H_0 \subset \mathrm{Sp}(\frac{n}{4})\mathrm{Sp}(1)$, the splitting, $\mathfrak{sp}(\frac{n}{4}) \oplus \mathfrak{sp}(1)$, gives rise to orthogonal parallel sub-bundles of $\bigwedge^2(M^n)$. The second of these, which we denote by E, is spanned locally by I, J, K.

As above, let R denote the curvature tensor of M^n, and let R denote the curvature of E. It follows directly that $\mathrm{R}(L) = [R, L]$, for all $L \in E$.

In fact, for $n > 4$,

$$\begin{aligned}
\mathrm{R}_{J,K} &= \frac{8}{n+8} \cdot \mathrm{Ric}_{M^n}(I(x), y)\,, \\
\mathrm{R}_{I,K} &= -\frac{8}{n+8} \cdot \mathrm{Ric}_{M^n}(J(x), y)\,, \\
\mathrm{R}_{I,J} &= \frac{8}{n+8} \cdot \mathrm{Ric}_{M^n}(K(x), y)\,. \qquad (10.3)
\end{aligned}$$

To see this, fix an almost complex structure, say I, and attempt to repeat (10.2), which in the hyper-Kähler case, gave $\mathrm{Ric}_{M^n} \equiv 0$. Using (10.3) in place of $\langle \omega_I, R\rangle(x,y) = \mathrm{Ric}_{M^n}(x, I(y))$ yields

$$\langle \omega_I, R\rangle(x,y) = \mathrm{Ric}_{M^n}(x, I(y)) - \mathrm{R}_{I,J}(x, J(y)) + \mathrm{R}_{I,K}(x, K(y))\,, \qquad (10.4)$$

and instead of (10.2), we get

$$\mathrm{Ric}_{M^m}(x,y) = -\frac{n}{4}(x,y)\mathrm{R}_{J,K}\,. \qquad (10.5)$$

Relations (10.4), (10.5) imply

$$\mathrm{Ric}_{M^m}(x,y) = \frac{n}{4}\mathrm{R}_{J,K}(x, I(y)) - \mathrm{R}_{I,K}(x, J(y)) + \mathrm{R}_{I,J}(x, K(y))\,, \qquad (10.6)$$

which together with the corresponding relations for J, K, can be solved for $\mathrm{R}_{J,K}(x, I(y))$, $\mathrm{R}_{I,K}(x, J(y))$, $\mathrm{R}_{I,J}(x, K(y))$. This gives (10.3). (Note that for $n = 4$, the above system of equations is degenerate, but we are assuming $n > 4$.)

From the last equation in (10.4), we get

$$\mathrm{R}_{J,K}(x,y)|z|^2 = R(x,y,z,I(z)) + R(x,y,J(z),K(z))\,. \qquad (10.7)$$

Employing (10.7) four times gives

$$\begin{aligned}
&R(x, I(x), z, I(z)) + R(x, I(x), J(z), K(z)) + R(J(x), K(x), z, I(z)) \\
&\quad + R(J(x), K(x), J(z), K(z)) \\
&= \frac{16}{n+8}\mathrm{Ric}_{M^n}(x,x)|z|^2 \\
&= \frac{16}{n+8}\mathrm{Ric}_{M^n}(z,z)|x|^2\,. \qquad (10.8)
\end{aligned}$$

Relation (10.8) implies that M^n is Einstein.

In (10.2)–(10.8), we have followed [**Be**], which is based on [**Ish**].

Recall that the twistor space associated to M^n is by definition, the unit sphere bundle, $S(E)$, of E. When endowed with the natural riemannian submersion metric induced by the connection on E, the fibration,

$$\mathbf{S}^2 \to S(E) \xrightarrow{\pi} M^n ,$$

has totally geodesic fibres and, by (10.2), apriori bounded parallel integrability tensor.

Given $L \in \pi^{-1}(p)$, the tangent space at L splits as a direct sum of its horizontal and vertical parts. Since L defines an Hermitian almost complex structure on M^n_p, there is an induced Hermitian almost complex structure on the horizontal subspace of $S(E)$ at L. The tangent space to the fibre is an oriented 2-dimensional inner product space. Thus, it carries a natural Hermitian almost complex structure as well. Hence, there is an induced almost complex structure, $\mathcal{I}$, on $S(E)$ which, as can be checked directly, is actually integrable.

In general, the almost complex structure, $\mathcal{I}$ is *not* parallel with respect to the riemannian connection on the tangent bundle to $S(E)$. However, there is a naturally associated Hermitian orthogonal connection for which $\mathcal{I}$ *is* parallel. This connection is characterized by the condition that its torsion tensor is given by $T = -\frac{1}{2}d^\nabla \mathcal{I}$, where $\mathcal{I}$ is viewed as 1-form with values in the tangent space and the exterior derivative, d^∇, is defined by using the riemannian connection, ∇, on $S(E)$.

The case: $\mathbf{G_2}$.

Let V^7 denote a real 7-dimensional inner product space, equipped with a 3-form, ϕ, for which there exists an orthonormal basis, $e_1, \dots, e_7$, such that for $e_1^*, \dots, e_7^*$ the corresponding dual basis,

$$\begin{aligned}\phi =& e_1^* \wedge e_2^* \wedge e_3^* + +e_1^*\wedge e_4^*\wedge e_5^* + e_1^*\wedge e_6^*\wedge e_7^* \\ &+ e_2^*\wedge e_4^*\wedge e_6^* - e_2^*\wedge e_5^*\wedge e_7^* - e_3^*\wedge e_4^*\wedge e_7^* - e_3^*\wedge e_5^*\wedge e_6^* . \end{aligned} \tag{10.9}$$

The Lie group, G_2, is the subgroup of $GL(V^7)$ that fixes ϕ. (A calculation shows that up to normalization, ϕ can be identified with $*\Omega$ of (0.8).)

Let $a = a_1e_1 + \cdots + a_7e_7$, $b = b_1e_1^* + \cdots + b_7e_7^*$, and put $\mathrm{Vol} = e_1^*\wedge \cdots \wedge e_7^*$. It is easy to check that

$$b\wedge i_a\phi\wedge i_a\phi\wedge i_a\phi = b(a) \cdot |a|^2 \cdot \mathrm{Vol} . \tag{10.10}$$

Since, g fixes ϕ, by choosing b such that $b(a) = 1$, it follows that $|a|^2 = |g^{-1}(a)|^2 \det(g)$, for all $g \in \mathrm{G}_2$. Since g^i fixes ϕ for all i, by replacing g by g^i and letting $i \to \infty$, it follows $\det(g) = 1$. Thus, $\mathrm{G}_2 \subset \mathrm{SO}(7)$.

Let i_e denote interior product with e. When restricted to the orthogonal complement, $[e_i]^\perp$, of the span of e_i, the form, $\omega_{e_i} := i_{e_i}\phi$, is clearly symplectic. Define the corresponding almost complex structure, I_{e_i}, on $[e_i]^\perp$, by $\omega_{e_i}(x,y) = \langle I_{e_i}(x), y\rangle$.

Let $H_e \subset G_2$ denote the subgroup fixing the unit vector e and put $H_{e_i} = H_i$. For all i, the subgroup $H_i \subset G_2$, consists of those elements of the unitary group, $U([e_i]^\perp)$, which also fix the 3-form, $\phi - e_i\wedge i_{e_i}\phi \in \bigwedge^3([e_i]^\perp)$. From this it is easily checked $H_{e_i} = \mathrm{SU}([e_i]^\perp)$. In particular, $\dim H_i = 8$.

It is not difficult to verify that the action of G_2 is transitive on the unit sphere. Hence, we get the fibration,

$$\mathrm{SU}(3) \to \mathrm{G}_2 \to \mathbf{S}^6 ,$$

from which it follows that G_2 is simply connected.

Let M^n denote a riemannian manifold with restricted holonomy contained in G_2.

To see that $\mathrm{Ric}_{M^7} \equiv 0$, note that since ϕ is G_2 invariant, it follows that ϕ is in the kernel of the action (by derivation) of the Lie algebra $\mathfrak{g}_2$. Writing this out in terms of the basis, $e_1, \ldots, e_7$, yields the 7 independent equations which characterize $\mathfrak{g}_2$. If we view $R \in \bigwedge^2(V^7) \otimes \mathfrak{g}_2$, then (written asymmetrically) these equations are

$$\begin{aligned} R_{e_1,e_2} &= R_{e_4,e_7} + R_{e_5,e_6}\,, \\ R_{e_1,e_3} &= R_{e_4,e_6} - R_{e_5,e_7}\,, \\ R_{e_1,e_4} &= R_{e_2,e_7} - R_{e_3,e_6}\,, \\ R_{e_1,e_5} &= -R_{e_2,e_6} + R_{e_3,e_7}\,, \\ R_{e_1,e_6} &= R_{e_2,e_5} + R_{e_3,e_4}\,, \\ R_{e_1,e_7} &= R_{e_2,e_4} - R_{e_3,e_5}\,. \end{aligned} \tag{10.11}$$

Let both sides of the i-th equation in (10.11) act on e_i, take the innner product with e_1, and apply the Jacobi identity to the right-hand of the resulting equation. If we sum over i, then terms the right-hand side cancel in pairs. Thus, $\mathrm{Ric}_{M^7}(e_1, e_1) \equiv 0$. Since G_2 acts transitively on the unit sphere, this shows that $\mathrm{Ric}_{M^n} \equiv 0$.

The case: Spin(7).

Let V^8 denote a real 8-dimensional inner product space, equipped with a 4-form, λ, for which there exists an orthonormal basis, $e_1, \ldots, e_8$, such that for $e_1^*, \ldots, e_8^*$ the corresponding dual basis,

$$\lambda = \phi_{e_8} \wedge e_8^* + *_{e_8} \phi_{e_8}\,, \tag{10.12}$$

where ϕ_{e_8} is the form in (10.9) and $*_{e_8}$ denotes the $*$ operator on the subspace spanned by $e_1^*, \ldots, e_7^*$. It is easy to check that the 4-form, λ, has a corresponding description with respect to any of the basis vectors, $e_1, \ldots, e_8$. The subgroup, $D \subset \mathrm{GL}(V^8)$, for which the adjoint action fixes λ, turns out to be $\mathrm{Spin}(7) \subset \mathrm{SO}(8)$ (and a calculation shows that up to normalization, λ can be identified with Ω of (0.8)). This can be seen as follows.

Let K_i denote the subgroup fixing e_i and let $\mathfrak{k}_i$ denote its Lie algebra. Clearly, K_i is isomorphic to G_2, for all i.

The Lie algebra generated by $\mathfrak{k}_8$ and the standard almost complex structure, I, given by $I(e_{2i-1}) = e_{2i}$, $I(e_{2i}) = -e_{2i-1}$, is $\mathfrak{su}(4)$ (where dim $\mathfrak{su}(4) = 15$). It is easy to check that λ is in the kernel of the action of $\mathfrak{su}(4)$ (by derivation). Hence, $\mathrm{SU}(4) \subset D$.

Suppose there exists $g \in D$ such that $g(e_8) = c \cdot e_8$, with $c \neq 1$. The action of g^* on $(V^8)^*$ fixes the subspace spanned by $e_1^*, \ldots, e_7^*$. In addition, $g^*(e_8^*) = c^{-1} e_8^* + v^*$, where v^* is in the subspace spanned by $e_1^*, \cdots, e_7^*$. From this it is trivial to check that g fixes $*_{e_8} \phi_{e_8}$. Hence $g \in K_8$ and $c = 1$.

Since the subgroup, $\mathrm{SU}(4) \subset D$ acts isometrically and transitively on S^7, it follows that $D \subset \mathrm{O}(8)$; otherwise, there would exist g as above with $c \neq 1$. Since D preserves $\lambda^2 \neq 0$, this shows that $D \subset \mathrm{SO}(8)$.

The existence of the fibration,

$$\mathrm{G}_2 \to D \to S^7\,,$$

implies that D is simply connected, with dim D=21. (This fibration plays a role in the proof of Theorem 9.4, as does the corresponding fibration for the group G_2.)

The subalgebra spanned by $\mathfrak{k}_7, \mathfrak{k}_8$ and I, is easily checked to be isomorphic to $\mathfrak{so}(7) = \mathfrak{spin}(7)$. Thus, D is isomorphic to Spin(7).

The remainder of the discussion of the Spin(7) case can be completed in a manner strictly analogous to the case G_2.

References

[AbGl] U. Abresch, D. Gromoll, On complete manifolds with nonnegative Ricci curvature, J. Am. Math. Soc. 3 N.2 (1990) 355–374.

[Al] D. V. Alekseevsky, Riemannian spaces with exceptional holonomy, Functional Analysis and its Applications, 2, (1968) 297–339.

[AmSing] W. Ambrose, I. Singer, A theorem on holonomy, Trans. Amer. Math. Soc. 75 (1953) 428–443.

[An1] M.T. Anderson, Ricci curvature bounds and Einstein metrics, J. Am. Math. Soc. 2, N.3 (1989) 455–490.

[An2] M.T. Anderson, Convergence and rigidity of metrics under Ricci curvature bounds, Invent. Math. 102 (1990) 429-445.

[An3] M.T. Anderson, Hausdorff perturbations of Ricci flat manifolds and the splitting theorem, Duke Math. J., V. 68, N.1 (1992) 67–82.

[An4] M.T. Anderson, Einstein metrics and metrics with bounds on Ricci curvature, Proceedings of International Congress of Mathematicians, Vol. 1, 2 (Zürich, 1994) Birkhäuser, 443–452.

[AnCh] Diffeomorphism finiteness for manifolds with Ricci curvature and $L^{n/2}$-norm of curvature bounded, GAFA, Geom. Funct. Anal. 1 (1991) 231–252.

[AtWit] M. Atiyah, E. Witten, M-theory dynamics on a manifold of G_2 holonomy, arXiv:hep-th/0107177, 16 Oct 2001.

[BaHiSing1] L. Baulieu, K. Hiroaki, I. M. Singer, Cohomological Yang-Mills theory in eight dimensions, Dualities in gauge and string theories, Seoul/Sokcho (1997) 365–373.

[BaHiSing2] L. Baulieu, K. Hiroaki, I. M. Singer, Special quantum field theories in eight and other dimensions, Comm. Math. Phys. 194 (1998) N. 1, 149–175.

[Be] A.L. Besse, Einstein manifolds, Springer Verlag, New York (1987).

[Ber] M. Berger, Sur les goupes d'holonomie homogène des variétés à connexion affines et des variétés riemanniennces. Bull. Soc. Math. France 83 (1955) 279–330.

[Bish] R.L. Bishop, A relation between volume, mean curvature and diameter, Notices Am. Math. Soc. 10 364 (1956).

[BrGr] R. Brown, A. Gray, Riemannian manifolds with holonomy groups Spin(9). In S. Kobayashi et al., editors, Differential Geometry (In honour of Kentaro Yano), Kinokuniya, Tokyo, (1972) 279–330.

[Bry] Metrics with exceptional holonomy, Ann. Math. 126 (1987) 525–576.

[BrySa] R. Bryant, S. Salamon, On the construction of some complete metrics with special holonomy, Duke Math. J. 58 (1989) 829–850.

[B] P. Buser, A note on the isoperimetric constant, Ann. Sci. Éc. Norm. Sup., Paris 15 (1982) 213–230.

[Ca1] E. Calabi, On Kähler manifolds with vanishing canonical class, Algebraic geometry and topology, A symposium in honor of S. Lefschetz, Princeton Univ. Press, Princeton, N.J., (1957) 78–89.

[Ca2] E. Calabi, An extension of E. Hopf's maximum principle with application to Riemannian geometry. Duke Math. J. 25 (1957) 45–56.

[Ca3] E. Calabi, Métriques káhleriennes et fibrés holomorphes, Ann. Scient. É. Norm. Sup. 12 (1979) 269–294.

[Ch1] J. Cheeger, Finiteness theorems for riemannian manifolds, Amer. J. Math., XCII, N. 1 (1970) 61–74.

[Ch2] J. Cheeger, Differentiability of Lipschitz functions on metric measure spaces GAFA, Geom. Funct. Anal. 9 (1999) 428–517.

[Ch3] J. Cheeger, L_p-bounds on curvature, elliptic estimates and rectifiability of singular sets, C.R. Acad. Sci Paris, Ser. I 334 (2002) 195–198.

[Ch4] J. Cheeger, Integral bounds on curvature, elliptic estimates and rectifiability of singular sets, GAFA, Geom. Funct. Anal., 13, (2003) 20–72..

[Ch5] J. Cheeger, Degeneration of riemannian metrics under Ricci curvature bounds, Lezione Fermiane, Pisa 2001, Scuola Normale Superiore, Accademia Nazional dei Lincei, (2003).

[ChCo1] J. Cheeger and T.H. Colding, Almost rigidity of warped products and the structure of spaces with Ricci curvature bounded below, C.R. Acad. Sci. Paris t. 320, Série 1 (1995) 353–357.

[ChCo2] J. Cheeger and T.H. Colding, Lower bounds on the Ricci curvature and the almost rigidity of warped products, Ann. of Math., 144 (1996) 189–237.

[ChCo3] J. Cheeger and T.H. Colding, On the structure of spaces with Ricci curvature bounded below. I, J. Diff. Geom. 46 (1997) 406–480.

[ChCo4] J. Cheeger and T.H. Colding, On the structure of spaces with Ricci curvature bounded below. II, J. Diff. Geom. 52 (1999) 13–35.

[ChCo5] J. Cheeger and T.H. Colding, On the structure of spaces with Ricci curvature bounded below. III, J. Diff. Geom. 52 (1999) 37–74.

[ChCoTi1] J. Cheeger, T.H. Colding, and G. Tian, Constraints on singularities under Ricci curvature bounds, C.R. Acad. Sci. Paris, t. 324, Série 1 (1997) 645–649.

[ChCoTi2] J. Cheeger, T.H. Colding, and G. Tian, On the singularities of spaces with bounded Ricci curvature, GAFA, Geom. Funct. Anal., 12, N. 5 (2002) 873–914.

[ChGl1] J. Cheeger, D. Gromoll, The splitting theorem for manifolds of nonegative Ricci curvature, J. Diff. Geom. 6, N. 1 (1971) 119–128.

[ChGl2] J. Cheeger, D. Gromoll, On the structure of complete manifolds with nonnegative curvature, Ann. of Math. 96, N. 3 (1972) 413–443.

[ChSi] J. Cheeger, J. Simons, Differential characters and geometric invariants, Lecture Notes distributed at Amer. Math. Soc. conference on Differenetial Geometry, Stanford (1973); published in Geometry and Topology, Lecture Notes in Math. 1167, Springer-Verlag, 1985.

[ChTi1] J. Cheeger, G. Tian, On the cone structure an infinity of Ricci flat manifolds with Euclidean volume growth and quadratic curvature decay, Invent. Math. 118 (1994) 493–571.

[ChTi2] J. Cheeger, G. Tian, Anti-self-duality of curvature and degeneration of metrics with special holonomy (in preparation).

[CgYau] S.Y. Cheng and S.T. Yau, Differential equations on Riemannian manifolds and their geometric applications, Comm. Pure Appl. Math. 28 (1975) 333–354.

[Co1] T.H. Colding, Stability and Ricci curvature, C.R. Acad. Sci. Paris, t. 320, Série 1 (1995) 1343–1347.

[Co2] T.H. Colding, Shape of manifolds with positive Ricci curvature, Invent. Math. V.1, 124, 1–3 (1996) 175–191.

[Co3] T.H. Colding, Large manifolds with positive Ricci curvature, Invent. Math. vol. 124, 1–3 (1996) 193–214.

[Co4] T.H. Colding, Ricci curvature and volume convergence, Ann. of Math. 145 (1997) 477–501.

[DetKa] D. Deturck, J. Kazdan, Some regularity theorems in Riemannian geometry, Ann. Sci. École Norm. Sup. 14 (1980) 249–260.

[EschHei] J.-H. Eschenburg, E. Heintze, An elementary proof of the Cheeger-Gromoll splitting theorem, Ann. Global Anal. Geom 2 (1984) N. 2, 141–151.

[EgHan] T. Eguchi, A. Hanson, Asymptotically flat self-dual solutions to Euclidean gravity, Phy. Lett. B 74 (1978) 249–251.

[Fe] H. Federer, Geometric measure theory, Springer Verlag (Berlin-New York) 1969.

[GaLaw] K. Galicki, H.B. Lawson, Quaternionic reduction and quaternionic orbifolds, Math. Ann. 282 N. 1 (1988) 1–21.

[GibHaw] G.W. Gibbons, S.W. Hawking, Gravitational multi-instantons, Phys. Lett. 78B (1978) 430–432.

[Gi] E. Giusti, Minimal surfaces and functions of bounded variation, Birkhäuser, 1984.

[GlMey] D. Gromoll and W. Meyer, On complete open manifolds of positive curvature, Ann. of Math. 90 (1969) 75–90.

[GvLP] M. Gromov, J. Lafontaine, and P. Pansu, Structures métriques pour les variétiés riemanniennes, Cedic/Fernand Nathan, Paris, 1981.

[GsWil] M. Gross, P. Wilson, Large complex structure limits of $K3$ surfaces, J. Diff. Geom. 55 N. 3 (2000) 475–546.

[Ham] R. Hamilton, Three manifolds with positive Ricci curvature, J. Diff. Geom. 17 (9182) 255–306.

[HarLaw] R. Harvey, H. B. Lawson, Calibrated Geometries, Acta. Math. 148 (1982) 47–157.

[HarShiff] R. Harvey, B. Schiffman, A characterization of holomorphic chains, Ann. of Math. 99 (1974) 553–587.

[HerHer] H. Herrera, R. Herrera, A-genus on non-spin manifolds with S^1-actions and the classification of positive quaternion-Kähler 12-manifolds, J. Diff. Geom. (to appear).

[HiSa] N. Hitchin, J. Sawon, Curvature and characteristic numbers of hyper-Kähler manifolds, Duke Math. J. 106, N. 3 (2001) 599–615.

[Ish] S. Ishihara, Quaternion Kähler manifolds, J. Diff. Geom. 4 (1974) 483–500.

[JoKar] J. Jost and H. Karcher, Geometrische Methoden zur Gewinnung von a-priori Schranken für harmonische Abbildungen, Manuscripta Math. 40 (1982) 27–77.

[Joy1] D. Joyce, Compact Riemannian 7-manifolds with holonomy G_2. I, J. Diff. Geom. V. 43, N. 2 (1996) 291–328.

[Joy2] D. Joyce,Compact Riemannian 7-manifolds with holonomy G_2. II, J. Diff. Geom. V. 43, N. 2 (1996) 329–375.

[Joy3] D. Joyce, Compact manifolds with special holonomy, Oxford Mathematical Monographs, Oxford University Press, Oxford, 2000.

[KonSoi] M. Kontsevich, Y. Soibelman, Homological mirror symmetry and torus fibrations, Symplectic geometry and mirror symmetry (Seoul, 2000) World Sci. River Edge, N. J. 2001, 203–263.

[Kron1] P. Kroneheimer, On self-dual Einstein manifolds, (Harmonic mappings, twistors and σ-models) Luminy (1986), 113–123, Adv. Ser. Math. Phys. 4, World Sci. Publishing, Singapore 1988.

[Kron2] P. Kroneheimer, The construction of ALE spaces as hyper-Kähler quotients, J. Diff. Geom. 29 N. 3 (1989) 665–683.

[LebWa] C. LeBrun, M. Wang, Essays on Einstein manifolds, Surveys in Differential Geometry, VI, International Press, Boston, MA, 1999.

[Li] P. Li, Lecture notes on geometric analysis, Lecture Notes Series N. 6, Research Institute of Mathematics, Global Analysis Research Center, Seoul National University, Seoul Korea.

[LiSch] P. Li, R. Schoen, L_p and mean value properties of subharmonic functions on riemannian manifolds, Acta Math. 153 N. 3-4 (1984) 279–301.

[Ma] P. Matilla, Geometry of sets and measures in Euclidean spaces; Fractals and rectifiability, Cambridge studies in advanced mathematics, 1995.

[Men1] X. Menguy, Noncollapsing examples with positive Ricci curvature and infinite topological type, GAFA, Geom. Funct. Anal. V. 10, N.3 (2000) 600–627.

[Men2] X. Menguy, Examples with bounded diameter growth and infinite topological type, Duke Math. J. 102, N. 3 (2000) 403–412.

[My] S. Myers, Riemannian manifolds with positive mean curvature, Duke Math. J., 8 (1941) 401–404.

[Nak1] H. Nakajima, Hausdorff convergence of Einstein 4-manifolds, J. Fac. Sci. Univ. Toky Sect. 1A Math. 35 N. 2 (1988) 411-424.

[Nak2] H. Nakajima, A convergence theorem for Einstein metrics and the ALE spaces, Selected papers on number theory, algbraic geometry and differential geometry, Amer. Math. Soc. Transl. Ser. 2, 160 (1994) 79–94.

[Pa] D. N. Page, A physical picture of the $K3$ gravitational instanton, Phys. Lett. 80B (1978) 55-57.

[Per1] G. Perelman, A complete Riemannian manifold of positive Ricci curvature with Euclidean volume growth and nonunique asymptotic cone, Comparison geometry, (Berkely CA 1993–94) 165–166, Math. Sci. Res. Inst. Publ., 30, Cambridge Univ. Press, Cambridge, 1997.

[Per2] G. Perelman, Construction of manifolds of positive Ricci curvature and large Betti numbers, Comparison geometry, (Berkely CA 1993–94) 157–163, Math. Sci. Res. Inst. Publ., 30, Cambridge Univ. Press, Cambridge, 1997.

[Reif] E. Reifenberg, Solution of the Plateau problem for m-dimensional surfaces of varying topological type, Acta Math. 104 (1960) 1–92.

[Sal] S. Salamon, Riemannian geometry and holonomy groups, Pitman Research Notes in Mathematics, Longman Scientrific and Technical, Essex, 1987.

[Sa2] S. Salamon, Quaternion-Kähler geometry, Surveys in Differential Geometry, VI: Essays on Einstein Manifolds, C. LeBrun, M. Wang eds., International Press, 1999.

[Si] J. Simons, On the transitivity of holonomy systems, Ann. Math. 76 (1962) 213–234.

[Sim1] L. Simon, Lectures on geometric measure theory, Proceedings of the Centre for Mathematical Analysis, Australian National University, 3 1983.

[Sim2] L. Simon, Rectifiability of the singular sets of multiplicity 1 minimal surfaces and energy minimizing maps, Surveys in Differential Geometry V. 2 (Cambridge, MA, 1993) International Press, Cambridge MA (1995) 246–305.

[Sim3] L. Simon, Rectifiability of singular sets of energy minimzing maps, Calc. Var. Partial Differential Equations 3:1 (1995) 1–65.

[StYauZa] A. Strominger, S-T. Yau, E. Zaslow, Mirror symmetry is T-duality, Nuclear Phys. B., 479 N. 1-2 (1996) 243–259.

[Ti1] G. Tian, On Calabi's conjecture for complex surfaces with positive first Chern class, Inv. Math., V. 101, N.1 (1990) 101–172.

[Ti2] Kähler Einstein metrics on algebraic manifolds, Proc. ICM, V. 1, 2 (Kyoto 1990) 587–598, Math. Soc. Japan, Tokyo, 1991.

[Ti3] G. Tian, Gauge theory and calibrated geometry, I, Ann. Math. 151 (2000) 193–168.

[TiYau1] G. Tian, S-T. Yau, Kähler-Einstein metrics on complex surfaces with $C_1 > 0$. Comm. Math. Phys. 112 N. 1 (1987) 175–203.

[TiYau2] G.Tian, S-T. Yau, Existence of Kähler-Einstein metrics on complete Kähler manifolds and their applications to algebraic geometry. Mathematical aspects of string theory (San Diego, Calif., 1986), 574–628, Adv. Ser. Math. Phys. 1, World Sci. Publishing, Singapore, 1987.

[TiYau3] G. Tian, S-T. Yau, Complete Kähler manifolds with zero Ricci curvature. I. J. Amer. Math. Soc. 3 N. 3, (1990) 579–609.

[TiYau4] G. Tian, S-T. Yau, Complete Khler manifolds with zero Ricci curvature. II. Invent. Math. 106 N.1, (1991) 27–60.

[Topi1] P. Topiwala, A new proof of the existence of Kähler-Einstein metrics on $K3$. I, Invent. Math. 89, N.2 (1987) 425–448.

[Topi2] P. Topiwala, A new proof of the existence of Kähler-Einstein metrics on $K3$. II. Invent. Math. 89, N.2 (1987) 449–454.

[Top1] V. Toponogov, Riemannian spaces with curvature bounded below by a positive number, Uspekhi Math. Nauk. 14 N. 1 (1959) 87–135.

[Top2] V. Toponogov, The metric structure of Riemannian spaces with nonnegative curvature which contain straight lines, Sibirsk Math. Z. 5 (1964) 1358–1369.

[Yam] H. Yamabe, On an arcwise connected subgroup of a Lie group, Osaka Math. J. 2 (1950) 13–14.

[Yau1] S-T. Yau, Calabi's conjecture and some new results in algebraic geometry. Proc. Nat. Acad. Sci. U.S.A. 74 N. 5, (1977) 1798–1799.

[Yau2] S-T. Yau, On the Ricci curvature of a compact Kähler manifold and the complex Monge-Ampère equation. I. Comm. Pure Appl. Math. 31 N. 3 (1978) 339–411.

[Yau3] S-T. Yau, A survey on Kähler-Einstein metrics. Complex analysis of several variables (Madison, Wis., 1982), 285–289, Proc. Sympos. Pure Math., 41, Amer. Math. Soc., Providence, RI, 1984.

[Yau4] S-T. Yau, Compact three-dimensional Kähler manifolds with zero Ricci curvature. Symposium on anomalies, geometry, topology (Chicago, Ill., 1985), 395–406, World Sci. Publishing, Singapore, 1985.

[Yau5] S-T. Yau, Einstein manifolds with zero Ricci curvature. Surveys in differential geometry: essays on Einstein manifolds, 1–14, Surv. Diff. Geom., VI, Int. Press, Boston, MA, 1999.

[Wh] B. White, Stratification of minimal surfaces, mean curvature flows and harmonic maps, J. Reine Angew. Math. 488 N. 3 (1997) 1–35.

J.C.: Courant Institute of Mathematical Sciences, 251 Mercer Street, New York, NY 10012

E-mail address: cheeger@cims.nyu.edu

The min–max construction of minimal surfaces

Tobias H. Colding and Camillo De Lellis

Dedicated to Eugenio Calabi on occasion of his eightieth birthday

1. Introduction

In this paper we survey with complete proofs some well–known, but hard to find, results about constructing closed embedded minimal surfaces in a closed 3-dimensional manifold via min–max arguments. This includes results of J. Pitts, F. Smith, and L. Simon and F. Smith.

The basic idea of constructing minimal surfaces via min–max arguments and sweep-outs goes back to Birkhoff, who used such a method to find simple closed geodesics on spheres. In particular when M^2 is the 2-dimensional sphere we can find a 1–parameter family of curves starting and ending at a point curve in such a way that the induced map $F : \mathbf{S}^2 \to \mathbf{S}^2$ (see Fig. 1) has nonzero degree. Birkhoff's argument (or the min-max argument) allows us to conclude that M has a nontrivial closed geodesic of length less than or equal to the length of the longest curve in the 1-parameter family. A curve shortening argument gave that the geodesic obtained in this way is simple.

The difficulty in generalizing this method to get embedded minimal surfaces in 3–manifolds is three fold. The first problem is getting regularity of the min–max surface obtained. In Birkhoff's case (curves in surfaces) this was almost immediate. The second key difficulty is to show that the min–max surface is embedded. Using the technical tools of Geometric Measure Theory (mostly the theory of varifolds), these two problems are tackled at the same time. The third key difficulty is to get a good genus bound for the embedded minimal surface obtained.

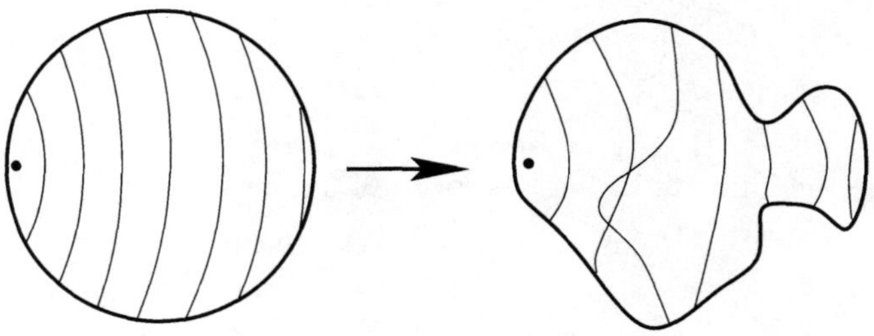

FIGURE 1. A 1–parameter family of curves on a 2–sphere which induces a map $F : \mathbf{S}^2 \to \mathbf{S}^2$ of degree 1.

The first author was partially supported by NSF Grant DMS 0104453.

1.1. The min–max construction in 3–manifolds. In the following M denotes a closed 3–dimensional Riemannian manifold, Diff_0 is the identity component of the diffeomorphism group of M, and $\mathfrak{Is}$ is the set of smooth isotopies. Thus $\mathfrak{Is}$ is the set of maps $\psi \in C^\infty([0,1] \times M, M)$ such that $\psi(0,\cdot)$ is the identity and $\psi(t,\cdot) \in \text{Diff}_0$ for every t.

We will first give a version of 1–parameter families of surfaces in 3-manifolds. The most direct way of doing this is to let $F : [0,1] \times \Sigma \to M$ be a smooth map such that $F(t,\cdot)$ is an embedding of the surface Σ for every $t \in [0,1]$. If we let $\Sigma_t = F(\{t\} \times \Sigma)$, then $\{\Sigma_t\}_{t\in[0,1]}$ is a smooth 1–parameter family of surfaces in M. This notion can be generalized in two directions. The first one is to relax the regularity required in the t–variable:

DEFINITION 1.1. *A family $\{\Sigma_t\}_{t\in[0,1]}$ of surfaces of M is said to be* continuous *if*

(c1) *$\mathcal{H}^2(\Sigma_t)$ is a continuous function of t;*
(c2) *$\Sigma_t \to \Sigma_{t_0}$ in the Hausdorff topology whenever $t \to t_0$.*

A second generalization allows the family of surfaces to degenerate in finitely many points:

DEFINITION 1.2. *A family $\{\Sigma_t\}_{t\in[0,1]}$ of subsets of M is said to be a* generalized family *of surfaces if there are a finite subset T of $[0,1]$ and a finite set of points P in M such that*

1. *(c1) and (c2) hold;*
2. *Σ_t is a surface for every $t \notin T$;*
3. *For $t \in T$, Σ_t is a surface in $M \setminus P$.*

Figure 1 gives (in one dimensions less) an example of a generalized 1–parameter family with $T = \{0,1\}$. To avoid confusion, families of surfaces will be denoted by $\{\Sigma_t\}$. Thus, when referring to a surface a subscript will denote a real parameter, whereas a superscript will denote an integer as in a sequence.

Given a generalized family $\{\Sigma_t\}$ we can generate new generalized families via the following procedure. Take an arbitrary map $\psi \in C^\infty([0,1] \times M, M)$ such that $\psi(t,\cdot) \in \text{Diff}_0$ for each t and define $\{\Sigma'_t\}$ by $\Sigma'_t = \psi(t, \Sigma_t)$. We will say that a set Λ of generalized families is *saturated* if it is closed under this operation.

REMARK 1.3. *For technical reasons we will require that any of the saturated sets Λ that we consider has the additional property that there exists some $N = N(\Lambda) < \infty$ such that for any $\{\Sigma_t\} \subset \Lambda$, the set P in Definition 1.2 consists of at most N points. This additional property will play a crucial role in the proof of Theorem 1.6.*

Given a family $\{\Sigma_t\} \subset \Lambda$ we denote by $\mathcal{F}(\{\Sigma_t\})$ the area of its maximal slice and by $m_0(\Lambda)$ the infimum of $\mathcal{F}$ taken over all families of Λ; that is,

$$\mathcal{F}(\{\Sigma_t\}) = \max_{t\in[0,1]} \mathcal{H}^2(\Sigma_t) \qquad \text{and} \tag{1}$$

$$m_0(\Lambda) = \inf_{\Lambda} \mathcal{F} = \inf_{\{\Sigma_t\}\in\Lambda} \left[\max_{t\in[0,1]} \mathcal{H}^2(\Sigma_t)\right]. \tag{2}$$

If $\lim_n \mathcal{F}(\{\Sigma_t\}^n) = m_0(\Lambda)$, then we say that the sequence of generalized families of surfaces $\{\{\Sigma_t\}^n\} \subset \Lambda$ is a *minimizing sequence.* Assume $\{\{\Sigma_t\}^n\}$ is a minimizing sequence and let $\{t_n\}$ be a sequence of parameters. If the areas of the slices $\{\Sigma^n_{t_n}\}$

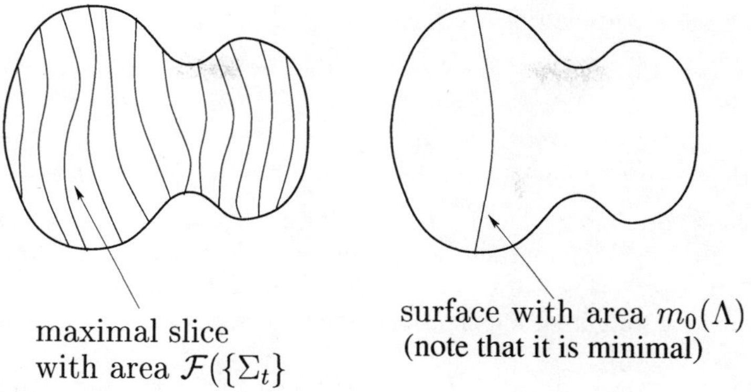

FIGURE 2. $\mathcal{F}(\{\Sigma_t\})$ and $m_0(\Lambda)$.

converge to m_0, i.e. if $\mathcal{H}^2(\Sigma^n_{t_n}) \to m_0(\Lambda)$, then we say that $\{\Sigma^n_{t_n}\}$ is a *min–max sequence.*

An important point in the min–max construction is to find a saturated set Λ of generalized families of surfaces with $m_0(\Lambda) > 0$. This can for instance be done by using the following elementary proposition proven in Appendix A; see Fig. 3:

PROPOSITION 1.4. *Let M be a closed 3-manifold with a Riemannian metric and let $\{\Sigma_t\}$ be the level sets of a Morse function. The smallest saturated set Λ containing the family $\{\Sigma_t\}$ has $m_0(\Lambda) > 0$.*

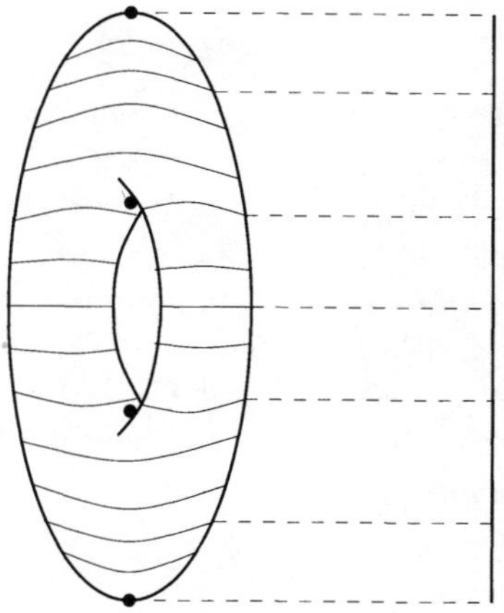

FIGURE 3. A sweep–out of the torus by level sets of a Morse function. In this case there are four degenerate slices in the 1–parameter family.

The following example of sweep–outs of the 3–sphere is a direct generalization of the families of curves on the 2–sphere considered by Birkhoff:

EXAMPLE 1.5. *Let x_4 be the coordinate function on the 3–sphere coming from its standard embedding into $\mathbf{R}^4$. By Proposition 1.4, for any fixed metric on $\mathbf{S}^3$ the level sets of x_4 generate a saturated set of generalized families of surfaces with $m_0 > 0$.*

In this survey we will prove the following theorem:

THEOREM 1.6. *[Simon–Smith] Let M be a closed 3-manifold with a Riemannian metric. For any saturated set of generalized families of surfaces Λ, there is a min–max sequence obtained from Λ and converging in the sense of varifolds to a smooth embedded minimal surface with area $m_0(\Lambda)$ (multiplicity is allowed).*

An easy corollary of Proposition 1.4 and Theorem 1.6 is the existence of a smooth embedded minimal surface in any closed Riemannian 3–manifold (Pitts proved that in any closed Riemannian manifold of dimension at most 7 there is a closed embedded minimal hypersurface; see theorem A and the final remark of the introduction of [**P**]).

For Λ as in Example 1.5 (where M is topologically a 3-sphere but could have an arbitrary metric and the sweep–outs are by 2-spheres) Simon and Smith proved that the min–max sequence given by Theorem 1.6 converges to a disjoint union of embedded minimal 2–spheres, possibly with multiplicity. The following generalization of this result was announced by Pitts and Rubinstein in [**PR1**] (in this theorem $\mathbf{g}(\Sigma)$ is the genus of the surface Σ):

THEOREM 1.7. *If $\{\Sigma^k_{t_k}\}$ is the min–max sequence of Theorem 1.6 and Σ^∞ its limit, then*

$$\mathbf{g}(\Sigma^\infty) \leq \liminf_{k\to\infty} \mathbf{g}(\Sigma^k_{t_k})\,. \tag{3}$$

We plan to address the proof of Theorem 1.7 elsewhere.

Part 1. Overview of the proof

2. Preliminaries

2.1. Notation. We begin by fixing some notation which will be used throughout. When speaking of an isotopy ψ, that is, of a map $\psi : [0,1]\times M \to M$ such that $\psi(t,\cdot) \in \mathrm{Diff}_0$ for every t, then if not otherwise specified we assume that $\psi \in \mathfrak{Is}$. Recall that $\mathfrak{Is}$ is the set of smooth isotopies that start at the identity. Moreover, we say that ψ *is supported in U* if $\psi(t,x) = x$ for every $(t,x) \in [0,1]\times(M\setminus U)$.

Most places Γ and Σ will either denote smooth closed surfaces in M (multiplicity allowed) or smooth surfaces in some subset $U \subset M$ with $\overline{\Sigma}\setminus\Sigma \subset \partial U$. However, there are a few places where Σ and Γ denote surfaces which are smooth away from finitely many (singular) points.

Below is a list of our notation:

T_xM	the tangent space of M at x
TM	the tangent bundle of M.
$\mathrm{Inj}\,(M)$	the injectivity radius of M.
$\mathcal{H}^2$	the 2–d Hausdorff measure in the metric space (M,d).
$B_\rho(x)$	open ball
$\overline{B}_\rho(x)$	closed ball
$\partial B_\rho(x)$	distance sphere of radius ρ in M.
$\mathrm{diam}(G)$	diameter of a subset $G \subset M$.
$d(G_1, G_2)$	the Hausdorff distance between the subsets G_1 and G_2 of M.
$\mathcal{D}$, $\mathcal{D}_\rho$	the unit disk and the disk of radius ρ in $\mathbf{R}^2$.
$\mathcal{B}$, $\mathcal{B}_\rho$	the unit ball and the ball of radius ρ in $\mathbf{R}^3$.
$\exp_x$	the exponential map in M at $x \in M$.
$\mathfrak{Is}(U)$	smooth isotopies supported in U.
$G^2(U)$, $G(U)$	grassmannian of (unoriented) 2–planes on $U \subset M$.

$\mathrm{An}(x,\tau,t)$	the open annulus $B_t(x) \setminus \overline{B}_\tau(x)$.
$\mathcal{AN}_r(x)$	the set $\{\mathrm{An}(x,\tau,t)$ where $0<\tau<t<r\}$.
$C^\infty(X,Y)$	smooth maps from X to Y.
$C^\infty_c(X,Y)$	smooth maps with compact support from X to the vector space Y.

2.2. Varifolds. We will need to recall some basic facts from the theory of varifolds; see for instance chapter 4 and chapter 8 of [**Si**] for further information. Varifolds are a convenient way of generalizing surfaces to a category that has good compactness properties. An advantage of varifolds, over other generalizations (like currents), is that they do not allow for cancellation of mass. This last property is fundamental for the min–max construction.

If U is an open subset of M, any finite nonnegative measure on the Grassmannian of unoriented 2–planes on U is said to be a 2*–varifold in* U. The Grassmannian of 2–planes will be denoted by $G^2(U)$ and the vector space of 2–varifolds is denoted by $\mathcal{V}^2(U)$. With the exception of Appendix C, throughout we will consider only 2–varifolds; thus we drop the 2.

We endow $\mathcal{V}(U)$ with the topology of the weak convergence in the sense of measures, thus we say that a sequence V^k of varifolds converge to a varifold V if for every function $\varphi \in C_c(G(U))$

$$\lim_{k\to\infty} \int \varphi(x,\pi)\, dV^k(x,\pi) \;=\; \int \varphi(x,\pi)\, dV(x,\pi)\,.$$

Here π denotes a 2–plane of T_xM. If $U' \subset U$ and $V \in \mathcal{V}(U)$, then we denote by $V \llcorner U'$ the restriction of the measure V to $G(U')$. Moreover, $\|V\|$ will be the unique measure on U satisfying

$$\int_U \varphi(x)\, d\|V\|(x) \;=\; \int_{G(U)} \varphi(x)\, dV(x,\pi) \qquad \forall \varphi \in C_c(U)\,.$$

The support of $||V||$, denoted by supp $(||V||)$, is the smallest closed set outside which $||V||$ vanishes identically. The number $||V||(U)$ will be called the *mass of V in U*. When U is clear from the context, we say briefly the *mass of V*.

Recall also that a 2–dimensional rectifiable set is a countable union of closed subsets of C^1 surfaces (modulo sets of $\mathcal{H}^2$–measure 0). Thus, if $R \subset U$ is a 2–dimensional rectifiable set and $h : R \to \mathbf{R}^+$ is a Borel function, then we can define a varifold V by

$$\int_{G(U)} \varphi(x,\pi)\, dV(x,\pi) = \int_R h(x)\varphi(x, T_x R)\, d\mathcal{H}^2(x) \qquad \forall \varphi \in C_c(G(U))\,. \tag{4}$$

Here $T_x R$ denotes the tangent plane to R in x. If h is integer–valued, then we say that V is an *integer rectifiable varifold.* If $\Sigma = \bigcup n_i \Sigma_i$, then by slight abuse of notation we use Σ for the varifold induced by Σ via (4).

2.3. Pushforward, first variation, monotonicity formula. If V is a varifold induced by a surface $\Sigma \subset U$ and $\psi : U \to U'$ a diffeomorphism, then we let $\psi_\sharp V \in \mathcal{V}(U')$ be the varifold induced by the surface $\psi(\Sigma)$. The definition of $\psi_\sharp V$ can be naturally extended to *any* $V \in \mathcal{V}(U)$ by

$$\int \varphi(y,\sigma)\, d(\psi_\sharp V)(y,\sigma) \;=\; \int J\psi(x,\pi)\, \varphi(\psi(x), d\psi_x(\pi))\, dV(x,\pi)\,;$$

where $J\psi(x,\pi)$ denotes the Jacobian determinant (i.e. the area element) of the differential $d\psi_x$ restricted to the plane π; cf. equation (39.1) of [**Si**].

Given a smooth vector field χ, let ψ be the isotopy generated by χ, i.e. with $\frac{\partial \psi}{\partial t} = \chi(\psi)$. The first variation of V with respect to χ is defined as

$$[\delta V](\chi) \;=\; \frac{d}{dt}(||\psi(t,\cdot)_\sharp V||)\Big|_{t=0}\;;$$

cf. sections 16 and 39 of [**Si**]. When Σ is a smooth surface we recover the classical definition of first variation of a surface:

$$[\delta\Sigma](\chi) \;=\; \int_\Sigma \mathrm{div}_\Sigma \chi\, d\mathcal{H}^2 \;=\; \frac{d}{dt}(\mathcal{H}^2(\psi(t,\Sigma)))\Big|_{t=0}\,.$$

If $[\delta V](\chi) = 0$ for every $\chi \in C^\infty_c(U, TU)$, then V is said to be *stationary in U*. Thus stationary varifolds are natural generalizations of minimal surfaces.

Stationary varifolds in Euclidean spaces satisfy the monotonicity formula (see sections 17 and 40 of [**Si**]):

$$\text{For every } x \text{ the function } f(\rho) = \frac{||V||(B_\rho(x))}{\pi\rho^2} \text{ is non–decreasing.} \tag{5}$$

When V is a stationary varifold in a Riemannian manifold a similar formula with an error term holds. Namely, there exists a constant $C(r) \geq 1$ such that

$$f(s) \;\leq\; C(r) f(\rho) \qquad \text{whenever } 0 < s < \rho < r. \tag{6}$$

Moreover, the constant $C(r)$ approaches 1 as $r \downarrow 0$. This property allows us to define the *density* of a stationary varifold V at x, by

$$\theta(x,V) \;=\; \lim_{r\downarrow 0} \frac{||V||(B_r(x))}{\pi r^2}.$$

Thus $\theta(x,V)$ corresponds to the upper density θ^{*2} of the measure $||V||$ as defined in section 3 of [**Si**]. The following theorem gives a useful condition for rectifiability in terms of density:

THEOREM 2.1. *(Theorem 42.4 of* **[Si]***). If V is a stationary varifold with $\theta(V,x) > 0$ for $\|V\|$–a.e. x, then V is rectifiable.*

2.4. Tangent cones, Constancy Theorem. Tangent varifolds are the natural generalization of tangent planes for smooth surfaces. In order to define tangent varifolds in a 3–dimensional manifold we need to recall what a dilation in a manifold is. If $x \in M$ and $\rho < \text{Inj}\,(M)$, then the dilation around x with factor ρ is the map $T^x_\rho : B_\rho(x) \to \mathcal{B}_1$ given by $T^x_\rho(z) = (\exp_x^{-1}(z))/\rho$; thus if $M = \mathbf{R}^3$, then T^x_ρ is the usual dilation $y \to (y-x)/\rho$.

DEFINITION 2.2. *If $V \in \mathcal{V}(M)$, then we denote by V^x_ρ the dilated varifold in $\mathcal{V}(\mathcal{B}_1)$ given by $V^x_\rho = (T^x_\rho)_\sharp V$. Any limit $V' \in \mathcal{V}(\mathcal{B}_1)$ of a sequence $V^x_{s_n}$ of dilated varifolds, with $s_n \downarrow 0$, is said to be a* tangent varifold at x. *The set of all tangent varifolds to V at x is denoted by $T(x,V)$.*

It is well known that if the varifold V is stationary, then any tangent varifold to V is a stationary *Euclidean* cone (see section 42 of [**Si**]); that is a stationary varifold in $\mathbf{R}^3$ which is invariant under the dilations $y \to y/\rho$. If V is also integer rectifiable and the support of V is contained in the union of a finite number of disjoint connected surfaces Σ_i, i.e. $\text{supp}\,(\|V\|) \subset \bigcup \Sigma_i$, then the Constancy Theorem (see theorem 41.1 of [**Si**]) gives that $V = \bigcup m_i \Sigma^i$ for some natural numbers m_i.

2.5. Curvature estimates for stable minimal surfaces. In many of the proofs we will use Schoen's curvature estimate (see [**Sc**] or [**CM2**]) for stable minimal surfaces. Recall that this estimate asserts that if $U \subset\subset M$, then there exists a universal constant, $C(U)$, such that for every stable minimal surface $\Sigma \subset U$ with $\partial\Sigma \subset \partial U$ and second fundamental form A

$$|A|^2(x) \leq \frac{C(U)}{d^2(x,\partial U)} \qquad \forall x \in \Sigma\,. \tag{7}$$

In fact, what we will use is not the actual curvature estimate, rather it is the following consequence of it:

$$\text{If } \{\Sigma^n\} \text{ is a sequence of stable minimal surfaces in } U\text{, then a subsequence converges to a stable minimal surface } \Sigma^\infty\,. \tag{8}$$

3. Overview of the proof of Theorem 1.6

In the following we fix a saturated set Λ of generalized 1-parameter families of surfaces and denote by $m_0 = m_0(\Lambda)$ the infimum of the areas of the maximal slices in Λ; cf. (1). The proof of Theorem 1.6, which we will outline in this section, follows by combining two results, Proposition 5.1 and Theorem 7.1. The proofs of these two results will involve all the material presented in Sections 3, 4, 5, and 6.

3.1. Stationarity. If $\{\{\Sigma_t\}^k\} \subset \Lambda$ is a minimizing sequence, then it is easy to show the existence of a min–max sequence which converge (after possibly passing to subsequences) to a stationary varifold. However, as Fig. 4 illustrate, a general minimizing sequence $\{\{\Sigma_t\}^k\}$ can have slices $\Sigma^k_{t_k}$ with area converging to m_0 but not "clustering" towards stationary varifolds.

In the language introduced above, this means that a given minimizing sequence $\{\{\Sigma_t\}^k\}$ can have min–max sequences which are not clustering to stationary varifolds. This is a source of some technical problems and forces us in Section 4 to

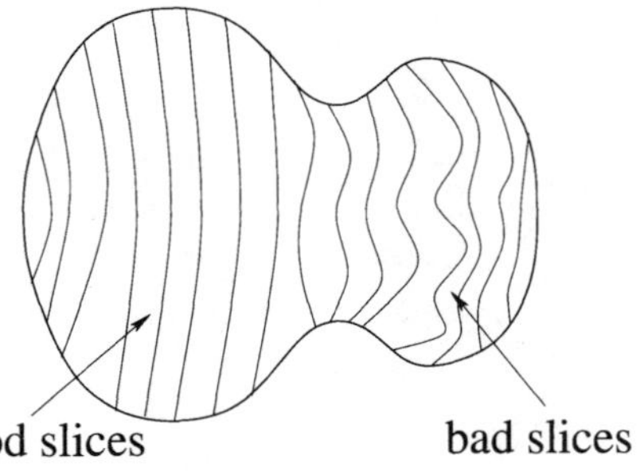

FIGURE 4. Slices with area close to m_0. The good ones are very near to a minimal surface of area m_0, whereas the bad ones are far from any stationary varifold.

show how to choose a "good" minimizing sequence $\{\{\Sigma_t\}^k\}$. This is the content of the following proposition:

PROPOSITION 3.1. *There exists a minimizing sequence* $\{\{\Sigma_t\}^n\} \subset \Lambda$ *such that every min–max sequence* $\{\Sigma^n_{t_n}\}$ *clusters to stationary varifolds.*

A result similar to Proposition 3.1 appeared in [**P**] (see theorem 4.3 of [P]). The proof follows from ideas of [**Alm**] (cf. 12.5 there).

3.2. Almost minimizing. A stationary varifold can be quite far from an embedded minimal surface. The key point for getting regularity for varifolds produced by min–max sequences is the concept of "almost minimizing surfaces" or a.m. surfaces. Roughly speaking a surface Σ is almost minimizing if any path of surfaces $\{\Sigma_t\}_{t\in[0,1]}$ starting at Σ and such that Σ_1 has small area (compared to Σ) must necessarily pass through a surface with large area. That is, there must exist a $\tau \in]0,1[$ such that Σ_τ has large area compared with Σ; see Fig. 5.

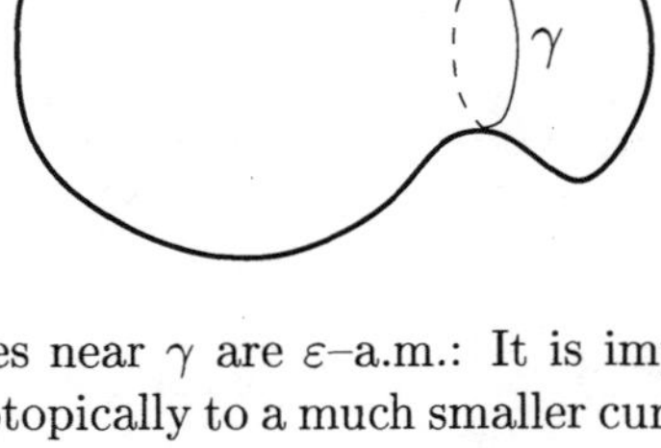

FIGURE 5. Curves near γ are ε–a.m.: It is impossible to deform any such curve isotopically to a much smaller curve without passing through a large curve.

The precise definition of a.m. surfaces is the following:

DEFINITION 3.2. *Given* $\varepsilon > 0$*, an open set* $U \subset M^3$*, and a surface* Σ*, we say that* Σ *is* ε–a.m. in U *if there* DOES NOT *exist any isotopy* ψ *supported in* U *such that*

$$\mathcal{H}^2(\psi(t,N)) \leq \mathcal{H}^2(N) + \varepsilon/8 \text{ for all } t; \tag{9}$$

$$\mathcal{H}^2(\psi(1,N)) \leq \mathcal{H}^2(N) - \varepsilon. \tag{10}$$

A sequence $\{\Sigma^n\}$ *is said to be* a.m. in U *if each* Σ^n *is* ε_n*–a.m. in* U *for some* $\varepsilon_n \downarrow 0$.

This definition first appeared in Smith's dissertation, [**Sm**], and was inspired by a similar one of Pitts (see the definition of almost minimizing varifolds in 3.1 of [**P**]). In section 4 of his book, Pitts used combinatorial arguments (some of which were based on ideas of Almgren, [**Alm**]) to prove a general existence theorem for almost minimizing varifolds. The situation we deal with here is much simpler, due to the fact that we only consider 1–parameter families of surfaces and not general multi–parameter families. Using a version of the combinatorial arguments of Pitts, we will prove in Section 5 the following proposition:

PROPOSITION 3.3. *There exists a function* $r : M \to \mathbf{R}^+$ *and a min–max sequence* $\{\Sigma^j\}$ *such that:*

- $\{\Sigma^j\}$ *is a.m. in every annulus* An *centered at* x *and with outer radius at most* $r(x)$*;*
- *In any such annulus,* Σ^j *is smooth when* j *is sufficiently large;*
- Σ^j *converges to a stationary varifold* V *in* M*, as* $j \uparrow \infty$.

The reason why we work with annuli is two fold. The first is that we allow the generalized families to have slices with point–singularities. The second is that even if any family of Λ were made of smooth surfaces, then the combinatorial proof of Proposition 3.3 would give a point $x \in M$ in which we are forced to work with annuli (cf. the proof of Proposition 5.1).

For a better understanding of this point consider the following example, due to Almgren ([**Alm**], p. 15–18; see also [**P**], p. 20–21). The surface M in Fig. 6 is diffeomorphic to $\mathbf{S}^2$ and metrized as a "three–legged starfish". The picture shows a sweep–out with a unique maximal slice, which is a geodesic figure–eight (cf. Fig. 5 of [**P**]). The slices close to the figure–eight are *not* almost minimizing in balls centered at its singular point P. But they are almost minimizing in every sufficiently small annulus centered at P.

If Λ is the saturated set generated by the sweep–out of Fig. 6, then no min–max sequence generated by Λ converges to a simple closed geodesic. However, there are no similar examples of sweep–outs of 3–dimensional manifolds by 2–dimensional objects: The reason for this is that point–singularities of (2–dimensional) minimal surfaces are removable.

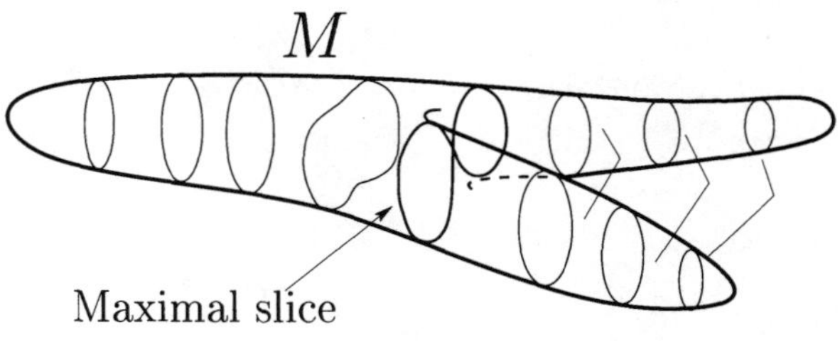

FIGURE 6. A sweep–out of the three–legged starfish, which can be realized as level–sets of a Morse function.

3.3. Gluing replacements and regularity. The task of the last sections is to prove that the stationary varifold V of Proposition 3.3 is a smooth surface. In

Section 7 we will see that if An is an annulus in which $\{\Sigma^j\}$ is a.m., then there exists a stationary varifold V', referred to as a *replacement*, such that

$$V \text{ and } V' \text{ have the same mass and } V = V' \text{ on } M \setminus \mathrm{An}. \tag{11}$$

$$V' \text{ is a stable minimal surface inside An.} \tag{12}$$

In Lemma 6.4 we use this "replacement property" and (8) to show that the stationary varifold V of Proposition 3.3 is integer rectifiable. The properties of (smooth) minimal surfaces would naturally lead to the following unique continuation–type conjecture, cf. [**SW**]:

CONJECTURE 3.4. *Let $\rho > 0$ be smaller than the convexity radius and let V and V' be stationary integer rectifiable in M with the same mass. If the outer radius of the annulus* An *is less than ρ, $V = V'$ in $M \setminus \mathrm{An}$, and V' is a stable minimal surface in* An*, then $V = V'$.*

An affirmative answer to Conjecture 3.4 would immediately yield the regularity of the stationary varifold V of Proposition 3.3 in sufficiently small annuli. By letting the inner radius of such annuli go to zero, we would be able to conclude that V is a stable minimal surface in $B_{\rho(x)}(x) \setminus \{x\}$, provided that $\rho(x)$ is sufficiently small. Hence, after showing that x is a removable singularity we would get that V is an embedded minimal surface.

Unluckily we are not able to argue in this way. In fact, in Appendix C we give an example of two distinct integer rectifiable 1–varifolds V_1 and V_2 in $\mathbf{R}^2$ which have the same mass and coincide outside a disk. This example does not disprove Conjecture 3.4; because, besides the dimensional difference, in the disk where $V_1 \neq V_2$, both the varifolds are singular. It does however show that a proof of Conjecture 3.4 could be rather delicate.

In [**Sm**] this problem of unique continuation was overcome by showing that for V as in Proposition 3.3, one can construct "secondary" replacements V'' also for the replacements V'. This idea goes back to [**P**]. In Section 6 we follow [**Sm**] and show that if we can replace sufficiently many times, then V is regular (cf. Definition 6.2 and Proposition 6.3 for the precise statement).

3.4. Replacements. As discussed in the previous subsection, to prove the regularity of V, we need to construct (sufficiently many) replacements. This task is accomplished in two steps in Section 7.

Step 1: Fix an annulus An in M in which Σ^k is ε_k–a.m. In this annulus we deform Σ^k into a further sequence of surfaces $\{\Sigma^{k,l}\}^l$ with the following properties:

- $\Sigma^{k,l}$ is the image of Σ^k under some isotopy ψ which satisfies (9) (with $\varepsilon = \varepsilon_k$ and $U = \mathrm{An}$);
- If we denote by $\mathcal{S}^k$ the family of all such isotopies, then

$$\lim_{l\to\infty} \mathcal{H}^2(\Sigma^{k,l}) = \inf_{\psi\in\mathcal{S}^k} \mathcal{H}^2(\psi(1,\Sigma^k))\,. \tag{13}$$

After possibly passing to a subsequence, then $\Sigma^{k,l} \to V^k$ and $V^k \to V'$, where V^k is a varifold which is stationary *in* An. By the a.m. property of V, it follows that V' is stationary *in all of* M and satisfies (11).

The second step is to prove that V^k is a (smooth) stable minimal surface in An. Thus, (8) will give that also V' is a stable minimal surface in An. After checking some details we show that V meets the technical requirements of Proposition 6.3.

Step 2: It remains to prove that V^k is a stable minimal surface. Stability is a trivial consequence of (13). For the regularity we use again Proposition 6.3. The key is proving the following property:

(P) If $B \subset \mathrm{An}$ is a sufficiently small ball and l is a sufficiently large number, then *any* $\psi \in \mathfrak{Js}(B)$ with $\mathcal{H}^2(\psi(1,\Sigma^{k,l})) \leq \mathcal{H}^2(1,\Sigma^k)$ can be replaced by a $\Psi \in \mathfrak{Js}(\mathrm{An})$ with

$$\Psi(1,\cdot) = \psi(1,\cdot) \quad \text{and} \quad \mathcal{H}^2(\Psi(t,\Sigma^{k,l})) \leq \mathcal{H}^2(\Sigma^{k,l}) + \varepsilon_k/8 \quad \text{for all } t. \tag{14}$$

We will now discuss how (P) gives the regularity of V^k.

Fix a sufficiently small ball B and a large number l so that the property (P) above holds. Take a sequence of surfaces $\Gamma^j = \Sigma^{k,l,j}$ which are isotopic to $\Sigma^{k,l}$ in B and such that $\mathcal{H}^2(\Gamma^j)$ converges to

$$\inf_{\psi \in \mathfrak{Js}(B)} \mathcal{H}^2(1,\psi(\Sigma^{k,l}))\,.$$

By a result of Meeks–Simon–Yau, [**MSY**], Γ^j converges to a varifold $V^{k,l}$ which is a stable minimal surface in B. Thus, by (8), the sequence of varifolds $\{V^{k,l}\}^l$ converges to a varifold W^k which is a stable minimal surface in B. The property (P) is used to show that, for j and l sufficiently large, $\Sigma^{k,l,j}$ is a good competitor with respect to the ε_k–a.m. property of Σ^k. This is then used to show that W^k is a replacement for V^k in B. Again it is only a technical step to check that we can apply Proposition 6.3, and hence get that V^k is a stable minimal surface in An.

Part 2. Proof of Theorem 1.6

4. Limits of suitable min–max sequences are stationary

This section is devoted to the proof of Proposition 3.1. For simplicity we metrize the weak topology on the space of varifolds and restate Proposition 3.1 using this metric.

Denote by X the set of varifolds $V \in \mathcal{V}(M)$ with mass bounded by $4\,m_0$, i.e., with $\|V\|(M) \leq 4m_0$. Endow X with the weak* topology and let $\mathcal{V}_\infty$ be the set of stationary varifolds contained in X. Clearly, $\mathcal{V}_\infty$ is a closed subset of X. Moreover, by standard general topology theorems, X is compact and metrizable. Fix one such metric and denote it by $\mathfrak{d}$. The ball of radius r and center V in this metric will be denoted by $U_r(V)$.

Proposition 4.1. *There exists a minimizing sequence $\{\{\Sigma_t\}^n\} \subset \Lambda$ such that, if $\{\Sigma^n_{t_n}\}$ is a min–max sequence, then* $\mathfrak{d}\,(\Sigma^n_{t_n}, \mathcal{V}_\infty) \to 0$.

Proof. The key idea of the proof is building a continuous map $\Psi : X \to \mathfrak{Js}$ such that :

- If V is stationary, then Ψ_V is the trivial isotopy;
- If V is not stationary, then Ψ_V decreases the mass of V.

Since each Ψ_V is an isotopy, and thus is itself a map from $[0,1] \times M \to M$, to avoid confusion we use the subscript V to denote the dependence on the varifold V. The map Ψ will be used to deform a minimizing sequence $\{\{\Sigma_t\}^n\} \subset \Lambda$ into another minimizing sequence $\{\{\Gamma_t\}^n\}$ such that :

For every $\varepsilon > 0$, there exist $\delta > 0$ and $N \in \mathbb{N}$ such that

$$\text{if} \quad \left\{ \begin{array}{rc} & n > N \\ \text{and} & \mathcal{H}^2(\Gamma^n_{t_n}) > m_0 - \delta \end{array} \right\}, \quad \text{then} \quad \mathfrak{d}\,(\Gamma^n_{t_n}, \mathcal{V}_\infty) < \varepsilon. \tag{15}$$

Such a $\{\{\Gamma_t\}^n\}$ would satisfy the requirement of the proposition.

The map Ψ_V should be thought of as a natural "shortening process" of varifolds which are not stationary. If the mass (considered as a functional on the space of varifolds) were smoother, then a gradient flow would provide a natural shortening process like Ψ_V. However, this is not the case; even if we start with smooth initial datum, in very short time the motion by mean curvature, i.e. the gradient flow of the area functional on smooth submanifolds, gives surfaces which are not isotopic to the initial one.

Step 1: A map from X to the space of vector fields.

The isotopies Ψ_V will be generated as 1–parameter families of diffeomorphisms satisfying certain ODE's. In this step we associate to any V a suitable vector field, which in Step 2 will be used to construct Ψ_V.

For $k \in \mathbb{Z}$ define the annular neighborhood of $\mathcal{V}_\infty$

$$\mathcal{V}_k = \left\{ V \in X | 2^{-k+1} \geq \mathfrak{d}\,(V, \mathcal{V}_\infty) \geq 2^{-k-1} \right\}.$$

There exists a positive constant $c(k)$ depending on k such that to every $V \in \mathcal{V}_k$ we can associate a smooth vector field χ_V with

$$||\chi_V||_\infty \leq 1 \qquad \text{and} \qquad \delta V(\chi_V) \leq -c(k).$$

Our next task is choosing χ_V with continuous dependence on V. Note that for every V there is some radius r such that $\delta W(\chi_V) \leq -c(k)/2$ for every $W \in U_r(V)$. Hence, for any k we can find balls $\{U^k_i\}_{i=1,\dots,N(k)}$ and vector fields χ^k_i such that :

$$\text{The balls } \tilde{U}^k_i \text{ concentric to } U^k_i \text{ with half the radii cover } \mathcal{V}_k; \tag{16}$$

$$\text{If } W \in U^k_i, \text{ then } \delta W(\chi) \leq -c(k)/2; \tag{17}$$

$$\text{The balls } U^k_i \text{ are disjoint from } \mathcal{V}_j \text{ if } |j-k| \geq 2. \tag{18}$$

Hence, $\{U^k_i\}_{k,i}$ is a locally finite covering of $X \setminus \mathcal{V}_\infty$. To this family we can subordinate a continuous partition of unit φ^k_i. Thus we set $H_V = \sum_{i,k} \varphi^k_i(V)\chi^k_i$. The map $H : X \to C^\infty(M, TM)$ which to every V associates H_V is continuous. Moreover, $||H_V||_\infty \leq 1$ for every V.

Step 2: A map from X to the space of isotopies.

For $V \in \mathcal{V}_k$ we let $r(V)$ be the radius of the smaller ball $\tilde{U}^j_i$ which contains it. We find that $r(V) > r(k) > 0$, where $r(k)$ only depends on k. Moreover, by (17) and (18), for every W contained in the ball $U_{r(V)}(V)$ we have that

$$\delta W(H_V) \leq -\frac{1}{2}\min\{c(k-1), c(k), c(k+1)\}.$$

Summarizing there are two continuous functions $g : \mathbf{R}^+ \to \mathbf{R}^+$ and $r : \mathbf{R}^+ \to \mathbf{R}^+$ such that

$$\delta W(H_V) \leq -g(\mathfrak{d}\,(V, \mathcal{V}_\infty)) \qquad \text{if} \qquad \mathfrak{d}\,(W, V) \leq r(\mathfrak{d}\,(V, \mathcal{V}_\infty)). \tag{19}$$

Now for every V construct the 1–parameter family of diffeomorphisms

$$\Phi_V : [0, +\infty) \times M \to M \qquad \text{with} \qquad \frac{\partial \Phi_V(t,x)}{\partial t} = H_V(\Phi_V(t,x)).$$

For each t and V, we denote by $\Phi_V(t,\cdot)$ the corresponding diffeomorphism of M. We claim that there are continuous functions $T : \mathbf{R}^+ \to [0,1]$ and $G : \mathbf{R}^+ \to \mathbf{R}^+$ such that

- If $\gamma = \mathfrak{d}\,(V, \mathcal{V}_\infty) > 0$ and we transform V into V' via the diffeomorphism $\Phi_V(T(\gamma),\cdot)$, then $\|V'\|(M) \leq \|V\|(M) - G(\delta)$;
- $G(s)$ and $T(s)$ both converge to 0 as $s \downarrow 0$.

Indeed fix V. For every $r > 0$ there is a $T > 0$ such that the curve of varifolds

$$\{V(t) = (\Phi_V(t,\cdot))_\sharp V\,, \quad t \in [0,T]\}$$

stays in $U_r(V)$. Thus

$$\begin{aligned}\|V(T)\|(M) - \|V\|(M) &= \|V(T)\|(M) - \|V(0)\|(M)\\ &\leq \int_0^T [\delta V(t)](H_V)\,dt\,,\end{aligned}$$

and therefore if we choose $r = r(\mathfrak{d}\,(V, \mathcal{V}_\infty))$ as in (19), then we get the bound

$$\|\Gamma(T)\|(M) - \|V\|(M) \;\leq\; -Tg(\mathfrak{d}\,(V, \mathcal{V}_\infty)).$$

Using a procedure similar to that of Step 1 we can choose T depending continuously on V. It is then trivial to see that we can in fact choose T so that at the same time it is continuous and depends only on $\mathfrak{d}\,(V, \mathcal{V}_\infty)$.

Step 3: Constructing the competitor and the conclusion.

For each V, set $\gamma = \mathfrak{d}\,(V, \mathcal{V}_\infty)$ and

$$\Psi_V(t,\cdot) \;=\; \Phi_V([T(\gamma)]\,t,\cdot) \qquad \text{for } t \in [0,1].$$

Ψ_V is a "normalization" of Φ_V. From Step 2 we know that there is a continuous function $L : \mathbf{R} \to \mathbf{R}$ such that

- L is strictly increasing and $L(0) = 0$;
- $\Psi_V(1,\cdot)$ deforms V into a varifold V' with $\|V'\| \leq \|V\| - L(\gamma)$.

Choose a sequence of families $\{\{\Sigma_t\}^n\} \subset \Lambda$ with $\mathcal{F}(\{\Sigma_t\}^n) \leq m_0 + 1/n$ and define $\{\Gamma_t\}^n$ by

$$\Gamma_t^n = \Psi_{\Sigma_t^n}(1, \Sigma_t^n) \qquad \text{for all } t \in [0,1] \text{ and all } n \in \mathbb{N} \tag{20}$$

Thus

$$\mathcal{H}^2(\Gamma_t^n) \leq \mathcal{H}^2(\Sigma_t^n) - L(\mathfrak{d}\,(\Sigma_t^n, \mathcal{V}_\infty)). \tag{21}$$

Note that $\{\Gamma_t\}^n$ does not necessarily belong to Λ, since the families of diffeomorphisms $\psi_t(\cdot) = \Psi_{\Sigma_t^n}(1,\cdot)$ may not depend smoothly on t. In order to overcome this technical obstruction fix n and note that $\Psi_t = \Psi_{\Sigma_t^n}$ is the 1–parameter family of isotopies generated by the 1–parameter family of vector fields $h_t = T(\Sigma_t^n)H_{\Sigma_t^n}$. Think of h as a continuous map

$$h : [0,1] \to C^\infty(M, TM) \quad \text{with the topology of } C^k \text{ seminorms.}$$

Thus h can be approximated by a *smooth* map $\tilde{h} : [0,1] \to C^\infty(M, TM)$. Consider the *smooth* 1–parameter family of isotopies $\tilde{\Psi}_t$ generated by the vector fields $\tilde{h}_t$ and the family of surfaces $\{\Gamma_t\}^n$ given by $\Gamma_t^n = \tilde{\Psi}_t(1, \Sigma_t^n)$. If $\sup_t \|h_t - \tilde{h}_t\|_{C^1}$ is sufficiently small, then we easily get (by the same calculations of the previous steps)

$$\mathcal{H}^2(\Gamma_t^n) \leq \mathcal{H}^2(\Sigma_t^n) - L(\mathfrak{d}\,(\Sigma_t^n, \mathcal{V}_\infty))/2. \tag{22}$$

Moreover, since $\tilde{\Psi}_t(1,\cdot)$ is a smooth map, this new family belongs to Λ.

Clearly $\{\{\Gamma_t\}^n\}$ is a minimizing sequence. We next show that $\{\{\Gamma_t\}^n\}$ satisfies (15). Note first that the construction yields a continuous and increasing function $\lambda : \mathbf{R}^+ \to \mathbf{R}^+$ such that

$$\lambda(0) = 0 \qquad \text{and} \qquad \mathfrak{d}\,(\Sigma_t^n, \mathcal{V}_\infty) \geq \lambda(\mathfrak{d}\,(\Gamma_t^n, \mathcal{V}_\infty))\,. \tag{23}$$

Fix $\varepsilon > 0$ and choose $\delta > 0$, $N \in \mathbb{N}$ such that $L(\lambda(\varepsilon))/2 - \delta > 1/N$. We claim that (15) is satisfied with this choice. Suppose not; then there are $n > N$ and t such that $\mathcal{H}^2(\Gamma_t^n) > m_0 - \delta$ and $\mathfrak{d}\,(\Gamma_t^n, \mathcal{V}_\infty) > \varepsilon$. Hence, from (22) and (23) we get

$$\mathcal{H}^2(\Sigma_t^n) \;\geq\; \mathcal{H}^2(\Gamma_t^n) + \frac{L(\lambda(\varepsilon))}{2} - \delta \;>\; m_0 + \frac{1}{N} \geq m_0 + \frac{1}{n}\,.$$

This contradicts the assumption that $\mathcal{F}(\{\Sigma_t\}^n) \leq m_0 + 1/n$. Thus (15) holds and the proof is completed. □

5. Almost minimizing min–max sequences

As above, Λ is a fixed saturated set of 1–parameter families $\{\Sigma_t\}$ in M. In the previous section we showed that there exists a family $\{\Sigma_t\}$ such that every min–max sequence is clustering towards stationary varifolds. We will now prove that one of these min–max sequences is a.m. in sufficiently many annuli.

PROPOSITION 5.1. *There exists a function $r : M \to \mathbf{R}^+$ and a min–max sequence $\{\Sigma^j\}$ such that:*

$$\{\Sigma^j\} \textit{ is a.m. in every } \mathrm{An} \in \mathcal{AN}_{r(x)}(x),\textit{ for all } x \in M\,. \tag{24}$$

$$\textit{In every such } \mathrm{An},\ \Sigma^j \textit{ is a smooth surface when } j \textit{ is sufficiently large}\,. \tag{25}$$

$$\Sigma^j \textit{ converges to a stationary varifold } V \textit{ as } j \uparrow \infty. \tag{26}$$

We first fix some notation.

DEFINITION 5.2. *Given a pair of open sets (U^1, U^2) we say that a surface Σ is ε–a.m. in (U^1, U^2) if it is ε–a.m. in at least one of the two open sets. We denote by $\mathcal{CO}$ the set of pairs (U^1, U^2) of open sets with*

$$\mathrm{d}\,(U^1, U^2) \geq 2\min\{\mathit{diam}(U^1), \mathit{diam}(U^2)\}\,.$$

Proposition 5.1 will be an easy corollary of the following:

PROPOSITION 5.3. *There is a min–max sequence $\{\Sigma^L\} = \{\Sigma^{n(L)}_{t_{n(L)}}\}$ which converges to a stationary varifold and such that*

$$\textit{each } \Sigma^L \textit{ is } 1/L\textit{–a.m. in every } (U^1, U^2) \in \mathcal{CO}\,. \tag{27}$$

Note that the Σ^L's in the previous proposition may be degenerate slices (that is, they may have a finite number of singular points). The key point for proving Proposition 5.3 is the following obvious lemma:

LEMMA 5.4. *If (U^1, U^2) and $(V^1, V^2) \in \mathcal{CO}$, then there are $i, j \in \{1, 2\}$ with $\mathrm{d}\,(U^i, V^j) > 0$.*

Before giving a rigorous proof of Proposition 5.3 we will explain the ideas behind it.

5.1. Outline of the proof of Proposition 5.3. First of all note that if a slice $\Sigma^n_{t_0}$ is not ε–a.m. in a given open set U, then we can decrease its area by an isotopy ψ satisfying (9) and (10). Now fix an open interval I around t_0 and choose a smooth bump function $\varphi \in C^\infty_c(I, [0,1])$ with $\varphi(t_0) = 1$. Define $\{\Gamma_t\}^n$ by

$$\Gamma^n_t \;=\; \psi(\varphi(t), \Sigma^n_t)\,.$$

If the interval I is sufficiently small, then by (9), for any $t \in I$, the area of Γ^n_t will not be much larger than the area of Σ^n_t. Moreover, for t very close to t_0 (say, in a smaller interval $J \subset I$) the area of Γ^n_t will be much less than the area of Σ^n_t.

We will show Proposition 5.3 by arguing by contradiction. So suppose that the proposition fails; we will construct a better competitor $\{\{\Gamma_t\}^n\}$. Here the pairs $\mathcal{CO}$ will play a crucial role. Indeed when the area of Σ^n_t is sufficiently large (i.e. close to m_0), we can find *two* disjoint open sets U_1 and U_2 in which Σ^n_t is not almost minimizing. Consider the set $K_n \subset [0,1]$ of slices with sufficiently large area. Using Lemma 5.4 (and some elementary considerations), we find a finite family of intervals I_j, open sets U_j, and isotopies $\psi_j : I_j \times M \to M$ satisfying the following conditions; see Fig. 7:

ψ_j is supported in U_j and is the identity at the ends of I_j. (28)

If $I_j \cap I_k \neq \emptyset$, then $U_j \cap U_k = \emptyset$. (29)

No point of $[0,1]$ belong to more than two I_j's. (30)

$\mathcal{H}^2(\psi_j(t, \Sigma^n_t))$ is never much larger than $\mathcal{H}^2(\Sigma^n_t)$. (31)

For every $t \in K_n$, there is j s.t. $\mathcal{H}^2(\psi_j(t, \Sigma^n_t))$ is much smaller than $\mathcal{H}^2(\Sigma^n_t)$. (32)

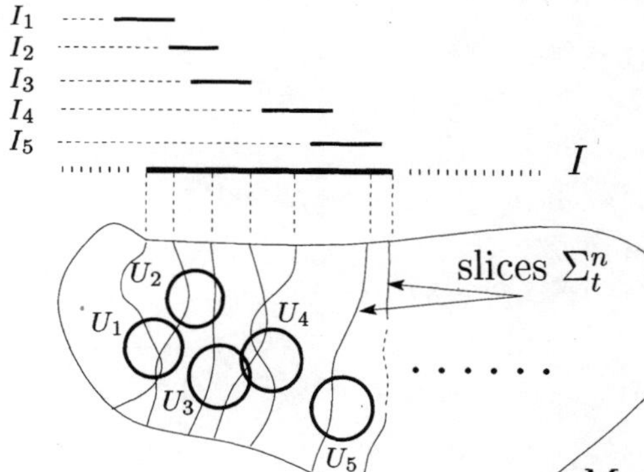

FIGURE 7. The covering I_j and the sets U_j. No point of I is contained in more than two I_j's. The intersection $U_j \cap U_k = \emptyset$ if I_j and I_k overlap.

Conditions (28) and (29) allow us to "glue" the ψ_j's in a unique $\psi \in \mathfrak{Js}$ such that $\psi = \psi_j$ on $I_j \times U_j$. The family $\{\Gamma_t\}^n$ given by $\Gamma^n_t = \psi(t, \Sigma^n_t)$ is our competitor. Indeed for every t, there are at most two ψ_j's which change Σ^n_t. If $t \notin K_n$, then none of them increases the area of Σ^n_t too much. Whereas, if $t \in K_n$, then one ψ_j decreases the area of Σ^n_t a definite amount, and the other increases the area of Σ^n_t a small amount. Thus, the area of the "small–area" slices are not increased much and the area of "large–area" slices are decreased. This yields that $\mathcal{F}(\{\Gamma_t\}^n) - \mathcal{F}(\{\Sigma_t\}^n) < 0$. We will now give a rigorous bound for this (negative) difference.

5.2. Proof of Proposition 5.3.

PROOF OF PROPOSITION 5.3. We choose $\{\{\Sigma_t\}^n\} \subset \Lambda$ such that $\mathcal{F}(\{\Sigma_t\}^n) < m_0 + 1/n$ and satisfying the requirements of Proposition 4.1. Fix $L \in \mathbb{N}$. To prove the proposition we claim there exist $n > L$ and $t_n \in [0,1]$ such that $\Sigma^n = \Sigma^n_{t_n}$ satisfies (27) and $\mathcal{H}^2(\Sigma^n) \geq m_0 - 1/L$. We define the sets

$$K_n = \left\{ t \in [0,1] : \mathcal{H}^2(\Sigma^n_t) \geq m_0 - \frac{1}{L} \right\}$$

and argue by contradiction. Suppose not; then for every $t \in K_n$ there exists a pair of open subsets (U^1_t, U^2_t) such that Σ^n_t is not $1/L$–a.m. in either of them. So for every $t \in K_n$ there exists isotopies ψ^i_t such that

(1) ψ^i_t is supported on U^i_t;
(2) $\mathcal{H}^2(\psi^i_t(1, \Sigma^n_t)) \leq \mathcal{H}^2(\Sigma^n_t) - 1/L$;
(3) $\mathcal{H}^2(\psi^i_t(\tau, \Sigma^n_t)) \leq \mathcal{H}^2(\Sigma^n_t) + 1/(8L)$ for every $\tau \in [0,1]$.

In the following we fix n and drop the subscript from K_n. Since $\{\Sigma^n_t\}$ is continuous in t, if $t \in K$ and $|s-t|$ is sufficiently small, then

(2') $\mathcal{H}^2(\psi^i_t(1, \Sigma^n_s)) \leq \mathcal{H}^2(\Sigma^n_s) - 1/(2L)$;
(3') $\mathcal{H}^2(\psi^i_t(\tau, \Sigma^n_s)) \leq \mathcal{H}^2(\Sigma^n_s) + 1/(4L)$ for every $\tau \in [0,1]$.

By compactness we can cover K with a finite number of intervals satisfying (2') and (3'). This covering $\{I_k\}$ can be chosen so that I_k overlaps only with I_{k-1} and I_{k-2}. Summarizing we can find

closed intervals	$I_1, \ldots I_r$
pairs of open sets	$(U^1_1, U^2_1), \ldots, (U^1_r, U^2_r) \in \mathcal{CO}$
and pairs of isotopies	$(\psi^1_1, \psi^2_1), \ldots, (\psi^1_r, \psi^2_r)$

such that

(A) the interiors of I_j cover K and $I_j \cap I_k = \emptyset$ if $|k-j| \geq 2$;
(B) ψ^i_j is supported in U^i_j;
(C) $\mathcal{H}^2(\psi^i_j(1, \Sigma^n_s)) \leq \mathcal{H}^2(\Sigma^n_s) - 1/(2L)$ $\forall s \in I_j$;
(D) $\mathcal{H}^2(\psi^i_j(\tau, \Sigma^n_s)) \leq \mathcal{H}^2(\Sigma^n_s) + 1/(4L)$ $\forall s \in I_j$ and $\tau \in [0,1]$.

In Step 1 we refine this covering. In Step 2 we use the refined covering to construct a competitor $\{\Gamma_t\}^n \in \Lambda$ with

$$\mathcal{F}(\{\Gamma_t\}^n) \leq \mathcal{F}(\{\Sigma_t\}^n) - 1/(2L). \tag{33}$$

The arbitrariness of n will give that $\liminf_n \mathcal{F}(\{\Gamma_t\}^n) < m_0$. This is the desired contradiction which yields the proposition.

Step 1: Refinement of the covering.

First we want to find

a covering $\{J_1, \ldots, J_R\}$	which is a refinement of $\{I_1, \ldots I_r\}$,
open sets $V_1, \ldots V_R$	among $\{U^i_j\}$,
and isotopies $\varphi_1, \ldots, \varphi_R$	among $\{\psi^i_j\}$,

such that:

(A1) The interiors of J_i cover K and $J_i \cap J_k = \emptyset$ for $|k-i| \geq 2$;

(A2) If $J_i \cap J_k \neq \emptyset$, then $d(V_i, V_k) > 0$;
(B') φ_i is supported in V_i;
(C') $\mathcal{H}^2(\varphi_i(1, \Sigma_s^n)) \leq \mathcal{H}^2(\Sigma_s^n) - 1/(2L)$ $\forall s \in J_i$;
(D') $\mathcal{H}^2(\varphi_i(\tau, \Sigma_s^n)) \leq \mathcal{H}^2(\Sigma_s^n) + 1/(4L)$ $\forall s \in J_i$ and $\tau \in [0,1]$.

We start by setting $J_1 = I_1$ and we distinguish two cases.

- **Case a1**: $I_1 \cap I_2 = \emptyset$; we set $V_1 = U_1^1$, and $\varphi_1 = \psi_1^1$.
- **Case a2**: $I_1 \cap I_2 \neq \emptyset$; by Lemma 5.4 we can choose $i, k \in \{1, 2\}$ such that $\mathrm{d}\,(U_1^i, U_2^k) > 0$ and we set $V_1 = U_1^i$, $\varphi_1 = \psi_1^i$.

We now come to the choice of J_3. If we come from case a1 then:

- **Case b1:** We make our choice as above replacing I_1 and I_2 with I_2 and I_3;

If we come from case a2, then we let i and k be as above and we further distinguish two cases.

- **Case b21**: $I_2 \cap I_3 = \emptyset$; we define $J_2 = I_2$, $V_2 = U_2^k$, $\varphi_2 = \psi_2^k$.
- **Case b22**: $I_2 \cap I_3 \neq \emptyset$; by Lemma 5.4 there exist $l, m \in \{1, 2\}$ such that $d(U_2^l, U_3^m) > 0$. If $l = k$, then we define $J_2 = I_2$, $V_2 = U_2^k$, $\varphi_2 = \psi_2^k$. Otherwise we choose two closed intervals $J_2, J_3 \subset I_2$ such that
  - their interiors cover the interior of I_2,
  - J_2 does not overlap with any I_h for $h \neq 1, 2$,
  - J_3 does not overlap with any I_h for $h \neq 2, 3$.

  Thus we set $V_2 = U_2^k$, $\varphi_2 = \psi_2^k$, and $V_3 = U_2^l$, $\varphi_3 = \psi_2^l$.

An inductive argument using this procedure gives the desired covering. Note that the cardinality of $\{J_1, \ldots, J_R\}$ is at most $2r - 1$.

Step 2: Construction.

Choose C^∞ functions η_i on $\mathbf{R}$ taking values in $[0,1]$, supported in J_i, and such that for every $s \in K$, there exists η_i with $\eta_i(s) = 1$. Fix $t \in [0,1]$ and consider the set $\mathrm{Ind}_t \subset \mathbb{N}$ of all i containing t; thus Ind_t consists of at most two elements. Define subsets of M by

$$\Gamma_t^n = \begin{cases} \varphi_i(\eta_i(t), \Sigma_t^n) & \text{in the open sets } V^i,\ i \in \mathrm{Ind}_t\ , \\ \Sigma_t^n & \text{outside.} \end{cases} \tag{34}$$

In view of (A1), (A2) and (B'), then $\{\Gamma_t\}^n$ is well defined and belongs to Λ.

Step 3: The contradiction.

We now want to bound the energy $\mathcal{F}(\{\Gamma_t\}^n)$ and hence we have to estimate $\mathcal{H}^2(\Gamma_t^n)$. Note that by (A1) every Ind_t consists of at most two integers. Assume for the sake of argument that Ind_t consists of *exactly* two integers. From the construction, there exist $s_i, s_k \in [0,1]$ such that Γ_t^n is obtained from Σ_t^n via the diffeomorphisms $\varphi_i(s_i, \cdot)$, $\varphi_k(s_k, \cdot)$. By (A2) these diffeomorphisms are supported on disjoint sets. Thus if $t \notin K$, then (D') gives

$$\mathcal{H}^2(\Gamma_t^n) \leq \mathcal{H}^2(\Sigma_t^n) + \frac{2}{4L} \leq m_0 - \frac{1}{2L}.$$

If $t \in K$, then at least one of s_i, s_k is equal to 1. Hence (C) and (D) give

$$\mathcal{H}^2(\Gamma_t^n) \leq \mathcal{H}^2(\Sigma_t^n) - \frac{1}{L} + \frac{1}{4L} \leq \mathcal{F}(\{\Sigma_t^n\}) - \frac{3}{4L}.$$

Therefore $\mathcal{F}(\{\Gamma_t\}^n) \leq \mathcal{F}(\{\Sigma_t\}^n) - 1/(2L)$. This is the desired bound (33). □

We now come to Proposition 5.1.

PROOF OF PROPOSITION 5.1. We claim that a subsequence of the Σ^k's of Proposition 5.3 satisfies the requirements of Proposition 5.1. Indeed fix $k \in \mathbb{N}$ and r such that $\text{Inj}(M) > 4r > 0$. Since $(B_r(x), M \setminus B_{4r}(x)) \in \mathcal{CO}$ we then know that Σ^N is $1/k$–a.m. in $M \setminus B_{4r}(x)$. Thus we have that

$$\text{either } \Sigma^k \text{ is } 1/k\text{–a.m. on } B_r(y) \text{ for every } y \tag{35}$$

$$\text{or there is } x^k_r \in M \text{ s.t. } \Sigma^k \text{ is } 1/k\text{–a.m. on } M \setminus B_{4r}(x^k_r). \tag{36}$$

If for some $r > 0$ there exists a subsequence $\{\Sigma^{k(n)}\}$ satisfying (35), then we are done. Otherwise we may assume that there are two sequences of natural numbers $n \uparrow \infty, j \uparrow \infty$ and points x^n_j such that

- For every j, and for n large enough, Σ^n is $1/n$–a.m. in $M \setminus B_{1/j}(x^n_j)$.
- $x^n_j \to x_j$ for $n \uparrow \infty$ and $x_j \to x$ for $j \uparrow \infty$.

Thus for every j, the sequence $\{\Sigma^n\}$ is a.m. in $M \setminus B_{2/j}(x)$. Of course if $U \subset V$ and N is ε–a.m. in V, then N is ε–a.m. in U. This proves that there exists a subsequence $\{\Sigma^j\}$ which satisfies conditions (24) and (26) for some positive function $r : M \to \mathbf{R}^+$.

It remains to show that an appropriate further subsequence satisfies (25). Each Σ^j is smooth except at finitely many points. We denote by P_j the set of singular points of Σ^j. After extracting another subsequence we can assume that P_j is converging, in the Hausdorff topology, to a finite set P. If $x \in P$ and An is any annulus centered at x, then $P_j \cap \text{An} = \emptyset$ for j large enough. If $x \notin P$ and An is any (small) annulus centered at x with outer radius less than $\mathrm{d}(x, P)$, then $P_j \cap \text{An} = \emptyset$ for j large enough. Thus, after possibly modifying the function r above, the sequence $\{\Sigma^j\}$ satisfies (24), (25), and (26). □

6. Regularity for the replacements

We will now define a notion of a "good replacement" for stationary varifolds and prove that the existence of (sufficiently many) replacements for a stationary varifold implies that it is a smooth minimal surface; see Proposition 6.3. In section 6 we will show that the varifold V of Proposition 5.1 satisfies the hypotheses of Proposition 6.3 and thus is smooth.

DEFINITION 6.1. *Let $V \in \mathcal{V}(M)$ be stationary and $U \subset M$ be an open subset. A stationary varifold $V' \in \mathcal{V}(M)$ is said to be a* replacement for V in U *if* (37) *and* (38) *below hold.*

$$V' = V \text{ on } G(M \setminus \overline{U}) \text{ and } \|V'\|(M) = \|V\|(M). \tag{37}$$

$$V \llcorner U \text{ is a stable minimal surface } \Sigma \text{ with } \overline{\Sigma} \setminus \Sigma \subset \partial U. \tag{38}$$

DEFINITION 6.2. *Let V be a stationary varifold and $U \subset M$ be an open subset. We say that V has the* good replacement property *in U if (a), (b), and (c) below hold.*

(a) *There is a positive function $r : U \to \mathbf{R}$ such that for every annulus* An $\in \mathcal{AN}_{r(x)}(x)$ *there is a replacement for V' in* An.
(b) *The replacement V' has a replacement V'' in any* An $\in \mathcal{AN}_{r(x)}(x)$ *and in any* An $\in \mathcal{AN}_{r'(y)}(y)$ *(where r' is positive).*
(c) *V'' has a replacement V''' in any* An $\in \mathcal{AN}_{r''(y)}(y)$ *(where $r'' > 0$).*

If V and V' are as above, then we will say that V' is a good replacement *and V'' a* good further replacement.

This section is devoted to prove the following:

PROPOSITION 6.3. *Let G be an open subset of M. If V has the good replacement property in G, then V is a (smooth) minimal surface in G.*

In the proof Proposition 6.3 we need the two technical Lemmas B.1 and B.2, stated and proved in Appendix B. Note that Lemma B.1 is just a weak version (in the framework of varifolds) of the classical maximum principle for minimal surfaces. As a first step towards the proof of Proposition 6.3 we have the following:

LEMMA 6.4. *Let U be an open subset of M and V a stationary varifold in U. If there exists a positive function r on M such that V has a replacement in any annulus $\mathrm{An} \in \mathcal{AN}_{r(x)}(x)$, then V is integer rectifiable. Moreover, $\theta(x,V) \geq 1$ for any $x \in U$ and any tangent cone to V in x is an integer multiple of a plane.*

PROOF. Since V is stationary, the monotonicity formula (6) gives a constant C_M such that

$$\frac{\|V\|(B_\sigma(x))}{\sigma^2} \leq C_M \frac{\|V\|(B_\rho(x))}{\rho^2} \qquad \forall \sigma < \rho < \mathrm{Inj}\,(M) \text{ and } \forall x \in M\,. \tag{39}$$

Fix $x \in \mathrm{supp}\,(\|V\|)$ and $r < r(x)$ so that $4r$ is smaller than the convexity radius. Replace V with V' in $\mathrm{An}(x,r,2r)$. We claim that $\|V'\|$ cannot be identically 0 on $\mathcal{AN}(x,r,2r)$. Assume it was; since $x \in \mathrm{supp}\,(\|V'\|)$, there would be a $\rho \leq r$ such that V' "touches" ∂B_ρ from the interior. More precisely, there would exist ρ and ε such that $\mathrm{supp}\,\|V'\| \cap \partial B_\rho(x) \neq \emptyset$ and $\mathrm{supp}\,\|V'\| \cap \mathcal{AN}(x,\rho,\rho+\varepsilon) = \emptyset$. Since $B_\rho(x)$ is convex this would contradict Lemma B.1. Thus $V' \llcorner \mathrm{An}(x,r,2r)$ is a non–empty smooth surface and so there is $y \in \mathrm{An}(x,r,2r)$ with $\theta(V',y) \geq 1$. Using (39) we get

$$\frac{\|V\|(B_{4r}(x))}{16r^2} = \frac{\|V'\|(B_{4r}(x))}{16r^2} \geq \frac{C_M\|V'\|(B_{2r}(y))}{16r^2} \overset{(39)}{\geq} \frac{\pi C_M}{4}. \tag{40}$$

Hence, $\theta(x,V)$ is bounded uniformly from below on $\mathrm{supp}\,(\|V\|)$ and applying Theorem 2.1 we conclude that V is rectifiable.

We next prove that V is *integer* rectifiable. We use the notation of Definition 2.2. Fix $x \in \mathrm{supp}\,(\|V\|)$, a stationary cone $C \in TV(x,V)$, and a sequence $\rho_n \downarrow 0$ such that $V^x_{\rho_n} \to C$. Replace V by V'_n in $\mathrm{An}(x,\rho_n/4,3\rho_n/4)$ and set $W'_n = (T^x_{\rho_n})_\sharp V'_n$. After possibly passing to a subsequence, we can assume that $W'_n \to C'$, where C' is a stationary varifold. The following properties of C' are trivial consequences of the definition of replacements:

$$C' = C \text{ in } \mathcal{B}_{1/4}(x) \cup \mathrm{An}(x,3/4,1), \tag{41}$$

$$\|C'\|(\mathcal{B}_\rho) = \|C\|(\mathcal{B}_\rho) \text{ if } \rho \in]0,1/4[\cup]3/4,1[. \tag{42}$$

Since C is a cone, in view of (42) we have

$$\frac{\|C'\|(\mathcal{B}_\sigma)}{\sigma^2} = \frac{\|C'\|(\mathcal{B}_\rho)}{\rho^2} \qquad \forall \sigma,\rho \in]0,1/4[\cup]3/4,1[. \tag{43}$$

Hence, the stationarity of C' and the monotonicity formula imply that C' is a cone. By (8), W'_n converge to a stable embedded minimal surface in $\mathrm{An}(x,1/4,3/4)$. This means that $C' \llcorner \mathrm{An}(x,1/4,3/4)$ is an embedded minimal cone in the classical sense and hence it is supported on a disk containing the origin. This forces C' and C to coincide and be an integer multiple of the same plane. □

Proof of Proposition 6.3. The strategy of the proof is as follows. Fix $x \in M$, a good replacement V' for V in $\mathrm{An}(x, \rho, 2\rho)$, and let Σ' be the stable minimal surface given by V' in $\mathrm{An}(x, \rho, 2\rho)$. Consider $t \in]\rho, 2\rho[$, $s < \rho$ and the replacement V'' of V' in $\mathrm{An}(x, s, t)$, which in this annulus coincides with a smooth minimal surface Σ''. In step 2 we will prove that, for ρ sufficiently small and for an appropriate choice of t, then $\Sigma'' \cup \Sigma'$ is a smooth surface. Letting $s \downarrow 0$ we get a minimal surface $\Sigma \subset B_\rho(x) \setminus \{x\}$ such that every Σ'' constructed as above is a subset of Σ. Loosely speaking, any replacement of V' will coincide with Σ in the annular region where it is smooth.

Now, fix a z which belongs to $\mathrm{supp}\,(\|V\|)$ and such that V intersects $\partial B_s(x)$ "transversally" in z. If we consider a replacement V'' of V' in $\mathrm{An}(x, s, \rho)$, then z will belong to the closure of the minimal surface $\Sigma'' = V'' \llcorner \mathrm{An}(x, s, \rho)$. The discussion above gives that $z \in \Sigma$. Lemma B.2 implies that "transversality" to the spheres centered at x in a dense subset of $(\mathrm{supp}\,(\|V\|)) \cap B_\rho(x)$. Thus in step 3 we conclude that

$$(\mathrm{supp}\,(\|V\|)) \cap B_\rho(x) \setminus \{x\} \subset \Sigma.$$

Since $\mathcal{H}^2(\Sigma \cap B_\rho(x)) = \|V\|(B_\rho(x))$, then $V = \Sigma$ in $B_\rho(x)$. Step 4 concludes the proof by showing that x is a removable singularity for Σ.

The key fact that Σ'' and Σ' can be "glued" smoothly together is a consequence of the curvature estimates for stable minimal surfaces combined with the characterization of the tangent cones given in Lemma 6.4. These two ingredients will be used to prove that Σ'' is (locally) a Lipschitz graph nearby $\partial B_t(x)$; thus allowing us to apply standard theory of Elliptic PDE.

Step 1: The set up.

Fix x, V, V', V'', Σ', Σ'', ρ, s, and t as above. We require that 2ρ is less than the convexity radius of M and that Σ' intersects $\partial B_t(x)$ transversally. Fix a point $y \in \Sigma' \cap \partial B_t(x)$ and a sufficiently small radius r, so that $\Sigma' \cap B_r(y)$ is a disk and $\gamma = \Sigma' \cap \partial B_t(x) \cap B_r(y)$ is a smooth arc.

Let $\zeta : B_r(y) \to \mathcal{B}_1$ be a diffeomorphism such that

$$\zeta(\partial B_t(x)) \subset \{z_1 = 0\} \qquad \text{and} \qquad \zeta(\Sigma'') \subset \{z_1 > 0\},$$

where z_1, z_2, z_3 are orthonormal coordinates on $\mathcal{B}_1$; see Fig. 8. We will also assume that $\zeta(\gamma) = \{(0, z_2, g'(0, z_2))\}$ and $\zeta(\Sigma') \cap \{z_1 \leq 0\} = \{(z_1, z_2, g'(z_1, z_2))\}$ where g' is smooth.

Note the following elementary facts:

- Any kind of estimates (like curvature or area bounds or monotonicity) for a minimal surface $\Sigma \subset B_r(y)$ translates into similar estimates for the surface $\zeta(\Sigma)$.
- Varifolds in $B_r(y)$ are push–forwarded to varifolds in $\mathcal{B}_1$ and there is a natural correspondence between tangent cones to V in ξ and tangent cones to $\zeta_\sharp V$ in $\zeta(\xi)$.

By slight abuse of notation, we use the same symbols (e.g. γ, V', Σ') for both the objects of $B_r(y)$ and their images under ζ.

Step 2: Graphicality; gluing Σ' and Σ'' smoothly together.

The varifold V'' consists of $\Sigma'' \cup \Sigma'$ in $B_r(y)$. Moreover, Lemma 6.4 applied to V'' gives that $TV(z, V'')$ is a family of (multiples of) 2–planes. Fix $\overline{z} \in \gamma$. Since Σ' is regular and transversal to $\{z_1 = 0\}$ in $\overline{z}$, each plane $P \in TV(\overline{z}, V'')$ coincides

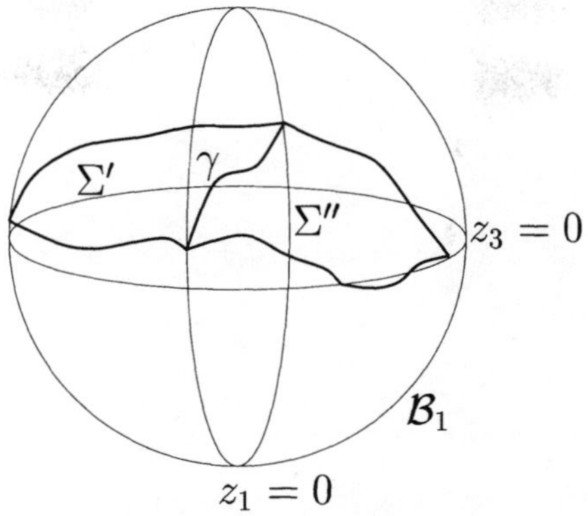

FIGURE 8. The surfaces Σ' and Σ'' and the curve γ in $\mathcal{B}_1$.

with the half plane $T_{\overline{z}}\Sigma'$ in $\{z_1 < 0\}$. Hence $TV(\overline{z}, V'') = \{T_{\overline{z}}\Sigma'\}$. Let $\tau(\overline{z})$ be the unit normal to the graph of g'

$$\tau(\overline{z}) \;=\; \frac{(-\partial_1 g'(0,\overline{z}_2), -\partial_2 g'(0,\overline{z}_2), 1)}{\sqrt{1+|\nabla g'0,\overline{z}_2)|^2}}$$

and let $R^{\overline{z}}_r : \mathbf{R}^3 \to \mathbf{R}^3$ be the dilatation of 3-space defined by

$$R^{\overline{z}}_r(z) \;=\; \frac{z-\overline{z}}{r}\,.$$

Since $TV(\overline{z}, V'') = \{T_{\overline{z}}\Sigma'\}$, the surfaces $\Sigma_r = R^{\overline{z}}_r(\Sigma'')$ converge to the half plane $HP = \{\tau(\overline{z})\cdot v = 0, v_1 > 0\}$ — half of the plane $\{\tau(\overline{z})\cdot v = 0\}$. This convergence implies that

$$\lim_{z\to\overline{z}, z\in\Sigma''} \frac{|(z-\overline{z})\cdot\tau(\overline{z})|}{|\overline{z}-z|} = 0\,. \tag{44}$$

Indeed assume that (44) fails; then there is a sequence $\{z_n\}\subset\Sigma''$ such that $z_n\to\overline{z}$ and $|(z_n-\overline{z})\cdot\tau(\overline{z})| \geq k|z_n-\overline{z}|$ for some $k>0$. Set $r_n = |z_n - \overline{z}|$. There exists a constant k_2 such that $\mathcal{B}_{2k_2r_n}(z_n)\cap HP = \emptyset$. Thus $\text{dist}(HP, \mathcal{B}_{k_2r_n}(z_n)) \geq k_2 r_n$. Since Σ'' is regular in z_n we get by the monotonicity formula that

$$||V''||(\mathcal{B}_{k_2r_n}(z_n)) \geq Ck_2^2r_n^2 \qquad \text{where } C \text{ depends on } \zeta.$$

This contradicts the fact that HP is the only element of $TV(\overline{z}, V'')$. Note also that the convergence of (44) is uniform for $\overline{z}$ in compact subsets of γ. The argument is explained in Fig. 9.

Let ν denote the smooth unit vector field to Σ'' such that $\nu\cdot(0,0,1)\geq 0$. We next use the stability of Σ'' to show that

$$\lim_{z\to\overline{z}, z\in\Sigma''} \nu(z) \;=\; \tau(\overline{z})\,. \tag{45}$$

Indeed let σ be the plane $\{(0,\alpha,\beta), \alpha,\beta\in\mathbf{R}\}$, assume that $z_n\to\overline{z}$ and set $r_n = \text{dist}(z_n,\sigma)$. Define the rescaled surfaces $\Sigma^n = R^{z_n}_{r_n}(\Sigma''_n\cap\mathcal{B}_{r_n}(z_n))$. Each Σ^n is a stable minimal surface in $\mathcal{B}_1$, and hence, after possibly passing to a subsequence, Σ^n converges smoothly in $\mathcal{B}_{1/2}$ to a minimal surface Σ^∞ (by (8)). By (44), we have that Σ^∞ is the disk $T_{\overline{z}}\Sigma'\cap\mathcal{B}_{1/2}$. Thus the normals to Σ^n in 0, which are given by

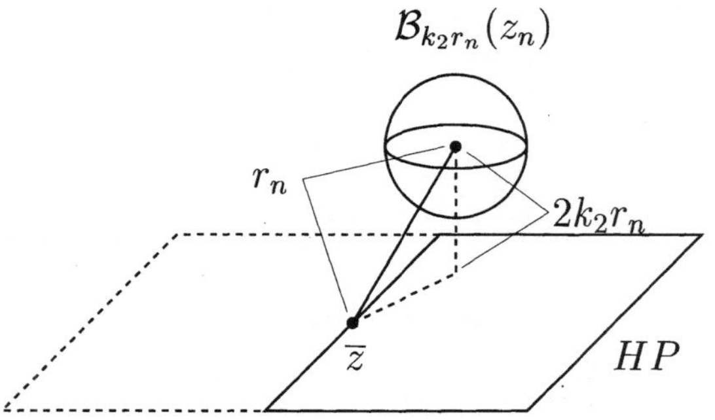

FIGURE 9. If $z_n \in \Sigma''$ is far from the plane HP, the monotonicity formula gives a "good amount" of the varifold V'' which lives far from HP.

$\nu(z_n)$, converge to $\tau(\overline{z})$; see Fig. 10. It is easy to see that the convergence in (45) is uniform on compact subsets of γ.

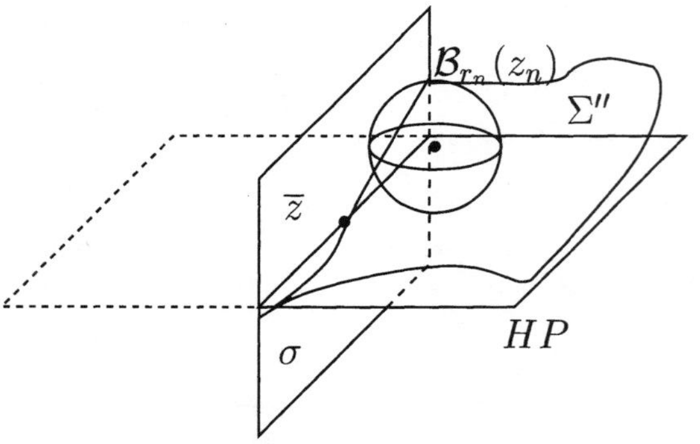

FIGURE 10. If we rescale $\mathcal{B}_{r_n}(z_n)$, then we find a sequence of stable minimal surfaces Σ^n which converge to the half-plane HP.

Hence, for each $\overline{z} \in \gamma$, there exists $r > 0$ and a function $g'' \in C^1(\{z_1 \geq 0\})$ such that

$$\begin{aligned} \Sigma'' \cap B_r(\overline{z}) &= \{(z_1, z_2, g''(z_1, z_2)), z_1 > 0\}, \\ g''(0, z_2) = g'(0, z_2), &\qquad \text{and} \qquad \nabla g''(0, z_2) = \nabla g'(0, z_2). \end{aligned}$$

In the coordinates z_1, z_2, z_3, the minimal surface equation yields a second order uniformly elliptic equation for g' and g''. Thus the classical theory of elliptic PDE gives that g' and g'' are restrictions of a unique smooth function g.

Step 3: Regularity of V in the punctured ball.

Let Σ' and Σ'' be as in the previous step. We will now show that :

$$\text{If } \Gamma \text{ is a connected component of } \Sigma'', \text{ then } \overline{\Gamma} \cap \Sigma' \cap \partial B_t(y) \neq \emptyset. \tag{46}$$

Indeed assume that for some Γ equation (46) fails. Since t is assumed to be less than the convexity radius we have by the maximum principle that $\overline{\Gamma} \cap \partial B_t(x) \neq \emptyset$.

Fix z in $\overline{\Gamma} \cap \partial B_t(x)$. If (46) were false, then the varifold V'' would "touch" $\partial B_t(x)$ in z from the interior. More precisely, there would be an $r > 0$ such that

$$z \in \operatorname{supp}(\|V''\|) \qquad \text{and} \qquad (B_r(y) \cap \operatorname{supp}(\|V''\|)) \subset \overline{B}_t(x)\,.$$

This contradicts Lemma B.1; thus (46) holds.

Let t, ρ be as in the first paragraph of Step 1. Step 2 and (46) imply the following:

$$\text{if } s < \rho, \text{ then } \Sigma' \text{ can be extended to}$$
$$\text{a surface } \Sigma_s \text{ in } \operatorname{An}(x, s, 2\rho) \tag{47}$$
$$\text{if } s_1 < s_2 < \rho, \text{ then } \Sigma_{s_1} = \Sigma_{s_2} \text{ in } \operatorname{An}(x, s_2, 2\rho). \tag{48}$$

Thus $\Sigma = \bigcup_s \Sigma_s$ is a stable minimal surface Σ with $\overline{\Sigma} \setminus \Sigma \subset (\partial B_{2\rho}(x) \cup \{x\})$, i.e. Σ' can be continued up to x (which, in principle, could be a singular point).

We will next show that V coincides with Σ in $B_\rho(x) \setminus \{x\}$. Recall that $V = V'$ in $B_\rho(x)$. Fix

$$y \in (\operatorname{supp}(\|V\|)) \cap B_\rho(x) \setminus \{x\} \qquad \text{and set } s = d(y, x).$$

We first prove that if $TV(y, V)$ consists of a (multiple of a) plane π transversal to $\partial B_s(x)$, then y belongs to Σ. Consider the replacement V'' of V' in $\operatorname{An}(x, s, t)$ and split V'' into the three varifolds

$$\begin{aligned} V_1 &= V'' \llcorner B_s(x) &&= V \llcorner B_s(x)\,, \\ V_2 &= V'' \llcorner \operatorname{An}(x, s, 2\rho) &&= \Sigma \cap \operatorname{An}(x, s, 2\rho)\,, \\ V_3 &= V'' - V_1 - V_2\,. \end{aligned}$$

By Lemma 6.4, the set $TV(y, V'')$ consists of planes and since $V_1 = V \llcorner B_s(x)$, all these planes have to be multiples of π. Thus y is in the closure of $(\operatorname{supp}(\|V''\|)) \setminus \overline{B}_s(x)$, which implies $y \in \overline{\Sigma}_t \subset \Sigma$.

Let T be the set of points $y \in B_\rho(x)$ such that $TV(y, V)$ consists of a (multiple of) a plane transversal to $\partial B_{d(y,x)}(x)$. Lemma B.2 gives that T is dense in $\operatorname{supp}(\|V\|)$. Thus

$$(\operatorname{supp}(\|V\|)) \cap B_\rho(x) \setminus \{x\} \subset \Sigma\,.$$

Property (37) of replacements implies $\mathcal{H}^2(\Sigma \cap B_\rho(x)) = \|V\|(B_\rho(x))$. Hence $V = \Sigma$ on $B_\rho(x) \setminus \{x\}$.

Step 4: Regularity in x.

We will next show that Σ is smooth also in x, i.e. that x is a removable singularity for Σ. If $x \notin \operatorname{supp}(\|V\|)$, then we are done. So assume that $x \in \operatorname{supp}(\|V\|)$. In the following we will use that, by Lemma 6.4, every $C \in TV(x, V)$ is a multiple of a plane.

Map $B_t(x)$ into $\mathcal{B}_t(0)$ by the exponential map, use the notation of Step 1, and set $\Sigma_r = R_r^x(\Sigma)$. Every convergent subsequence $\{\Sigma_{r_n}\}$ converges to a plane in the sense of varifolds. The curvature estimates for stable minimal surfaces (see (8)) gives that this convergence is actually smooth in $\mathcal{B}_1 \setminus \mathcal{B}_{1/2}$. Thus, for r sufficiently small, there exist natural numbers $N(\rho)$ and $m_i(\rho)$ such that

$$\Sigma \cap \operatorname{An}(x, \rho/2, \rho) = \bigcup_{i=1}^{N(\rho)} m_i(\rho) \Sigma_\rho^i\,,$$

where each Σ_ρ^i is a Lipschitz graph over a (planar) annulus. Note also that the Lipschitz constants are uniformly bounded, independently of ρ.

By continuity, the numbers $N(r)$ and $m_i(r)$ do not depend on r. Moreover, if $s \in]\rho/2, \rho[$, then each Σ^i_ρ can be continued through $\mathrm{An}(s/2, \rho/2, x)$ by a Σ^j_s. Repeating this argument a countable number of times, we get N minimal punctured disks Σ^i with

$$\Sigma \cap B_\rho(x) \setminus \{x\} = \bigcup_{i=1}^{N} m_i \Sigma^i .$$

Note that x is a removable singularity for each Σ^i. Indeed, Σ^i is a stationary varifold in $B_\rho(x)$ and $TV(x, \Sigma^i)$ consists of planes with multiplicity one. This means that

$$\lim_{r\downarrow 0} \frac{\|V\|(B_r(x))}{\pi r^2} = 1 .$$

Hence we can apply Allard's regularity theorem (see section 8 of [**All**]) to conclude that Σ^i is a graph in a sufficiently small ball around x. Standard elliptic PDE theory gives that x is a removable singularity.

Finally, the maximum principle for minimal surfaces implies that N must be 1. This completes the proof. □

REMARK 6.5. *In the case at hand, there are other ways of proving that x is a removable singularity. For example one could use the existence of a conformal parameterization $u : \mathbf{C} \setminus \{0\} \to \Sigma^i \cap B_\rho(x) \setminus \{x\}$. The minimality of Σ^i gives that u is an harmonic map. Since the energy of u is finite, we can use theorem 3.6 in* [**SU**] *to conclude that u is smooth in* 0.

7. Construction of the replacements

In this section we conclude the proof of Theorem 1.6 by showing that the varifold V of Proposition 5.1 is a smooth minimal surface.

THEOREM 7.1. *Let $\{\Sigma^j\}$ be a sequence of compact surfaces in M which converge to a stationary varifold V. If there exists a function $r : M \to \mathbf{R}^+$ such that*

- *in every annulus of $\mathcal{AN}_{r(x)}(x)$ and for j large enough Σ^j is a $1/j$–a.m. smooth surface in* An,

then V is a smooth minimal surface.

To prove this theorem, we will show that V satisfies the requirements of Proposition 6.3. Thus we need to construct good replacements for V, using the strategy outlined in Section 3. In subsection 7.1 we fix some notation and recall a theorem of Meeks–Simon–Yau. In subsection 7.2 we show how to construct the varifolds V^* which are our candidates for replacements. Subsections 7.3 and 7.4 prove the regularity of the V^*'s constructed in subsection 7.2. Finally, in subsection 7.5 we prove the last details needed to show that V meets the requirements of Proposition 6.3.

7.1. The result of Meeks–Simon–Yau.

DEFINITION 7.2. *Let $\mathcal{I}$ be a class of isotopies of M and $\Sigma \subset M$ a smooth embedded surface. If $\{\varphi^k\} \subset \mathcal{I}$ and*

$$\lim_{k\to\infty} \mathcal{H}^2(\varphi^k(1,\Sigma)) = \inf_{\psi\in\mathcal{I}} \mathcal{H}^2(\psi(1,\Sigma)) ,$$

then we say that $\varphi^k(1,\Sigma)$ is a minimizing sequence for Problem $(\Sigma, \mathcal{I})$.

We will need the following result from[**MSY**]:

THEOREM 7.3. [Meeks–Simon–Yau [**MSY**]] *If* $\{\Sigma^k\}$ *is a minimizing sequence for Problem* $(\Sigma, \mathfrak{Is}(U))$ *which converges to a varifold* V*, then there exists a stable minimal surface* Γ *with* $\overline{\Gamma} \setminus \Gamma \subset \partial U$ *and* $V = \Gamma$ *in* U.

In [**MSY**] Theorem 7.3 is proved for $U = M$. However, the theory developed there is local and can be extended in a straightforward way to cover the case at hand.

7.2. Construction of replacements. Let V be as in Theorem 7.1 and fix an annulus $\mathrm{An} \in \mathcal{AN}_{r(x)}(x)$. Set

$$\begin{aligned} \mathfrak{Is}_j(\mathrm{An}) &= \\ &\left\{\psi \in \mathfrak{Is}(\mathrm{An}) \middle| \, \mathcal{H}^2(\psi_j(\tau, \Sigma^j)) \leq \mathcal{H}^2(\Sigma_j) + 1/(8j) \quad \forall \tau \in [0,1]\right\}, \\ m_j &= \inf_{\psi \in \mathfrak{Is}_j} \mathcal{H}^2(\psi(1, \Sigma^j)). \end{aligned}$$

Fix j. The following lemma implies that we can deform Σ^j into a sequence $\Sigma^{j,k}$ which is minimizing for Problem $(\Sigma^j, \mathfrak{Is}_j(\mathrm{An}))$ and converges, as $k \to \infty$, to a stable minimal surface in An.

LEMMA 7.4. *If a sequence* $\{\Sigma^{j,k}\}^k$ *is minimizing for Problem* $(\Sigma^j, \mathfrak{Is}_j(\mathrm{An}))$ *and converges to a varifold* V^j*, then* V^j *is a stable minimal surface in* An.

Lemma 7.4 will be proved in the next subsection. Here we use it for constructing a replacement for V in An.

PROPOSITION 7.5. *Let* V^j *be the varifold of Lemma 7.4. Any* V^* *which is the limit of a subsequence of* $\{V^j\}$ *is a replacement for* V.

PROOF. Without loss of generality, we can assume that the sequence $\{V^j\}$ converges to V. Note that every V^j coincides with V in $M \setminus \overline{\mathrm{An}}$; thus the same is true for V^*. Moreover, $\|V^j\|(M) \geq \mathcal{H}^2(\Sigma^j) - 1/j$ since Σ^j is a.m. This gives that $\|V^*\|(M) = \|V\|(M)$. By Lemma 7.4 and (8) we have that V^* is a stable minimal surface in An.

To complete the proof we need to show that V^* is stationary. Since $V = V^*$ in $M \setminus \overline{\mathrm{An}}$, then V^* is stationary in this open set. Hence it suffices to prove that V^* is stationary in an open annulus $\mathrm{An}' \in \mathcal{AN}_r$ containing $\overline{\mathrm{An}}$. Choose such an An$'$ and suppose that V^* is not stationary in An$'$; we will show that this contradict that $\{\Sigma^j\}$ is a.m. in An$'$. Namely, suppose that for some vector field χ supported in An$'$ we have $\delta V^*(\chi) \leq -C < 0$. Let ψ be the isotopy given by that $\frac{\partial \psi(t,x)}{\partial t} = \chi(\psi(t,x))$ and set

$$\begin{aligned} V^*(t) &= \psi(t)_\sharp V^*, \\ V^j(t) &= \psi(t)_\sharp V^j, \\ \Sigma^{j,k}(t) &= \psi(t, \Sigma^{j,k}). \end{aligned}$$

For ε sufficiently small, we have that

$$[\delta V^*(t)](\chi) \leq -\frac{C}{2} \qquad \text{for all } t < \varepsilon.$$

Since $V^j(t) \to V^*(t)$, there exists J such that

$$[\delta V^j(t)](\chi) \leq -\frac{C}{4} \qquad \text{for every } j > J \text{ and every } t < \varepsilon.$$

Moreover, since $\Sigma^{j,k}(t) \to V^j(t)$, for each $j > J$ there exists $K(j)$ with

$$[\delta\Sigma^{j,k}(t)](\chi) \leq -\frac{C}{8} \qquad \text{for all } t < \varepsilon \text{ and all } k > K(j). \tag{49}$$

Integrating both sides of (49) we get

$$\mathcal{H}^2(\psi(t,\Sigma^{j,k})) - \mathcal{H}^2(\Sigma^{j,k}) \leq -\frac{tC}{8} \tag{50}$$

Choose j and k sufficiently large so that $\varepsilon C/8 > 1/j$ and (50) holds. Each $\Sigma^{j,k}$ is isotopic to Σ^j via an isotopy $\varphi^{j,k} \in \mathfrak{Js}_j(\text{An})$. By gluing $\varphi^{j,k}$ and ψ smoothly together, we find a smooth isotopy $\Phi : [0, 1+\varepsilon] \times M \to M$ supported on An$'$. In view of (50), Φ satisfies

$$\begin{aligned} \mathcal{H}^2(\Phi(t,\Sigma^j)) &\leq \mathcal{H}^2(\Sigma^j) + 1/(8j) \qquad \forall t \in [0, 1+\varepsilon], \\ \mathcal{H}^2(\Phi(1+\varepsilon,\Sigma^j) &< \mathcal{H}^2(\Sigma^j) - 1/j, \end{aligned}$$

which give the desired contradiction and prove the proposition. □

7.3. Proof of Lemma 7.4. Without loss of generality we may assume that $j = 1$ and use V', Σ and Σ^k in place of V^j, $\Sigma^{j,k}$ and Σ^j. Clearly V' is stationary and stable in An, by its minimizing property. Thus we need only prove that V' is regular. The proof of this uses Theorem 7.3 and the following:

LEMMA 7.6. *Let $x \in$ An and assume that $\{\Sigma^k\}$ is minimizing for Problem $(\Sigma, \mathfrak{Js}_1(\text{An}))$. There exists $\varepsilon > 0$ such that, for k sufficiently large, the following holds:*

(Cl) *For any $\varphi \in \mathfrak{Js}(B_\varepsilon(x))$ with $\mathcal{H}^2(\varphi(1,\Sigma^k)) \leq \mathcal{H}^2(\Sigma^k)$, there exists an isotopy $\Phi \in \mathfrak{Js}(B_\varepsilon)$ such that*

$$\Phi(1,\cdot) = \varphi(1,\cdot), \tag{51}$$

$$\mathcal{H}^2(\Phi(t,\Sigma^k)) \leq \mathcal{H}^2(\Sigma^k) + 1/8. \tag{52}$$

Moreover, $\varepsilon > 0$ can be chosen so that (Cl) holds for any *sequence $\{\tilde{\Sigma}^k\}$ which is minimizing for Problem $(\Sigma, \mathfrak{Js}_1(\text{An}))$ and with $\Sigma^j = \tilde{\Sigma}^j$ on $M \setminus \overline{B}_\varepsilon(x)$.*

Lemma 7.6 will be proved in the next subsection. We now return to the proof of Lemma 7.4. We will use Proposition 6.3. Hence, once again we need to construct replacements for a varifold, which this time is V'. We divide the proof into two steps. The first one is the basic construction of replacements for V'. The second shows that the replacements satisfy (b) and (c) in Definition 6.2

Step 1 Fix $x \in$ An and $\varepsilon > 0$ such that Lemma 7.6 holds. Fix any annulus

$$\text{An}^* = \text{An}(x,\tau,t) \subset B_\varepsilon(x) \subset \text{An}$$

and consider a minimizing sequence $\{\overline{\Sigma}^{k,l}\}^l$ for Problem $(\Sigma^k, \mathfrak{Js}(\text{An}^*))$. Lemma 7.6 implies that, for k sufficiently large, $\overline{\Sigma}^{k,l}$ can be constructed from Σ via an isotopy of $\mathfrak{Js}_1(\text{An})$. Thus if we let W^k be the varifold limit of $\overline{\Sigma}^{k,l}$ and W the limit of W^k, then we have that $\|W\|(M) = \|V'\|(M)$.

Let $\{\overline{\Sigma}^{k,l(k)}\}$ be a subsequence which converges to W. By the discussion above the subsequence is a minimizing sequence for Problem $(\Sigma, \mathfrak{Js}_1(\text{An}))$. Hence W is stationary in An. Moreover, by Theorem 7.3, every W^k is a stable minimal surface in the annulus An*: the curvature estimates (see (8)) give that W is a stable minimal surface in An*. Hence W is a replacement for V'.

Step 2 Summarizing we have proven in Step 1 that for any $y \in \mathrm{An}$ there exists $r(y) > 0$ such that in the class of annuli $\mathcal{AN}_{r(y)}(y)$ we can construct replacements. In order to complete the proof we have to check that the replacements so constructed satisfy all the technical requirements of Proposition 6.3. Thus define W as in Step 1. Since $\{\overline{\Sigma}^{k,l(k)}\}$ is a minimizing sequence for Problem $(\Sigma, \mathfrak{Is}_1(\mathrm{An}))$, in all the arguments of Step 1 we can use W in place of V'. Thus for every $y \in \mathrm{An}$, W has the replacement property for a class of annuli centered at y. This shows the second part of (b) in Definition 6.2

We still have to settle the first part of (b) in Definition 6.2, i.e. that if the W constructed in Step 1 replaces V' in an annulus of $\mathcal{AN}_{r(x)}(x)$, then W has the replacement property on the *whole* collection of annuli $\mathcal{AN}_{r(x)}(x)$. Note that our $r(x) = \varepsilon$, where ε is given by Lemma 7.6. Every $\overline{\Sigma}^{k,l(k)}$ coincides with Σ^k in $M \setminus B_\varepsilon(x)$ and $\{\overline{\Sigma}^{k,l(k)}\}$ is minimizing for Problem $(\Sigma, \mathfrak{Is}_1(\mathrm{An}))$. Thus the last line of Lemma 7.6 applies and again we can argue as in Step 1 with W in place of V'. We conclude that also W has a replacement in every annulus of $\mathcal{AN}_\varepsilon(x)$.

Condition (c) in Definition 6.2 follows from similar arguments. Summarizing, V' and all its replacements just constructed satisfy the requirements of Proposition 6.3. Hence V' is a smooth surface in An. □

7.4. Proof of Lemma 7.6. Step 1: Small area slices. Let x, ε, An, Σ^k and φ be as in the statement of the lemma. We know that Σ^k converges to a varifold V' which is stationary in An and has mass m_0. Thus the monotonicity formula gives that

$$\|V'\|(\mathrm{An}(x,\varepsilon,2\varepsilon)) < 4m_0\varepsilon^2 .$$

Therefore, if k is sufficiently large we get that $\mathcal{H}^2(\Sigma^k \cap B_{2\varepsilon}(\delta)) < 5m_0\varepsilon^2$. Applying the coarea formula we have that, for every such k, there exists an interval $I_k \subset]\varepsilon, 2\varepsilon[$ such that

$$\mathcal{L}^1(I_k) > 0 \quad \text{and} \quad \mathcal{H}^1(\Sigma^k \cap \partial B_\tau(x)) \le 10m_0\varepsilon \quad \text{for all } \tau \in I_k . \tag{53}$$

Thus, applying Sard's Theorem to the function $\mathrm{d}\,(\cdot, x)$ on Σ^k, we can find $\tau_k \in]\varepsilon, 2\varepsilon[$ such that

$$\mathcal{H}^1(\Sigma^k \cap \partial B_{\tau_k}(x)) < 10m_0\varepsilon \quad \text{and} \quad \Sigma^k \text{ is transversal to } \partial B_{\tau_k}(x) . \tag{54}$$

Moreover, the smoothness of Σ^k implies that we can choose a small interval $]\sigma_k, s_k[$ with $\varepsilon < \sigma_k$ and so that (54) holds for every $\tau_k \in]\sigma_k, s_k[$.

Step 2: Radial deformations.

For $\eta > 0$ we denote by O_η the usual radial deformation of Euclidean 3–space given by $O_\eta(x) = \eta x$. If both r and $r\eta$ are less than $\mathrm{Inj}\,(M)/2$, then we define the diffeomorphism

$$I_\eta : B_r(x) \to B_{\eta r}(x) \qquad \text{by } \exp \circ O_\eta \circ \exp^{-1} .$$

By the smoothness of M there exists $\mu > 0$ such that, for any surface $\Gamma \subset M$,

$$\mathcal{H}^2(I_\eta(\Gamma) \cap B_{\eta r}(x)) \le \mu \eta^2 \mathcal{H}^2(\Gamma \cap B_r(x)) , \tag{55}$$

$$\mathcal{H}^1(I_\eta(\Gamma) \cap \partial B_{\eta r}(x)) \le \mu \eta \mathcal{H}^1(\Gamma \cap \partial B_r(x)) \quad \text{if } \Gamma \text{ is transversal to } \partial B_r(x). \tag{56}$$

$$\mathcal{H}^2(\Gamma \cap B_r(x)) \le \mu \int_0^r \mathcal{H}^1(\Gamma \cap \partial B_\rho(x))\, d\rho. \tag{57}$$

Fix k and choose ε, σ_k and s_k as in the previous step. In the current step we use I_η to construct a smooth $\psi : [0,1] \times M \to M$ such that

- For every $\delta > 0$, $\psi|_{[0,1-\delta]\times M}$ is a smooth isotopy supported in $\mathfrak{Is}(B_{s_k}(x))$;
- $\psi|_{\{1\}\times M}$ "squeezes" the ball $B_{\sigma_k}(x)$ to the point $\{x\}$ and "stretches" the annulus $\mathrm{An}(x, \sigma_k, s_k)$ to the ball $B_{s_k}(x)$;
- For some constant C depending on μ we have (see Fig. 11)

$$\mathcal{H}^2(\psi(t, \Sigma^k)) \le \mathcal{H}^2(\Sigma^k) + C\varepsilon^2 . \tag{58}$$

We construct ψ explicitly. We first choose a nondecreasing smooth $f : [0,1] \times [0,1] \to [0,1]$ such that

$$f(t,r) = \begin{cases} (1-t) & \text{if } r \in [0, \sigma_k], \\ 1 & \text{if } r \in [s_k, 1], \end{cases}$$

and then set

$$\psi(t,y) = \begin{cases} y & \text{if } \mathrm{d}\,(y,x) \ge s_k\,, \\ I_{f(t,\mathrm{d}\,(x,y))}(y) & \text{otherwise.} \end{cases}$$

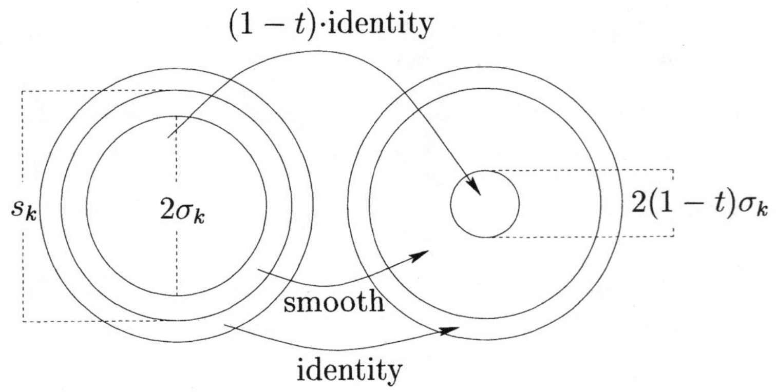

FIGURE 11. The map $\psi(t, \cdot) : M \to M$

We will only prove (58), since the other properties are easy to check. First of all, since $\psi(t,y) = y$ on $M \setminus B_{\sigma_k}(x)$, we have

$$\mathcal{H}^2(\psi(t, \Sigma^k) \cap (M \setminus B_{2\varepsilon}(x)) = \mathcal{H}^2(\Sigma^k \cap (M \setminus B_{2\varepsilon}(x)). \tag{59}$$

By (55)

$$\mathcal{H}^2(\psi(t, \Sigma^k) \cap B_{(1-t)\sigma_k}(x)) \le \mu(1-t)^2 \mathcal{H}^2(\Sigma^k \cap B_{\sigma_k}) \le 10\mu m_0 \varepsilon\,. \tag{60}$$

To estimate the remaining portion of $\psi(t, \Sigma^k)$ we use (57) and get

$$\mathcal{H}^2(\psi(t, \Sigma^k) \cap \mathrm{An}(x, (1-t)\sigma_k, s_k)) \le \mu \int_{(1-t)\sigma_k}^{s_k} \mathcal{H}^1(\psi(t, \Sigma^k) \cap \partial B_\rho(x)) d\rho. \tag{61}$$

Note that for $\rho \in ((1-t)\sigma_k, s_k)$ there exists $\tau \in (\sigma_k, s_k)$ and $\eta \le 1$ such that

$$\psi(t, \Sigma^k) \cap \partial B_\rho = I_\eta(\Sigma^k \cap \partial B_\tau(x))\,.$$

Thus, by (56) and (54) we have

$$\mathcal{H}^1(\psi(t, \Sigma^k) \cap \partial B_\rho) \le \mu \mathcal{H}^1(\Sigma^k \cap \partial B_\tau(x)) \le 10\mu m_0 \varepsilon\,. \tag{62}$$

Inserting (62) in (61) we find that

$$\mathcal{H}^2(\psi(t,\Sigma^k)\cap \mathrm{An}(x,(1-t)\sigma_k,s_k)) \;\leq\; 20\mu^2 m_0\varepsilon^2. \tag{63}$$

Equations (59), (60), and (63) yield

$$\mathcal{H}^2(\psi(t,\Sigma^k)) \;\leq\; \mathcal{H}^2(\Sigma^k)+C\varepsilon^2\,. \tag{64}$$

Step 3: The conclusion.

Choose ε such that $\mu C\varepsilon^2 < 1/32$ and K such that :

- We can construct the ψ of the previous step with (64) valid for any $k > K$;
- $\mathcal{H}^2(\Sigma^k) \leq \mathcal{H}^2(\Sigma) + 1/32$ for any $k > K$.

We want to prove that ε satisfies the requirement of the lemma. Indeed choose any smooth isotopy φ which is supported on $B_\varepsilon(x)$ and such that $\mathcal{H}^2(\varphi(1,\Sigma^k)) \leq \mathcal{H}^2(\Sigma^k)$. Set

$$K \;=\; \sup_t \mathcal{H}^2(\varphi(t,\Sigma^k)\cap B_{\sigma_k}(x))$$

and choose t such that $\mu(1-t)^2K \leq 1/32$.

Define isotopies ψ^- and $\tilde{\varphi}$ by

$$\psi^-(s,\cdot) \;=\; \psi(1-s,\cdot) \qquad\text{and}\qquad \tilde{\varphi} \;=\; I_{(1-t)}\circ\varphi\circ I_{(1-t)^{-1}}\,,$$

and note that ψ^- is the "backward" of ψ and hence instead of "squeezing", it magnifies, whereas $\tilde{\varphi}$ is the "$(1-t)$–shrunk" version of φ. We now define a (piecewise) smooth isotopy $\Psi : I_1\cup I_2\cup I_3\times M \to M$ by gluing ψ, $\tilde{\phi}$, and ψ^- together. Namely, set

- $\Psi(s,\cdot) = \psi(s,\cdot)$ for $s \in I_1 = [0,1-t]$;
- $\Psi(s,\cdot) = \tilde{\varphi}(s-(1-t))$ for $s \in I_2 = [1-t,2-t]$;
- $\Psi(s,\cdot) = \psi^-((s-(2-t))+t,\tilde{\varphi}(1,\cdot))$ for $s\in I_3 = [2-t,3-2t]$.

Loosely speaking, the isotopy Ψ performs the following operations (see Fig. 12):

First it "shrinks" the ball $B_{s_k}(x)$ to the ball $B_{(1-t)s_k}(x)$;
Then it applies the "$(1-t)$–shrunk" version of φ to the ball $B_{(1-t)s_k}(x)$;
Finally it magnifies back $B_{(1-t)s_k}(x)$ to the ball $B_{s_k}(x)$.

By changing parameter we can assume that Ψ is smooth and that $I_i = [(i-1)/3, i/3]$. Note that (58) implies that

$$\mathcal{H}^2(\Psi(s,\Sigma^k)) \;\leq\; \mathcal{H}^2(\Sigma^k)+1/32 \qquad \text{for } s\in[0,1/3]. \tag{65}$$

Since $\mu(1-t)^2K \leq 1/32$ we have

$$\mathcal{H}^2(\Psi(s,\Sigma^k)) \;\leq\; \mathcal{H}^2(\Sigma^k)+\mu(1-t)^2K \;\leq\; \frac{1}{32} \qquad \text{for } s\in[1/3,2/3]. \tag{66}$$

Again by (58), and since $\mathcal{H}^2(\varphi(1,\Sigma^k)) \leq \mathcal{H}^2(\Sigma^k)$, we have

$$\begin{aligned}\mathcal{H}^2(\Psi(s,\Sigma^k)) \;&\leq\; \mathcal{H}^2(\Sigma^k)+1/32+\mu\mathcal{H}^2(\varphi(1,\Sigma^k)\cap B_{\sigma_k}(x))\\ &\leq\; \mathcal{H}^2(\Sigma^k)+1/32+\mu C\varepsilon^2\\ &\leq\; \mathcal{H}^2(\Sigma^k)+1/16 \qquad \text{for } s\in[2/3,1].\end{aligned} \tag{67}$$

Thus for every s

$$\mathcal{H}^2(\Psi(s,\Sigma^k)) \;\leq\; \mathcal{H}^2(\Sigma^k)+1/16 \;\leq\; \mathcal{H}^2(\Sigma)+3/32. \tag{68}$$

Note also that $\Psi(1,\Sigma^k) \;=\; \varphi(1,\Sigma^k)$. Finally, recall that Σ^k was obtained via an isotopy φ^k such that

$$\mathcal{H}^2(\varphi^k(t,\Sigma)) \;\leq\; \mathcal{H}^2(\Sigma)+1/8.$$

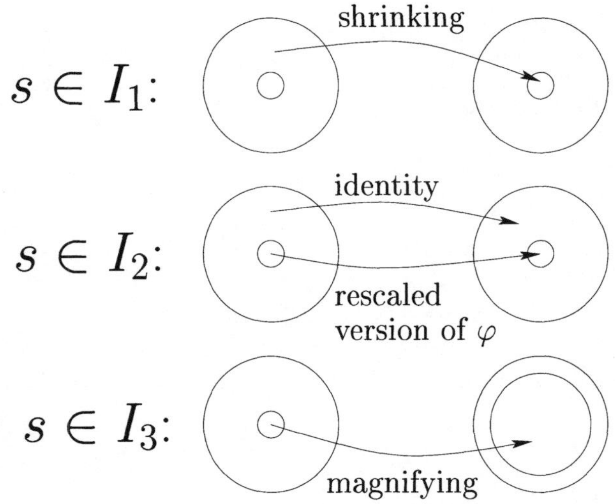

FIGURE 12. The isotopy Ψ.

Gluing together φ_k and Ψ we easily obtain a Φ which satisfies both (51) and (52). Clearly the ε found in this proof satisfies the last requirement of the statement of the lemma. □

7.5. Proof of Theorem 7.1. We will apply Proposition 6.3. From Proposition 7.5 we know that in every annulus $\text{An} \in \mathcal{AN}_{r(x)}(x)$ there is a replacement V^* for V. We still need to show that V satisfies (a), (b), and (c) in Definition 6.2. Consider the family of surfaces $\Sigma^{j,k}$ of Lemma 7.4. By a diagonal argument we can extract a subsequence $\Sigma^{j,k(j)}$ converging to V^*. Note the following consequence of the way we constructed $\{\Sigma^{j,k(j)}\}^j$. If U is open and

- either $U \cup \text{An}$ is contained in some annulus $\mathcal{AN}_{r(x)}(x)$
- or $U \cap \text{An} = \emptyset$ and U is contained in some annulus of $\mathcal{AN}_{r(y)}(y)$ with $y \neq x$,

then $\Sigma^{j,k(j)}$ is a.m. in U. Thus $\{\Sigma^{j,k(j)}\}$ is still a.m. in

- every annulus of $\mathcal{AN}_{r(x)}(x)$;
- every annulus of $\mathcal{AN}_{\rho(y)}(y)$ for $y \neq x$, provided $\rho(y)$ is sufficiently small.

This shows that (b) in Definition 6.2 holds for V. Similarly, we can show that also condition (c) of that Definition holds. Hence Proposition 6.3 applies and we conclude that V is a smooth surface. □

Appendix A. Proof of Proposition 1.4

Let M^3 be a closed Riemannian 3–manifold with a Morse function $f : M \to [0,1]$. Denote by Σ_t the level set $f^{-1}(\{t\})$ and let Λ be the saturated set of families

$$\Big\{\{\Gamma_t\}\Big|\ \Gamma_t = \psi(t,\Sigma_t) \text{ for some } \psi \in C^\infty([0,1]\times M, M)$$
$$\text{with } \psi_t \in \text{Diff}_0 \text{ for every } t\Big\}$$

To prove Proposition 1.4 we need to show that $m_0(\Lambda) > 0$. To do that set $U_t = f^{-1}([0,t[)$ and $V_t = \psi(t, U_t)$. Clearly $\Gamma_t = \partial V_t$ and if we let Vol denote the volume on M, then $\text{Vol}(U_t)$ is a continuous function of t. Since V_0 is a finite set of points and $V_1 = M$, then there exists an s such that $\mathrm{Vol}(V_s) = \mathrm{Vol}(M)/2$. By the isoperimetric inequality there exists a constant $c(M)$ such that

$$\frac{\mathrm{Vol}(M)}{2} = \mathrm{Vol}(V_s) \leq c(M)\mathcal{H}^2(\Gamma_s)^{3/2}\,.$$

Hence,

$$\mathcal{F}(\{\Gamma_t\}) = \max_{t\in[0,1]} \mathcal{H}^2(\Gamma_t) \geq \left(\frac{\mathrm{Vol}(M)}{2c(M)}\right)^{\frac{2}{3}} > 0\,, \tag{69}$$

and the proposition follows.

Appendix B. Two lemmas about varifolds

LEMMA B.1. *Let U be an open subset of a 3–manifold M and W a stationary 2–varifold in $\mathcal{V}(U)$. If $K \subset\subset U$ is a smooth strictly convex set and $x \in (\mathrm{supp}\,(\|W\|)) \cap \partial K$, then*

$$(B_r(x)\setminus \overline{K}) \cap \mathrm{supp}\,(\|W\|) \neq \emptyset \qquad \textit{for every } r>0.$$

PROOF. For simplicity assume that $M = \mathbf{R}^3$. The proof can be easily adapted to the general case. Let us argue by contradiction; so assume that there are $x \in \mathrm{supp}\,(\|W\|)$ and $B_r(x)$ such that $(B_r(x)\setminus\overline{K}) \cap \mathrm{supp}\,(\|W\|) = \emptyset$. Given a vector field $\chi \in C^\infty_c(U,\mathbf{R}^3)$ and a 2–plane π we set

$$\mathrm{Tr}\,(D\chi(x),\pi) = D_{v_1}\chi(x)\cdot v_1 + D_{v_2}\chi(x)\cdot v_2$$

where $\{v_1, v_2\}$ is an orthonormal base for π. Recall that the first variation of W is given by

$$\delta W(\chi) = \int_{G(U)} \mathrm{Tr}\,(D\chi(x),\pi)\, dW(x,\pi)\,.$$

Take an increasing function $\eta \in C^\infty([0,1])$ which vanishes on $[3/4,1]$ and is identically 1 on $[0,1/4]$. Denote by φ the function given by $\varphi(x) = \eta(|y-x|/r)$ for $y \in B_r(x)$. Take the interior unit normal ν to ∂K in x, and let z_t be the point $x+t\nu$. If we define vector fields ψ_t and χ_t by

$$\psi_t(y) = -\frac{y-z_t}{|y-z_t|} \qquad \text{and} \qquad \chi_t = \varphi\psi_t\,,$$

then χ_t is supported in $B_r(x)$ and $D\chi_t = \varphi D\psi_t + \nabla\varphi\otimes\psi_t$. Moreover, by the strict convexity of the subset K,

$$\nabla\varphi(y)\cdot\nu > 0 \qquad \text{if } y\in\overline{K}\cap B_r(x) \text{ and } \nabla\varphi(y)\neq 0.$$

Note that ψ_t converges to ν uniformly in $B_r(x)$, as $t\uparrow\infty$. Thus, $\psi_T(y)\cdot\nabla\varphi(y)\geq 0$ for every $y\in\overline{K}\cap B_r(x)$, provided T is sufficiently large. This yields that

$$\mathrm{Tr}\,(\nabla\varphi(y)\otimes\psi_T(y),\pi)\geq 0 \qquad \text{for all } (y,\pi)\in G(B_r(x)\cap\overline{K})\,. \tag{70}$$

Note that $\mathrm{Tr}\,(D\psi_t(y),\pi) > 0$ for all $(y,\pi)\in G(B_r(x))$ and all $t>0$. Thus

$$\begin{aligned}
\delta W(\chi_T) &= \int_{G(B_r(x)\cap\overline{K})} \mathrm{Tr}\,(D\chi_T(y),\pi)\, dW(y,\pi)\\
&\stackrel{(70)}{\geq} \int_{G(B_r(x)\cap\overline{K})} \mathrm{Tr}\,(\varphi(y)D\psi_T(y),\pi)\, dW(y,\pi)\\
&\geq \int_{G(B_{r/4}(x)\cap\overline{K})} \mathrm{Tr}\,(D\psi_T(y),\pi)\, dW(y,\pi) > 0.
\end{aligned}$$

This contradicts that W is stationary and completes the proof. □

LEMMA B.2. *Let $x \in M$ and V be a stationary integer rectifiable varifold in M. Assume T is the subset of the support of $\|V\|$ given by*

$$T = \{T(y,V) \text{ consists of a plane transversal to } \partial B_{d(x,y)}(x)\}.$$

If $\rho < Inj(M)$, then T is dense in $(\mathrm{supp}\,(\|V\|)) \cap B_\rho(x)$.

PROOF. Since V is integer rectifiable, then V is supported on a rectifiable 2–dimensional set R and there exists a Borel function $h : R \to \mathbb{N}$ such that $V = hR$. Assume the lemma is false; then there exists $y \in B_\rho(x) \cap \mathrm{supp}\,(\|V\|)$ and $t > 0$ such that

- the tangent plane to R in z is tangent to $\partial B_{d(z,x)}(x)$, for any $z \in B_t(y)$.

We choose t so that $B_t(y) \subset B_\rho(x)$. Take polar coordinates (r, θ, φ) in $B_\rho(x)$ and let f be a smooth nonnegative function in $C^\infty_c(B_t(y))$ with $f = 1$ on $B_{t/2}(y)$. Denote by χ the vector field $\chi(\theta, \varphi, r) = f(\theta, \varphi, r)\frac{\partial}{\partial r}$. We use the notation of the proof of Lemma B.1. For every $z \in R \cap B_t(x)$, the plane π tangent to R in z is also tangent to the sphere $\partial B_{d(z,x)}(x)$. Hence, an easy computation yields that $\mathrm{Tr}\,(\chi, \pi)(z) = 2\psi(z)/d(z,x)$. This gives

$$[\delta V](\chi) = \int_{R\cap B_t(y)} \frac{2h(z)\psi(z)}{\mathrm{d}\,(z,x)} d\mathcal{H}^2(z) \;>\; C\|V\|(B_{t/2}(y)),$$

for some positive constant C. Since $y \in \mathrm{supp}\,(\|V\|)$, we have

$$\|V\|(B_{t/2}(y)) \;>\; 0.$$

This contradicts that V is stationary. □

Appendix C. An example

Let $V_1 \in \mathcal{V}^1(\mathcal{D}_2)$ be the 1–dimensional varifold given by three straight lines ℓ_1, ℓ_2, ℓ_3 which meet in the origin at angles of 60 degrees and let V_2 be the 1–dimensional varifold given by (see Fig. 13):

- $V_2 = V_1$ in $\mathcal{D}_2 \setminus \mathcal{D}_1$;
- In $\mathcal{D}_1$, V_2 is given by the regular hexagon Hex with sides of Length 1 and vertices lying on the l_i's.

Note that both V_1 and V_2 are stationary in $\mathcal{D}_2$, they have the same mass, and they coincide in $\mathcal{D}_2 \setminus \mathcal{D}_1$.

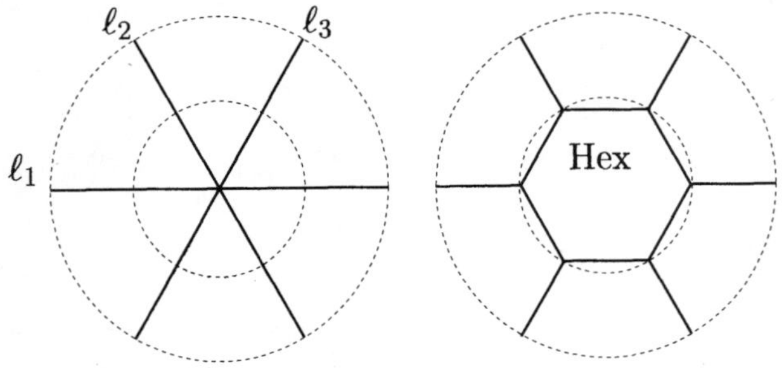

FIGURE 13. The varifolds V_1 and V_2.

References

[All] W.K. Allard, On the first variation of a varifold, *Ann. of Math.(2)* 95 (1972) no. 3, 417–491.

[Alm] F.J. Almgren, The theory of varifolds, Mimeographed notes, Princeton, 1965.

[CM1] T.H. Colding and W.P. Minicozzi II, Minimal surfaces, Courant Lecture Notes in Mathematics, 4. New York University, Courant Institute of Mathematical Sciences, New York, 1999.

[CM2] T.H. Colding and W.P. Minicozzi II, Estimates for parametric elliptic integrands, *International Mathematics Research Notices*, no. 6 (2002) 291-297.

[Cr] C.B. Croke, Area and length of the shortest closed geodesic, *Jour. of Diff. Geom.*, vol. 27 (1988) 1–21.

[J] J. Jost, Unstable solutions of two–dimensional geometric variational problems *Proceedings of Symposia in Pure Mathematics*, v. 54, Part I (1993).

[MSY] W. Meeks III, L. Simon, and S.T. Yau, Embedded minimal surfaces, exotic spheres, and manifolds with positive ricci curvature *Ann. of Math.(2)* 116 (1982) no. 3, 621–659.

[Mi] J. Milnor, Morse Theory, *Ann. of Math. Studies*, v. 51, Princeton University Press (1969).

[P] J.T. Pitts, Existence and regularity of minimal surfaces on Riemannian manifolds. Princeton University Press, Princeton, N.J.; University of Tokyo Press, Tokyo (1981).

[PR1] J.T. Pitts and J.H. Rubinstein, Existence of minimal surfaces of bounded topological type in three–manifolds, Miniconference on Geometry and Partial Differential Equations, *Proceedings of the Centre for Mathematical Analysis,* Australian National University (1985).

[PR2] J.T. Pitts and J.H. Rubinstein, Application of minmax to minimal surfaces and the topology of 3–manifolds, Miniconference on Geometry and Partial Differential Equations, 2 *Proceedings of the Centre for Mathematical Analysis,* Australian National University (1986).

[SU] J. Sacks and K. Uhlenbeck, The existence of minimal immersions of 2–spheres, *Ann. of Math.(2)* 113 (1981), no. 1 1–24.

[SW] B. Solomon and B. White, A strong maximum principle for varifolds that are stationary with respect to even parametric functionals, *Indiana Univ. Math. J.* 38 (1989) no. 3, 683–691.

[Sc] R. Schoen, Estimates for stable minimal surfaces in three-dimensional manifolds, Seminar on Minimal submanifolds, *Ann. of Math. Studies*, v. 103, Princeton University Press (1983).

[Si] L. Simon, Lectures on geometric measure theory, *Proceedings of the Centre for Mathematical Analysis,* Australian National University (1983).

[Sm] F. Smith, On the existence of embedded minimal 2–spheres in the 3–sphere, endowed with an arbitrary Riemannian metric, supervisor L. Simon, University of Melbourne (1982).

Courant Institute of Mathematical Sciences, 251 Mercer Street, New York, NY 10012

Max-Planck-Institute for Mathematics in the Sciences, Inselstr. 22 - 26, 04103 Leipzig / Germany

E-mail address: colding@cims.nyu.edu and delellis@mis.mpg.de

Universal volume bounds in Riemannian manifolds

Christopher B. Croke and Mikhail Katz

ABSTRACT. In this survey article we will consider universal lower bounds on the volume of a Riemannian manifold, given in terms of the volume of lower dimensional objects (primarily the lengths of geodesics). By 'universal' we mean without curvature assumptions. The restriction to results with no (or only minimal) curvature assumptions, although somewhat arbitrary, allows the survey to be reasonably short. Although, even in this limited case the authors have left out many interesting results.

CONTENTS

1. Introduction

In the present article, we will consider n-dimensional Riemannian manifolds $(X^n, \mathbf{g})$, for the most part compact. We plan to survey the results and open questions concerning volume estimates that do not involve curvature (or involve it only very weakly). Our emphasis will be on open questions, and we intend the article to be accessible to graduate students who are interested in exploring these questions. In choosing what to include here, the authors have concentrated on results that have influenced their own work and on recent developments. In particular, we are only able to mention some of the highlights of M. Gromov's seminal paper [**Gr83**]. His survey devoted specifically to systoles appeared in [**Gr96**]. The interested reader is encouraged to explore those papers further, as well as his recent book [**Gr99**].

Supported by NSF grant DMS 02-02536.
Supported by ISF grant no. 620/00-10.0.

The estimates we are concerned with are lower bounds on the Riemannian volume (*i.e.* the n-dimensional volume), in terms of volume-minimizing lower dimensional objects. For example, one lower dimensional volume we will consider will be the infimum of volumes of representatives of a fixed homology or homotopy class. This line of research was, apparently, originally stimulated by a remark of René Thom, in a conversation with Marcel Berger in the library of Strasbourg University in the 1960's, not long after the publication of [**Ac60, Bl61**]. Having been told of the latter results (discussed in sections 3 and 1), Professor Thom reportedly exclaimed: "Mais c'est fondamental! [These results are of fundamental importance!]." Ultimately this lead to the so called isosystolic inequalities, *cf.* section 4. An intriguing historical discussion appears in section "Systolic reminiscences" [**Gr99**, pp. 271-272].

We will devote section 2 to the study of lower bounds on the area of surfaces (two dimensional manifolds) in terms of the length of closed geodesics. This is where the subject began, with the theorems of Loewner and Pu, and where the most is known.

In section 3 we discuss some of the results and questions that come from [**Gr83**]. In particular, inequality (3.1), which provides a lower bound for the total volume in terms of an invariant called the Filling Radius, is one of the main tools used to obtain isosystolic inequalities in higher dimensions. We also discuss Gromov's notion of Filling Volume, as well as some of the open questions and recent results relating the volume of a compact Riemannian manifold with boundary, to the distances among its boundary points [**B-C-G96, Cr01, C-D-S00, C-D, Iv02**].

In section 4, we see a collection of results that show that the inequalities envisioned in M. Berger's original question are often violated. Unless we are dealing with one-dimensional objects, such isosystolic inequalities are systematically violated by suitable families of metrics. This line of work was stimulated by M. Gromov's pioneering example (4.5), *cf.* [**Gr96, Ka95, Pitte97, BabK98, BKS98, KS99, KS01, Bab02, Ka02**].

Section 5 discusses how the systolic inequalities that we saw failed in Section 4, can in fact hold if we pass to stable versions. We survey the known results of this nature [**Gr83, He86, BanK03, BanK2**], and also examine the related conformal systolic invariants and their asymptotic behavior, studied in [**BuSa94, Ka3**].

In section 6, we discuss bounds on the volume in terms of the injectivity radius of a compact Riemannian manifold. We discuss the (sharp) isoembolic inequality (6.3) of M. Berger, as well as the local version, *i.e.* volumes of balls, by Berger and the first author . We survey some of the extensions of these results, with an emphasis on the open questions and conjectures.

2. Area and length of closed geodesics in 2 dimensions

In this section, we will restrict ourselves to 2-dimensional Riemannian manifolds $(X^2, \mathbf{g})$ and discuss lower bounds on the total area, $area(\mathbf{g})$, in term of the least length of a closed geodesic $L(\mathbf{g})$. In any dimension, the shortest loop in every nontrivial homotopy class is a closed geodesic. We will denote by $sys\pi_1(\mathbf{g})$, the least length (often referred to as the "Systole" of $\mathbf{g}$) of a noncontractible loop γ in a compact, non-simply-connected Riemannian manifold $(X, \mathbf{g})$:

$$sys\pi_1(\mathbf{g}) = \min_{[\gamma]\neq 0 \in \pi_1(X)} length(\gamma). \tag{2.1}$$

	$\sup_{\mathbf{g}} \frac{sys\pi_1(\mathbf{g})^2}{area(\mathbf{g})}$	numerical value	where to find it
$X = \mathbb{R}P^2$	$= \frac{\pi}{2}$ **[Pu52, Ta92]**	≈ 1.5707	(2.5)
infinite $\pi_1(X)$	$< \frac{4}{3}$ **[Gr83]**	< 1.3333	(2.10)
$X = \mathbb{T}^2$	$= \frac{2}{\sqrt{3}}$ (Loewner)	≈ 1.1547	(2.3)
$X = \mathbb{R}P^2 \# \mathbb{R}P^2$	$= \frac{\pi}{2^{3/2}}$ **[Bav86, Sak88]**	≈ 1.1107	(2.14)
X of genus 2	$> \left(27 - 18\sqrt{2}\right)^{-\frac{1}{2}}$	> 0.8047	(2.8)
X of genus 3	$\geq \frac{8}{7\sqrt{3}}$ **[Ca96]**	> 0.6598	section 5
X of genus 4	$> \frac{9\sqrt{7}}{40}$ **[Ca03, Ca3]**	> 0.5953	section 5
X of genus s	$< \frac{64}{4\sqrt{s} + 27}$ **[Gr83, Ko87]**		(2.9)

FIGURE 2.1. Values for optimal systolic ratio of surface $(X, \mathbf{g})$, in decreasing order

Hence $sys\pi_1(\mathbf{g}) \geq L(\mathbf{g})$. In this section (with the exception of section 4 where there is no homotopy) we will consider isosystolic inequalities of the following form: $sys\pi_1(\mathbf{g})^2 \leq const\ area(\mathbf{g})$. This leads naturally to the notion of the systolic ratio of an n-dimensional Riemannian manifold $(X, \mathbf{g})$, which is defined to be the scale-invariant quantity $\frac{sys\pi_1(\mathbf{g})^n}{vol_n(\mathbf{g})}$. We also define the *optimal systolic ratio* of a manifold X to be the quantity

$$\sup_{\mathbf{g}} \frac{sys\pi_1(\mathbf{g})^n}{vol_n(\mathbf{g})}, \tag{2.2}$$

where the supremum is taken over all Riemannian metrics $\mathbf{g}$ on the given manifold X. Note that in the literature, the reciprocal of this quantity is sometimes used instead. In this section, we will discuss the optimal systolic ratio for $\mathbb{T}^2$, $\mathbb{R}P^2$ (section 1), and the Klein bottle $\mathbb{R}P^2 \# \mathbb{R}P^2$ (section 5). These are the only surfaces for which the optimal systolic ratio is known. However, upper bounds for the systolic ratio, namely isosystolic inequalities, are available for all compact surfaces (section 3) except of course for S^2 (but see section 4). Figure 2.1 contains a chart showing the known upper and lower bounds on the optimal systolic ratios of surfaces.

1. Inequalities of Loewner and Pu. The first results of this type were due to C. Loewner and P. Pu. Around 1949, Carl Loewner proved the first systolic inequality, *cf.* **[Pu52]**. He showed that for every Riemannian metric $\mathbf{g}$ on the torus $\mathbb{T}^2$, we have

$$sys\pi_1(\mathbf{g})^2 \leq \frac{2}{\sqrt{3}} area(\mathbf{g}), \tag{2.3}$$

while a metric satisfying the boundary case of equality in (2.3) is necessarily flat, and is homothetic to the quotient of $\mathbb{C}$ by the lattice spanned by the cube roots of unity.

Two distinct optimal generalisations of (2.3) are available, *cf.* (5.17) and (5.14). It follows from Gromov's estimate (2.9) that aspherical surfaces satisfy Loewner's inequality (2.3) if the genus is bigger than 50. It is an open question whether the genus assumption can be removed, but for genus up to 30, some information can be deduced from the Buser-Sarnak inequality (5.8).

We give a slightly modified version of M. Gromov's proof [**Gr96**], using conformal representation and Cauchy-Schwartz, of Loewner's theorem for the 2-torus. We present the following slight generalisation: there exists a pair of closed geodesics on $(\mathbb{T}^2, \mathbf{g})$, of respective lengths λ_1 and λ_2, such that

$$\lambda_1 \lambda_2 \leq \tfrac{2}{\sqrt{3}} area(\mathbf{g}), \tag{2.4}$$

and whose homotopy classes form a generating set for $\pi_1(\mathbb{T}^2) = \mathbb{Z} \times \mathbb{Z}$.

PROOF. The proof relies on the conformal representation $\phi : \mathbb{T}_0 \to (\mathbb{T}^2, \mathbf{g})$, where $\mathbb{T}_0$ is flat. Here ϕ may be chosen in such a way that $(\mathbb{T}^2, \mathbf{g})$ and $\mathbb{T}_0$ have the same area. Let f be the conformal factor of ϕ.

Let ℓ_0 be any closed geodesic in $\mathbb{T}_0$. Let $\{\ell_s\}$ be the family of geodesics parallel to ℓ_0. Parametrize the family $\{\ell_s\}$ by a circle S^1 of length σ, so that $\sigma \ell_0 = area(\mathbb{T}_0)$. Thus $\mathbb{T}_0 \to S^1$ is a Riemannian submersion. Then $area(\mathbb{T}^2) = \int_{\mathbb{T}_0} Jac_\phi = \int_{\mathbb{T}_0} f^2$. By Fubini's theorem, $area(\mathbb{T}^2) = \int_{S^1} ds \int_{\ell_s} f^2 dt$. By the Cauchy-Schwartz inequality,

$$area(\mathbb{T}^2) \geq \int_{S^1} ds \frac{\left(\int_{\ell_s} f dt\right)^2}{\ell_0} = \frac{1}{\ell_0} \int_{S^1} ds \, (length \phi(\ell_s))^2 .$$

Hence there is an s_0 such that $area(\mathbb{T}^2) \geq \frac{\sigma}{\ell_0} length\phi(\ell_{s_0})^2$, so that $length\phi(\ell_{s_0}) \leq \ell_0$. This reduces the proof to the flat case. Given a lattice in $\mathbb{C}$, we choose a shortest lattice vector λ_1, as well as a shortest one λ_2 not proportional to λ_1. The inequality is now obvious from the geometry of the standard fundamental domain for the action of $PSL(2, \mathbb{Z})$ in the upper half plane of $\mathbb{C}$. □

We record here a slight generalization of Pu's theorem from [**Pu52**]. The generalization follows from Gromov's inequality (2.10). Namely, every surface $(X, \mathbf{g})$ which is not a 2-sphere satisfies

$$sys\pi_1(\mathbf{g})^2 \leq \frac{\pi}{2} area(\mathbf{g}), \tag{2.5}$$

where the boundary case of equality in (2.5) is attained precisely when, on the one hand, the surface X is a real projective plane, and on the other, the metric $\mathbf{g}$ is of constant Gaussian curvature.

In both Loewner's and Pu's proofs, the area of two pointwise conformal metrics were compared to the minimum lengths of paths in a fixed family of curves (the family of all non contractible closed curves). This is sometimes called the conformal-length method and analyzed in [**Gr83**, section 5.5]. C. Bavard [**Bav92a**] was able to show that in a given conformal class, on any n-dimensional manifold, there is at most one metric with maximum systolic ratio, and was able to give a characterization of such metrics.

We conjecture a generalisation, 2.7, of Pu's inequality (2.5). Let S be a nonorientable surface and let $\phi : \pi_1(S) \to \mathbb{Z}_2$ be a homomorphism from its fundamental group to $\mathbb{Z}_2$, corresponding to a map $\hat{\phi} : S \to \mathbb{R}P^2$ of absolute degree one. We define the "1-systole relative to ϕ", denoted $\phi sys_1(\mathbf{g})$, of a metric $\mathbf{g}$ on S, by minimizing length over loops γ which are not in the kernel of ϕ, *i.e.* loops whose image under ϕ is not contractible in the projective plane:

$$\phi sys_1(\mathbf{g}) = \min_{\phi([\gamma]) \neq 0 \in \mathbb{Z}_2} length(\gamma). \tag{2.6}$$

The question is whether this systole of $(S, \mathbf{g})$ satisfies the following sharp inequality, related to Gromov's inequality $(*)_{inter}$ from [**Gr96**, 3.C.1], see also [**Gr99**, Theorem 4.41].

Conjecture 2.7. *For any nonorientable surface S and map $\hat{\phi} : S \to \mathbb{R}P^2$ of absolute degree one, we have $\phi sys_1(\mathbf{g})^2 \leq \frac{\pi}{2} area(\mathbf{g})$.*

In the above, the ordinary 1-systole of Pu's inequality (2.5) is replaced by the one relative to ϕ. The example of the connected sum of a standard $\mathbb{R}P^2$ with a little 2-torus shows that such an inequality would be optimal in every topological type S. Conjecture 2.7 is closely related to the the filling area of the circle (Conjecture 3.8). Given a filling X^2, one identifies antipodal points on the boundary circle to get a nonorientable surface S. One can define a degree one map f from S to $\mathbb{R}P^2$ by taking all points outside of a tubular neighborhood of the original boundary circle to a point. This gives a natural choice of ϕ above which relates the two conjectures, *cf.* Remarks $(e), (e')$ following [**Gr83**, Theorem 5.5.B′].

2. A surface of genus 2 with octahedral Weierstrass set. The optimal systolic ratio in genus 2 is unknown. Here we discuss a lower bound for the optimal systolic ratio in genus 2. The example of M. Berger (see [**Gr83**, Example 5.6.B′], or [**Be83**]) in genus 2 is a singular flat metric with conical singularities. Its systolic ratio is 0.6666, which is not as good as the two examples we will now discuss.

C. Bavard [**Bav92b**] and P. Schmutz [**Sch93**, Theorem 5.2] identified the hyperbolic genus 2 surface with the optimal systolic ratio among all hyperbolic genus 2 surfaces. The surface in question is a triangle surface (2,3,8). It admits a regular hyperbolic octagon as a fundamental domain, and has 12 systolic loops of length $2x$, where $x = \cosh^{-1}(1 + \sqrt{2})$. It has $sys\pi_1 = 2\log\left(1 + \sqrt{2} + \sqrt{2 + 2\sqrt{2}}\right)$, area 4π, and systolic ratio 0.7437.

This ratio can be improved by a singular flat metric $\mathbf{g}_{\mathcal{O}}$, described below. Note that it is an Hadamard space in a generalized sense, *i.e.* a CAT(0) space.

Start with a triangulation of the real projective plane with 3 vertices and 4 faces, corresponding to the octahedral triangulation of the 2-sphere. The fact that the metric is pulled back from $\mathbb{R}P^2$ is significant, to the extent that if we can show that an extremal metric satisfies this, we would go a long way toward identifying it explicitly.

Every conformal structure on the surface of genus 2 is hyperelliptic [**FK92**, Proposition III.7.2], *i.e.* it admits a ramified double cover over the 2-sphere with 6 ramification points, called Weierstrass points. We take the 6 vertices of the regular octahedral triangulation, corresponding to the Riemann surface which has an equation of the form

$$y^2 = x^5 - x.$$

This produces a triangulation of the genus 2 surface Σ_2 consisting of 16 triangles. If each of them is flat equilateral with side x, then the total area is $16\left(\frac{1}{2}x^2 \sin\frac{\pi}{3}\right) = 4\sqrt{3}x^2$. Meanwhile, $sys\pi_1 = 2x$, corresponding to the inverse image of an edge under the double cover $\Sigma_2 \to S^2$. The systolic ratio is $\frac{(2x)^2}{4\sqrt{3}x^2} = \frac{1}{\sqrt{3}} = 0.5773$, which is not as good as the Bavard-Schmutz example.

Notice that the systolic loop remains locally minimizing if the angle of the triangle with side x is decreased from $\frac{\pi}{3}$ to $\frac{\pi}{4}$ (and the space remains CAT(0)). Therefore we replace the flat equilateral triangles by singular flat equilateral "triangles", whose singularity at the center has a total angle of $\frac{9\pi}{4} = 2\pi\left(1+\frac{1}{8}\right)$, while the angle at each of the vertices is $\frac{\pi}{4}$. Its barycentric subdivision consists of 6 copies of a flat right angle triangle, denoted $\measuredangle\left(\frac{x}{2},\frac{\pi}{8}\right)$, with side $\frac{x}{2}$ and adjacent angle $\frac{\pi}{8}$. Notice that this results in a smooth ramification point (with a total angle of 2π) in the ramified cover. We thus obtain a decomposition of Σ_2 into 96 copies of the triangle $\measuredangle\left(\frac{x}{2},\frac{\pi}{8}\right)$.

The resulting singular flat metric $\mathbf{g}_{\mathcal{O}}$ on Σ_2 (here $\mathcal{O}$ stands for "octahedron") has 16 singular points, with a total angle $\frac{9\pi}{4}$ around each of them, 4 of them over each center of a face of the $\mathbb{R}P^2$ decomposition. This calculation is consistent with M. Troyanov's [**Tro86**] Gauss-Bonnet formula $\sum_\sigma \alpha(\sigma) = 2s-2$, where s is the genus, where the cone angle at singularity σ is $2\pi(1+\alpha(\sigma))$.

Dually, the metric $\mathbf{g}_{\mathcal{O}}$ can be viewed as glued from six flat regular octagons, centered on the Weierstrass points. The hyperelliptic involution is the 180 degree rotation on each of them. The 1-skeleton projects to that of the inscribed cube in the 2-sphere. The systolic ratio of the resulting metric $\mathbf{g}_{\mathcal{O}}$ is

$$\begin{aligned}\frac{(sys\pi_1(\mathbf{g}_{\mathcal{O}}))^2}{area(\mathbf{g}_{\mathcal{O}})} &= \frac{(2x)^2}{96\ area\left(\measuredangle\left(\frac{x}{2},\frac{\pi}{8}\right)\right)} \\ &= \frac{x^2}{24\left(\frac{1}{2}\left(\frac{x}{2}\right)^2\tan\frac{\pi}{8}\right)} \\ &= \left(3\sqrt{3-2\sqrt{2}}\right)^{-1} \\ &= 0.8047.\end{aligned} \tag{2.8}$$

The genus 2 case is further investigated in [**Ca3**].

3. Gromov's area estimates for surfaces of higher genus. The earliest work on isosystolic inequalities for surfaces of genus s is by R. Accola [**Ac60**] and C. Blatter [**Bl61**]. Their bounds on the optimal systolic ratio went to infinity with the genus, *cf.* (5.8). J. Hebda [**He82**] and independently Yu. Burago and V. Zalgaller [**B-Z80, B-Z88**] showed that for $s > 1$ the optimal systolic ratio is bounded by 2 (which is not as good as (2.5), which came later). M. Gromov [**Gr83**, p. 50] (*cf.* [**Ko87**, Theorem 4, part (1)]) proved a general estimate which implies that if Σ_s is a closed orientable surface of genus s with a Riemannian metric, then

$$\frac{sys\pi_1(\Sigma_s)^2}{area(\Sigma_s)} < \frac{64}{4\sqrt{s}+27}. \tag{2.9}$$

Thus, as expected, the optimal systolic ratio goes to 0 as the genus goes to infinity (see (5.9) for the correct asymptotic behavior).

Another helpful estimate is found in [**Gr83**, Corollary 5.2.B]. Namely, every aspherical closed surface $(\Sigma, \mathbf{g})$ admits a metric ball $B = B_p\left(\frac{1}{2}sys\pi_1(\mathbf{g})\right) \subset \Sigma$ of radius $\frac{1}{2}sys\pi_1(\mathbf{g})$ which satisfies

$$sys\pi_1(\mathbf{g})^2 \leq \frac{4}{3}area(B). \tag{2.10}$$

Furthermore, whenever a point $x \in \Sigma$ lies on a two-sided loop which is minimizing in its free homotopy class, the metric ball $B_x(r) \subset \Sigma$ of radius $r \leq \frac{1}{2}sys\pi_1(\mathbf{g})$ satisfies the estimate $2r^2 < area(B_x(r))$.

4. Simply connected and noncompact surfaces. One of the motivating questions for this section is from [**Gr83**, p. 135], see also problem 87 in [**Ya82**] (or [**S-Y94**]).

Question 2.11. For an n-dimensional compact manifold X, is there a constant $C(X)$ such that for every Riemannian metric $\mathbf{g}$ on X, we have

$$vol(\mathbf{g}) \geq C(X)L(\mathbf{g})^n, \tag{2.12}$$

where $L(\mathbf{g})$ is the length of the shortest nontrivial closed geodesic.

This is still open for many manifolds X, *e.g.* for $X = S^n, n \geq 3$. One could also ask the (stronger) question whether $C(X)$ depends only on n. Since $sys\pi_1(\mathbf{g}) \geq L(\mathbf{g})$, upper bounds on the optimal systolic ratio give upper bounds on the constant $C(X)$ in (2.12). Thus we have a positive answer for all closed surfaces except for the two-sphere, S^2, which of course can have no nontrivial systolic inequalities. However, Question 2.11 does have an affirmative answer in this case. It was shown in [**Cr88B**] that every metric $\mathbf{g}$ on S^2 satisfies the bound

$$\frac{1}{31^2}L(\mathbf{g})^2 \leq area(\mathbf{g}). \tag{2.13}$$

Another estimate in that paper was $L(\mathbf{g}) \leq 9D(\mathbf{g})$, where $D(\mathbf{g})$ represents the diameter. Neither of these constants are best possible, and both have been improved recently [**Mae94, N-R02, Sab1**]. The best known bounds are $L(\mathbf{g}) \leq 4D(\mathbf{g})$ and $\frac{1}{64}L(\mathbf{g})^2 \leq area(\mathbf{g})$. The conjectured best constant $C(S^2)$ in (2.12) for the 2-sphere (suggested to the first author by E. Calabi) is $\frac{1}{2\sqrt{3}}$, attained by the singular metric obtained by gluing two equilateral triangles along their edges.

A natural way to find closed geodesics on a non simply connected closed Riemannian manifold is to look for the shortest curve in a nontrivial homotopy class, as we have done in our consideration of $sys\pi_1(\mathbf{g})$. However, when $\pi_1(X)$ is trivial, the standard technique is to used minimax arguments on nontrivial families of curves. For example, G. Birkhoff [**Bi27**] considered 1-parameter families of closed curves starting and ending in point curves, which pass over S^2 in the sense that the induced map from S^2 to S^2 has nonzero degree. He found a closed geodesic on S^2 by taking minimum over all these families of the maximum length curve in the family. These minimax geodesics are stationary in the sense that they are critical points (but not necessarily local minima) for the length functional on the space of curves. Thus one can find a short closed geodesic by finding such a family where every curve in the family is short. This idea played an important part in the proof of (2.13) above.

Instead of homotopy classes we could consider homology classes (leading to the notion of $sys_1(X)$ *cf.* (4.1)). That is, we could look at 1-cycles of minimal mass in nontrivial homology classes. When $H_1(X)$ is trivial one can instead use a minimax method on 1-cycles, similar to the one described in the previous paragraph, to get stationary 1-cycles (also see section 1). In some important cases these also turn out to be closed geodesics but in any event they are natural objects. The basis for this is the work Almgren [**Al62**] and Pitts [**Pitts81**, Theorem 4.10], who get stationary varifolds (in all dimensions) via minimax techniques. The case of 1-cycles (where things are easier) was exploited by Calabi and Cao [**CaCa92**] in their proof that the shortest closed geodesic on a convex surface is simple. They use the fact that on S^2 this minimax technique produces a closed geodesic. The estimates of A. Nabutovsky, R. Rotman, and S. Sabourau [**N-R02, Sab1**] improving (2.13) mentioned above exploited these techniques, *cf.* (3.6).

Although, as mentioned above, Question 2.11 is still open for general metrics on S^n, if one considers only convex hypersurfaces of $\mathbb{R}^{n+1}$ then such a result was shown independently in [**Tre85**] and [**Cr88B**]. The sharp constants are still not known in this case.

Finally one can ask Question 2.11 for noncompact surfaces and complete metrics (of finite area). In fact, the question has a positive answer for all surfaces. Most of the cases were dealt with as a special case of [**Gr83**, Theorem 4.4A], while the other cases (the plane and the cylinder) were dealt with in [**Cr88B**].

5. Optimal surfaces, existence of optimal metrics, fried eggs. In the case of $\mathbb{T}^2$ or $\mathbb{R}P^2$, we saw in (2.3) and (2.5) that there is a particular smooth metric which achieves the optimal systolic ratio. In general this will not be the case unless one admits metrics with singularities. In fact, due to Gromov's compactness theorem [**Gr83**, theorem 5.6.C′], there will always be a singular metric achieving the maximal systolic ratio. Such metrics (as well as those that have locally maximal systolic ratio) are called "extremal isosystolic metrics". See [**Gr83**] or [**Ca96**] for a description of these singular metrics.

For the Klein bottle, C. Bavard [**Bav86**] (also see [**Sak88**]) found the maximal systolic ratio:

$$sys\pi_1(\mathbf{g})^2 \leq \frac{\pi}{2\sqrt{2}} area(\mathbf{g}). \tag{2.14}$$

He also identified the extremal isosystolic metric achieving it. This metric is singular, and is built out of two möbius strips of constant curvature +1 (where the central curve has length π, and width is $\frac{\pi}{2}$). They are then glued together along their boundaries. The metric is singular since the geodesic curvatures of both boundaries point outward.

An extensive study of such extremal metrics was undertaken by E. Calabi [**Ca96**]. He derived a number of properties of such extremal metrics using variation methods. For points in such a space at least two systoles must pass through each point. (We slightly abuse terminology by referring to noncontractible curves that have length equal to $sys\pi_1$ "systoles".) One consequence of his analysis is that any such extremal metric must be flat at points where exactly two systoles pass through each point in a neighborhood. Furthermore, these systoles must intersect orthogonally.

The study of solutions to the Euler - Lagrange equations for this problem was futher taken up by R. Bryant in [**Br96**].

Some very interesting examples of piecewise flat (singular) metrics in genus 3 that satisfy these criteria were presented in [**Ca96**]. They do not have the same systolic ratio, one has ratio $\frac{9}{8\sqrt{3}} \sim .6495$ and the other $\frac{8}{7\sqrt{3}} \sim .6598$, but Calabi stated in that paper that these were probably both local maxima (or at least stationary points) of the systolic ratio. Today **[Ca03]** Calabi feels that the example with the smaller systolic ratio (a ramified 4-fold cover of the octahedron with ratio $\frac{9}{8\sqrt{3}}$) is probably not locally a maximum. The example with the larger systolic ratio is still conjectured to be not only a local maximum but the best possible. Calabi **[Ca03, Ca3]** also has an example on a surface of genus 4 of a piecewise flat (with conical singularities) metric (having a symmetry group of order 120) with a systolic ratio of $\frac{(18\sqrt{7})^2}{1440\sqrt{7}} = \frac{9\sqrt{7}}{40} \approx 0.5953$. It is built out of 60 flat rhombi with angles arccos(1/8) and arccos(-1/8). It is suspected that it is close to the optimum systolic ratio but it is not optimum since there are points through which only one systole passes.

An important local solution to the Euler-Lagrange equations for extremal isosystolic metrics, the "Fried Egg", was also described in **[Ca96]**. This example is a (non flat) metric on a disk whose boundary is a regular hexagon all of whose angles are $\frac{\pi}{2}$. It arises in consideration of the problem of finding the least area Riemannian metric in a hexagon (having the symmetries of a hexagon - i.e. the dihedral group of order 12 generated by reflections) such that each point on a given side has distance exactly 2 to the opposite side. (The above additional symmetry assumption did not appear in the original paper but should be included **[Ca03]**.) A solution to related problems for an octagon and a triangle might be useful in further increasing the systolic ratio in the example of section 2 by replacing the flat octagons (in the dual picture) or the singular triangles with such fried eggs.

The example (piecewise flat with conical singularities) of Calabi above in genus 3 with systolic ratio $\frac{8}{7\sqrt{3}}$ is the only candidate that exists for metrics on surfaces of higher genus achieving the maximal systolic ratio. In contrast, Sabourau in **[Sab02c]** shows that in genus 2, no flat metric with conical singularities (such as the example of section 2) can achieve the maximal systolic ratio.

3. Gromov's Filling Riemannian Manifolds

In this section, we discuss some of the main results in **[Gr83]**. The reader should look at that paper for many interesting results in this area.

1. Filling radius + main estimate. Let X^n be a closed, smooth, n-dimensional manifold with a metric d_X (not necessarily Riemannian). Let A be a coefficient ring (either $\mathbb{Z}$ or $\mathbb{Z}/2\mathbb{Z}$). In **[Gr83]**, Gromov introduced the notion of the filling radius, *FillRad*(X, A), with respect to A, of (X, d_X). There is a natural strong isometric embedding (in the metric space sense!) $i : X \to L^\infty X$, defined by $i(x)(\cdot) = d_X(x, \cdot)$. The filling radius is the infimum of r such that $i(X)$ bounds in the tubular r-neighborhood

$$T_r(i(X)) \subset L^\infty X,$$

in the sense that the image $i([X])$ of the fundamental class vanishes in $H_n(T_r(i(X)); A)$.

The only Riemannian manifolds for which the precise value of the filling radius is known **[Ka83]** are spheres and real projective spaces of constant curvature K, as well as a single additional case of $\mathbb{C}P^2$ **[Ka91B]**. Thus, $\mathit{FillRad}(S^n) = \frac{1}{2}\arccos\left(-\frac{1}{n+1}\right) K^{-\frac{1}{2}}$. Meanwhile, $\mathit{FillRad}(\mathbb{R}P^n) = \frac{1}{3}\mathit{diam}(\mathbb{R}P^n)$, and thus "round" real projective spaces are extremal for the optimal inequality $\mathit{FillRad}(X) \leq \frac{1}{3}\mathit{diam}(X)$, valid for all Riemannian manifolds [**Ka83**]. Partial results in the direction of calculating the filling radius for other two-point homogeneous spaces were obtained in [**Ka91A, Ka91B, Ka91C**]. An optimal inequality for the filling radius appears in [**Wi92**].

One of the fundamental estimates that allows universal inequalities in higher dimensions is the following theorem, due to Gromov [**Gr83**, Section 1.2, Main Theorem].

Theorem 3.1. *For any closed Riemannian n-manifold X we have*

$$vol(X)^{\frac{1}{n}} > \left((n+1)(n^n)\sqrt{(n+1)!}\right)^{-1} FillRad(X).$$

Many of the higher dimensional universal inequalities use this estimate along with an estimate for the filling radius. The first of these was Gromov's systolic theorem (below) for essential manifolds.

Definition 3.2. The manifold X^n is called *essential* (over A) if it admits a map, $F : X \to K$, to an aspherical space K such that the induced homomorphism $F_* : H_n(X, A) \to H_n(K, A)$ sends the fundamental class $[X]$ to a nonzero class: $F_*([X]) \neq 0 \in H_n(K, A)$.

Here we must take $A = \mathbb{Z}/2\mathbb{Z}$ if X is nonorientable. In [**Gr83**, Theorem 0.1.A] Gromov showed the following.

Theorem 3.3. *Assume X is essential over A. Then every Riemannian metric* **g** *on X satisfies the inequality*

$$FillRad(\mathbf{g}) \geq \frac{1}{6} sys\pi_1(\mathbf{g}).$$

Combined with Theorem 1, this yields the inequality

$$sys\pi_1(\mathbf{g})^n < \left(6(n+1)(n^n)\sqrt{(n+1)!}\right)^n vol_n(\mathbf{g}). \tag{3.4}$$

This theorem generalizes both the Loewner (2.3) and the Pu (2.5) theorems to higher dimensions (with non-sharp constants) since both $\mathbb{T}^n$ and $\mathbb{R}P^n$ are essential, and in particular answers Question 2.11 in the affirmative for these spaces. Since the proof of the theorem is the model for most other estimates of *FillRad*(X) we give it now.

Proof. We will represent a filling as continuous map $\sigma : \Sigma \to T_r(i(X))$ from an $(n+1)$-dimensional simplicial complex Σ such that the restriction $\sigma|_{\partial\Sigma} : \partial\Sigma \to i(X)$ represents a generator in $H_n(X; A)$. We note that for any fixed $\epsilon > 0$, by taking barycentric subdivisions as needed, we may assume that the σ-image of any simplex has diameter less than ϵ in $L^\infty(X)$. We note that there can be no continuous map $f : \Sigma \to K$ which agrees with $F \circ \sigma$ on $\partial\Sigma$, since $F \circ \sigma|_{\partial\Sigma}$ represents $F_*[X]$ (so is not a boundary in K by the hypothesis of Definition 3.2). Since F is defined on X, we can also think of it as defined on $i(X) \subset L^\infty(X)$ because i is an embedding.

The proof will be by contradiction. We assume that $FillRad(X) < \frac{1}{6} sys\pi_1(X)$, let $0 < \epsilon$ be such that $2FillRad(X) + 3\epsilon < \frac{1}{3} sys\pi_1(X)$. Take a filling $\sigma : \Sigma \to T_r(i(X))$ (for $r = FillRad(X) + \epsilon$) where the image of the simplices have diameter less than ϵ, and show that $F \circ \sigma : \partial\Sigma \to K$ extends to a continuous $f : \Sigma \to K$ giving the desired contradiction.

We construct the extension by first choosing, for every vertex v of $\Sigma - \partial\Sigma$, a point $x(v) \in X$ such that $d(i(x(v)), \sigma(v)) < r$ (which we can do since $\sigma(\Sigma) \subset T_r(i(X))$). (For $v \in \partial\Sigma$ just take $x(v) = \sigma(v)$) Now we define $f(v) \equiv F(x(v))$. Let e be an edge of a simplex in Σ with endpoints v_1 and v_2. The strongly isometric nature of the imbedding $X \to L^\infty X$ and the triangle inequality in $L^\infty X$ imply that

$$d(x(v_1), x(v_2)) = d(i(x(v_1)), i(x(v_2))) \leq 2r + \epsilon < \frac{1}{3} sys\pi_1(X).$$

Choose a shortest path $x(e)(t)$ from $x(v_1)$ to $x(v_2)$ in X and let $f(e(t)) = F(x(e)(t))$. This extends f to the one-skeleton of Σ. Now for each two-simplex of Σ with edges e_1, e_2, and e_3, the closed curve formed by the $x(e_i)$ has length less than $sys\pi_1(X)$. Therefore it can be contracted to a point in X. Hence we can define a map of the two-simplex to X, and then composing with F to K which agrees with f on the edges. Thus we can extend f to the two-skeleton. Now since K is aspherical, there is no obstruction to extending f to the rest of Σ. This gives the desired contradiction. □

In general, proofs of lower bounds on the filling radius take this form. That is, one assumes the filling radius is small, takes a nice filling, and uses it to construct some map (usually skeleton by skeleton) that one knows does not exist.

The following theorem from [Gro2, 3.C_1] can be thought of as a generalisation of Loewner's inequality (2.3), and a homological analogue of (3.4), see also Gromov's sharp stable inequality (5.14). Let X be a smooth compact n-dimensional manifold. Assume that there is a field F and classes $\alpha_1, \ldots, \alpha_n \in H^1(X, F)$ with a nonvanishing cup product $\alpha_1 \cup \ldots \cup \alpha_n \neq 0$. Then every Riemannian metric $\mathbf{g}$ on X satisfies the inequality

$$sys_1(\mathbf{g})^n \leq C_n vol_n(\mathbf{g}), \tag{3.5}$$

where C_n is a constant depending only on the dimension.

In section 4 we introduced the notion of a stationary 1-cycle. These are usually found by the minimax techniques of Almgren and Pitts (see [**Al62**] and [**Pitts81**, Theorem 4.10]). A stationary one cycle is built out of finitely many geodesic segments and the mass is just the sum of the lengths of the segments. For a given Riemannian metric $\mathbf{g}$ we will let $m_1(\mathbf{g})$ be the minimal mass of a stationary 1-cycle. Since a closed geodesic is a stationary 1-cycle we have $L(\mathbf{g}) \geq m_1(\mathbf{g})$.

Recently these techniques have been combined with the Filling techniques to get isosystolic type estimates in all dimensions for all manifolds. In [**N-R2**] Nabutovsky and Rotman showed that

$$m_1(\mathbf{g}) \leq 2(n+2)! \mathit{FillRad}(\mathbf{g}) \tag{3.6}$$

and hence, by Theorem 3.1, we have $m_1(\mathbf{g})^n \leq C(n) vol(\mathbf{g})$ (for an explicit constant $C(n)$).

In [**Sab1**] Sabourau gives a lower bound on the Filling radius of a generic metric on the two-sphere in terms of the minimal mass of a 1-cycle of index 1. The minimal mass is achieved by either a simple closed geodesic or a figure 8 geodesic. The advantage of this estimate is that short geodesics around thin necks (which, in nearby generic metrics, will have index 0 as 1-cycles) can be ignored to give better bounds on the area.

2. Filling volume and chordal metrics. In [**Gr83**] Gromov also introduced the notion of Filling Volume, $FillVol(N^n, d)$, for a compact manifold N with a metric d (here d is a distance function which is not necessarily Riemannian). For the actual definition one should see [**Gr83**], but it is shown in [**Gr83**] that when $n \geq 2$

$$FillVol(N^n, d) = \inf_{\mathbf{g}} vol(X^{n+1}, \mathbf{g}) \tag{3.7}$$

where X is any fixed manifold such that $\partial X = N$ (one can even take $X = N \times [0, \infty)$), the infimum is taken over all Riemannian metrics $\mathbf{g}$ on X for which the boundary distance function is $\geq d$. In the case $n = 1$, the topology of the filling X^2 could affect the infimum, as is shown by example in [**Gr83**, Counterexamples 2.2.B].

In fact, the filling volume is not known for any Riemannian metric. However, Gromov does conjecture the following in [**Gr83**], immediately after Proposition 2.2.A.

Conjecture 3.8. *$Fillvol(S^n, can) = \frac{1}{2}\omega_{n+1}$, where ω_{n+1} represents the volume of the unit $(n+1)$-sphere.*

This is still open in all dimensions. In the case that $n = 1$, as we pointed out earlier, this is closely related to Conjecture 2.7. The filling area of the circle of length 2π with respect to the simply connected filling (by a disk) is indeed $\frac{\pi}{2}$, by Pu's inequality (2.5) applied to the projective plane obtained by identifying opposite points of the circle.

In many cases it is more natural to consider "chordal metrics" than Riemannian metrics when discussing Filling volume. Consider a compact manifold X with boundary ∂X with a Riemannian metric $\mathbf{g}$. Then there is a (typically not Riemannian) metric $d_{\mathbf{g}}$ on ∂X, where $d_{\mathbf{g}}(x, y)$ represents the distance in X with respect to the the metric $\mathbf{g}$, *i.e.* the length of the $\mathbf{g}$-shortest path in X between boundary points. We call $d_{\mathbf{g}}$ the *chordal* metric on ∂X induced by $\mathbf{g}$.

Sharp filling volume estimates for chordal metrics are related to a universal length versus volume question for a given compact manifold X with boundary ∂X. This question compares the volumes, $vol(\mathbf{g}_0)$ and $vol(\mathbf{g}_1)$, of two Riemannian metrics $\mathbf{g}_0$ and $\mathbf{g}_1$ on X if we know that for every pair of points $x, y \in \partial X$ we have $d_{\mathbf{g}_0}(x, y) \leq d_{\mathbf{g}_1}(x, y)$. Of course, without some further assumptions on the metrics (such as some minimizing property of geodesics) there is no general comparison between the volumes. For a fixed $\mathbf{g}_0$ (and $n \geq 2$) the statement that for all such $\mathbf{g}_1$ we have $vol(\mathbf{g}_0) \leq vol(\mathbf{g}_1)$ is just the statement that $FillVol(\partial X, d_{\mathbf{g}_0}) = vol(\mathbf{g}_0)$. Note that the standard metric on the n-sphere is just the chordal metric of the $(n+1)$-dimensional hemisphere that it bounds. So Conjecture 3.8 is in fact such a question.

In the computation of the Filling volume via formula (3.7) above when d is the chordal distance function of some Riemannian manifold $(X^{n+1}, \partial X, \mathbf{g}_0)$ $(n \geq 2)$ one can not only fix the topology of X^{n+1} but also restrict to metrics $\mathbf{g}$ such that the Riemannian metrics on ∂X gotten by restricting $\mathbf{g}$ and $\mathbf{g}_0$ to ∂X are the same (see [**Cr01**]).

The filling volume is known for some chordal metrics. Gromov in [**Gr83**] proved this for X^{n+1} a compact subdomain of $\mathbb{R}^{n+1}$ (or in fact for some more general flat manifolds with boundary). The minimal entropy theorem of Besson, Courtois, and Gallot [**B-C-G96, B-C-G95**] can be used to prove the result for compact subdomains of symmetric spaces of negative curvature (see [**Cr01**]). For general convex simply connected manifolds $(X, \partial X, \mathbf{g}_0)$ of negative curvature, there is a C^3 neighborhood in the space of metrics such that any metric $\mathbf{g}_1$ in that neighborhood with $\mathbf{g}_1|_{\partial X} = \mathbf{g}_0|_{\partial X}$ and $d_{\mathbf{g}_1}(x,y) \geq d_{\mathbf{g}_0}(x,y)$ has $vol(\mathbf{g}_1) \geq vol(\mathbf{g}_0)$ and equality of the volumes implies $\mathbf{g}_1$ is isometric to $\mathbf{g}_0$ (see [**C-D-S00**]). This leads to the conjecture:

Conjecture 3.9. *For any compact subdomain of a simply connected space of negative (nonpositive?) curvature of dimension ≥ 3, the filling volume of the boundary with the chordal metric is just the volume of the domain. Furthermore, the domain is the unique (up to isometry) volume minimizing filling.*

The uniqueness part of this question has applications to the boundary rigidity problem. In that problem one considers the case when $d_{\mathbf{g}_0}(x,y) = d_{\mathbf{g}_1}(x,y)$ for all boundary points x and y and asks if $\mathbf{g}_0$ must be isometric to $\mathbf{g}_1$. In some natural cases (see *SGM* below) the volumes can be shown to be equal. For example, Gromov's result for subdomains of Euclidean space showed that they were boundary rigid. Again this cannot hold in general. A survey of what is known about the boundary rigidity problem can be found in [**Cr01**]. There are a few natural choices for assumptions in this case. The most general natural assumption of this sort is *SGM.* The *SGM* condition (which is given in terms of the distance function $d_{\mathbf{g}} : \partial X \times \partial X \to \mathbb{R}$ alone) would take some space to define precisely (see [**Cr91**]), but loosely speaking (*i.e.* the definitions coincide except in a few cases) it means the following:

Definition 3.10. A metric is "loosely" *SGM* if all nongrazing geodesic segments are strongly minimizing.

By a nongrazing geodesic segment we mean a segment of a geodesic which lies in the interior of X except possibly for the endpoints. A segment is said to minimize if its length is the distance between the endpoints, and to strongly minimize if it is the unique such path. (This loose definition seems to rely on more than $d_{\mathbf{g}}$ but the relationship is worked out in [**Cr91**].) Examples of such $(X, \partial X, \mathbf{g})$ are given by compact subdomains of an open ball, B, in a Riemannian manifold where all geodesics segments in B minimize. The only reason not to use the "loose" definition above is that using a definition (such as *SGM*) given in terms only of $d_{\mathbf{g}}$ guarantees that if $d_0(x,y) = d_1(x,y)$ and $\mathbf{g}_0$ is *SGM*, then $\mathbf{g}_1$ will be as well. In fact, the questions and results stated here for the *SGM* case also hold for manifolds satisfying the loose definition, so the reader can treat that as a definition of *SGM* for the purpose of this paper.

The most general result one could hope for would be of the form:

Question 3.11. If $\mathbf{g}_0$ is an *SGM* metric on $(X, \partial X)$ and $\mathbf{g}_1$ is another metric with $d_{\mathbf{g}_0}(x,y) \leq d_{\mathbf{g}_1}(x,y)$ for $x, y \in \partial X$ then $vol(\mathbf{g}_0) \leq vol(\mathbf{g}_1)$ with equality of volumes implying $\mathbf{g}_0$ is isometric to $\mathbf{g}_1$.

This is still very much an open question, which as stated includes the boundary rigidity problem.

The case $\mathbf{g}_1 = f^2(x)\mathbf{g}_0$ (*i.e.* $\mathbf{g}_1$ is pointwise conformal to $\mathbf{g}_0$) was answered positively in **[C-D]** (also see **[Cr91]**).

In two dimensions more is known. Recently Ivanov **[Iv02]** considered the case of compact metrics $\mathbf{g}_0$ and $\mathbf{g}_1$ on a disk. He assumes that $\mathbf{g}_0$ is a convex metric in the sense that every pair of interior points can be joined by a unique geodesic, and proves that if $d_{\mathbf{g}_1} \geq d_{\mathbf{g}_0}$ then $vol(\mathbf{g}_1) \geq vol(\mathbf{g}_0)$. He also says that equality in the area would imply that $d_{\mathbf{g}_1} = d_{\mathbf{g}_0}$. Of course, going from here to isometry of $\mathbf{g}_0$ and $\mathbf{g}_1$ is the boundary rigidity problem. This boundary rigidity problem (in 2 dimensions) is solved **[Cr91, Ot90B]** in the case that the metric $\mathbf{g}_0$ has negative curvature.

Using different methods **[C-D]** proves a similar though somewhat different result where the metrics $\mathbf{g}_0$ and $\mathbf{g}_1$ are both assumed to be SGM but the surfaces are not assumed to be simply connected. The estimate comes from a formula for the difference between the areas of two SGM surfaces. Since the surfaces are not simply connected one needs to worry about the homotopy class of a path between two boundary points. We will let $L_{\mathbf{g}}(x, y, [\alpha])$ be the length of the $\mathbf{g}$-shortest curve from x to y in a homotopy class $[\alpha]$ of curves from x to y. $\mathcal{A}$ will represent the space of such triples $(x, y, [\alpha])$ The formula is:

$$area(\mathbf{g}_1) - area(\mathbf{g}_0) = \frac{1}{2\pi} \int_{\mathcal{A}} \Big(L_{\mathbf{g}_1}(x, y, [\alpha]) - L_{\mathbf{g}_0}(x, y, [\alpha]) \Big) \left(\mu_{\mathbf{g}_1} + \mu_{\mathbf{g}_0} \right), \tag{3.12}$$

where the measures $\mu_{\mathbf{g}_i}$ on $\mathcal{A}$ are the push forwards of the standard Liouville measure on the space of geodesics. Since for each pair (x, y) there is at most one geodesic segment in each metric from x to y, we see that for most $(x, y, [\alpha])$ there will be neither a $\mathbf{g}_1$-geodesic nor a $\mathbf{g}_2$-geodesic segment from x to y in the class $[\alpha]$ and hence the term for $(x, y, [\alpha])$ will contribute nothing to the integral (see **[C-D]** for details). This formula easily leads to a result relating lengths of paths to area.

Another powerful result with filling volume consequences is the Besicovitch lemma which was exploited and generalized in **[Gr83**, section 7**]**. It says that for any Riemannian metric on a cube, the volume is bounded below by the product of the distances between opposite faces. Equality only holds for the Euclidean cube.

3. Marked length spectrum and volume. The question for compact manifolds N without boundary corresponding to Question 3.11 involves the marked length spectrum. The marked length spectrum for a Riemannian metric $\mathbf{g}$ on N is a function, $MLS_{\mathbf{g}} : \mathcal{C} \to \mathbb{R}^+$, from the set $\mathcal{C}$ of free homotopy classes of the fundamental group $\pi_1(N)$ to the nonnegative reals. For each $\langle\gamma\rangle \in \mathcal{C}$, $MLS_{\mathbf{g}}(\langle\gamma\rangle)$ is the length of the shortest curve in $\langle\gamma\rangle$ (always a geodesic). We consider two Riemannian metrics $\mathbf{g}_0$ and $\mathbf{g}_1$ on N such that $MLS_{\mathbf{g}_1}(\langle\gamma\rangle) \geq MLS_{\mathbf{g}_0}(\langle\gamma\rangle)$ for all free homotopy classes $\langle\gamma\rangle$ (we then say $MLS_{\mathbf{g}_1} \geq MLS_{\mathbf{g}_0}$) and ask if $vol(\mathbf{g}_1)$ must be greater than or equal to $vol(\mathbf{g}_0)$. Again this is hopeless without further assumptions.

The natural setting for this is in negative curvature. Here there are lots of closed geodesics but exactly one for each free homotopy class (achieving the minimum length in that class). The following was conjectured in **[C-D-S00]**:

Conjecture 3.13. *For two negatively curved metrics,* $\mathbf{g}_0$ *and* $\mathbf{g}_1$*, on a manifold* N *the inequality* $MLS_{\mathbf{g}_1} \geq MLS_{\mathbf{g}_0}$ *implies* $vol(\mathbf{g}_1) \geq vol(\mathbf{g}_0)$*. Furthermore,* $vol(\mathbf{g}_1) = vol(\mathbf{g}_0)$ *if and only if* $\mathbf{g}_0$ *and* $\mathbf{g}_1$ *are isometric.*

This conjecture was proved in dimension 2 in **[C-D]**. The higher dimensional version of the above was shown when $\mathbf{g}_0$ and $\mathbf{g}_1$ are pointwise conformal. This used ideas developed in **[Bo91]** and **[Si90]**.

A consequence of this conjecture is the conjecture for negatively curved $\mathbf{g}_0$ and $\mathbf{g}_1$ that $MLS_{\mathbf{g}_1} = MLS_{\mathbf{g}_0}$ would imply $\mathbf{g}_0$ is isometric to $\mathbf{g}_1$. See **[Cr01]** for a survey of this problem and **[Ban94, C-F-F92, Cr91, Ot90A]** for results (stronger than the conjecture) in two dimensions. However, it is not even known if $MLS_{\mathbf{g}_1} = MLS_{\mathbf{g}_0}$ implies $vol(\mathbf{g}_1) = vol(\mathbf{g}_0)$. In fact this is itself an important open question. Hamenstädt **[Ha99]** (also see **[Ha92]**) proved this in the special case that $\mathbf{g}_0$ is further assumed to be locally symmetric. This was the case that had the most important immediate applications. For example, it allows one to drop the volume assumption in the conjugacy rigidity result in **[B-C-G95]** and hence to see that if $\mathbf{g}_0$ is locally symmetric and $\mathbf{g}_1$ is a metric of negative curvature with $MLS_{\mathbf{g}_1} = MLS_{\mathbf{g}_0}$, then $\mathbf{g}_0$ is isometric to $\mathbf{g}_1$.

4. Systolic freedom for unstable systoles

In **[Gr83**, p. 5], M. Gromov, following M. Berger, asks the following basic question. What is the best constant C, possibly depending on the topological type of the manifold, for which the k-systolic inequality (4.2) holds? It was shown by the second author in collaboration with A. Suciu **[KS99, KS01]** that such a constant typically does not exist whenever $k > 1$, in sharp contrast with the inequalities of Loewner, Pu, and Gromov discussed in sections 2 and 3.

Similar non-existence results were obtained for a pair of complementary dimensions (4.4), culminating in the work of I. Babenko **[Bab02]**, *cf.* section 4. These results are placed in the context of other systolic results in the table of Figure 4.1.

1. Table of systolic results, definitions. Let $(X, \mathbf{g})$ be a Riemannian manifold, and let $k \in \mathbb{N}$. Given a homology class $\alpha \in H_1(X; A)$, denote by $len(\alpha)$ the least length (in the metric $\mathbf{g}$) of a 1-cycle with coefficients in A representing α. The *homology 1-systole* $sys_1(\mathbf{g}, A)$ is defined by setting

$$sys_1(\mathbf{g}, A) = \min_{\alpha \neq 0 \in H_1(X,A)} len(\alpha), \tag{4.1}$$

and we let $sys_1(\mathbf{g}) = sys_1(\mathbf{g}, \mathbb{Z})$. The other important case is $sys_1(\mathbf{g}, \mathbb{Z}_2)$. In other words, $sys_1(\mathbf{g})$ is the length of a shortest loop which is not nullhomologous in a compact Riemannian manifold $(X, \mathbf{g})$ with a nontrivial group $H_1(X, \mathbb{Z})$.

More generally, let $sys_k(\mathbf{g})$ be the infimum of k-volumes of integer (Lipschitz) k-cycles which are not boundaries in X, *cf.* formula (4.1). Note that if X is orientable of dimension n, then the total volume is a systolic invariant: $vol_n(\mathbf{g}) = sys_n(\mathbf{g})$.

The term "systolic freedom" refers to the absence of a systolic inequality, *e.g.* violation of the inequality relating a single systole to the total volume, namely

$$sys_k(\mathbf{g})^{\frac{n}{k}} < C vol_n(\mathbf{g}), \tag{4.2}$$

or similarly the inequality involving a pair of complementary dimensions, k and $(n-k)$, namely $sys_k(\mathbf{g}) sys_{n-k}(\mathbf{g}) < C vol_n(\mathbf{g})$, by a suitable sequence of metrics, *cf.* formula (4.5). Thus, we say that an n-manifold is k-systolically free if

$$\inf_{\mathbf{g}} \frac{vol_n(\mathbf{g})}{sys_k(\mathbf{g})^{n/k}} = 0, \tag{4.3}$$

	1 $k=1$	2 $k \geq 2$
A homotopy k-systole	Loewner's inequality for $\mathbb{T}^2$ (2.3), Pu's for $\mathbb{R}P^2$ (2.5), Gromov's for essential X (3.4)	
B homology k-systole	if cuplength$(X^n) = n$ then Gromov proves inequality (3.5)	freedom reigns: Gromov (4.5), Babenko (4.3), (4.4); Katz, Suciu (4.3); freedom of $S^1 \times S^2$ over $\mathbb{Z}_2$: Freedman (4.6)
C stable homology k-systole	sharp inequality if $b_1(X^n) =$ cuplength$(X^n) = n$ (Gromov, based on Burago-Ivanov) (5.14)	multiplicative relations in $H^*(X)$ entail inequalities: Gromov, Hebda (5.15), Bangert-Katz (5.13), (5.17)
D interpolation homotopy–homology	"fiberwise" inequality if Abel-Jacobi map "surjective" (Katz-Kreck-Suciu) [**KKS**], *cf.* (5.12)	
E conformal k-systole	Accola, Blatter (section 3); for a surface Σ_s, logarithmic in genus s: Buser-Sarnak (5.8)	X^4, indefinite case: polynomial in $\chi(X)$, modulo surjectivity of period map: Katz (5.10)
F contractible k-systole	short contractible geodesics: Croke (2.13), Maeda, Nabutovsky and Rotman (3.6), Sabourau (6.5)	

FIGURE 4.1. A 2-D map of systolic geometry

and $(k, n-k)$-systolically free if

$$\inf_{\mathbf{g}} \frac{vol_n(\mathbf{g})}{sys_k(\mathbf{g}) sys_{n-k}(\mathbf{g})} = 0, \tag{4.4}$$

where the infimum is over all smooth metrics $\mathbf{g}$ on the manifold. See Section 4 for results in this direction. Note that we use the reciprocals of our convention (2.2) for the systolic ratio, when discussing systolic freedom.

2. Gromov's homogeneous (1,3)-freedom. M. Gromov first described a $(1,3)$-systolically free family of metrics on $S^1 \times S^3$ in 1993, *cf.* [**Gr96**, section 4.A.3], [**Gr99**, p. 268]. His construction of (1,3)-systolic freedom on the manifold $S^1 \times S^3$ exhibits a family $\mathbf{g}_\epsilon$ of homogeneous metrics with the following asymptotic behavior:

$$\frac{vol_4(\mathbf{g}_\epsilon)}{sys_1(\mathbf{g}_\epsilon) sys_3(\mathbf{g}_\epsilon)} \to 0 \text{ as } \epsilon \to 0. \tag{4.5}$$

Due to the exceptional importance of this example, we present a proof. Let $U(2)$ be the unitary group acting on the unit sphere $S^3 \subset \mathbb{C}^2$. In particular, we have the scalar matrices $e^{i\theta} I_2 \in U(2)$. Consider the direct sum metric on $\mathbb{R} \times S^3$, and let α be the pullback to $\mathbb{R} \times S^3$ of the volume form of the second factor. Let $Trans(\mathbb{R})$ be the group of translations of the real line. Let $\epsilon > 0$ be a real parameter. We will define an infinite cyclic subgroup $\Gamma_\epsilon \subset Trans(\mathbb{R}) \times U(2)$ of the group of isometries of $\mathbb{R} \times S^3$, in such a way that the quotient

$$(\mathbb{R} \times S^3)/\Gamma_\epsilon \simeq (S^1 \times S^3, \mathbf{g}_\epsilon)$$

produces a $(1,3)$-systolically free family of metrics $\mathbf{g}_\epsilon$ on $S^1 \times S^3$ as $\epsilon \to 0$. Namely, we may take the generator $\gamma_\epsilon \in \Gamma_\epsilon$ to be the element $\gamma_\epsilon = \left(\epsilon, e^{i\sqrt{\epsilon}} I_2\right) \in \mathit{Trans}(\mathbb{R}) \times U(2)$. Then the quantity $sys_1(\mathbf{g}_\epsilon)$ is dominated by the second factor $e^{i\sqrt{\epsilon}}$, and is asymptotic to $\sqrt{\epsilon}$. Meanwhile, $sys_3(\mathbf{g}_\epsilon) = \omega_3$, where ω_3 the volume of the unit 3-sphere, by a calibration argument. Namely, every nontrivial 3-cycle C in $S^1 \times S^3$ satisfies $vol_3(C) \geq \int_C \alpha = n\omega_3 \geq \omega_3$, where n is the order of $[C] \in H_3(S^1 \times S^3) = \mathbb{Z}$.

Note that we have a fundamental domain with closure $[0, \epsilon] \times S^3 \subset \mathbb{R} \times S^3$ in the universal cover. Hence

$$\frac{vol_4(\mathbf{g}_\epsilon)}{sys_1(\mathbf{g}_\epsilon) sys_3(\mathbf{g}_\epsilon)} \sim \frac{\epsilon \omega_3}{\sqrt{\epsilon}\, \omega_3} = \sqrt{\epsilon} \to 0,$$

proving (1,3)-systolic freedom of $S^1 \times S^3$.

Note that a calculation of the stable 1-systole, *cf.* formula (5.5), reveals the behavior of systolic constraint instead of freedom. Thus, the n-th power of the generator $\gamma_\epsilon \in \pi_1(S^1 \times S^3) = \Gamma_\epsilon$, where $n = \left[\frac{2\pi}{\sqrt{\epsilon}}\right]$ is the integer part, also contains a closed geodesic whose length is on the order of $2\pi\sqrt{\epsilon}$, as

$$\gamma_\epsilon^n = \left(n\epsilon, e^{i\sqrt{\epsilon} n} I_2\right) \sim \left(2\pi\epsilon^{-\frac{1}{2}}\epsilon, e^{2\pi i} I_2\right) = \left(2\pi\epsilon^{\frac{1}{2}}, I_2\right).$$

It follows that the stable norm satisfies

$$\|\gamma_\epsilon\| \sim n^{-1} \mathit{length}(\gamma_\epsilon^n) \sim \frac{2\pi\sqrt{\epsilon}}{n} \sim \epsilon,$$

resulting in a stable 1-systole on the order of ϵ instead of $\sqrt{\epsilon}$, which is consistent with Hebda's inequality (5.15).

Remarkably, M. Freedman [**Fr99**] proves the $(1,2)$-systolic freedom of $S^1 \times S^2$ even when one works with homology with coefficients in $\mathbb{Z}_2$, *i.e.* nonorientable surfaces are allowed to compete in the definition of the 2-systole. He showed that

$$\inf_{\mathbf{g}} \frac{vol_3(\mathbf{g})}{sys_1(\mathbf{g}, \mathbb{Z}_2) sys_2(\mathbf{g}, \mathbb{Z}_2)} = 0, \tag{4.6}$$

where the infimum is over all metrics $(S^1 \times S^2, \mathbf{g})$, *cf.* formula (4.1). His technique relies on a common starting point with the lower bound of the Buser-Sarnak Theorem [**BuSa94**] (*cf.* (5.8)), namely the existence of Riemann surfaces of arbitrarily high genus with a uniform lower bound for the positive eigenvalues of the Laplacian. Whether or not $(\mathbb{R}P^3, \mathbf{g})$ satisfies (4.6) is unknown.

3. Pulling back freedom by (n, k)-morphisms. The early papers in the subject [**Ka95, Pitte97**] contain explicit constructions of systolically free metrics, using geometrically controlled surgery. Here parts of the manifold X^n look like a *lower dimensional* manifold

$$M^m, \text{ where } m \leq n - 1, \tag{4.7}$$

or more concretely, like circle bundle over such an M, with *short circle fibers*. In particular, such constructions yield free families of metrics on *products of spheres*.

Later, [**BabK98**] developed a more general framework for constructing metrics well adapted for the systolic problem. This was done in the context of simplicial maps f, where some of the top dimensional simplices are collapsed to walls of *positive codimension*. A positive quadratic form is obtained by "pullback" by f, suitably interpreted. The form can then be inflated a little to make it definite, yielding a true smooth metric. This can be thought of as a generalisation of the *short circle fiber* construction, *cf.* (4.7). The next step was to realize that metrics can be pulled back by an even more general morphism, defined below.

Let X and Y be n-dimensional simplicial complexes. Notions of volume, as well as systole with coefficients in A, can be defined for such objects by using piecewise smooth metrics.

A *morphism*$_{n,k}$ "from X to Y" (over A) is a continuous map $f : X \to W$ satisfying the following two conditions:

(a) the simplicial complex W is obtained from the "target" Y by attaching cells of dimension at most $n-1$;

(b) the map f induces a monomorphism in k-dimensional homology, $H_k(X, A) \hookrightarrow H_k(W, A)$.

Note that the positive codimension of the attached cells parallels the positive codimension in (4.7) above.

Morphisms of this type were referred to as "topological meromorphic maps" in [**KS99**]. Such a notion encompasses on the one hand, *surgeries* (*cf.* item 2 of subsection 4), and on the other, *complex blow ups*, in suitable contexts. The importance of these *morphisms*$_{n,k}$ stems from the following observation [**BabK98, KS99**].

Proposition 4.8. *If X admits a morphism$_{n,k}$ to Y, then the k-systolic freedom of X follows from that of Y.*

In the problem of systolic freedom, the appropriate *objects* are not manifolds but CW-complexes X, even though piecewise smooth metrics can only be defined on simplicial complexes X' homotopy equivalent to X. To establish the k-systolic freedom for a general X, one first maps X, by a map inducing a monomorphism on $H_k(X, \mathbb{R})$, to a product of $b_k(X)$ copies of the loop space $\Omega(S^{k+1})$. The space $\Omega(S^{k+1})$ is of the rational homotopy type of the Eilenberg Maclane space $K(\mathbb{Z}, k)$.

One uses the existence of a cell structure, due to Morse, which is *sparse* (in the sense that only cells in an arithmetic progression of dimensions occur) in the homotopy type of $\Omega(S^{k+1})$. This reduces the problem to a single $\Omega(S^{k+1})$, rather than a Cartesian product, and thus eliminates the dependence on the Betti number. The k-systolic freedom of the skeleta of the latter is established by finding a morphism$_{n,k}$ to a product of k-spheres, *cf.* Proposition 4.8.

This is done by analyzing the cell structure of a suitable rationalisation, where the cases $k = 2$ and $k \geq 4$ must be treated separately (k odd is easy). In the former case, one relies on the *telescope model* of Sullivan, and in the latter, on the existence of rational models with cells of dimension at most $n - k + 2$ added (where $n = \dim X$) in the construction of the *morphism*$_{n,k}$, *cf.* Proposition 4.8 and [**KS99, KS01**] for details.

4. Systolic freedom of complex projective plane, general freedom. The 2-systolic freedom of $\mathbb{C}P^2$ [**KS99**] answers in the negative the question from [**Gr83**, p. 136]. Such freedom results through the following 4 steps.

1. The space $\mathbb{C}P^2$ admits a *morphism*$_{4,2}$ to $S^2 \times S^2$. Here a 3-cell is attached along a diagonal class in $\pi_2(S^2 \times S^2)$.

2. The space $S^2 \times S^2$ admits a $morphism_{4,2}$ to $S^1 \times S^1 \times S^2$. Here a pair of 2-cells is attached along the generators of $\pi_1(S^1 \times S^1 \times S^2)$, corresponding to a surgery that allows one to pass from $S^1 \times S^1 \times S^2$ to $S^2 \times S^2$.
3. There exists a $(1, 2)$-systolically free sequence of metrics $\mathbf{g}_j$ on the 3-manifold $S^1 \times S^2$ (*cf.* [**Ka95, Pitte97**] and formula (4.5) above).
4. The 2-systolic freedom of $S^1 \times S^1 \times S^2$ results from the Kunneth formula, by taking Cartesian product of $\mathbf{g}_j$ with a circle S^1 of length $sys_2(\mathbf{g}_j)/sys_1(\mathbf{g}_j)$.

Part of the procedure described above can be carried out with torsion coefficients. Starting with Freedman's metrics (4.6) on $S^1 \times S^2$, we can construct in this way 2-systolically free metrics on $S^2 \times S^2$ even in the sense of $\mathbb{Z}_2$ coefficients. However, the 2-systolic freedom of $\mathbb{C}P^2$ with $\mathbb{Z}_2$ coefficients is still open. The technique described above fails over $\mathbb{Z}_2$ because the map in step 1 has even degree, violates condition (b) above, and hence does not define a $morphism_{4,2}$ over $A = \mathbb{Z}_2$.

The series of papers [**Ka95, Pitte97, BabK98, BKS98, KS99, KS01, Bab02**] proved general k-systolic freedom, as well as freedom in a pair of complementary dimensions, as soon as any systole other than the first one is involved. Thus, every closed n-manifold is k-systolically free in the sense of formula (4.3) whenever $2 \leq k < n$ and $H_k(X, \mathbb{Z})$ is torsionfree [**KS99, KS01**]. This answers the basic question of [**Gr83**, p. 5]. Similarly, every n-dimensional polyhedron is $(k, n-k)$-systolically free in the sense of formula (4.4) whenever $1 \leq k < n-k < n$ and $H_{n-k}(X, \mathbb{Z})$ is torsionfree [**Bab02**]. Simultaneous $(k, n-k)$-systolic freedom for a pair of adjacent values of k is explored in [**Ka02**].

5. Stable systolic and conformal inequalities

M. Gromov showed in [**Gr83**, 7.4.C] that to multiplicative relations in the cohomology ring of a manifold X, are associated stable systolic inequalities, satisfied by an arbitrary metric on X, *cf.* (5.13). These inequalities follow from the analogous, stronger inequalities for conformal systoles. He points out that the dependence of the constants in these inequalities on the Betti numbers is unsatisfactory. The invariants involved are defined in (5.5) and (5.7). We first point out the following open question.

Conjecture 5.1. *Every closed orientable smooth 4-manifold X satisfies the inequality*

$$stsys_2(\mathbf{g})^2 \leq C\,vol_4(\mathbf{g}),\ \forall \mathbf{g}, \tag{5.2}$$

for a suitable numerical constant C (independent of X).

The lower bound of (5.10) shows that one is unlikely to prove such an inequality via conformal systoles, except in the case of the connected sum of copies of $\mathbb{C}P^2$, when (5.2) holds with $C = 6$, *cf.* [**Ka3**]. The best one has available in general is $C = 6b_2(X)$ [**BanK03**].

Let $H_1(X; \mathbb{Z})_\mathbb{R}$ denote the lattice (*i.e.* free $\mathbb{Z}$-module) obtained as the quotient of $H_1(X; \mathbb{Z})$ by its torsion subgroup. The function $i \mapsto len(i\alpha)$ defines a norm $\|\ \|$, called *stable norm* [**Fe74**, 4.10], [**Gr99**, 4.35], on $H_1(X; \mathbb{Z})_\mathbb{R}$ by setting

$$\|\alpha_{\mathbb{R}}\| = \lim_{i \to \infty} i^{-1} len(i\alpha). \tag{5.3}$$

Denote by $\lambda_1(L, \|\ \|)$ the least length of a nonzero vector of a lattice L with respect to a norm $\|\ \|$. More generally, let i be an integer satisfying $1 \leq i \leq rk(L)$. The

i-th successive minimum $\lambda_i(L, \| \ \|)$ is the least $\lambda > 0$ such that there exist i linearly independent vectors in L of norm at most λ:

$$\lambda_i(L, \| \ \|) = \inf_{\lambda} \{\lambda \in \mathbb{R} \mid \exists v_1, \dots, v_i (l.i.) \ \forall k = 1, \dots, i : \|v_k\| \leq \lambda\} \tag{5.4}$$

The *stable homology k-systole*, denoted $stsys_k(\mathbf{g})$, is the least norm of a nonzero element in the lattice $H_k(X;\mathbb{Z})_{\mathbb{R}}$ with respect to the stable norm $\| \ \|$:

$$stsys_k(\mathbf{g}) = \lambda_1 \left(H_k(X;\mathbb{Z})_{\mathbb{R}}, \| \ \|\right). \tag{5.5}$$

It can be shown that if $(X, \mathbf{g})$ is an orientable surface then $sys_1(\mathbf{g}, \mathbb{Z}) = stsys_1(\mathbf{g})$.

The conformally invariant norm $\| \ \|_{L^n}$ in $H_1(X^n, \mathbb{R})$ is by definition dual to the conformally invariant L^n-norm in de Rham cohomology. The latter norm on $H^1(X, \mathbb{R})$ is the quotient norm of the corresponding norm on closed forms. Thus, given $\alpha \in H^1(X, \mathbb{R})$, we set

$$\|\alpha\|_{L^n} = \inf_{\omega} \left\{ \left(\int_X |\omega_x|^n dvol(x) \right)^{\frac{1}{n}} \middle| \ \omega \in \alpha \right\},$$

where ω runs over closed one-forms representing α. On a surface, or more generally for a middle-dimensional class $\alpha \in H^p(X, \mathbb{R})$ where $\dim(X) = 2p$, we may write

$$\|\alpha\|_{L^2}^2 = \int_X \omega \wedge *\omega, \tag{5.6}$$

where ω is the harmonic representative of α and $*$ is the Hodge star operator of the metric $\mathbf{g}$. The conformal 1-systole of $(X^n, \mathbf{g})$ is the quantity

$$confsys_1(\mathbf{g}) = \lambda_1 \left(H_1(X, \mathbb{Z})_{\mathbb{R}}, \| \ \|_{L^n}\right), \tag{5.7}$$

satisfying $stsys_1(\mathbf{g}) \leq confsys_1(\mathbf{g}) vol(\mathbf{g})^{\frac{1}{n}}$ on an n-manifold $(X, \mathbf{g})$.

1. Asymptotic behavior of conformal systole as function of $\chi(X)$. The inequality of Accola and Blatter is valid for the conformal systole. Their bound was improved by P. Buser and P. Sarnak [**BuSa94**], who proved that if Σ_s is a closed orientable surface of genus s, the conformal 1-systole satisfies the bounds

$$C^{-1} \log s < \sup_{\mathbf{g}} \left\{\lambda_1 \left(H^1(\Sigma_s, \mathbb{Z}), | \ |_{L^2}\right)\right\}^2 < C \log s, \ \forall s = 2, 3, \dots \tag{5.8}$$

where $C > 0$ is a numerical constant, the supremum is over all conformal structures $\mathbf{g}$ on Σ_s, and $| \ |_{L^2}$ is the associated L^2-norm, *cf.* formula (5.6). An explicit upper bound of $\frac{3}{\pi} \log(4s + 3)$ in (5.8) is provided in [**BuSa94**, formula (1.13)]. Note that by Poincaré duality, $\lambda_1 \left(H^1(\Sigma_s, \mathbb{Z}), | \ |_{L^2}\right) = \lambda_1 \left(H_1(\Sigma_s, \mathbb{Z}), | \ |_{L^2}\right)$.

It should be kept in mind that the asymptotic behavior of the 1-systole as a function of the genus is completely different from the conformal length. Indeed, M. Gromov [**Gr96**, 2.C] reveals the existence of a universal constant C such that we have an asymptotically *vanishing* upper bound

$$\frac{sys_1(\Sigma_s)^2}{area(\Sigma_s)} \leq C \frac{(\log s)^2}{s}, \tag{5.9}$$

whenever Σ_s is a closed, orientable surface of genus $s \geq 2$ with a Riemannian metric.

Inequality (5.8) admits the following higher-dimensional analogue [**Ka02**]. Let $n \in \mathbb{N}$ and consider the complex projective plane blown up at n points, $\mathbb{C}P^2 \# n \overline{\mathbb{C}P}^2$, where bar denotes reversal of orientation, while $\#$ is connected sum. Assume that the Surjectivity conjecture for the period map is satisfied for such manifolds, namely every line in the positive cone of the intersection cone occurs as the selfdual direction of a suitable metric. Then the conformal 2-systole satisfies the bounds

$$C^{-1}\sqrt{n} < \sup_{\mathbf{g}} \left\{ \lambda_1 \left(H^2(\mathbb{C}P^2 \# n\overline{\mathbb{C}P}^2, \mathbb{Z}), |\ |_{L^2} \right) \right\}^2 < Cn, \tag{5.10}$$

as $n \to \infty$, where $C > 0$ is a numerical constant, the supremum is taken over all smooth metrics $\mathbf{g}$ on $\mathbb{C}P^2 \# n\overline{\mathbb{C}P}^2$, and $|\ |_{L^2}$ is the norm associated with $\mathbf{g}$ by formula (5.6).

It would be interesting to eliminate the dependence of inequality (5.10) on the surjectivity conjecture. Moreover, can one improve the lower bound in (5.10) to linear dependence on n? Here one could try to apply an averaging argument, using Siegel's formula as in [**MH73**], over integral vectors satisfying $q_{n,1}(v) = -p$. Here one seeks a vector $v \in \mathbb{R}^{n,1}$ such that the integer lattice $\mathbb{Z}^{n,1} \subset \mathbb{R}^{n,1}$ has the Conway-Thompson behavior with respect to the positive definite form $SR(q_{n,1}, v)$.

2. A conjectured Pu-times-Loewner inequality. We conjecture an optimal inequality for certain 4-manifolds X with first Betti number $b_1(X) = 2$ which can be thought of as a product of the inequalities of Loewner (2.3) and Pu (2.5). The main obstacle to its proof is the absence of a generalized Pu's inequality (2.7) (or conjecture 3.8 for $n = 1$). For simplicity, we first state it for the manifold $\mathbb{T}^2 \times \mathbb{R}P^2$, and later discuss the relevant topological hypothesis.

Conjecture 5.11. *Assume* (2.7). *Then every metric* $\mathbf{g}$ *on* $\mathbb{T}^2 \times \mathbb{R}P^2$ *satisfies the inequality*

$$sys\pi_1(\mathbf{g})^2 stsys_1(\mathbf{g})^2 \leq \frac{2}{\sqrt{3}}\frac{\pi}{2} vol_4(\mathbf{g}), \tag{5.12}$$

while a metric which satisfies the boundary case of equality must admit a Riemannian submersion onto a Loewner-extremal torus, with fibers which are Pu-extremal, i.e. real projective planes of constant Gaussian curvature.

The relevant topological hypothesis is the following: X should be a 4-manifold with $b_1(X) = 2$, such that moreover the universal free abelian cover $\overline{X}$ is essential in dimension 2, *cf.* [**KKS**], so that in particular X satisfies the nonvanishing condition $[\overline{X}] \neq 0 \in H_2(\overline{X}, \mathbb{Z}_2)$, where $[\overline{X}]$ is the Poincaré dual of the pullback by $\overline{AJ_X}$ of the fundamental class with compact support of the universal cover $\mathbb{R}^2$ of the Jacobi torus of X.

We conjecture that in the boundary case of equality, both the topology and the metrics must be right, namely: (a) the manifold X must fiber smoothly over $\mathbb{T}^2$ with fiber $\mathbb{R}P^2$; (b) the Abel-Jacobi map $AJ_X : X \to \mathbb{T}^2$ is a Riemannian submersion with fibers of constant volume.

3. An optimal inequality in dimension and codimension 1. Recently, V. Bangert and the second author [**BanK03**] clarified the constants involved in Gromov's inequalities mentioned at the beginning of section 5, and showed in particular the following. Let $k > 0$, and assume $H^k(X, \mathbb{R}) \neq 0$. Let $k = \sum k_j$ and assume that the group $H^k(X,\mathbb{R})$ is spanned by cup products of classes of dimensions k_j. Then

$$\prod_j stsys_{k_j}(\mathbf{g}) \le C(k) \left(\prod_j b_{k_j}(X)\left(1+\log b_{k_j}(X)\right)\right) stsys_k(\mathbf{g}),\ \forall \mathbf{g}, \tag{5.13}$$

where the constant $C(k)$ depends only on k (and not on the metric or the Betti numbers). On the other hand, the optimal constants in such inequalities are generally unknown, unlike Loewner's classical inequality (2.3).

The following sharp inequality generalizing Loewner's is proved in [**Gr99**, pp. 259-260], based on the techniques of [**Bu-Iv95**]. Assume that the dimension, first Betti number, and real cuplength of X are all equal to n. Then

$$stsys_1(\mathbf{g})^n \le (\gamma_n)^{\frac{n}{2}} vol_n(\mathbf{g}),\ \forall \mathbf{g} \tag{5.14}$$

where γ_n denotes the classical Hermite constant, while equality in (5.14) is attained precisely by flat tori whose deck transformations define the densest sphere packings in dimension n. An optimal stable systolic inequality, for n-manifolds X with first Betti number $b_1(X,\mathbb{R})$ equal to one, is due to J. Hebda [**He86**]:

$$stsys_1(\mathbf{g}) sys_{n-1}(\mathbf{g}) \le vol_n(\mathbf{g}), \tag{5.15}$$

with equality if and only if $(X,\mathbf{g})$ admits a Riemannian submersion with connected minimal fibers onto a circle. An optimal inequality, involving the conformal 1-systole, is proved in [**BanK2**], namely equation (5.17) below. The new inequality generalizes simultaneously Loewner's inequality (2.3), Hebda's inequality (5.15), the inequality [**BanK03**, Corollary 2.3], as well as certain results of G. Paternain [**Pa01**]. We define the Bergé-Martinet constant, γ'_b, by setting

$$\gamma'_b = \sup_L \left\{ \lambda_1(L)\lambda_1(L^*) \,\middle|\, L \subset \left(\mathbb{R}^b, |\ |\right) \right\}, \tag{5.16}$$

where the supremum is over all lattices L in $\mathbb{R}^b$, *cf.* [**BeM89, CS94**]. Here L^* denotes the lattice dual to L, while $|\ |$ is a Euclidean norm. The supremum defining γ'_b is attained, and the lattices realizing it are called *dual-critical* [**BeM89**]. The following is proved in [**BanK2**]. Let X be a compact, oriented, n-dimensional manifold with positive first Betti number $b_1(X) \ge 1$. Then every metric $\mathbf{g}$ on X satisfies

$$confsys_1(\mathbf{g}) sys_{n-1}(\mathbf{g}) \le \gamma'_{b_1(X)}\ vol_n(\mathbf{g})^{\frac{n-1}{n}}, \tag{5.17}$$

where equality occurs if and only if there exists a dual-critical lattice L in Euclidean space $\mathbb{R}^b$ (*cf.* formula (5.16)), and a Riemannian submersion of X onto the flat torus $\mathbb{R}^b/L$, such that all fibers are connected minimal submanifolds of X.

6. Isoembolic Inequalities

Let $inj(X)$ represent the injectivity radius of a Riemannian manifold X. All geodesics of length $\le inj(X)$ will thus be minimizing. This section has to do with "isoembolic inequalities" which are estimates of volume in terms of the injectivity radius. In the closed ball, $B_p(r)$, centered at p of radius $r < \frac{inj(X)}{2}$ all geodesic segments will minimize and it will be *SGM* as a manifold with boundary (which we denote $S(r)$). We will let ω_n represent the volume of the standard unit sphere (of injectivity radius and diameter π) while β_n will be the volume of the Euclidean n-ball (so that $n\beta_n = \omega_{n-1}$). One major open question is:

Conjecture 6.1.
a) For any $r \leq \frac{inj(X)}{2}$, $vol(B(r)) \geq \frac{\omega_n}{2}(\frac{2}{\pi})^n r^n$ with equality holding only if the ball is isometric to a hemisphere.
b) For any $r \leq inj(X)$, $vol(B(r)) \geq \frac{\omega_n}{\pi^n} r^n$, where equality holds if and only if X is isometric to the round sphere of injectivity radius r (i.e. extrinsic radius $\frac{r}{\pi}$).

In part a) above, the hemisphere in the equality case will be that of a sphere of (extrinsic) radius $\frac{2r}{\pi}$ (so the hemisphere is a metric ball of intrinsic radius r). In fact a) is a stronger conjecture than b) since by the triangle inequality $B(r)$ contains two balls of radius $\frac{r}{2}$ with disjoint interiors.

1. Berger's 2 and 3 dimensional estimates. The early results on this conjecture and the conjecture itself are due to Berger (see [**Be77**]). The known results take the form of proving inequalities $vol(B(r)) \geq c(n)r^n$ for non-sharp constants $c(n)$. In 2 dimensions it is not hard to get such an estimate when $r \leq \frac{inj(X)}{2}$. Simply notice that for all $t \leq r$, $vol(S_p(t)) \geq 4t$ (*i.e.* the length of the boundary of $B_p(t)$ is $\geq 4t$) since "antipodal" points on $S(t)$ must be at least $2t$ apart in X and hence also along $S(t)$. This says that $vol(B_p(r)) = \int_0^r vol(S_p(t))dt \geq 2r^2$. The conjecture would say $vol(B(r)) \geq \frac{8}{\pi}r^2$ and hence this simple argument leads to a constant not too far from the best possible. The conjecture with the sharp constant is still open. (Using ideas in [**C-D**] - *cf.* section 2 - one can show in this 2-dimensional setting that $vol(B_p(t)) \geq 4\pi - vol(S_p(t))$ and then use this to get the better estimate $vol(B_p(r)) \geq \frac{16}{4+\pi}r^2$, but since this argument is a little involved and does not yield the sharp constant we will not pursue it.)

An interesting example to keep in mind in the "Mercedes-Benz" example. Consider three unit length line segments in the plane emanating from the origin making angles of $\frac{2\pi}{3}$. Thin tubular neighborhoods have arbitrarily small area but every boundary point is at a bounded distance from its antipodal point (the point halfway around the boundary). Thus one needs more than just bounded distance between "antipodal boundary points" to get area lower bounds.

One might hope to generalize the above simple argument for the area of 2-balls by showing:

Question 6.2. Let $\mathbf{g}$ be a Riemannian metric on the n-sphere S^n such that the $\mathbf{g}$-distance between antipodal points is $\geq t$.
a) Is $vol((S^n, \mathbf{g})) \geq \frac{\omega_n}{\pi^n} t^n$?
b) Is there any constant $c(n)$ such that $vol((S^n, \mathbf{g})) \geq c(n)t^n$?

In the above one can either think of the standard antipodal map when S^n is embedded in $\mathbb{R}^n$ in the standard way, or simply take the antipodal map to be any order two fixed point free diffeomorphism. Metric balls of radius $\frac{t}{2}$ satisfy this when $t \leq inj(X)$.

The easy argument we gave above for the volume of 2-balls consists of answering a) above in the case $n = 1$. This is the only case where a) is known. Berger in [**Be77**] gave an answer to b) above for $n = 2$ and hence showed by integrating (as in the 2 dimensional argument) that there is a constant c such that for 3 dimensional manifolds when $r \leq \frac{inj(X)}{2}$, then $vol(B_p(r)) \geq cr^3$.

Though part a) of Question 6.2 in 2 dimensions is still open (and very interesting), there are enough reasons to believe it to be true that it would warrant being called a conjecture. One approach would be to use the uniformization theorem and conformal length techniques (but note that the antipodal map need not be conformal).

In higher dimensions there is not enough evidence one way or the other even for question b). However (see [**Cr02**]) there are constants such that for any given metric g on the 3-sphere either Question 6.2 b) holds or Question 2.11 holds. This follows from a lower bound on the filling radius (*cf.* section 1) in terms of the infimum of the length of closed curves who link their antipodal images along with an application of Theorem 3.1. In the other direction, Ivanov (see [**Iv97**] or [**Iv98**]) has given examples of a sequence of metrics on S^3 that Gromov-Housdorff converge to the standard metric but whose volumes go to zero. Although this does not give a counter example it does show that the topological properties of the antipodal map are important in the above question. (Even in two dimensions long thin cigar shapes show that simply because every point is far from some other point the area need not be large.)

2. Higher dimensions. The two ways that higher dimensional isoembolic type inequalities have been proved are via Gromov's estimate of section 1 and via an estimate of Berger and Kazdan [**Be-Ka78**]. The Berger-Kazdan inequality first appeared in the proof of the Baschke conjecture for spheres (see [**Bes78**] appendices D and E). Berger then used the result to give the sharp estimate which holds for any compact Riemannian manifold:

$$vol(X) \geq \frac{\omega_n}{\pi^n} inj(X)^n \tag{6.3}$$

with equality holding only for round spheres. If you know that X is not homeomorphic to a sphere then you can do better [**Cr88A**], showing $vol(X) \geq c(n)inj(X)^n$ where $c(n)$ is an explicit constant larger than $\frac{\omega_n}{\pi^n}$.

The Berger-Kazdan inequality along with Santaló's formula [**San52, San76**, Chapter 19] was used in [**Cr80**] to give a general sharp isoperimetric inequality for compact Riemannian manifolds, X, with boundary, ∂X where all geodesic minimize (*e.g.* *SGM* manifolds):

$$\frac{vol(\partial X)}{vol(X)^{\frac{n-1}{n}}} \geq C(n)$$

where the constant $C(n)$ is just $\omega_{n-1}(\frac{2}{\omega_n})^{\frac{n-1}{n}}$. Equality holds if and only if X is isometric to a hemisphere. (The inequality has a version for all manifolds with boundary which involves a term measuring the fraction of geodesics that minimize to the boundary.) One gets a (non-sharp) version of conjecture 6.1 a) in all dimensions by applying this to the balls $B_p(t)$ and integrating t from 0 to r:

$$vol(B_p(r)) \geq C(n)^n r^n. \tag{6.4}$$

Further, applying the isoperimetric inequality again we see: $vol(S_p(r)) \geq C(n)^n r^{n-1}$. The failure of this estimate to be sharp comes from the fact that the isoperimetric inequality is not sharp for spherical caps (*i.e.* metric balls in the round sphere) unless they are hemispheres. We mentioned before (Conjecture 6.1 part a) that, for $r \leq \frac{inj}{2}$, we do not know the optimal lower bound for $vol(B_p(r))$. Except in case of dimension 2, we also don't know the optimal lower bound for $vol(S_p(r))$, however the above isoperimetric inequality will imply the sharp estimate on $vol(S_p(r))$ if Conjecture 6.1 part a is proven.

Although we are unable to prove Conjecture 6.1 for all balls there are some results about the "average" volume of balls, $AveVol(r)$, of a compact Riemannian manifold X. By this we mean

$$AveVol(r) \equiv \frac{1}{vol(X)} \int_X vol(B_p(r))dp$$

where dp is the Riemannian volume form. In **[Cr84]** it is shown for any $r \leq inj(X)$ that $AveVol(r) \geq \frac{\omega_n}{\pi^n} r^n$ with equality holding only for the round sphere of injectivity radius r. Thus showing that part b) of Conjecture 6.1 is true "on the average" (and giving another proof of Berger's isoembolic inequality 6.3). If one further knows that X has no conjugate points, then one can improve this to $AveVol(r) \geq \beta_n r^n$, where equality holds if and only if X is flat **[Cr92]**. In fact, this holds for all r, if one interprets $vol(B_p(r))$ to mean the volume of a ball in the universal cover centered at a lift of p.

Gromov in **[Gr83]**, showed (in an argument similar to the one presented in section 1) that the filling radius of a compact Riemannian manifold is always bounded from below by a constant times the injectivity radius. This, along with Theorem 3.1, gave an alternative proof to Berger's isoembolic inequality 6.3 (albeit with a non-sharp constant). This idea was extended in **[G-P92]** to get non-sharp universal estimates on the volumes of balls in terms of the local geometric contractibility function. As a special case, it gives an alternative proof of inequality 6.4. Sabourau also used these filling radius ideas in **[Sab2]** to give improved versions of both (6.3) and (6.4) (with worse constants) where instead of injectivity radius he is able to substitute the length of the shortest geodesic loop. A geodesic loop is a closed curve which is a geodesic at all but one point. Geodesic loops are much more abundant than closed geodesics and the shortest one will have length, $sgl(\mathbf{g})$, which satisfies

$$2inj(\mathbf{g}) \leq sgl(\mathbf{g}) \leq L(\mathbf{g}) \tag{6.5}$$

Sabourau's version of (6.4), which holds for all $r \leq \frac{1}{2}sgl(\mathbf{g})$, also used the ideas of **[G-P92]**.

7. Acknowledgments

We are grateful to E. Calabi, I. Chavel, A. Nabutovsky, and S. Sabourau for reading a draft version of the manuscript and for their comments.

References

[Ac60] R. D. M. Accola, *Differential and extremal lengths on Riemannian surfaces*, Proc. Math. Acad. Sci. USA **46** (1960), 83–96.

[Ak02] H. Akrout, *Théorème de Voronoï dans les espaces symétriques.* [Voronoi theorem in symmetric spaces] Canad. J. Math. **54** (2002), no. 3, 449–467.

[Al62] F. Almgren Jr., *The homotopy groups of the integral cycle groups*, Topology **1** (1962),257–299.

[Ar79] M. A. Armstrong, Basic topology. McGraw-Hill Book Co. (UK), Ltd., London-New York, 1979.

[Bab02] I. Babenko, *Forte souplesse intersystolique de variétés fermées et de polyèdres*, Annales de l'Institut Fourier **52** 4 (2002), 1259-1284.

[BabK98] I. Babenko and M. Katz, *Systolic freedom of orientable manifolds*, Annales Scientifiques de l'E.N.S. (Paris) **31** (1998), 787-809.

[BKS98] I. Babenko, M. Katz, and A. Suciu, *Volumes, middle-dimensional systoles, and Whitehead products*, Math. Res. Lett. **5** (1998), no. 4, 461–471.

[Ban94] V. Bangert, *Geodesic rays, Busemann functions and monotone twist maps*, Calc. Var. Part. Diff. Eq. **2** (1994), no. 1, 49–63.

[BanK03] V. Bangert and M. Katz, *Stable systolic inequalities and cohomology products*, Comm. Pure Appl. Math. **56** (2003), in press. `math.DG/0204181`

[BanK2] V. Bangert, and M. Katz, *Riemannian manifolds with harmonic* 1*-forms of constant norm*, in preparation.

[Bav86] C. Bavard, *Inégalité isosystolique pour la bouteille de Klein*, (French) [*Isosystolic inequality for the Klein bottle*] Math. Ann. **274** (1986), no. 3, 439–441.

[Bav92a] C. Bavard, *Ingalits isosystoliques conformes*, Comment. Math. Helv. **67** (1992), no. 1, 146–166.

[Bav92b] C. Bavard, *La systole des surfaces hyperelliptiques*, Prepubl. Ec. Norm. Sup. Lyon 71 (1992).

[BeM89] A.-M. Bergé, and J. Martinet, *Sur un problème de dualité lié aux sphères en géométrie des nombres*, J. Number Theory **32** (1989), 14–42.

[Be77] M. Berger, *Volume et rayon d'injectivité dans les variétés riemanniennes de dimension 3*, Osaka J. Math., **14** (1977), 191- 200.

[Be80] M. Berger, *Une bourne inférieure pour le volume d'une variété riemannienne en fonction du rayon d'injectivité*, Ann. Inst. Fourier (Grenoble) **30** (1980) no.3, 259-265.

[Be83] M. Berger, *Filling Riemannian manifolds or isosystolic inequalities*, Global Riemannian Geometry, Ellis Horwood Ser. Math. Appl. (1983), 75–84.

[Be-Ka78] M. Berger and J. L. Kazdan, *A Sturm-Liouville inequality with applications to an isoperimetric inequality for volume in terms of injectivity radius, and to Wiedersehen manifolds*, in General Inequalities 2 (Proceedings of the second International Conference on General Inequalities, 1978), E. F. Beckenbach (ed.), ISNMA47, Birkhauser Verlag, no.3, 251-254.

[Bes78] A. Besse. *Manifolds all of whose geodesics are closed*, Ergebisse Grenzgeb. Math., no. 93, Springer, Berlin, 1978.

[B-C-G95] G. Besson, G. Courtois, and S. Gallot, *Entropies et rigidités des espaces localement symétriques de courbure strictement négative*, Geom. Funct. Anal. **5** (1995) no. 5, 731-799.

[B-C-G96] G. Besson, G. Courtois, and S. Gallot, *Minimal entropy and Mostow's rigidity theorems*, Ergodic Theory Dynam. Systems **16** (1996) no. 4, 623-649.

[Bi27] G. D. Birkhoff, *Dynamical systems*, Amer. Math. Soc. Colloq. Publ. Vol. 9, Providence, RI, 1927.

[Bl61] C. Blatter, *Über extremallängen auf geschlossen flächen*, Comment. Math. Helv. **35** (1961), no.3, 153–168.

[Bo91] F. Bonahon, *The geometry of Teichmüller space via geodesic currents*, Inven. Math. **33** (1991), 445–464.

[Bo93] F. Bonahon, *Surfaces with the same marked length spectrum*, Topology Appl. **50** (1993) no. 1, 55-62.

[Br96] R. Bryant, *On extremals with prescribed Lagrangian densities* Manifolds and geometry (Pisa, 1993), 86–111, Sympos. Math., XXXVI, Cambridge Univ. Press, Cambridge, 1996.

[Bu-Iv95] Burago, D.; Ivanov, S.: *On asymptotic volume of tori*, Geom. Funct. Anal. **5** (1995), no. 5, 800–808.

[Bu-Iv2] D. Burago and S. Ivanov, *On asymptotic volume of Finsler tori, minimal surfaces in normed spaces, and symplectic filling volume*, (preprint).

[B-Z80] Yu. Burago and V. Zalgaller, Geometric Inequalities, "Nauka"(1980), Russian.

[B-Z88] Yu. Burago and V. Zalgaller, Geometric Inequalities, Springer (1988).

[BuSa94] P. Buser, and P. Sarnak, *On the period matrix of a Riemann surface of large genus. With an appendix by J. H. Conway and N. J. A. Sloane*, Invent. Math. **117** (1994) no. 1, 27–56.

[Ca96] E. Calabi, *Extremal isosystolic metrics for compact surfaces*, Actes de la table ronde de geometrie differentielle, Sem. Congr. 1 (1996), Soc. Math. France 165–204.

[Ca03] E. Calabi, private conversations.

[Ca3] E. Calabi, On the Extremal Isosystolic Problems, in preparation.

[CaCa92] E. Calabi and J. Cao, *simple closed geodesics on convex surfaces*, J. Diff. Geom., **36** (1992), 517-549

[Ch93] I. Chavel, Riemannian geometry—a modern introduction. Cambridge Tracts in Mathematics, 108. Cambridge University Press, Cambridge, 1993.

[CS94] J. H. Conway; N. J. A. Sloane: On lattices equivalent to their duals. J. Number Theory 48 (1994), no. 3, 373–382.

[Cr80] C. Croke, *Some isoperimetric inequalities and eigenvalue estimates*, Ann. Scient. Ec. Norm. Sup., 4e serie, t.**13** (1980), 419-435.

[Cr84] C. Croke, *Curvature Free Volume Estimates*, Inventiones Mathematicae **76** (1984), 515-521.

[Cr88A] C. Croke, *An Isoembolic Pinching Theorem*, Invent. Math., **92** (1988), 385–387.

[Cr88B] C. Croke, *Area and the Length of the Shortest Closed Geodesic*, J. Diff. Geom. **27** (1988), 1–21.

[Cr90] C. Croke, *Rigidity for surfaces of non-positive curvature*, Comm. Math. Helv. **65** (1990) no.1, 150-169.

[Cr91] C. Croke, *Rigidity and the distance between boundary points*, J. Diff. Geom. **33** (1991), 445-464.

[Cr92] C. Croke, *Volumes of balls in manifolds without conjugate points*, International J. Math. **3** (1992) no.4, 455-467.

[Cr01] C. Croke, *Rigidity Theorems in Riemannian geometry*, (preprint – to appear in "Geometry, Inverse Problems and Control of PDE" based on the conference at IMA, July 2001).

[Cr02] C. Croke, *the volume and lengths on a three sphere*, Comm. Anal. Geom., **10** (2002) no. 3, 467–474.

[C-D] C. Croke, N. Dairbekov, *Lengths and volumes in Riemannian manifolds*, preprint.

[C-D-S00] C. Croke, N. Dairbekov, and V. Sharafutdinov, *Local boundary rigidity of a compact Riemannian manifold with curvature bounded above*, Trans. Amer. Math. Soc. **325** (2000) no.9, 3937-3956.

[C-F-F92] C. Croke, A. Fathi, and J. Feldman, *The marked length-spectrum of a surface of nonpositive curvature*, Topology **31** (1992) no. 4, 847–855.

[FK92] H. M. Farkas, I. Kra, *Riemann surfaces*. Second edition. Graduate Texts in Mathematics, 71. Springer-Verlag, New York, 1992.

[Fe74] H. Federer, Real flat chains, cochains, and variational problems. *Indiana Univ. Math. J.* **24** (1974), 351–407.

[Fr99] M. Freedman, $\mathbb{Z}_2$-*systolic-freedom*, Proceedings of the Kirbyfest (Berkeley, CA, 1998), 113–123, Geom. Topol. Monogr., vol. 2, Geom. Topol, Coventry, 1999. `http://www.maths.warwick.ac.uk/gt/GTMon2/paper6.abs.html`.

[Gr81] M. Gromov, *Structures métriques pour les variétés riemanniennes*. (French) [Metric structures for Riemann manifolds] Edited by J. Lafontaine and P. Pansu. Textes Mathématiques [Mathematical Texts], 1. CEDIC, Paris, 1981.

[Gr83] M. Gromov, *Filling Riemannian manifolds*, J. Diff. Geom. **18** (1983), 1-147.

[Gr96] M. Gromov, *Systoles and intersystolic inequalities Actes de la Table Ronde de Géométrie Différentielle (Luminy, 1992)*, 291–362, Sémin. Congr., vol. 1, Soc. Math. France, Paris, 1996. `www.emis.de/journals/SC/1996/1/ps/smf_sem-cong_1_291-362.ps.gz`

[Gr99] M. Gromov, *Metric structures for Riemannian and non-Riemannian spaces*, Progr. in Mathematics, vol 152, Birkhäuser, Boston, 1999.

[G-P92] R. Green and P. Petersen, *Little topology, big volume*, Duke Math. J. **67** (1992), no. 2, 273–290.

[Ha92] U. Hamenstädt, *Time-preserving conjugacies of geodesic flows*, Ergod. Th. & Dynam. Sys. **12** (1992), 67-74.

[Ha99] U. Hamenstädt, *Cocycles, symplectic structures and intersection*, Geom. Funct. Anal. **9** (1999) no. 1, 90-140.

[He82] J. Hebda, *Some lower bounds for the area of surfaces*, Invent. Math. **65** (1982), 485–491.

[He86] J. Hebda, *The collars of a Riemannian manifold and stable isosystolic inequalities*, Pacific J. Math. **121** (1986), 339–356.

[Iv02] S. Ivanov, *On two-dimensional minimal fillings*, St. Petersbg. Math. J. **13** (2002), no. 1, 17–25.

[Iv97] S. Ivanov *Gromov-Hausdorff Convergence and volumes of manifolds*, Algebra i Analiz, **9** (1997) No.5, 65-83 (Russian).

[Iv98] S. Ivanov, *Gromov-Hausdorff Convergence and volumes of manifolds*, St. Petersburg Math. J., **9** (1998) No.5, 945-959.

[Ka83] M. Katz, *The filling radius of two-point homogeneous spaces*, J. Diff. Geom. 18 (1983) 505-511.

[Ka91A] M. Katz, *On neighborhoods of the Kuratowski imbedding beyond the first extremum of the diameter functional*, Fundamenta Mathematicae 137 (1991) 15-29.

[Ka91B] M. Katz, *The rational filling radius of complex projective space*, Topology and Its Applications 42 (1991) 201-215.

[Ka91C] M. Katz, *Pyramids in the complex projective plane*, Geometriae Dedicata 40 (1991) 171-190.

[Ka95] M. Katz, *Counterexamples to isosystolic inequalities*, Geom. Dedicata **57** (1995), 195–206.

[Ka02] M. Katz, *Local calibration of mass and systolic geometry*, Geometric and Functional Analysis **12**, issue 3 (2002), 598–621.

[Ka3] M. Katz, *Four-manifold systoles and surjectivity of period map*, Comment. Math. Helvetici (2003), conditionally accepted for publication.

[KS99] M. Katz, and A. Suciu, *Volume of Riemannian manifolds, geometric inequalities, and homotopy theory*, in *Tel Aviv Topology Conference: Rothenberg Festschrift* (M. Farber, W. Lück, and S. Weinberger, eds.), Contemp. Math., vol. 231, AMS (1999), pp. 113–136.

[KS01] M. Katz, and A. Suciu, *Systolic freedom of loop space*, Geometric and Functional Analysis **11** (2001), 60-73.

[KKS] M. Katz, M, Kreck, & A. Suciu, *Free abelian covers, short loops, stable length, and systolic inequalities.* `arXiv:math.DG/0207143`

[Ko87] S. Kodani, *On two-dimensional isosystolic inequalities* Kodai Math. J. **10** (1987) no. 3, 314–327.

[Mae94] M. Maeda, *The length of a closed geodesic on a compact surface*, Kyushu J. Math., **48:1** (1994), 9–18.

[Mas97] D. Massart, *Stable norms of surfaces: local structure of the unit ball of rational directions.* Geom. Funct. Anal. **7** (1997), no. 6, 996–1010.

[MH73] J. Milnor and D. Husemoller, Symmetric bilinear forms. Springer, 1973.

[N-R02] A. Nabutovsky and R. Rotman, *The length of the shortest closed geodesic on a 2-dimensional sphere*, Int. Math. Res. Not. (2002) no. 23, 1211–1222.

[N-R1] A. Nabutovsky and R. Rotman, *Upper bounds on the length of a shortest closed geodesic and quantitative Hurewicz theorem*, (preprint).

[N-R2] A. Nabutovsky and R. Rotman, *Volume, diameter, and the minimal mass of a stationary 1-cycle*, (preprint).

[Ot90A] J.-P. Otal, *Le spectre marqué des longueurs des surfaces à courbure négative*, Ann. of Math. **131** (1990), 151-162.

[Ot90B] J.-P. Otal, *Sur les longueurs des géodésiques d'une métrique à courbure négative dans le disque*, Comment. Math. Helv, **65** (1990) no. 2, 334-347.

[Pa01] G. P. Paternain, *Schrödinger operators with magnetic fields and minimal action functionals*, Israel J. Math. **123** (2001), 1–27.

[Pitte97] C. Pittet, *Systoles on* $S^1 \times S^n$, Differential Geom. Appl. **7** (1997), 139–142.

[Pitts81] J. Pitts, *Existence and regularity of minimal surfaces on Riemannian manifolds*, Ann. Math. studies **27**(1981), Princeton University Press.

[Pu52] P.M. Pu, *Some inequalities in certain nonorientable Riemannian manifolds*, Pacific J. Math. **2** (1952), 55–71.

[Sab1] S. Sabourau, *Filling radius and short closed geodesics of the sphere*, preprint (2002).

[Sab2] S. Sabourau, *Global and local volume bounds and the shortest geodesic loop*, preprint (2002).

[Sab02c] S. Sabourau, *Systoles des surfaces plates singulières de genre deux*, preprint (2002).

[Sak88] T. Sakai, *A proof of the isosystolic inequality for the Klein bottle*, Proc. Amer. Math. Soc. **104** (1988) no. 2, 589–590.

[San52] L. A. Santaló, *Measure of sets of geodesics in a Riemannian space and applications to integral formulas in elliptic and hyperbolic spaces*, Summa Brasil. Math. **3**, (1952). 1–11.

[San76] L. A. Santaló, Integral Geometry and Geometric Probability. With a Foreword by Mark Kac. Encyclopedia of Mathematics and its Applications, Vol. 1. Addison-Wesley Publishing Co., Reading, Mass.-London-Amsterdam, 1976.

[Sch93] P. Schmutz, Riemann surfaces with shortest geodesic of maximal length. Geom. Funct. Anal. 3 (1993), no. 6, 564–631.

[S-Y94] R Schoen and S.-T. Yau, *Lectures on Differential Geometry*, International Press (1994).

[Si90] K. Sigmund, *On the space of invariant measures for hyperbolic flows*, Ann. of Math. **131**(1990), 151–162.

[Ta92] P. M. Tamrazov, *Methods for studying extremal metrics and moduli in a twisted Riemannian manifold* (Russian. Russian summary) Mat. Sb. **183** (1992), no. 3, 55–75; translation in Russian Acad. Sci. Sb. Math. **75** (1993), no. 2, 333–351.

[Tre85] A. Treibergs, *Estimates of volume by the length of shortest closed geodesics on a convex hypersurface*, Invent. Math., **80** (1985), 481–488.

[Tro86] M. Troyanov, *Les surfaces euclidiennes á singularités coniques.* (French) [Euclidean surfaces with conical singularities] Enseign. Math. (2) **32** (1986), no. 1-2, 79–94.

[Wi92] F. Wilhelm, *On the filling radius of positively curved manifolds.* Invent. Math. **107** (1992), no. 3, 653–668.

[Ya82] S.-T. Yau, *Problem section in "Seminar in Differential Geometry"*, ed. by S.-T. Yau, Princeton University Press (1982).

Christopher B Croke:
Department of Mathematics
University of Pennsylvania
Philadelphia, PA 19104-6395
ccroke@math.upenn.edu

Mikhail G. Katz
Department of Mathematics and Statistics
Bar Ilan University
Ramat Gan 52900, Israel
katzmik@math.biu.ac.il

A Kawamata-Viehweg Vanishing Theorem on compact Kähler manifolds

Jean-Pierre Demailly*, Thomas Peternell**

ABSTRACT. We prove a Kawamata-Viehweg vanishing theorem on a normal compact Kähler space X: if L is a nef line bundle with $L^2 \neq 0$, then the group $H^q(X, K_X + L)$vanishes for $q \geq \dim X - 1$. As an application we complete a part of the abundance theorem for minimal Kähler threefolds: if X is a minimal Kähler threefold, then the Kodaira dimension $\kappa(X)$ is nonnegative.

0. Introduction

In this paper we establish the following Kawamata-Viehweg type vanishing theorem on a compact Kähler manifold or, more generally, a normal compact Kähler space.

CLAIM 0.1. *Let X be a normal compact Kähler space of dimension n and L a nef line bundle on X. Assume that $L^2 \neq 0$. Then*

$$H^q(X, K_X + L) = 0$$

for $q \geq n - 1$.

In general, one expects a vanishing

$$H^q(X, K_X + L) = 0$$

for $q \geq n + 1 - \nu(L)$, where $\nu(L)$ is the numerical Kodaira dimension of the nef line bundle L, i.e. $\nu(L)$ is the largest integer ν such that $L^\nu \neq 0$.

Of course, when X is projective, Theorem 0.1 is contained in the usual Kawamata-Viehweg vanishing theorem, but the methods of proof in the algebraic case clearly fail in the general Kähler setting. Instead we proceed in the following way. Clearly we may assume that X is smooth and by Serre duality, only the cohomology group H^{n-1} is of interest. Take a singular metric h on L with positive curvature current T with local weight function h. By [Si74, De93a] there exists a decomposition

$$T = \sum \lambda_j D_j + G, \tag{D}$$

where $\lambda_j \geq 1$ are irreducible divisors, and G is a pseudo-effective current such that $G|D_i$ is pseudo-effective for all i. Consider the multiplier ideal sheaf $\mathcal{I}(h)$. We associate to h another, "upper regularized" multiplier ideal sheaf $\mathcal{I}_+(h)$ by setting

$$\mathcal{I}_+(h) := \lim_{\varepsilon \to 0_+} \mathcal{I}(h^{1+\varepsilon}) = \lim_{\varepsilon \to 0_+} \mathcal{I}\big((1+\varepsilon)\varphi\big).$$

It is unknown whether $\mathcal{I}(h)$ and $\mathcal{I}_+(h)$ actually differ; in all known examples they are equal. Then in Section 2 the following vanishing theorem is proved.

CLAIM 0.2. *Let (L, h) be a holomorphic line bundle over a compact Kähler n-fold X. Assume that L is nef and has numerical Kodaira dimension $\nu(L) = \nu \geq 0$, i.e. $c_1(L)^\nu \neq 0$ and ν is maximal. Then the morphism*

$$H^q(X, \mathcal{O}(K_X + L) \otimes \mathcal{I}_+(h)) \longrightarrow H^q(X, K_X + L)$$

induced by the inclusion $\mathcal{I}_+(h) \subset \mathcal{O}_X$ vanishes for $q > n - \nu$.

The strategy of the proof of Theorem 0.2 is based on a direct application of the Bochner technique with special hermitian metrics constructed by means of the Calabi-Yau theorem.

Now, coming back to the principles of the proof of Theorem 0.1, we introduce the divisor

$$D = \sum [\lambda_j] D_j.$$

Then Theorem 0.2 yields the vanishing of the map in cohomology

$$H^{n-1}(X, -D + L + K_X) \longrightarrow H^{n-1}(X, L + K_X).$$

Thus we are reduced to show that $H^{n-1}(D, L + K_X|D) = 0$, or dually that

$$H^0(D, -L + D|D) = 0.$$

This is now done by a detailed analysis of a potential non-zero section in $-L+D|D$; making use of the decomposition (D) and of a Hodge index type inequality.

The vanishing theorem 0.1 is most powerful when X is a threefold, and in the second part of the paper we apply 0.1 - or rather a technical generalization - to prove the following abundance theorem.

CLAIM 0.3. *Let X be a $\mathbb{Q}$-Gorenstein Kähler threefold with only terminal singularities, such that K_X is nef (a minimal Kähler threefold for short). Then $\kappa(X) \geq 0$.*

This theorem was established in the projective case by Miyaoka and in [Pe01] for Kähler threefolds, with the important exception that X is a simple threefold which is not Kummer. Recall that X is said to be *simple* if there is no proper compact subvariety through a very general point of X, and that X is said to be Kummer if X is bimeromorphic to a quotient of a torus. So our contribution here consists in showing that such a simple threefold X with K_X nef has actually $\kappa(X) = 0$. Needless to say that among all Kähler threefolds the simple non-Kummer ones (which conjecturally do not exist) are most difficult to deal with, since they do not carry much global information besides the fact that π_1 is finite and that they have a holomorphic 2-form. The first main ingredient in our approach is the inequality

$$K_X \cdot c_2(X) \geq 0$$

for a minimal simply connected Kähler threefold X with algebraic dimension $a(X) = 0$. Philosophically this inequality comes from Enoki's theorem that the tangent sheaf of X is K_X-semi-stable when $K_X^2 \neq 0$ resp. (K_X, ω)-semi-stable when $K_X^2 = 0$; here ω is any Kähler form on X. Now if this semi-stability with respect to a degenerate polarization would yield a Miyaoka-Yau inequality, then $K_X \cdot c_2(X) \geq 0$ would follow. However this type of Miyoka-Yau inequalities with respect to degenerate polarizations is completey unknown. In the projective case, the inequality follows from Miyaoka's generic nefness theorem and is based on char. p-methods.

Instead we approximate K_X (in cohomology) by Kähler forms ω_j. If T_X is still ω_j-semi-stable for sufficiently large j, then we can apply the usual Miyaoka-Yau inequality and pass to the limit to obtain $K_X \cdot c_2(X) \geq 0$. Otherwise we examine the maximal destabilizing subsheaf which essentially (because of $a(X) = 0$) is independent of the polarization.

The second main ingredient is the boundedness $h^2(X, mK_X) \leq 1$. If $K_X^2 \neq 0$, this is of course contained in Theorem 0.1. If $K_X^2 = 0$, we prove this boundedness under the additional assumption that $a(X) = 0$ and that $\pi_1(X)$ is finite (otherwise by a result of Campana X is already Kummer). The main point is that if $h^2(X, mK_X) \geq 2$, then we obtain "many" non-split extensions

$$0 \longrightarrow K_X \longrightarrow \mathcal{E} \longrightarrow mK_X \longrightarrow 0$$

and we analyze whether $\mathcal{E}$ is semi-stable or not. The assumption on π_1 is used to conclude that if $\mathcal{E}$ is projectively flat, then $\mathcal{E}$ is trivial after a finite étale cover.

From these two ingredients Theorem 0.3 immediately follows by applying Riemann-Roch on a desingularization of X.

The only remaining problem concerning abundance on Kähler threefolds is to prove that a simple Kähler threefold with K_X nef and $\kappa(X) = 0$ must be Kummer.

1. Preliminaries

We start with a few preliminary definitions.

DEFINITION 1.1. *A normal complex space X is said to be Kähler if there exists a Kähler form ω on the regular part of X such that the following holds. Every singular point $x \in X$ admits an open neighborhood U and a closed embedding $U \subset V$ into an open set $U \subset \mathbb{C}^N$ such that there is a Kähler form η on V with $\eta | U = \omega$.*

REMARK 1.2. Let X be a compact Kähler space and let $f : \hat{X} \longrightarrow X$ be a desingularization by a sequence of blow-ups. Then $\hat{X}$ is a Kähler manifold. More generally consider a holomorphic map $f : \hat{X} \longrightarrow X$ of a normal compact complex space to a normal compact Kähler space. If f is a projective morphism or, more generally, a Kähler morphism, then $\hat{X}$ is Kähler. For references to this and more informations on Kähler spaces, we refer to [Va84]. A Kähler form ω defines naturally a class $[\omega] \in H^2(X, \mathbb{R})$, see [Gr62] where Kähler metrics on singular spaces were first introduced. Therefore we also have a Kähler cone on a normal variety.

NOTATION 1.3. Let X be a normal compact complex space.
(1) Let A and B be reflexive sheaves of rank 1. Then we define $A \hat{\otimes} B := (A \otimes B)^{**}$. Moreover we let $A^{[m]} := A^{\hat{\otimes} m}$.
(2) A reflexive sheaf A is said to be a $\mathbb{Q}$-line bundle if there exists a positive integer m such that $A^{[m]}$ is locally free.
(3) X is $\mathbb{Q}$-Gorenstein if the canonical reflexive sheaf ω_X, also denoted K_X, is a $\mathbb{Q}$-line bundle. X is $\mathbb{Q}$-factorial, if every reflexive sheaf of rank 1 is a $\mathbb{Q}$-line bundle.

DEFINITION 1.4. *Let X be a normal compact Kähler threefold.*
(1) *X is simple if there is no proper compact subvariety through the very general point of X.*

(2) *X is Kummer, if X is bimeromorphic to a quotient T/G where T is a torus and G a finite group acting on T.*

It is conjectured that all simple threefolds are Kummer.

NOTATION 1.5.
(1) The *algebraic dimension* $a(X)$ of an irreducible reduced compact complex space is the transcendence degree of the field of meromorphic functions over $\mathbb{C}$. If $a(X) = 0$, i.e. all meromorphic functions on X are constant, then it is well known that X carries only finitely many irreducible hypersurfaces.
(2) A line bundle L on a compact Kähler manifold is *nef*, if $c_1(L)$ lies in the closure of the Kähler cone. For alternative descriptions see e.g. [DPS94,00]. If X is a normal compact Kähler, then L is nef if there exists a desingularization $\pi : \hat{X} \longrightarrow X$ such that $\pi^*(L)$ is nef. By [Pa98], this definition does not depend on the choice of π.

2. Hodge index type inequalities

We give here some generalizations of Hodge index inequalities for nef classes over compact Kähler manifolds. In this direction the main result is the Hovanskii-Teissier concavity inequality, which can be stated in the following way (see e.g. [De93b], Prop. 5.2 and Remark 5.3).

PROPOSITION 2.1. *Let $\alpha_1, \dots, \alpha_k$ and $\gamma_1, \dots \gamma_{n-k}$ be nef cohomology classes on a compact Kähler n-dimensional manifold X. Then*

$$\alpha_1 \cdots \alpha_k \cdot \gamma_1 \cdots \gamma_{n-k} \geq (\alpha_1^k \cdot \gamma_1 \cdots \gamma_{n-k})^{1/k} \cdots (\alpha_k^k \cdot \gamma_1 \cdots \gamma_{n-k})^{1/k}.$$

We want to derive from these a non vanishing property for intersection products of the form $\alpha^i \cdot \beta^j$. Let us fix a Kähler metric ω on X. By Proposition 2.1 applied with $k = i + j$ and the α_ℓ's being i copies of α followed by j copies of β and $\gamma_\ell = \omega$, we have

$$\alpha^i \cdot \beta^j \cdot \omega^{n-i-j} \geq (\alpha^k \cdot \omega^{n-k})^{i/k} \cdots (\beta^k \cdot \omega^{n-k})^{j/k}.$$

As all products α^k and analogues can be represented by closed positive currents, we have $\alpha^k \neq 0 \Rightarrow \alpha^k \cdot \omega^{n-k} > 0$, hence with $k = i + j$ we find

$$\alpha^{i+j} \neq 0 \text{ and } \beta^{i+j} \neq 0 \implies \alpha^i \cdot \beta^j \neq 0. \tag{2.2}$$

This is of course optimal in terms of the exponents if $\alpha = \beta$, but as we shall see in a moment, this is possibly not optimal in a dissymetric situation. Actually, we have the following additional inequalities which can be viewed as "differentiated" Hovanskii-Teissier inequalities.

THEOREM 2.3. *Let α and β be nef cohomology classes of type $(1,1)$ on a compact Kähler n-dimensional manifold X. Assume that $\alpha^p \neq 0$ and $\beta^q \neq 0$ for some integers $p, q > 0$. Then we have $\alpha^i \cdot \beta^j \neq 0$ as soon as there exists an integer $k \geq i + j$ such that*

$$i(k-p)_+ + j(k-q)_+ < k,$$

where x_+ means the positive part of a number x.

Proof. Assume that $\alpha^i \cdot \beta^j = 0$. We apply the Hovanskii-Teissier inequality respectively with $\alpha_\ell = \alpha + \varepsilon\omega$ (i terms), or $\alpha_\ell = \beta + \varepsilon\omega$ (j terms) or $\alpha_\ell = \omega$ ($k - i - j$ terms), and $\gamma_\ell = \omega$. This gives
$(*)$

$$(\alpha+\varepsilon\omega)^i \cdot (\beta+\varepsilon\omega)^j \cdot \omega^{n-i-j} \geq \big((\alpha+\varepsilon\omega)^k \cdot \omega^{n-k}\big)^{i/k} \big((\beta+\varepsilon)^k \cdot \omega^{n-k}\big)^{j/k} (\omega^n)^{1-i/k-j/k}.$$

By expanding the intersection form and using the assumption $\alpha^i \cdot \beta^j = 0$, we infer

$$(\alpha + \varepsilon\omega)^i \cdot (\beta + \varepsilon\omega)^j \cdot \omega^{n-i-j} \le O(\varepsilon)$$

as ε tends to zero. On the other hand $(\alpha + \varepsilon\omega)^k \cdot \omega^{n-k}$ is bounded away from 0 if $k \le p$ since then $\alpha^k \neq 0$, and $(\alpha + \varepsilon\omega)^k \cdot \omega^{n-k} \ge C\varepsilon^{k-p}$ for some constant $C > 0$ if $k \ge p$. Hence we infer from $(*)$ that

$$C\varepsilon^{(i/k)(k-p)_+ + (j/k)(k-q)_+} = O(\varepsilon),$$

and this is not possible if $i(k-p)_+ + j(k-q)_+ < k$. The theorem is proved. □

The special case $p = 2$, $q = 1$, $i = j = 1$, $k = 2$ provides the following result which will be needed later on several occasions.

COROLLARY 2.4. *Assume that α, β are nef with $\alpha^2 \neq 0$ and $\beta \neq 0$. Then $\alpha \cdot \beta \neq 0$.*

Finally, we state an extension of Proposition 2.1 in the case when one of the cohomology classes involved is not necessarily nef.

PROPOSITION 2.5. *Let α be a real $(1,1)$-cohomology class, and let $\beta, \gamma_1, \dots \gamma_{n-2}$ be nef cohomology classes. Then*

$$(\alpha \cdot \beta \cdot \gamma_1 \cdots \gamma_{n-2})^2 \ge (\alpha^2 \cdot \gamma_1 \cdots \gamma_{n-2})(\beta^2 \cdot \gamma_1 \cdots \gamma_{n-2}).$$

Proof. By proposition 2.1, the result is true when α is nef. If we replace β by $\beta + \varepsilon\omega$ and let $\varepsilon > 0$ tend to zero, we see that it is enough to consider the case when β is a Kähler class. Then $\alpha + \lambda\beta$ is also Kähler for $\lambda \gg 1$ large enough, and the inequality holds true with $\alpha + \lambda\beta$ in place of α. However, after making the replacement, the contributions of terms involving λ in the right and left hand side of the inequality are both equal to

$$2\lambda(\alpha \cdot \beta \cdot \gamma_1 \cdots \gamma_{n-2})(\beta^2 \cdot \gamma_1 \cdots \gamma_{n-2}) + \lambda^2(\beta^2 \cdot \gamma_1 \cdots \gamma_{n-2})^2.$$

Hence these terms cancel and the claim follows. □

3. Partial vanishing for multiplier ideal sheaf cohomology

Let (L, h) be a holomorphic line bundle over a compact Kähler n-fold X. Locally in a trivialization $L_{|U} \simeq U \times \mathbb{C}$, the metric is given by $\|\xi\|_x = |\xi| e^{-\varphi(x)}$ and we assume that the curvature $\Theta_h(L) := \frac{i}{\pi}\partial\overline{\partial}\varphi$ is a closed positive current (so that, in particular, L is pseudo-effective). We introduce as usual the *multiplier ideal sheaf* $\mathcal{I}(h) := \mathcal{I}(\varphi)$ where

$$\mathcal{I}(\varphi)_x := \{f \in \mathcal{O}_{X,x}\,;\ \exists V \ni x,\ \int_V |f(z)|^2 e^{-2\varphi(z)} < +\infty\}$$

and V is an arbitrarily small neighborhood of x. We also consider the *upper regularized multiplier ideal sheaf*

$$\mathcal{I}_+(h) := \lim_{\varepsilon \to 0_+} \mathcal{I}(h^{1+\varepsilon}) = \lim_{\varepsilon \to 0_+} \mathcal{I}\big((1+\varepsilon)\varphi\big).$$

It should be noticed that $\mathcal{I}\big((1+\varepsilon)\varphi\big)$ increases as ε decreases, hence the limit is locally stationary by the Noether property of coherent sheaves, and one has of course $\mathcal{I}_+(h) \subset \mathcal{I}(h)$. It is unknown whether these sheaves may actually differ (in all known examples they are equal). In any case, they coincide at least in codimension 1 (i.e., outside an analytic subset of codimension ≥ 2).

PROPOSITION 3.2. *Let*

$$\Theta_h(L) = \sum_{j=1}^{+\infty} \lambda_j D_j + G$$

be the Siu decomposition of the $(1,1)$*-current* $\Theta_h(L)$ *as a countable sum of effective divisors and of a* $(1,1)$*-current* G *such that the Lelong sublevel sets* $E_c(G)$, $c > 0$, *all have codimension* 2. *Then we have the inclusion of sheaves*

$$\mathcal{I}_+(h) \subset \mathcal{I}(h) \subset \mathcal{O}\big(-\sum [\lambda_j] D_j\big), \qquad [\lambda_j] := \text{integer part of } \lambda_j,$$

and equality holds on $X \smallsetminus Z$ *where* Z *is an analytic subset of* X *whose components all have codimension at least* 2.

Proof. The decomposition exists by [Siu74] (see also [De93a]). Now, if g_j is a local generator of the ideal sheaf $\mathcal{O}(-D_j)$, the plurisubharmonic weight φ of h can be written as

$$\varphi = \sum \lambda_j \log|g_j| + \psi$$

where ψ is plurisubharmonic and the $E_c(\psi)$ have codimension 2 at least. Since ψ is locally bounded from above, it is obvious that

$$\mathcal{I}(\varphi) \subset \mathcal{I}(\lambda_j \log|g_j|) \subset \mathcal{O}\big(-\sum [\lambda_j] D_j\big).$$

Now, let Y be the union of all sets $E_c(\psi)$ (with, say, $c = 1/k$), all pairwise intersections $D_j \cap D_k$ and all singular sets $D_{j\,\mathrm{sing}}$. This set Y is at most a countable union of analytic sets of codimension ≥ 2. Pick an arbitrary point $x \in X \smallsetminus Y$. Then x meets the support of $\bigcup D_j$ in at most one point which is then a smooth point of some D_k, and the Lelong number of $\psi_k = \psi + \sum_{j \neq k} \lambda_j \log|g_j|$ at x is zero. Then $\varphi = \lambda_k \log|g_k| + \psi_k$ and the inclusion

$$\mathcal{I}(h)_x \supset \mathcal{O}\big(-\sum_j [\lambda_j] D_j\big)_x = \mathcal{O}(-[\lambda_k] D_k)_x$$

holds true by Hölder's inequality. In fact, for every germ f in $\mathcal{O}(-[\lambda_k]D_k)_x$ we have

$$\int_{V \ni x} |f|^2 \exp\big(-(1+\varepsilon)\lambda_k \log|g_k|\big) < +\infty$$

for $\varepsilon > 0$ so small that $[(1+\varepsilon)\lambda_k] = [\lambda_k]$, while $e^{-\psi_k}$ is in $L^p(V_p)$ for some $V_p \ni x$, for every $p > 1$. Similarly, we have

$$\mathcal{I}_+(h)_x \supset \mathcal{O}\big(-[(1+\varepsilon)\lambda_k] D_k\big)_x = \mathcal{O}\big(-[\lambda_k] D_k\big)_x$$

for $\varepsilon > 0$ small enough. The analytic set Z where our sheaves differ $\big[$i.e. the union of supports of $\mathcal{I}(h)/\mathcal{I}_+(h)$ and $\mathcal{O}\big(-\sum_j [\lambda_j] D_j\big)/\mathcal{I}(h)\big]$ must be contained in Y, hence Z is of codimension ≥ 2. □

The main goal of this section is to prove the following partial vanishing theorem.

THEOREM 3.3. *Let* (L, h) *be a holomorphic line bundle over a compact Kähler* n*-fold* X, *equipped with a singular metric* h *such that* $\Theta_h(L) \geq 0$. *Assume that* L *is nef and has numerical dimension* $\nu(L) = \nu \geq 0$, *i.e.* $c_1(L)^\nu \neq 0$ *and* ν *is maximal. Then the morphism*

$$H^q(X, \mathcal{O}(K_X + L) \otimes \mathcal{I}_+(h)) \longrightarrow H^q(X, K_X + L)$$

induced by the inclusion $\mathcal{I}_+(h) \subset \mathcal{O}_X$ *vanishes for* $q > n - \nu$.

Of course, it is expected that the Kawamata-Viehweg vanishing theorem also holds for Kähler manifolds, in which case the whole group $H^q(X, K_X + L)$ vanishes and Theorem 3.3 would then be an obvious consequence. However, we will see in Section 4 that, conversely, Theorem 3.3 can be used to derive the Kawamata-Viehweg vanishing theorem in the first non trivial case $\nu = 2$. Using the same method for higher values of ν would probably be very hard, if not impossible.

Proof. Our strategy is based on a direct application of the Bochner technique with special hermitian metrics constructed by means of the Calabi-Yau theorem.

Let us fix a smooth hermitian metric h_∞ on L, which may have a curvature form $\Theta_{h_\infty}(L)$ of arbitrary sign, and let $\varepsilon > 0$. Then $c_1(L) + \varepsilon\omega$ is a Kähler class, hence by the Calabi-Yau theorem for complex Monge-Ampère equations there exists a hermitian metric $h_\varepsilon = h_\infty e^{-2\varphi_\varepsilon}$ such that

$$(3.4) \qquad \big(\Theta_{h_\varepsilon}(L) + \varepsilon\omega\big)^n = C_\varepsilon \omega^n.$$

Here $C_\varepsilon > 0$ is the constant such that

$$C_\varepsilon = \frac{\int_X (c_1(L) + \varepsilon\omega)^n}{\int_X \omega^n} \geq C\varepsilon^{n-\nu}.$$

Let $h = h_\infty e^{-2\psi}$ be a metric with $\Theta_h(L) \geq 0$ as given in the statement of the theorem, and let $\psi_\varepsilon \downarrow \psi$ be a regularization of ψ possessing only analytic singularities (i.e. only logarithmic poles), such that

$$\tilde{h}_\varepsilon := h_\infty e^{-2\psi_\varepsilon}$$

satisfies $\Theta_{\tilde{h}_\varepsilon}(L) \geq -\varepsilon\omega$ in the sense of currents. Such a metric exists by the general regularization results proved in [De92]. We consider the metric

$$\hat{h}_\varepsilon = (h_\varepsilon)^\delta (\tilde{h}_\varepsilon)^{1-\delta} = h_\infty \exp\big(-2(\delta\varphi_\varepsilon + (1-\delta)\psi_\varepsilon)\big)$$

where $\delta > 0$ is a sufficiently small number which will be fixed later. By construction,

$$\begin{aligned} \Theta_{\hat{h}_\varepsilon}(L) + 2\varepsilon\omega &= \delta\big(\Theta_{h_\varepsilon}(L) + \varepsilon\omega\big) + (1-\delta)\big(\Theta_{\tilde{h}_\varepsilon}(L) + \varepsilon\omega\big) + \varepsilon\omega \\ &\geq \delta\big(\Theta_{h_\varepsilon}(L) + \varepsilon\omega\big) + \varepsilon\omega. \end{aligned}$$

Denote by $\lambda_1 \leq \ldots \leq \lambda_n$ and $\hat{\lambda}_1 \leq \ldots \leq \hat{\lambda}_n$, respectively, the eigenvalues of the curvature forms $\Theta_{h_\varepsilon}(L) + \varepsilon\omega$ and $\Theta_{\hat{h}_\varepsilon}(L) + 2\varepsilon\omega$ at every point $z \in X$, with respect to the base Kähler metric $\omega(z)$. By the minimax principle we find $\hat{\lambda}_j \geq \delta\lambda_j + \varepsilon$. On the other hand, the Monge-Ampère equation (3.4) tells us that

$$(3.5) \qquad \lambda_1 \ldots \lambda_n = C_\varepsilon \geq C\varepsilon^{n-\nu}$$

everywhere on X. We apply the basic Bochner-Kodaira inequality to sections of type (n, q) with values in the hermitian line bundle $(L, \hat{h}_\varepsilon)$. As the curvature eigenvalues of $\Theta_{\hat{h}_\varepsilon}(L)$ are equal to $\hat{\lambda}_j - 2\varepsilon$ by definition, we find

$$(3.6) \qquad \|\overline{\partial}u\|^2_{\hat{h}_\varepsilon} + \|\overline{\partial}^\star u\|^2_{\hat{h}_\varepsilon} \geq \int_X (\hat{\lambda}_1 + \cdots + \hat{\lambda}_q - 2q\varepsilon)|u|^2_{\hat{h}_\varepsilon} dV_\omega$$

for every smooth (n, q)-form u with values in L. Actually this is formally true only if the metric $\hat{h}_\varepsilon$ is smooth on X. The metric h_ε is indeed smooth, but $\tilde{h}_\varepsilon$ may have poles along an analytic set $Z_\varepsilon \subset X$. In that case, we apply instead the inequality to forms u which are compactly supported in $X \smallsetminus Z_\varepsilon$, and replace the Kähler metric ω by a sequence of complete Kähler metrics $\omega_k \downarrow \omega$ on $X \smallsetminus Z_\varepsilon$, and pass to the limit as k tends to $+\infty$ (see e.g. [De82] for details about such techniques). In the limit we recover the same estimates as if we were in the smooth case, and we therefore allow ourselves to ignore this minor technical problem from now on.

Now, let us take a cohomology class $\{\beta\} \in H^q(X, K_X \otimes L \otimes \mathcal{I}_+(h))$. By using Čech cohomology and the De Rham-Weil isomorphism between Čech and Dolbeault cohomology (via a partition of unity and the usual homotopy formulas), we obtain a representative β of the cohomology class which is a smooth (n,q)-form with values in L, such that the coefficients of β lie in the sheaf $\mathcal{I}_+(h)\otimes_{\mathcal{O}_X}\mathcal{C}^\infty$. We want to show that β is a boundary with respect to the cohomology group $H^q(X, K_X \otimes L)$. This group is a finite dimensional Hausdorff vector space whose topology is induced by the L^2 Hilbert space topology on the space of forms (all Sobolev norms induce in fact the same topology on the level of cohomology groups). Therefore, it is enough to show that we can approach β by $\overline{\partial}$-exact forms in L^2 norm.

As in Hörmander [Hö65], we write every form u in the domain of the L^2-extension of $\overline{\partial}^*$ as $u = u_1 + u_2$ with

$$u_1 \in \operatorname{Ker}\overline{\partial} \quad \text{and} \quad u_2 \in (\operatorname{Ker}\overline{\partial})^\perp = \overline{\operatorname{Im}\overline{\partial}^*} \subset \operatorname{Ker}\overline{\partial}^*.$$

Therefore, since $\beta \in \operatorname{Ker}\overline{\partial}$,

$$\big|\langle\!\langle \beta, u\rangle\!\rangle\big|^2 = \big|\langle\!\langle \beta, u_1\rangle\!\rangle\big|^2 \le \int_X \frac{1}{\hat\lambda_1 + \cdots + \hat\lambda_q}|\beta|^2_{\hat h_\varepsilon}\,dV_\omega \int_X (\hat\lambda_1 + \cdots + \hat\lambda_q)|u_1|^2_{\hat h_\varepsilon}\,dV_\omega.$$

As $\overline{\partial}u_1 = 0$, an application of (3.6) to u_1 (together with an approximation of u_1 by compactly supported smooth sections on the corresponding complete Kähler manifold $X \smallsetminus Z_\varepsilon$) shows that the second integral in the right hand side is bounded above by

$$\|\overline{\partial}^* u_1\|^2_{\hat h_\varepsilon} + 2q\varepsilon\|u_1\|^2_{\hat h_\varepsilon} \le \|\overline{\partial}^* u\|^2_{\hat h_\varepsilon} + 2q\varepsilon\|u\|^2_{\hat h_\varepsilon},$$

so we finally get

$$\big|\langle\!\langle \beta, u\rangle\!\rangle\big|^2 \le \int_X \frac{1}{\hat\lambda_1 + \cdots + \hat\lambda_q}|\beta|^2_{\hat h_\varepsilon}\,dV_\omega \big(\|\overline{\partial}^* u\|^2_{\hat h_\varepsilon} + 2q\varepsilon\|u\|^2_{\hat h_\varepsilon}\big).$$

By the Hahn-Banach theorem (or rather a Hilbert duality argument in this situation), we can find elements v_ε, w_ε such that

$$\langle\!\langle \beta, u\rangle\!\rangle = \langle\!\langle v_\varepsilon, \overline{\partial}^* u\rangle\!\rangle + \langle\!\langle w_\varepsilon, u\rangle\!\rangle \quad \forall u, \qquad \text{i.e.} \quad \beta = \overline{\partial}v_\varepsilon + w_\varepsilon,$$

with

$$\|v_\varepsilon\|^2_{\hat h_\varepsilon} + \frac{1}{2q\varepsilon}\|w_\varepsilon\|^2_{\hat h_\varepsilon} \le \int_X \frac{1}{\hat\lambda_1 + \cdots + \hat\lambda_q}|\beta|^2_{\hat h_\varepsilon}\,dV_\omega.$$

As a consequence, the L^2 distance of β to the space of $\overline{\partial}$-exact forms is bounded by $\|w_\varepsilon\|_{\hat h_\varepsilon}$ where

$$\|w_\varepsilon\|^2_{\hat h_\varepsilon} = \int_X |w_\varepsilon|^2_{h_\infty} e^{-2(\delta\varphi_\varepsilon + (1-\delta)\psi_\varepsilon)}dV_\omega \le 2q\varepsilon \int_X \frac{1}{\hat\lambda_1 + \cdots + \hat\lambda_q}|\beta|^2_{\hat h_\varepsilon}\,dV_\omega.$$

We normalize the choice of the potentials φ_ε, ψ and ψ_ε so that

$$\sup_X \varphi_\varepsilon = 0, \qquad \sup_X \psi = -1, \qquad -1 \le \sup_X \psi_\varepsilon < 0\,;$$

in this way φ_ε, $\psi_\varepsilon \le 0$ everywhere on X (all inequalities can be achieved simply by adding suitable constants). From this we infer

$$\int_X |w_\varepsilon|^2_{h_\infty}dV_\omega \le 2\int_X \frac{q\varepsilon}{\hat\lambda_1 + \cdots + \hat\lambda_q}|\beta|^2_{\hat h_\varepsilon}\,dV_\omega,$$

and what remains to be shown is that the right hand side converges to 0 for a suitable choice of $\delta > 0$. By construction $\hat{\lambda}_j \geq \delta\lambda_j + \varepsilon$ and (3.5) implies

$$\lambda_q^q \lambda_{q+1} \ldots \lambda_n \geq \lambda_1 \ldots \lambda_n \geq C\varepsilon^{n-\nu},$$

hence

$$\frac{1}{\lambda_1 + \cdots + \lambda_q} \leq \frac{1}{\lambda_q} \leq C^{-1/q}\varepsilon^{-(n-\nu)/q}(\lambda_{q+1} \ldots \lambda_n)^{1/q}.$$

We infer

$$\gamma_\varepsilon := \frac{q\varepsilon}{\hat{\lambda}_1 + \cdots + \hat{\lambda}_q} \leq \min\left(1, \frac{q\varepsilon}{\delta\lambda_q}\right) \leq \min\left(1, C\delta^{-1}\varepsilon^{1-(n-\nu)/q}(\lambda_{q+1} \ldots \lambda_n)^{1/q}\right).$$

We notice that

$$\int_X \lambda_{q+1} \ldots \lambda_n \, dV_\omega \leq \int_X (\Theta_{h_\varepsilon}(L) + \varepsilon\omega)^{n-q} \wedge \omega^q = (c_1(L) + \varepsilon\{\omega\})^{n-q}\{\omega\}^q \leq C'',$$

hence the functions $(\lambda_{q+1} \ldots \lambda_n)^{1/q}$ are uniformly bounded in L^1 norm as ε tends to zero. Since $1 - (n - \nu)/q > 0$ by hypothesis, we conclude that γ_ε converges almost everywhere to 0 as ε tends to zero. On the other hand

$$|\beta|^2_{\tilde{h}_\varepsilon} = |\beta|^2_{h_\infty} e^{-2(\delta\varphi_\varepsilon + (1-\delta)\psi_\varepsilon)} \leq |\beta|^2_{h_\infty} e^{-2\delta\varphi_\varepsilon} e^{-2\psi}.$$

Our assumption that the coefficients of β lie in $\mathcal{I}_+(h)$ implies that there exists $p' > 1$ such that $\int_X |\beta|^2_{h_\infty} e^{-2p'\psi} dV_\omega < \infty$. Let $p \in]1, +\infty[$ be the conjugate exponent such that $\frac{1}{p} + \frac{1}{p'} = 1$. By Hölder's inequality, we have

$$\int_X \gamma_\varepsilon |\beta|^2_{\tilde{h}_\varepsilon} dV_\omega \leq \left(\int_X |\beta|^2_{h_\infty} e^{-2p\delta\varphi_\varepsilon} dV_\omega\right)^{1/p} \left(\int_X \gamma_\varepsilon^{p'} |\beta|^2_{h_\infty} e^{-2p'\psi} dV_\omega\right)^{1/p'}.$$

As $\gamma_\varepsilon \leq 1$, the Lebesgue dominated convergence theorem shows that

$$\int_X \gamma_\varepsilon^{p'} |\beta|^2_{h_\infty} e^{-2p'\psi} dV_\omega$$

converges to 0 as ε tends to 0. However, the family of quasi plurisubharmonic functions (φ_ε) is a bounded family with respect to the L^1 topology on the space of (quasi)-plurisubharmonic functions – we use here the fact that the currents

$$\Theta_{\tilde{h}_\varepsilon}(L) = \Theta_{h_\infty}(L) + \frac{i}{\pi}\partial\overline{\partial}\varphi_\varepsilon \geq 0$$

all sit in the same cohomology class; the boundedness of their normalized potentials then results from the continuity properties of the Green operator. By standard results of complex potential theory, we conclude that there exists a small constant $\eta > 0$ such that $\int_X e^{-2\eta\varphi_\varepsilon} dV_\omega$ is uniformly bounded. By choosing $\delta \leq \eta/p$, the integral $\int_X |\beta|^2_{h_\infty} e^{-2p\delta\varphi_\varepsilon} dV_\varepsilon$ remains bounded and we are done. □

4. Kawamata-Viehweg vanishing theorem for line bundles of numerical dimension 2

In this section we prove the Kawamata-Viehweg vanishing theorem for the cohomology group of degree $n - 1$ of nef line bundles L with $L^2 \neq 0$ on compact Kähler spaces of dimension n. Furthermore we will prove an extended version where L can be a reflexive sheaf. This will be needed for proving the abundance theorem for Kähler threefolds.

THEOREM 4.1. *Let X be a normal compact Kähler space of dimension n and L a nef line bundle on X. Assume that $L^2 \neq 0$. Then*

$$H^q(X, K_X + L) = 0$$

for $q \geq n-1$.

Proof. In a first step we reduce the proof to the case of a smooth space X (this is comfortable but not really necessary; all arguments would also work in the singular setting as well). In fact, let $\pi : \hat{X} \longrightarrow X$ be a Kähler desingularization. Then, assuming our claim in the smooth case, we have

$$H^q(\hat{X}, K_{\hat{X}} + \pi^*(L)) = 0.$$

By the projection formula and the Grauert-Riemenschneider vanishing theorem

$$R^j\pi_*(K_{\hat{X}}) = 0,$$

it follows

$$H^q(X, \pi_*(K_{\hat{X}}) \otimes L) = 0.$$

Since $\pi_*(K_{\hat{X}}) \subset K_X$ with cokernel supported in codimension at least 2, namely on the singular locus of X, the vanishing claim follows.

So from now on, we assume X smooth. In the case $q = n$, we have $H^n(X, K_X+L) = H^0(X, -L)^*$ by Serre duality, and for L nef, $-L$ has no section unless L is trivial. Therefore the only interesting case is $q = n-1$. We introduce a singular metric h on L with positive curvature current T. By [Siu74] and [De92, De93a] we obtain a decomposition

$$T = \sum \lambda_j D_j + G,$$

where $\lambda_j \geq 1$ are irreducible divisors, and G is a positive current such that G has Lelong numbers in codimension ≥ 2 only – so that in particular $G|D_i$ is pseudo-effective for all i. Consider the multiplier ideal sheaf $\mathcal{I}(h)$. By Proposition 3.2 we have

$$\mathcal{I}_+(h) \subset \mathcal{I}(h) \subset \mathcal{O}_X(-\sum[\lambda_j]D_j)$$

with equality in codimension 1. We put

$$D = \sum [\lambda_j] D_j.$$

We consider the canonical map in cohomology

$$H^{n-1}(X, -D + L + K_X) \longrightarrow H^{n-1}(X, L + K_X)$$

which is vanishing by (3.3). In order to prove our claim it is therefore sufficient to prove

$$H^{n-1}(D, L + K_X|D) = 0.$$

By Serre duality and the adjunction formula, this comes down to show

$$H^0(D, -L + D|D) = 0.$$

Supposing the contrary, we fix a non-zero section

$$\sigma \in H^0(D, -L + D).$$

We choose $p_1, \ldots, p_k$ maximal so that

$$\sigma \in H^0(D, -L + D - \sum p_j D_j),$$

i.e. we choose $\tilde{D} = \sum p_j D_j \subset D$ maximal such that $\sigma|\tilde{D} = 0$. [In this notation, we view $\tilde{D}$ as the subscheme of X defined by the structure sheaf $\mathcal{O}_X/\mathcal{O}_X(-\tilde{D})$]. Then $0 \le p_i \le [\lambda_i]$ for all $i \in I$, not all $p_i = [\lambda_i]$, and we shall always consider σ as a section of $-L + D - \sum p_j D_j$. Denote

$$c_i = \frac{\{\lambda_i\} + p_i}{\lambda_i}.$$

Then we have $0 \le c_i \le 1$. We introduce $c = \min c_i$ and

$$I_0 = \{i \in I \mid c_i = c\}.$$

Clearly $c < 1$. Notice that by construction $\sigma|D_i \ne 0$ unless $c_i = 1$. Let

$$E = -\big(\sum(\{\lambda_i\} + p_i)D_i\big) - G.$$

Since $L = \sum \lambda_i D_i + G$, we have

$$-L + D - \sum p_i D_i = -\big(\sum(\{\lambda_i\} + p_i)D_i\big) - G = E,$$

so E is effective (possibly zero) on every D_i with $c_i < 1$. Since L is nef, also the $\mathbb{R}$-divisor $cL = \sum \lambda_i c D_i + cG$ is nef. Adding this to the divisor E in the last equation, we deduce that

$$-\big(\sum(\{\lambda_i\} + p_i - c\lambda_i)D_i\big) - (1-c)G$$

is pseudo-effective on every D_i with $c_i < 1$. Since $\{\lambda_i\} + p_i - c\lambda_i = 0$ for $i \in I_0$, it follows that

$$-\big(\sum_{i \notin I_0}(\{\lambda_i\} + p_i - c\lambda_i)D_i\big) - (1-c)G$$

is pseudo-effective on every D_i with $c_i < 1$, in particular for every $i \in I_0$. Now $D_j|D_i$ is effective (possibly 0) for all $j \ne i$, and $G|D_i$ is always pseudo-effective, hence, having in mind $c < 1$ and $\{\lambda_i\} + p_i - c\lambda_i > 0$ for $i \notin I_0$, we conclude that

$$D_j|D_i \equiv 0 \tag{1}$$

for all (j, i) with $j \notin I_0$ and $i \in I_0$ and that

$$G|D_i \equiv 0 \tag{2}$$

for all $i \in I_0$. Introducing

$$D' = \sum_{i \in I_0} \lambda_i D_i$$

and

$$D'' = \sum_{i \notin I_0} \lambda_i D_i,$$

we have

$$L = D' + D'' + G$$

and $D'' \cdot D_i = G \cdot D_i = 0$ for all $i \in I_0$ by (1) and (2). Hence $L \cdot D_i = D' \cdot D_i$ for $i \in I_0$, so that $D'|D_i$ is nef, hence D' is nef by [Pa98]. In total

$$L \cdot D' = D' \cdot D'$$

and

$$D' \cdot D'' = D' \cdot G = 0.$$

As $L^2 \ne 0$ et $D' \ne 0$, Corollary 2.4 implies $L \cdot D' \ne 0$. First recall that

$$L|D_i = \sum_{j \in I_0} \lambda_j D_j|D_i$$

is nef. On the other hand

$$-\sum_{j\in I_0}\{\lambda_j\}+p_j)D_j=-c\sum_{j\in I_0}\lambda_j D_j$$

is of course pseudo-effective on every D_i for $i\in I_0$ (E is effective on those D_i). Combining these two facts, we deduce that either $c=0$ or that $L\cdot D_i=0$ for all $i\in I_0$, hence $L\cdot D'=0$, contradiction. So we have $c=0$. This means $p_j=0$ and $\lambda_j\in\mathbb{N}$ for all $j\in I_0$.

CLAIM 4.2. *The divisor $D''+G$ is nef, and in fact must be equal to zero.*

Proof of the claim. We consider the closed positive $(1,1)$-current $\Theta=[D'']+G$. By the results of [Pa98], the proof of nefness of $\{\Theta\}$ just amounts to showing that the restriction $\{\Theta\}_{|Z}$ of the $(1,1)$-cohomology class $\{\Theta\}$ to any component Z in the Lelong sublevel set $\bigcup_{c>0}E_c(\Theta)$ is nef. However Z is either a component of D'' or a component of $\bigcup_{c>0}E_c(G)$. In the first case, Z is contained in the support of D'', and as $D'\cdot D''=0$, Z must be disjoint from D'. Hence

$$\{\Theta\}_{|Z}=\{D''+G\}_{|Z}=\{L-D'\}_{|Z}=\{L\}_{|Z}$$

is nef. If Z is a component of $\bigcup_{c>0}E_c(G)$, then Z has codimension at least 2. Then we know by [De93a] that the intersection product $[D']\wedge G$ is well defined as a closed positive current. Since the cohomology class of this current is zero, we must have $[D']\wedge G=0$. However, we infer from [De93a] that

$$\nu([D']\wedge G,z)\geq\nu([D'],z)\,\nu(G,z)>0$$

at every point $z\in D'\cap Z$, hence Z must also be disjoint from D' in that case. We conclude as before that $\{\Theta\}_{|Z}=\{L\}_{|Z}$ is nef. Now we have $D'\cdot(D''+G)=0$, with D', $D''+G$ nef and $D'^2=L\cdot D'\neq 0$. Hence $\{D''+G\}=0$ by Corollary 2.4, and we conclude that $D''=0$, $G=0$ (both $[D'']$ and $[G]$ being positive currents). □From this we infer $L\equiv D'$ and $\mathcal{I}(h)=\mathcal{O}_X(-D')$.

Case 1 We assume that $L=D'$. Now the sequence

$$0\longrightarrow\mathcal{I}(h)\otimes K_X+L\longrightarrow K_X+L\longrightarrow K_X+L|D'=K_{D'}\longrightarrow 0$$

gives in cohomology

$$0\longrightarrow H^{n-1}(X,K_X+L)\longrightarrow H^{n-1}(D',K_{D'})\simeq H^0(D',\mathcal{O}_{D'})\longrightarrow H^n(K_X)=\mathbb{C}\longrightarrow 0.$$

Thus we need to show

$$h^0(D',\mathcal{O}_{D'})=1.$$

In order to verify this, we first observe that D' is connected. In fact otherwise write $D'=A+B$ with A and B effective and $A\cdot B=0$. But A and B are necessarily nef, hence the Hodge Index Theorem gives a contradiction to $L^2=(D')^2\neq 0$. So D' is connected and if $h^0(D',\mathcal{O}_{D'})\geq 2$, then $\mathcal{O}_{D'}$ contains a nilpotent section $t\neq 0$. Let $\sum_{j\in I}\mu_j D_j$ denote its vanishing divisor (notice that D' is Cohen-Macaulay!). Then $1\leq\mu_j\leq\lambda_j$ for all j. Let

$$J=\{j\in J|\frac{\lambda_j}{\mu_j}\text{maximal}\}$$

and let $c=\frac{\lambda_j}{\mu_j}$ be the maximal value. Notice that $-\sum_{j\in I}\mu_j D_j|D_i$ is effective (possibly 0) for all i. First we rule out the case that $c=\frac{\lambda_j}{\mu_j}$ for all $j\in I$. In fact, then $L|D_i = c\sum \mu_j D_j|D_i$ is nef and its dual is effective, hence $L|D_i \equiv 0$ for all i, whence $L^2 = 0$, contradiction. Thus we find some j such that

$$c > \frac{\lambda_j}{\mu_j}.$$

By connectedness of $D = D'$ we can choose $i_0 \in J$ in such a way that there exists $j_1 \in I \setminus J$ with $D_{i_0} \cap D_{j_1} \neq \emptyset$. Now

$$\sum_{j \in I} (\lambda_j - c\mu_j) D_j | D_{i_0}$$

is pseudo-effective as a sum of a nef and an effective line bundle (this has nothing to do with the choice of i_0). Since the sum, taken over I, is the same as the sum taken over $I \setminus \{i_0\}$, we conclude that

$$\sum_{j \neq i_0} (\lambda_j - \mu_j) D_j | D_{i_0}$$

is pseudo-effective, too. Now all $\lambda_j - c\mu_j \leq 0$ and $\lambda_{j_1} - c\mu_{j_1} < 0$ with $D_{j_1} \cap D_{i_0} \neq \emptyset$, hence the dual of

$$\sum_{j \neq i_0} (\lambda_j - \mu_j) D_j | D_{i_0}$$

is effective non-zero, a contradiction. □

Case 2. Now we deal with the case that $L \neq D'$. Then we can write

$$L = D' + L_0$$

where $L_0^m \in \mathrm{Pic}^0(X)$ (The exponent m is there because there might be torsion in $H^2(X, \mathbb{Z})$; we take m to kill the denominator of the torsion part). We may in fact assume that $m = 1$; otherwise we pass to a finite étale cover $\tilde{X}$ of X and argue there (the vanishing on $\tilde{X}$ clearly implies the vanishing on X). Then the sequence S is modified to

$$(S) \quad 0 \longrightarrow \mathcal{I}(h) \otimes (K_X + L) \longrightarrow K_X + L \longrightarrow (K_X + L)|D' = (K_{D'} + L_0)|D' \longrightarrow 0.$$

Taking cohomology as before, things come down to prove

$$(*) \qquad H^0(D', -L_0|D') = 0.$$

If $-L_0|D' \neq 0$, then we see as above that $-L_0|D'$ cannot have a nilpotent section. So if $(*)$ fails, then $-L_0|D'$ has a section s such that $s|\mathrm{red}D'$ has no zeroes, so that $-L_0|\mathrm{red}D'$ is trivial. But then $-L_0|D'$ is trivial. Now let $\alpha : X \longrightarrow A$ be the Albanese map with image Y. Then $L_0 = \alpha^*(L_0')$ with some line bundle L_0' on A which is topologically trivial but not trivial. Since $L_0|D'$ is trivial, we conclude that $\alpha(D') \neq Y$, and $\alpha(D')$ is contained in a proper subtorus B of A. Now consider the induced map

$$\beta : X \longrightarrow A/B$$

and denote its image by Z. Then $\beta(D')$ is a point; on the other hand D' is nef, so that $\dim Z = 1$ and D' consists of multiples of fibers of β. But this contradicts $D'^2 = L^2 \neq 0$. □

For applications to minimal Kähler 3-folds, 4.1 is still not good enough, because we need to know the vanishing property $H^2(X, mK_X) = 0$ on a $\mathbb{Q}$-Gorenstein 3-fold (with $K_X^2 \neq 0$). We would like to set $L = (m-1)K_X$ to apply 4.1 but this is no longer a line bundle. This difficulty is overcome by

PROPOSITION 4.3. *Let X be a normal $\mathbb{Q}$-Gorenstein compact Kähler 3-fold with at most terminal singularities. Let A be a $\mathbb{Q}$-line bundle. Suppose A is nef and $A^2 \neq 0$. Then*

$$H^2(X, A + K_X) = 0.$$

Proof. (A) In a first step we show that we may assume X to be $\mathbb{Q}$-factorial. (Actually, in our application in Section 5, it will be clear that we may always assume X to be $\mathbb{Q}$-factorial, so the reader only interested in the applications may skip (A)). In fact, if X is not $\mathbb{Q}$-factorial, there exists a bimeromorphic map $f : Y \longrightarrow X$ from a normal $\mathbb{Q}$-factorial Kähler space with at most terminal singularities ([Ka88,4.5′]). Moreover f is an isomorphism in codimension 1 and f is projective since X has only isolated singularities. Now consider the reflexive sheaf

$$\mathcal{H} = f^*(\mathcal{O}_X(A))^{**}.$$

Choose a number r such that $A^{[r]}$ is locally free. Then

$$\mathcal{H}^{[r]} = f^*(\mathcal{O}_X(rA)), \tag{1}$$

since both sheaves are reflexive and coincide in codimension 1. Thus $\mathcal{H}$ is nef (as $\mathbb{Q}$-line bundle) with $\mathcal{H}^2 \neq 0$. Once we know the result in the $\mathbb{Q}$-factorial case, we get

$$H^2(Y, \mathcal{H}\hat{\otimes}K_Y) = 0.$$

So by the Leray spectral sequence, we only have to show

$$R^1 f_*(\mathcal{H}\hat{\otimes}K_Y) = 0.$$

This however follows from [KMM87, 1-2-7]. Actually this citation deals with the algebraic case. However first notice that our statement is local around the isolated singularities of X. Now isolated singularities are algebraic by Artin's theorem, i.e. we can realize an open neighborhood of an isolated singularity as an open set in a normal algebraic variety. So locally on X the map $f : Y \longrightarrow X$ can be realized algebraically. Now we can approximate $\mathcal{H}$ by algebraic reflexive sheaves $\mathcal{H}_k$ up to high order k and then apply [KMM87, 1-2-7] to get the vanishing $R^1 f_*(\mathcal{H}_k) = 0$. This sheaf coincides with $R^1 f_*(\mathcal{H})$ to high order, so $R^1 f_*(\mathcal{H})$ vanishes to high order. For k approaching ∞, we obtain the vanishing we are looking for.

(B) From now on we assume X to be $\mathbb{Q}$-factorial. We proceed as in the proof of 4.1. First of all choose r such that $A^{[r]}$ is locally free. Then choose a singular metric h with positive curvature current on $A^{[r]}$. Now $\frac{1}{r}h$ is a metric at least on $A|X_{\mathrm{reg}}$ with positive curvature current T extending to all of X. We argue as in the first part of the proof of 4.1 to obtain the divisor D and the current R, however D is only an integral Weil divisor. By the same arguments as in 4.1 we can still reduce the problem to proving

$$H^2(D, \mathcal{O}_D(A + K_X)) = 0.$$

(Notice that $\mathcal{I}_D \otimes \mathcal{O}_X(A + K_X) = \mathcal{O}_X(-D + A + K_X)$ outside a finite set and that by definition $\mathcal{O}_D(A + K_X) = \mathcal{O}_X(A + K_X)|D)$). Now D is Cohen-Macaulay; here we need in an essential way that locally X is the quotient of a hypersurface by a finite group. To be more detailed, we can write locally $X = V/G$ with V a hypersurface singularity and G a finite group (see e.g. [Re87]). Let $\pi : V \longrightarrow X$ be the quotient map and let $\hat{D} = \pi^*(D)$. If we can prove that $\hat{D}$ is Cohen-Macaulay, then D will be Cohen-Macaulay, too, since this property is G-invariant. So we may assume that $X = V$. Now X is (locally) a compound Du Val singularity [Re87], i.e.

a 1-parameter deformation of a 2-dimensional rational double point. Hence we can find a *Cartier* divisor $H \subset X$ through x_0 which has just a rational double point at x_0. Now consider $D \cap H$. This is a Weil $\mathbb{Q}$-Cartier divisor on H. Since x_0 is a quotient singularity of H, we can argue as above to see that $D \cap H$ is Cohen-Macaulay. Hence D has a hyperplane section through x_0 which is Cohen-Macaulay. Thus D is Cohen-Macaulay at x_0 itself.

Therefore we have by Serre duality

$$H^2(D, \mathcal{O}_D(A + K_X)) \simeq \mathrm{Hom}(\mathcal{O}_D(A + K_X), \mathcal{O}_D(K_D)).$$

Suppose that H^2 does not vanish. Then we obtain a non-zero homomorphism $s : \mathcal{O}_D(A + K_X) \longrightarrow K_D$. This s must be generically non-zero. In fact, D is generically Gorenstein. Hence $\mathcal{O}_D(K_D)_x$ is isomorphic to an ideal in $\mathcal{O}_{X,x}$ for all x, in particular K_D has no torsion sections, D being Cohen-Macaulay; see [Ei95] in the algebraic case. Let X_0 be the regular part of X, this means that we eliminate a finite set from X, all singularities being terminal. Let denote $D_0 = D \cap X_0$ and let $s_0 = s|X_0$. Then by adjunction we have

$$0 \neq s_0 \in H^0(D_0, \mathcal{O}_D(-A + D)).$$

From now on we argue as in 4.1 just working on X_0 instead of X. The only exceptions are calculations of intersection numbers and Hodge index arguments. Here we still need to argue on X - we do not have any problems with singularities since all divisors are $\mathbb{Q}$-Cartier. □

5. The case $K_X^2 = 0, K_X \neq 0$

The second ingredient for the proof of the abundance theorem for Kähler threefolds is the following weak analogue of 4.3 in case $K_X^2 = 0$ (however one should have in mind that we are dealing with a cawe which does not exist a posteriori).

THEOREM 5.1. *Let X be a normal compact Kähler threefold with at most terminal singularities such that K_X is nef. Suppose that $K_X^2 = 0$, $K_X \neq 0$, and that X is simple and not Kummer. Then*

$$h^2(X, mK_X) \leq 1$$

for all $m \in \mathbb{N}$.

As already mentioned the essential property derived from X being simple non-Kummer is that $\pi_1(X)$ is finite [Ca94].

5.2 Start of the proof. Using Kawamata's $\mathbb{Q}$-factorialisation theorem (compare with proof of 4.5), we may assume that X is $\mathbb{Q}$-factorial. Suppose $h^2(X, mK_X) \geq 2$. Using Serre duality we get – following Miyaoka and Shepherd-Barron – (many) non-split extensions

$$0 \longrightarrow K_X \longrightarrow \mathcal{E} \longrightarrow mK_X \longrightarrow 0 \qquad (S)$$

with reflexive sheaves $\mathcal{E}$ of rank 2. We note

$$c_1(\mathcal{E}) = (m+1)K_X \qquad (5.2.1)$$

and

$$c_2(\mathcal{E}) = 0. \qquad (5.2.2)$$

5.3 The unstable case.

(5.3.a) Here we will assume that every non-split extension $\mathcal{E}$ as in (S) is ω- unstable for some fixed Kähler form ω independent of $\mathcal{E}$. Let $\mathcal{A}_S \subset \mathcal{E}$ be the ω-maximal destabilizing subsheaf. Then $\mathcal{A}_S$ is a $\mathbb{Q}$-line bundle and we determine a $\mathbb{Q}$-line bundle $\mathcal{B}_S$ such that $\mathcal{E}/\mathcal{A}_S = \mathcal{I}_Z\mathcal{B}_S$ with some subspace Z of codimension at least 2 (actually Z is generically (i.e. on the smooth part of X) locally a complete intersection or finite and supported in Sing$X$). Since $K_X \neq 0$, we obtain injective maps

$$\phi_S : \mathcal{A}_S \longrightarrow mK_X$$

and

$$\psi_S : K_X \longrightarrow \mathcal{I}\mathcal{B}_S.$$

Now there are (up to $\mathbb{C}^*$) only finitely many maps $\phi_S : \mathcal{A}_S \longrightarrow mK_X$ with some $\mathbb{Q}$-line bundle $\mathcal{A}_S$ arising as maximal ω-destabilizing subsheaf for some extension (S). In fact, fix $\phi = \phi_S : \mathcal{A} \longrightarrow mK_X$. Then by (6.13) there are only finitely many maximal reflexive subsheaves $\mathcal{A}' \subset mK_X$ such that $\mathcal{A}' \not\subset \mathcal{A}$. So we may suppose $\mathcal{A}' \subset \mathcal{A}$. Then

$$\mathcal{A}' \cdot \omega^2 < \mathcal{A} \cdot \omega^2.$$

Actually, putting

$$\epsilon := \min Y_j \cdot \omega^2,$$

where the minimum runs over the finitely many irreducible hypersurfaces $Y_j \subset X$, we have

$$\mathcal{A}' \cdot \omega^2 \leq \mathcal{A} \cdot \omega^2 - \epsilon.$$

On the other hand, restricting ourselves to $\mathcal{A}'$ of the form $\mathcal{A}' = \mathcal{A}_{S'}$, we have by the destabilizing property

$$\mathcal{A}' \cdot \omega^2 \geq \frac{c_1(\mathcal{E}) \cdot \omega^2}{2}. \qquad (*)$$

Having in mind that

$$\mathcal{A}' = \mathcal{A} \otimes \mathcal{O}_X(-\sum \lambda_j Y_j),$$

the finiteness of irreducible hypersurfaces in X gives the finiteness claim, since $(*)$ reads

$$(\mathcal{A} - \sum \lambda_j Y_j) \cdot \omega^2 \geq \frac{c_1(\mathcal{E}) \cdot \omega^2}{2}.$$

So we have only finitely many possible maps ϕ (up to $\mathbb{C}^*$). In the same way (by dualizing) we have only finitely many maps ψ (up to $\mathbb{C}^*$). In (2) we prove that (ϕ, ψ) and $(\lambda\phi, \lambda\psi)$ with $\lambda \in \mathbb{C}^*$ always define isomorphic extensions (S). Therefore in total $(\lambda\phi, \mu\psi)$ with $\lambda, \mu \in \mathbb{C}^*$ define just a 1-dimensional space of extensions, whence $h^2(X, mK_X) \leq 1$.

(5.3.b) We shall now prove that the extension class defining (S) is already determined by ϕ and ψ (modulo $\mathbb{C}^*$). So take another extension

$$0 \longrightarrow K_X \longrightarrow \mathcal{E}' \longrightarrow mK_X \longrightarrow 0 \qquad (S')$$

with the same destabilizing sheaves $\mathcal{A}$ and $\mathcal{B}$ and with the same morphisms ϕ and ψ (the case of $(\phi, \psi), (\lambda\phi, \lambda\psi)$ is exactly the same). Let D be the divisorial part of

$$\{\phi = 0\} \cup \{\psi = 0\} \cup \mathrm{Sing}(X);$$

then we obtain a splitting of the sequence (S) over $X \setminus D$ via ϕ :

$$\mathcal{E} \simeq mK_X \oplus K_X \simeq \mathcal{A} \oplus \mathcal{B},$$

and an analogous splitting of $\mathcal{E}'$ over $X \setminus D$. Observe also that $\mathcal{A} = mK_X - D$ and $\mathcal{B} = K_X + D$. Thus we obtain an isomorphism

$$f : \mathcal{E} \longrightarrow \mathcal{E}'$$

on $X \setminus D$ making the two extensions (S) and (S') isomorphic over $X \setminus D$:

$$\begin{array}{ccccccccc} 0 & \longrightarrow & K_X & \longrightarrow & \mathcal{E} & \longrightarrow & mK_X & \longrightarrow & 0 \\ & & \| & & \downarrow & & \| & & \\ 0 & \longrightarrow & K_X & \longrightarrow & \mathcal{E}' & \longrightarrow & mK_X & \longrightarrow & 0. \end{array}$$

It remains to extend the map f to X. Let us notice that we may assume $Z = \emptyset$. In fact let Z_1 be the codimension 1 part of Z. Restricting our two exact sequences describing $\mathcal{E}$ to D, we see that (modulo torsion at finitely many points)

$$\mathcal{A}|D = K_X|D,$$

hence

$$(m-1)K_X \cdot D = D \cdot D. \tag{5.3.1}$$

In particular we note that $D|D$ is nef, hence D itself is nef. Now (5.3.1) yields

$$0 = c_2(\mathcal{E}) = c_1(\mathcal{A}) \cdot c_1(\mathcal{B}) + c_2(\mathcal{I}_Z B) = (mK_X - D) \cdot (K_X + D) + Z_1 = Z_1,$$

hence $Z_1 = \emptyset$. In particular $Z \subset \operatorname{Sing} X$ has codimension at least 3. This shows that we may ignore Z in all our following considerations; in what follows restriction will always that we also divide by torsion. Take a local section $s \in \mathcal{E}(U)$ over a small disc U. We need to show that $f(s) \in \mathcal{E}'(U)$; a priori we only know $f(s) \in \mathcal{E}'(D)(U)$. Let

$$\kappa : \mathcal{E} \to mK_X,\ \kappa' : \mathcal{E}' \to mK_X$$

and

$$\lambda : \mathcal{E} \to \mathcal{B},\ \lambda' : \mathcal{E}' \to \mathcal{B}$$

be the canonical maps.

Now consider the exact sequence

$$0 \longrightarrow \mathcal{E} \overset{\kappa\oplus\lambda}{\longrightarrow} mK_X \oplus \mathcal{B} \overset{\rho}{\longrightarrow} mK_X|D \longrightarrow 0$$

Here $\rho(u \oplus v) = u_D - \tau(v_D)$, where $\tau : \mathcal{B}_D \to mK_X|D$ is the canonical sequence arising by restricting our sequences and the maps ϕ and ψ to D. Analogously for $\mathcal{E}'$. Suppose we know

(+) $$h^0(D_c, \mathcal{O}_{D_c}) = 1$$

for any connected component D_c of D. Actually it suffices to know this for $D_c \cap \operatorname{reg} X$. Then $\tau'^{-1} \circ \tau = a$ id, and of course we can normalize (in the extension class) to $a = 1$ (our arguments are local around every individual connected component of D). Therefore we can construct a diagram

$$\begin{array}{ccccccccc} 0 & \longrightarrow & \mathcal{E} & \longrightarrow & mK_X \oplus \mathcal{B} & \to & mK_X|D & \longrightarrow & 0 \\ & & \downarrow & & \| & & \downarrow & & \\ 0 & \longrightarrow & \mathcal{E}' & \longrightarrow & mK_X \oplus \mathcal{B} & \longrightarrow & mK_X|D & \longrightarrow & 0. \end{array}$$

Here the map $f : \mathcal{E} \to \mathcal{E}'$ is defined only on $X \setminus D$ and meromorphic on X, the left square is commutative on $X \setminus D$ and the right square is commutative on X. It is immediately checked that $mK_X|D \to mK_X|D$ is the identity (consider images of elements $u \oplus 0$), hence $f(s) \in \mathcal{E}'(U)$ and thus f extends to a global isomorphism making the two extensions isomorphic.

(5.3.c) It remains to prove (+) and we may assume $D = D_c$ for simplicity of notations. So suppose that $h^0(D, \mathcal{O}_D) \geq 2$, resp. $h^0(D \cap \mathrm{reg}X, c\mathcal{O}_{D\cap\mathrm{reg}X}) \geq 2$. Since $H^1(X, \mathcal{O}_X) = 0$, we obtain

$$H^1(X, \mathcal{O}_X(-D)) \neq 0.$$

Let X_0 be the regular part of X and $D_0 = D|X_0$. Since the singularities of X are at most finite,

$$H^1(X_0, \mathcal{O}(-D_0)) = H^1(X, \mathcal{O}(-D)) \neq 0.$$

Hence $\mathrm{Ext}^1(\mathcal{O}_{X_0}, \mathcal{O}(-D_0)) \neq 0$, and therefore there exists a non-split extension

$$(E_0) \qquad 0 \longrightarrow \mathcal{O}_{X_0} \longrightarrow \mathcal{F}_0 \longrightarrow \mathcal{O}(D_0) \longrightarrow 0,$$

with a locally free sheaf $\mathcal{F}_0$ over X_0. Now $\mathcal{F}_0$ has a unique reflexive extension to X : consider a singular point $x_0 \in X$ and let U be a Stein neighborhood of x_0. Then

$$H^1(U \setminus x_0, \mathcal{O}(-D_0)) = H^1(U, \mathcal{O}(-D)) = 0,$$

hence (E_0) splits over $U \setminus x_0$:

$$\mathcal{F}_0 \simeq \mathcal{O} \oplus \mathcal{O}_{U\setminus x_0}(D_0).$$

Hence $\mathcal{F}_0$ extends to a reflexive sheaf $\mathcal{F}$. Moreover (E_0) extends to

$$(E) \qquad 0 \longrightarrow \mathcal{O}_X \longrightarrow \mathcal{F} \longrightarrow \mathcal{O}_X(D) \longrightarrow 0.$$

In particular $\mathrm{Ext}^1(\mathcal{O}_X(D), \mathcal{O}_X) \neq 0$. This is easily seen to be equivalent to

$$\mathrm{Ext}^1(K_X + \mathcal{O}_X(D), \mathcal{O}_X(D)) \neq 0,$$

hence $H^2(X, K_X + \mathcal{O}_X(D)) \neq 0$ by Serre duality. Thus

$$D^2 = 0$$

by (4.3).

We observe $c_1(\mathcal{F})^2 = c_2(\mathcal{F}) = 0$, therefore $\mathcal{F}$ cannot be ω-stable. So let $\mathcal{A}'$ be the maximal ω-destabilizing subsheaf, and we obtain as before for $\mathcal{E}$ a sequence

$$(F) \qquad 0 \longrightarrow \mathcal{A}' \longrightarrow \mathcal{F} \longrightarrow \mathcal{I}_{Z'}\mathcal{B}' \longrightarrow 0,$$

where Z' has generically dimension 1 or is contained in the singular locus of X. As before, $\mathcal{A}' \subset \mathcal{O}_X(D)$ and $\mathcal{O}_X \subset \mathcal{B}'$ so that there is an effective divisor D' such that $\mathcal{B}' = \mathcal{O}_X(D')$ and $\mathcal{A}' = \mathcal{O}_X(D - D')$. By (E) and the fact that $a(X) = 0$ and $H^1(X, \mathcal{O}_X) = 0$, we deduce that $h^0(X, \mathcal{F}) = 2$. Hence (F) yields

$$h^0(X, \mathcal{A}') = 1.$$

So we can write $\mathcal{A}' = \mathcal{O}_X(E)$ with an effective divisor E. Thus $D = D' + E$. By restricting (E) and (F) to D', we obtain $\mathcal{A}'|D' = E|D' = \mathcal{O}_{D'}$ and that the 1-dimensional part Z_1' of Z' is empty, so that Z' is at most finite. Alternatively, we calculate

$$0 = c_2(\mathcal{F}) = E \cdot D' + Z_1'$$

and conclude together with the observation that $D'|D'$ and thus D' is nef.

So we have a decomposition $D = D' + E$ with D and D' nef. By instability, $E \neq 0$. Having in mind that D is connected, we are going to prove that D,D' and E are proportional (even numerical proportionality would be sufficient, which in our situation (X simply connected with $a(X) = 0$) gives equality). Assuming this proportionality for the moment, we obtain $D' = aE$ and $D = (a + 1)E$. Since $\mathcal{A}'$

destabilizes, we have $a \leq 1$. By restricting the sequence (E) and (F) to D' we obtain:

$$\mathcal{O}_{D'}(E) = \mathcal{O}_{D'} \qquad (*)$$

up to torsion. For the simplicity of notation we suppress the torsion and agree, when taking a restriction, that we also divide by the torsion. We may assume that $h^0(D', \mathcal{O}_{D'}) = 1$. In fact, if $h^0(D', \mathcal{O}_{D'}) \geq 2$, then we substitute D by D' and argue as before; of course this procedure has to terminate after finitely many steps. So $h^0(D', \mathcal{O}_{D'}) = 1$ and consequently

$$h^0(D', N^{*\mu}_{D'/D}) \neq 0$$

for some positive integer; hence $h^0(D', N^{*\mu}_{D'/X}) \neq 0$. (We may neglect the torsion, because we may compute over X_0). Now this non-vanishing implies that $\mu D'|D_0$ is trivial, where D_0 is the reduction of D'. In fact, let $\mu D' = \sum a_i Y_i$ and take a non-zero section s of $\mathcal{O}_{\mu D'}(\mu D')$. Let $\tilde{D} = \sum b_i Y_i \subset \mu D'$ be the maximal subdivisor such that $s|\tilde{D} = 0$. Introducing $c_i = a_i - b_i \geq 0$, we obtain a section

$$s' \in H^0(D_0, \mathcal{O}_{D_0}(-\sum c_i Y_i))$$

such that $s'|Y_i \neq 0$ for all i. Fix a Kähler form ω and let $\alpha_{ij} = Y_i \cdot Y_j \cdot \omega$ for $i \neq j$. Then we obtain for all j

$$-\sum_i c_i Y_i \cdot Y_j \cdot \omega = -\sum c_i \alpha_{ij} \geq 0.$$

Since $D^2 = 0$ and D is nef, we have $D \cdot Y_j = 0$ for all j and therefore

$$Y_j^2 = -\frac{1}{a_j} \sum_{i \neq j} a_i Y_i \cdot Y_j$$

so that we arrive at the inequalities (for each j)

$$\sum_{i \neq j} a_i c_j \alpha_{ij} \geq \sum_{i \neq j} a_j c_i \alpha_{ij}.$$

By simple algebraic considerations this is only possible if we have always equality. This means that the divisor $D^* = \sum c_i Y_i$ fulfills $D^* \cdot Y_j = 0$ for all j. Hence D^* is nef, and the proportionality arguments below shows that $D^* = cD'$ for some positive number c. Because of the non-vanishing of s' on Y_j, $D^*|Y_j$ is trivial, hence $D^*|D_0$ is trivial. Suppose for the moment that $\mu = 1$. Then $D^* \subset D'$ so that $c < 1$ and in total we have $D^*|D_0$ trivial, and $D^*|D^*$ torsion by $(*)$. Hence [Mi88a, 4.1] says that $D^*|D^*$ is trivial. Now the exact sequence

$$0 \longrightarrow \mathcal{O}_X \longrightarrow \mathcal{O}_X(D^*) \longrightarrow \mathcal{O}_X(D^*)|D^* \longrightarrow 0$$

implies by $(*)$ – keeping in mind $H^1(X, \mathcal{O}_X) = 0$ – that $h^0(X, \mathcal{O}_X(D')) \geq 2$, contradicting $a(X) = 0$. So we are left with the case $\mu \geq 2$. We deal with $\mu = 2$ and leave the trivial modifications in the general case to the reader. The difficulty here is that possibly $c > 1$ so that $D' \subset D^*$, otherwise we conclude as before. At least we know that $D^* \subset \mu D'$ and we are going to show that $\mu E|\mu D'$ is trivial; then we are done. This does not follows directly from restricting (E) and (F); instead we take S^2 and obtain an injection $\mathcal{O}_X(2E) \longrightarrow S^2(\mathcal{F})$. Restricting to $2D'$, we obtain a non-zero map $\mathcal{O}_{2D'}(2E) \to \mathcal{F}|2D'$. Then either the induced map $\mathcal{O}_{2D'}(2E) \to \mathcal{O}_D(D))$ is non-zero; this implies $H^0(D', \mathcal{O}_{D'}(-D')) \neq 0$ so that $h^0(X, \mathcal{O}_{2D'}) > 1$ and we may take $D = 2D'$ whence $\mu = 1$. Or this map vanishes;

then we get a non-zero map $\mathcal{O}_{2D'}(2E) \to \mathcal{O}_{2D'}$. This map is an isomorphism and settles our claim.

(5.3.d) It remains to settle the proportionality. After passing to a desingularization this comes down to prove the following statement. Let X be compact Kähler, $A = \sum_{i=1}^{s} a_i Y_i$ and $B = \sum_{i=1}^{s} b_i Y_i$ be effective divisors, a_i and b_i positive, with connected supports. Suppose that A and B are nef, and that $A \cdot Y_i = B \cdot Y_i = 0$ for all i (in particular $A^2 = B^2 = A \cdot B = 0$). Then $A = cB$ (not only for numerical equivalence). Observe that if X is a surface, then this is nothing than Zariski's lemma, which is usually formulated for fibers of maps to curves, but which works in this context; therefore the claim also follows for projective manifolds by taking hyperplane sections and applying Lefschetz. If X is merely Kähler, then we consider the vector space $V \subset H^2(X, \mathbb{R})$ generated by the classes of the hypersurfaces $Y_i \subset X$. Let W be the direct sum $\bigoplus \mathbb{R} \cdot Y_i$; and let Q be the bilinear form

$$Q(Y_i, Y_j) = -Y_i \cdot Y_j \cdot \omega.$$

In this situation we apply [BPV84, lemma I.2.10] to conclude. □

5.4 The stable case. By 5.3 we are reduced to the case that some extension $\mathcal{E}$ is ω-stable for some Kähler form ω. By (5.2.1) and (5.2.2) we have in particular

$$c_1^2(\mathcal{E}) \cdot \omega = 4c_2(\mathcal{E}) \cdot \omega,$$

which implies that $\mathcal{E}$ is projectively flat, at least on the regular part X_0. In fact, this is well-known if X is smooth and $\mathcal{E}$ is locally free. But the proof generalizes to our case since the singularities of X and $\mathcal{E}$ are in codimension at least 3. Now we follow the arguments in [Ko92,p.113/114].

Assume first that the degree of finite étale covers of X_0 is bounded: $\pi_1^{\text{alg}}(X_0)$ is finite. After performing a finite étale cover, we may assume $\pi_1^{\text{alg}}(X_0) = 0$. Since $\mathcal{E}|X_0$ is projectively flat, $\mathcal{E}^* \otimes \mathcal{E}|X_0$ is hermitian flat and therefore given by a unitary representation ρ of $\pi_1(X_0)$. Since $\rho(\pi_1(X_0))$ is residually finite, it follows that ρ is trivial, hence $\mathcal{E}^* \otimes \mathcal{E}$ is trivial. This implies, using the exact sequence

$$0 \longrightarrow K_X \longrightarrow \mathcal{E} \longrightarrow mK_X \longrightarrow 0$$

and dualizing that (over X_0) $h^0(\mathcal{E}^* \otimes K_X) = 0$, that $h^0(\mathcal{E}^* \otimes mK_X) \geq 4$ and that therefore $h^0((m-1)K_X) \geq 3$. This contradicts $a(X) = 0$.

If $\pi_1^{\text{alg}}(X_0)$ is infinite, we just take over the arguments of [Ko92p.114]: since the local fundamental groups of X at the singularities are finite, any finite étale cover h of X_0 of sufficiently large degree extends to a covering $h : \tilde{X} \to X$ which can be written in the form $h = g \circ f$, where $f : \tilde{X} \to X'$ is étale and $g : X' \to X$ is étale outside the singular locus. Therefore $\pi_1(X')$ is infinite, contradicting the fact that X' is simple non-Kummer. □

6. The inequality $K_X \cdot c_2(X) \geq 0$

The aim of this section is to prove

THEOREM 6.1. *Let X be a minimal Kähler 3-fold with $a(X) = 0$. Then we have $K_X \cdot c_2(X) \geq 0$.*

This inequality is an important step in the proof of abundance for Kähler threefolds. In the projective case, it follows from Miyaoka's inequality $K_X^2 \leq 3c_2(X)$ which in turn is a consequence of his generic nefness theorem for the cotangent bundle (relying on char p methods).

The rest of this section consists of the proof of 6.1 together with some auxiliary propositions (6.9, 6.10, 6.12/6.13).

6.2 Reduction to the unstable case. Suppose that there is a sequence (ω_j) of Kähler metrics converging in $H^2(X, \mathbb{R})$ to K_X such that T_X is ω_j-stable for all j. Then we have

$$c_1^2(X) \cdot \omega_j \leq 3c_2(X) \cdot \omega_j$$

for all j by Proposition 6.9. Taking limits, we obtain

$$K_X^3 \leq 3K_X \cdot c_2(X),$$

hence our claim results from $K_X^3 = 0$.

So from now on we shall assume that T_X *is* ω*-unstable for all* ω *near* K_X *(in* $H^2(X, \mathbb{R})$*)*.

6.3 The setup Let $\mathcal{S}_\omega \subset T_X$ be the maximal destabilizing subsheaf with respect to ω. Let r denote its rank. Then by Corollary 6.13 below, there are only finitely many choices for $\mathcal{S}_\omega$, hence there exists an open set in the Kähler cone of X having K_X as boundary point such that $\mathcal{S}_\omega$ does not depend on $[\omega]$ for $[\omega] \in U$. We shall write

$$S = \det \mathcal{S}$$

and let $Q = T_X/\mathcal{S}$, a torsion free sheaf of rank 1 or 2. We notice

$$c_1(Q) = -K_X - S \tag{6.3.1}$$

and, if $r = 1$,

$$c_2(Q) = c_2(X) + S \cdot (K_X + S). \tag{6.3.2}$$

$c_3(Q)$ will be irrelevant for us. The instability of T_X gives

$$S \cdot \omega^2 \geq \frac{-K_X \cdot \omega^2}{3}, \tag{6.3.3}$$

for $\omega \in U$ in case $r = 1$ and

$$\frac{S \cdot \omega^2}{2} \geq \frac{-K_X \cdot \omega^2}{3} \tag{6.3.3a}$$

in case $r = 2$. We claim

$$K_X^2 \cdot S = 0. \tag{6.3.4}$$

In fact, (6.3.3) resp. (6.3.3a) gives in the limit $K_X^2 \cdot S \geq 0$. Since we may assume $K_X^2 \neq 0$, the tangent sheaf T_X is K_X-semi-stable by Enoki [En87]. This implies $K_X^2 \cdot S \leq 0$, hence (6.3.4) follows.

The next lemma is a general statement on Kähler 3-folds with $a(X) = 0$, independent from our setup.

LEMMA 6.4. *Let* X *be a normal compact Kähler 3-fold. Let* A *and* B *be* $\mathbb{Q}$*-line bundles on* X *and let* A *be nef with* $A^2 \neq 0$*. If* $A^2 \cdot B = 0$*, then* $A \cdot B^2 \leq 0$*.*

Proof. By passing to a desingularization we may assume X smooth. Fix a Kähler class ω and apply (2.5) with $\alpha = c_1(B), \beta = K_X + \varepsilon\omega$ and $\gamma = K_X$. Then expand in terms of powers of ε to obtain the claim. □

COROLLARY 6.4*a* In our setup (6.3) we have $K_X \cdot S^2 \leq 0$ if $K_X^2 = 0$.

Proof. This follows from 6.4 via 6.3.4 □

Lemma 6.3 is of course not true in case $A^2 = 0$. Thus in order to obtain (6.4.a) also in case $K_X^2 = 0$ we need more specific arguments:

LEMMA 6.5. *Let X be a simply connected minimal Kähler 3-fold with $a(X) = 0$ and $K_X^2 = 0$. Let L be a $\mathbb{Q}$-line bundle on X. Then $K_X \cdot L^2 \leq 0$.*

Proof. Assume that $K_X \cdot L^2 > 0$. If a positive integer c satisfies the following condition:

$$2c^2 K_X \cdot L^2 > -K_X \cdot c_2(X) \qquad (*),$$

then by Riemann-Roch we easily get asymptotically

$$\chi(X, mK_X + cL) \sim m.$$

Observe also that $(*)$ is satisfied for large c since $K_X \cdot L^2 > 0$ by assumption. So let us fix such a number c. Then we conclude

$$h^2(mK_X + cL) \geq Cm. \qquad (**)$$

In fact, otherwise $h^0(mK_X + cL) \geq Cm$ by $(*)$, contradicting $a(X) = 0$.

Now, as in section 5, we obtain "many" extensions

$$0 \longrightarrow K_X + cL \longrightarrow \mathcal{E} \longrightarrow mK_X \to 0.$$

Observe that $\mathcal{E}$ cannot be ω-stable for ω near K_X. In fact, in that case we had

$$c_1(\mathcal{E})^2 \cdot \omega \leq 4c_2(\mathcal{E}) \cdot \omega,$$

hence $c_1(\mathcal{E})^2 \cdot K_X \leq 4c_2(\mathcal{E}) \cdot K_X$ in the limit. This comes down to

$$c^2 K_X \cdot L^2 \leq 0,$$

contradicting our assumption.

We proceed exactly in the same way as in section 5, introducing the divisors D_m, and now $(**)$ and the arguments in section 5 yield

$$h^0(\mathcal{O}_{D_m}) \geq Cm,$$

for large m. On the other hand,

$$(m-1)K_X + cL^*|D_m = D_m|D_m,$$

again referring to section 5, hence for large m, the normal bundle N_{D_m} gets more and more "nef". However, to have many functions on D_m means to have a tendency to negativity for the normal bundle. So we will show that (+) and (++) are contradictory. By passing to a subsequence - having in mind that X carries only finitely many irreducible hypersurfaces - we can suppose the following.

$$D_m = \sum_{i=1}^{s} a_{m,i} Y_i + \sum_{s+1}^{t} a_j Y_j,$$

where $a_{m,i} < a_{m+1,i}$ and the a_j are independent of m. Put $R = \sum_{s+1}^{t} a_j Y_j$ and $Y = \sum_1^s Y_i$. Then

$$N^*_{D_m/D_{m+1}} = (\sum_{i=1}^{s} a_{m,i} Y_i + R)|Y.$$

Since by (+), $h^0(N^*_{D_m/D_{m+1}}) > 0$, the sequence of divisors $-\sum a_{m,i}Y_i|Y$, suitably normalized, converges to an effective non-zero divisor on Y. Thus $N^*_{D_m|X}|Y$, suitably normalized converges to an effective non-zero divisor on Y. On the other hand, its dual is nef by (++). This is a contradiction. □

COROLLARY 6.5*a* *In our setup (6.3) we have* $K_X \cdot S^2 \le 0$.

6.6 The Case: rk$\mathcal{S}$ = 1 and Q stable. By "Q stable" we mean that there is a sequence of Kähler forms (ω_j) converging to K_X (as classes) such that Q is ω_j-stable for all j. Then by Proposition 6.9 we have

$$c_1^2(Q) \cdot \omega_j \le 4c_2(Q) \cdot \omega_j,$$

hence

$$c_1^2(Q) \cdot K_X \le 4c_2(Q) K_X.$$

Putting in (6.3.1) and (6.3.2) we obtain

$$K_X \cdot (K_X + S)^2 \le (4c_2(X) + 4S \cdot (K_X + S)) \cdot K_X,$$

which in turn yields

$$K_X \cdot c_2(X) \ge -\frac{3}{4} K_X \cdot S^2.$$

Thus 6.4.a gives $K_X \cdot c_2(X) \ge 0$.

6.7 The Case: rk$\mathcal{S}$ = 1 and Q is unstable. After the previous case it is clear what unstable has to mean: Q is ω-unstable for all ω near K_X. Then we obtain a destabilizing sequence

$$0 \longrightarrow L_1 \longrightarrow Q \longrightarrow \mathcal{I}_B L_2 \longrightarrow 0$$

where L_i are reflexive of rank 1 and $\dim B \le 1$. This sequence is - as usual - independent of ω, if ω is sufficiently near to K_X and contained in a suitable open set U as in (6.3). We first claim

$$K_X^2 \cdot L_1 \le 0. \tag{6.7.1}$$

To verify this, let $\mathcal{R}$ be the cokernel of

$$T_X \longrightarrow \mathcal{I}_B L_2 \longrightarrow 0.$$

Then we have an exact sequence

$$0 \longrightarrow S \longrightarrow \mathcal{R} \longrightarrow L_1 \longrightarrow 0.$$

Of course we may assume $K_X^2 \ne 0$. Then by Enoki [En87], T_X is K_X-semi-stable, hence

$$c_1(\mathcal{R}) \cdot K_X^2 \le 0.$$

This implies $K_X^2 \cdot (L_1 + S) \le 0$ by the last exact sequence. Now (6.3.4) gives our claim (6.7.1)

Next we show

$$K_X \cdot L_1 = 0. \tag{6.7.2}$$

In fact, the destabilizing property for L_1 reads

$$L_1 \cdot \omega^2 \ge \frac{c_1(Q) \cdot \omega^2}{2},$$

hence

$$L_1 \cdot K_X^2 \ge \frac{c_1(Q) \cdot K_X^2}{2} = \frac{1}{2}(-K_X - S) \cdot K_X^2 = 0.$$

We now conclude by (6.7.1).

Since $c_1(Q) \cdot K_X^2 = 0$, we also have

$$K_X^2 \cdot L_2 = 0. \tag{6.7.3}$$

Thus Lemma 6.4 applies:

$$K_X \cdot L_i^2 \leq 0 \tag{6.7.4}$$

for $i = 1, 2$.

The final preparation is

$$K_X \cdot L_1 \cdot L_2 = \frac{1}{2}(K_X \cdot S^2 - K_X \cdot L_1^2 - K_X \cdot L_2^2).$$

This follows from the two equations

$$K_X \cdot c_1^2(Q) = K_X \cdot (L_1 + L_2)^2$$

and

$$K_X \cdot c_1^2(Q) = K_X \cdot (K_X + S)^2 = K_X \cdot S^2.$$

After all these preparations we conclude using (6.7.5) as follows.

$$\begin{aligned} K_X \cdot c_2(X) &= K_X \cdot c_2(Q) + K_X \cdot S \cdot c_1(Q) \\ &= K_X \cdot c_2(\mathcal{I}_B) + K_X \cdot L_1 \cdot L_2 - K_X \cdot S^2 \\ &= K_X \cdot c_2(\mathcal{I}_B) + \tfrac{K_X \cdot S^2}{2} - \tfrac{K_X \cdot L_1^2}{2} - \tfrac{K_X \cdot L_2^2}{2} - K_X \cdot S^2. \end{aligned}$$

Since $K_X \cdot c_2(\mathcal{I}_B) \geq 0$ by nefness of K_X we conclude by virtue of (6.3) and (6.7.4).

6.8 The Case: $\operatorname{rk} \mathcal{S} = 2$.

In this case we consider the maximal destabilizing subsheaf $Q^* \subset \Omega_X^1 = (T_X)^*$. Here it is convenient to switch completely the notations: we denote the maximal destabilizing subsheaf of Ω_X^1 again by $\mathcal{S}$ and let Q denote the quotient. Then

$$c_1(Q) = K_X - S$$

and

$$c_2(Q) = c_2(X) - S \cdot (K_X - S).$$

Now (6.3) yields $K_X^2 \cdot S = 0$. Applying again 6.4 gives $K_X \cdot S^2 \leq 0$. Now (6.6) and (6.7) run in completely the same way; notice that some minus signs are irrelevant because $K_X^2 \cdot S = 0$.

PROPOSITION 6.9. *Let X be a normal compact Kähler n-fold with* $\operatorname{codim} \operatorname{Sing}(X) \geq 3$. *Suppose $a(X) = 0$. Let ω be a Kähler form on X and $\mathcal{E}$ a torsion free coherent sheaf on X of rank $r \geq 2$. If $\mathcal{E}$ is ω-stable, then*

$$c_1^2(\mathcal{E}) \cdot \omega^{n-2} \leq \frac{2r}{r-1} c_2(\mathcal{E}) \cdot \omega^{n-2}.$$

Proof. For simplicity of notations set $\mu = \frac{2r}{r-1}$.

(1) First we reduce the problem to the case "$\mathcal{E}$ reflexive". So suppose we know the assertion for reflexive sheaves and let $\mathcal{E}$ be torsion free. Then we consider the quotient sheaf

$$Q = \mathcal{E}^{**}/\mathcal{E},$$

which is supported on a complex subspace $Z \subset X$ of codimension at least 2. Now

$$c_2(Q) = -mc_2(\mathcal{I}_Z),$$

for some positive m, and $c_2(\mathcal{I}_Z)$ is an effective cycle supported on Z, hence

$$\omega^{n-2} \cdot c_2(Q) \leq 0. \tag{$*$}$$

Now $c_2(\mathcal{E}^{**}) = c_2(\mathcal{E}) + c_2(Q)$, hence $(*)$ implies

$$c_2(\mathcal{E})\omega^{n-2} \geq c_2(\mathcal{E}^{**}) \cdot \omega^{n-2}. \tag{$**$}$$

Notice that $\mathcal{E}^{**}$ is stable because $\mathcal{E}$ is ([Ko87,V.7.7]), hence by our assumption

$$c_1^2(\mathcal{E}^{**})\omega^{n-2} \leq \mu c_2(\mathcal{E}^{**}) \cdot \omega^{n-2}.$$

Since $c_1(\mathcal{E}^{**}) = c_1(\mathcal{E})$, the inequality $(**)$ implies our claim follows.

(2) From now on we shall assume $\mathcal{E}$ reflexive. Choose a desingularization $\pi : \hat{X} \longrightarrow X$ by a sequence of blow-ups whose centers all ly over the singularities of X and $\mathcal{E}$. Moreover we may assume that $\hat{\mathcal{E}} = \pi^*(\mathcal{E})^{**}$ is locally free (see [GR71]). Let $\hat{\omega} = \pi^*(\omega)$. By definition of Kähler forms on singular spaces $\hat{\omega}$ - which a priori exists only on a Zariski open part of $\hat{X}$ - extends to a semipositive $(1,1)$-form on all of $\hat{X}$. We claim

$$\mathcal{E} \text{ is } \hat{\omega} - \text{stable}. \tag{$+$}$$

Indeed, assume we have a subsheaf $\hat{\mathcal{S}} \subset \hat{\mathcal{E}}$ of rank s with

$$\frac{c_1(\hat{\mathcal{S}}) \cdot \hat{\omega}^{n-1}}{s} \geq \frac{c_1(\hat{\mathcal{E}}) \cdot \hat{\omega}^{n-1}}{r}.$$

Then consider

$$\mathcal{S} = \pi_*(\hat{\mathcal{S}}) \subset \pi_*(\hat{\mathcal{E}}).$$

Since $\pi_*(\hat{\mathcal{E}})$ is torsion free and since $\mathcal{E}$ is reflexive, we have $\pi_*(\hat{\mathcal{E}}) \subset \mathcal{E}$, hence $\mathcal{S} \subset \mathcal{E}$. Now

$$c_1(\hat{\mathcal{S}}) \cdot \hat{\omega}^{n-1} = c_1(\pi_*(\hat{\mathcal{S}})) \cdot \omega^{n-1} = c_1(\mathcal{S}) \cdot \omega^{n-1},$$

and

$$c_1(\hat{\mathcal{E}}) \cdot \hat{\omega}^{n-1} = c_1(\mathcal{E}) \cdot \omega^{n-1},$$

hence

$$\frac{c_1(\mathcal{S}) \cdot \omega^{n-1}}{s} \geq \frac{c_1(\mathcal{E}) \cdot \omega^{n-1}}{r},$$

contradicting the ω-stability of $\mathcal{E}$. This proves $(+)$.

Now $\hat{\omega}$ has the disadvantage not to be a Kähler form, but it is on the boundary of the Kähler cone. To circumvent this difficulty, let E_i denote the exceptional components of the exceptional set of π, then we can chose $a_i < 0$, such that $E := \sum a_i E_i$ is π-ample. Thus

$$\hat{\omega}_\epsilon := \hat{\omega} + \epsilon E$$

is a Kähler class for all small positive ϵ. We claim that $\hat{\mathcal{E}}$ is $\hat{\omega}_\epsilon$-stable for ϵ small enough. Indeed, suppose the contrary. Then there exists a sequence $\epsilon - k$ converging to 0 such that $\hat{\mathcal{E}}$ is not $\hat{\omega}_{\epsilon_k}$-stable. Let $\mathcal{S}_i \subset \hat{\mathcal{E}}$ be the maximal destabilizing subsheaf with respect to $\hat{\omega}_{\epsilon_i}$. Since $a(\hat{X}) = 0$, we find i_0 such that $\mathcal{S}_i = \mathcal{S}_j$ for all $i, j \geq i_0$ (Prop. 6.12), possibly after passing to a subsequence (but even this could be avoided). So let $\mathcal{S} = \mathcal{S}_i$, $i \geq i_0$. Then we have

$$\frac{c_1(\mathcal{S}) \cdot \hat{\omega}_{\epsilon_k}^{n-1}}{s} \geq \frac{c_1(\hat{\mathcal{E}}) \cdot \hat{\omega}_{\epsilon_k}^{n-1}}{r},$$

so passing to the limit,

$$\frac{c_1(\mathcal{S}) \cdot \hat{\omega}^{n-1}}{s} \geq \frac{c_1(\hat{\mathcal{E}}) \cdot \omega^{n-1}}{r}.$$

This contradicts (+).

Thus $\hat{\mathcal{E}}$ is $\hat{\omega}$-stable for small positive ϵ. Therefore $\hat{\mathcal{E}}$ is Hermite-Einstein with respect to $\hat{\omega}_\epsilon$ and hence

$$c_1^2(\hat{\mathcal{E}}) \cdot \hat{\omega}_\epsilon^{n-2} \leq \mu\, c_2(\hat{\mathcal{E}}) \cdot \hat{\omega}_\epsilon,$$

hence

$$c_1^2(\hat{\mathcal{E}}) \cdot \hat{\omega}^{n-2} \leq \mu\, c_2(\hat{\mathcal{E}}) \cdot \hat{\omega}^{n-2}.$$

Since $\operatorname{codim}(\operatorname{Sing}(X) \cup \operatorname{Sing}(\mathcal{E})) \geq 3$, we conclude

$$c_1^2(\mathcal{E})\omega^{n-2} \leq \mu\, c_2(\mathcal{E}) \cdot \omega^{n-2}.$$

□

PROPOSITION 6.10. *Let L be a line bundle or a reflexive sheaf of rank 1 on the normal compact complex space X. Suppose $a(X) = 0$. Let $\mathcal{S}_i \subset L$ be reflexive subsheaves, $i \in I$. Then for all i there are only finitely many j such that $\mathcal{S}_j \not\subset \mathcal{S}_i$.*

Proof. Of course we may assume X smooth. Since $a(X) = 0$, the complex space X has only finite many irreducible hypersurfaces $Y_1, \ldots Y_r$, therefore we can write

$$\mathcal{S}_i = L - \sum_{j=1}^{r} a_j^{(i)} Y_j$$

with $a_j^{(i)} \geq 0$. Thus the claim is clear. □

DEFINITION 6.11. *Let $\mathcal{F}$ be a torsion free coherent sheaf on a normal compact complex space and let $\mathcal{S} \subset \mathcal{F}$ be a reflexive subsheaf with $0 < \operatorname{rk}\mathcal{S} < \operatorname{rk}\mathcal{F}$. We say that $\mathcal{S}$ is maximal, if there is no proper reflexive subsheaf $\mathcal{S}' \subset \mathcal{F}$ of the same rank as $\mathcal{S}$ such that $\mathcal{S} \subset \mathcal{S}'$ and $\mathcal{S} \neq \mathcal{S}'$.*

If ω is a Kähler form on X and if $\mathcal{S}$ is the ω-maximal destabilizing subsheaf of the ω-unstable sheaf $\mathcal{F}$, then $\mathcal{S}$ is maximal. This is the way we will identify maximal subsheaves.

PROPOSITION 6.12. *Let X be a normal compact Kähler space with $a(X) = 0$ and $\mathcal{F}$ a reflexive coherent sheaf on X. Then $\mathcal{F}$ admits only finitely many maximal reflexive subsheaves of rank 1.*

Proof. Of course we may assume X smooth. Consider now the maximal subsheaves $\mathcal{S}_i \subset \mathcal{F}$ of rank 1, $i \in I = \mathbb{N}$. Choose $m \in \mathbb{N}$ and $i_1 < \ldots < i_m$ such that

$$\mathcal{S}' = \mathcal{S}_{i_1} + \ldots + \mathcal{S}_{i_m} \subset \mathcal{F}$$

has the following property: if j is different from the i_j, then $\operatorname{rk}\mathcal{S}' = \operatorname{rk}(\mathcal{S}' + \mathcal{S}_j)$. So things come down to show that there are only finitely many j such that

$$\operatorname{rk}(\mathcal{S}_j \cap \mathcal{S}') = 1.$$

In order to prove this, we assume to the contrary that there are infinitely many j such that $\operatorname{rk}(\mathcal{S}_j \cap \mathcal{S}') = 1$. Then we have infinitely many subsheaves

$$\mathcal{T}_j := \mathcal{S}_j \cap \mathcal{S}' \subset \mathcal{S}'$$

of rank 1 (use again the finiteness of hypersurfaces in X.) Now fix j_0. Then by (6.10) there are only finitely many j such that $\mathcal{T}_j \not\subset \mathcal{T}_{j_0}$. For all others we have $\mathcal{T}_j \subset \mathcal{T}_{j_0}$ and for those we write

$$\mathcal{T}_j = \mathcal{S}_j - A_j$$

and

$$\mathcal{T}_{j_0} = \mathcal{S}_{j_0} - A_{j_0}$$

with effective divisors A_j. Since X has only finitely many irreducible hypersurfaces, we have $A_{j_0} \subset A_j$ for almost all j, hence we obtain $\mathcal{S}_j \subset \mathcal{S}_{j_0}$ for almost all j, contradiction to maximality. □

COROLLARY 6.13. *Let X be an normal compact Kähler space with $a(X) = 0$ and $\mathcal{F}$ a torsion free sheaf of rank at most 3. Then $\mathcal{F}$ contains only finitely many maximal reflexive subsheaves.*

Proof. By 6.12 we have only to deal with the case of subsheaves of rank 2. This is done by dualizing and applying 6.12 to $\mathcal{F}^*$ using the following trivial remark: if $\mathcal{S} \subset \mathcal{F}$ is maximal with quotient Q, then $Q^* \subset \mathcal{F}^*$ is maximal. □

7. An abundance theorem for Kähler threefolds.

Here we want to solve (the remaining part of) the abundance problem for Kähler threefolds:

THEOREM 7.1. *Let X be a $\mathbb{Q}$-Gorenstein Kähler threefold with only terminal singularities such that K_X is nef (a minimal Kähler threefold for short). Then $\kappa(X) \geq 0$.*

Of course, more should be true:

CONJECTURE 7.2. *Let X be a minimal Kähler threefold. Then K_X is semi-ample, i.e. some multiple mK_X is spanned by global sections.*

REMARK 7.3.
(1) *In case X is projective, everything is proved by Miyaoka [Mi87,88] and Kawamata [Ka92].*
(2) *In the non-algebraic case, 7.1/7.2 is proved in [Pe00] with the important possible exception that X is simple and not Kummer (see 1.4). In particular in this remaining case we have algebraic dimension $a(X) = 0$ and $\pi_1(X)$ finite. In [DPS00] 7.1 is proved if K_X carries a sufficiently nice metric, e.g. if K_X is hermitian semipositive.*
(3) In case X is Gorenstein, we have the Riemann-Roch formula

$$\chi(X, \mathcal{O}_X) = -\frac{1}{24} K_X \cdot c_2(X).$$

Therefore the inequality (6.1) – recall we may assume that $a(X) = 0$ –

$$K_X \cdot c_2(X) \geq 0 \qquad (*)$$

implies $\chi(X, \mathcal{O}_X) \leq 0$ and therefore $h^0(X, K_X) = h^3(X, \mathcal{O}_X) \neq 0$, so that at least $\kappa(X) \geq 0$. In case X is not Gorenstein, this Riemann-Roch formula is not true; instead one has some positive correction term [Fl87] which might correct the negativity of $-K_X \cdot c_2(X)$ and therefore destroy the contradiction.

Proof of Theorem 7.1 As noticed in 7.3 we may assume that X is simple non-Kummer, in particular $q(X) = 0$. First we reduce ourselves to the case that X is $\mathbb{Q}$-factorial by applying Kawamata's factorialisation $f : \hat{X} \longrightarrow X$ as in the proof of 4.4. Since f is small, we have $K_{\hat{X}} = f^*(K_X)$, so $K_{\hat{X}}$ is nef. Hence we can work on $\hat{X}$ and thus may assume X to be $\mathbb{Q}$-factorial from the beginning. We consider a desingularization

$$\pi : \hat{X} \longrightarrow X$$

and compute by Riemann-Roch

$$\chi(\hat{X}, \pi^*(mK_X)) = \frac{m}{12} K_X \cdot c_2(X) + \chi(X, \mathcal{O}_X) \qquad (*)$$

for all m such that mK_X is Cartier. Assume $\kappa(X) = -\infty$, so $H^3(\mathcal{O}_X) = 0$. Since X is not projective, we have $H^2(\mathcal{O}_X) \neq 0$. In total we obtain:

$$\chi(X, \mathcal{O}_X) \geq 2.$$

If now $K_X \neq 0$, then by 4.3/5.1, we have $h^2(X, mK_X) \leq 1$, hence (6.1) and $(*)$ imply $h^0(X, mK_X) \geq 1$, a contradiction. If however $K_X = 0$, take a positive integer m such that mK_X is Cartier. If now mK_X is not a torsion line bundle, we must have $q(X) > 0$, contradiction. □

REMARK 7.4. In order to settle the abundance for Kähler threefolds completely, it remains to show that a simple threefold X with K_X nef and $\kappa(X) = 0$ must be Kummer. In the following we collect what we know about X. We shall assume that $q(X) = 0$, otherwise we consider the Albanese and are easily done. Thus we have $\chi(X, \mathcal{O}_X) \geq 1$. (1) $K_X \cdot c_2(X) = 0$ and $1 \leq \chi(X, \mathcal{O}_X) \leq 2$. The first part follows easily from equation $(*)$ in the proof of (7.3) together with 4.1/5.1. Hence

$$(*) \qquad \chi(X, mK_X) = \chi(X, \mathcal{O}_X)$$

for all integers m such that mK_X is Cartier. Then again 4.3/5.1 gives the inequality for $\chi(X, \mathcal{O}_X)$.
(2) X cannot be Gorenstein. In fact, then the Riemann-Roch formula

$$24\chi(X, \mathcal{O}_X) = -K_X \cdot c_2(X) = 0$$

gives a contradiction. (3) If $\chi(X, \mathcal{O}_X) = 2$, then $K_X^2 = 0$. This is a consequence of (1) via the vanishing 4.3. (4) If $\chi(X, \mathcal{O}_X) = 1$, then $h^0(X, K_X) = 1$ and $h^2(X, \mathcal{O}_X) = 1$.

8. Almost algebraic Kähler threefolds

In this section we show that simple non-Kummer threefolds are very far from projective threefolds, in a sense which is made precise in the following definition.

DEFINITION 8.1. *Let X be a normal Kähler variety with only terminal singularities. X is almost algebraic if there exists an algebraic approximation of X. This is a proper surjective flat holomorphic map $\pi : \mathcal{X} \to \Delta$ from a normal complex space $\mathcal{X}$ where $\Delta \subset \mathbb{C}^m$ is the unit disc, where $X \simeq X_0$, where all complex analytic fibers $X_t = \pi^{-1}(t)$ are normal Kähler spaces with at most terminal singularities such that there is a sequence (t_j) in Δ converging to 0 so that all $X_j := X_{t_j}$ are projective.*

Of course, in case X is smooth, all X_t will be smooth (after possibly shrinking Δ).

The following problem is attributed to Kodaira.

PROBLEM 8.2. *Is every compact Kähler manifold almost algebraic?*

From a point of view of algebraic geometry almost algebraic Kähler spaces seem to be the most interesting Kähler spaces. Therefore it is worthwile to notice

THEOREM 8.3. *Let X be a nearly algebraic Kähler threefold with only terminal singularities. If X is simple and additionally K_X nef or X smooth, then X is Kummer.*

Proof. Assume that X is not Kummer. Then $\pi_1(X)$ is finite by [Ca94] as already mentioned. Let $\pi : \mathcal{X} \to \Delta$ be an algebraic approximation of X. Let (t_j) be a sequence in Δ converging to 0 such that all $X_j = X_{t_j}$ are projective. Notice first that $\kappa(X_j) \geq 0$ for all j. In fact, otherwise X_j would be uniruled for some j and by standard arguments X_t would be uniruled for all t which is not possible, $X = X_0$ being simple.
(1) We show that $\kappa(X) = \kappa(X_0) \geq 0$. Fix a positive integer m. Then by [KM92, 1.6], every t_j admits an open neighborhood U_j such that

$$h^0(X_t, mK_{X_t}) = h^0(X_j, mK_{X_j})$$

for all $t \in U_j$. Now choose m such that $h^0(X_j, mK_{X_j}) > 0$ for some j. Then it follows that $h^0(X_t, mK_{X_t}) = h^0(X_j, mK_{X_j}) =: d > 0$ for all t in an open set in Δ. Let

$$A := \{t \in \Delta | h^0(X_t, mK_{X_t}) \geq d\}.$$

Then A is an analytic set in Δ (semi-continuity in the analytic Zariski topology), and it contains a non-empty open set, hence $A = \Delta$. Thus $\kappa(X_0) \geq 0$. Since X_0 is simple, we conclude $\kappa(X_0) = 0$. (2) Suppose that $\kappa(X_j) \geq 1$ for some j. Then fix m such that $h^0(X_j, mK_{X_j}) \geq 2$. Repeating the same arguments as in (1), we conclude $h^0(X, mK_X) \geq 2$, contradicting X being simple. So $\kappa(X_j) = 0$ for all j.
(3) Here we will show that X_j is Kummer for all j. Let X'_j be a minimal model of X_j. Observe that

$$h^2(X_j, \mathcal{O}_{X_j}) = h^2(X, \mathcal{O}_X) > 0;$$

in fact, $H^1(X, \mathcal{O}_X) = 0$, hence $H^1(X_j, \mathcal{O}_{X_j}) = 0$ for large j. Moreover $h^0(X_t, K_{X_t})$ is constant by [KM92], as shown above. Therefore the equality follows by Serre duality and the constancy of $\chi(X_t, \mathcal{O}_{X_t})$. Hence we also have $h^2(X'_j, \mathcal{O}_{X'_j}) > 0$.

Since $K_{X'_j} \equiv 0$, there exists a finite cover, the so-called canonical cover, $h : \tilde{X}_j \to X'_j$, étale in codimension 2, such that $K_{\tilde{X}_j} = \mathcal{O}_{\tilde{X}_j}$. In particular $\tilde{X}_j$ is Gorenstein and Riemann-Roch yields

$$\chi(\tilde{X}, \mathcal{O}_{\tilde{X}_j}) = 0.$$

Since $h^2(\mathcal{O}_{\tilde{X}_j}) > 0$, we must have $q(\tilde{X}_j) > 0$. Let $\alpha_j : \tilde{X}_j \longrightarrow A = A_j$ be the Albanese map. By [Ka85] there exists a finite étale cover $B \longrightarrow A$ such that

$$\hat{X}_j := \tilde{X}_j \times_A B \simeq F \times B.$$

In particular $\hat{X}_j$ and $\tilde{X}_j$ are smooth because of the isolatedness of singularities. We conclude that X'_j is Kummer unless F is a K3-surface. To exclude that case, consider the image $F' \subset \tilde{X}_j'$ of a general $F \times \{b\}$. Then F' is K3 or Enriques and does not meet the singularities of X_j'. Moreover the normal bundle $N_{F'}$ is numerically trivial. Since F' moves, it is actually trivial. Now consider the strict transform in X_j, again called F'. Then F' has the same normal bundle in X_j, so that $N_{F'/\mathcal{X}} = \mathcal{O}_{\mathcal{X}}^2$. Since $H^1(N) = 0$, the deformations of F' cover every X_t contradicting the simplicity of X_0. So F cannot be K3 and X_j is Kummer.

(4) Suppose K_{X_0} nef. Fix a positive number m such that

$$mK_{\mathcal{X}} = \mathcal{O}_{\mathcal{X}}(D)$$

with some effective divisor D. We may assume that D does not contain any fiber of π; denote $D_t = X_t \cap D$. We want to argue that K_{X_t} must be nef, therefore $K_{X_j} = 0$ so that $D_j = 0$ and $D = 0$ in total. So we will obtain $K_X = 0$. To see that X is Kummer, consider the canonical cover of X and argue as in the proof of (7.1). To prove nefness, we apply [KM92] to deduce that the sequence $\varphi_j : X_j \to X_j' = A_j/G$ appears in family $\mathcal{X}_{U_j} \longrightarrow \mathcal{X}'_{U_j}$ over a small neighborhood U_j of t_j. In particular some multiple $N_{D_t}^{*\mu}$ has many sections for $t \in U_j$ on an at least 1-dimensional family of curves. By semicontinuity, also $N_{D_0}^{*\mu}$ has many sections on such a family, contradicting the nefness of D_0. Alternatively, $K_{\mathcal{X}}$ is negative on a family of rational curves over U_j, which converges to a family of rational curves in X_0 and therefore forces K_{X_0} to be non-nef.

(5) Now suppose X_0 smooth, i.e. π is smooth after shrinking Δ. Take a sequence of blow-ups of smooth subvarieties of $\mathcal{X}$ such that the preimage of red$D$ has normal crossings. After shrinking Δ we may assume that the only points and compact curves blown up lying over X_0 so that all fibers over $\Delta \setminus 0$ are smooth. Then take a covering $h : \tilde{\mathcal{X}} \longrightarrow \mathcal{X}$ such that $K_{\tilde{\mathcal{X}}} = \mathcal{O}(\tilde{D})$ with $\tilde{\mathcal{X}}$ smooth. This is possible e.g. by applying [Ka81]. Then $\mathcal{X}_{t_j}$ is Kummer and admits a 3-form and therefore must be bimeromorphically a torus (if A/G admits a 3-form, then it is a torus covered by A. This is a consequence of the simplicity of A and the fact that G acts without fix points). Hence every $\tilde{\mathcal{X}}_t$, $t \neq 0$, has 3 holomorphic 1-forms which are independent at the general point and therefore every X_t, $t \neq 0$, is Kummer. In order to show that X_0 is Kummer, consider the central fiber $\tilde{\mathcal{X}}_0$ which contains the preimage of the strict transform X_0' of X_0. More precisely, we have

$$\tilde{\mathcal{X}}_0 = X_0' + \sum a_i E_i$$

where the E_i are smooth threefolds contracted to points or curves. By semicontinuity, $h^2(\mathcal{O}_{\tilde{\mathcal{X}}_0}) \geq 3$. Now we check easily that

$$H^2(\tilde{\mathcal{X}}_0, \mathcal{O}_{\tilde{\mathcal{X}}_0}) = H^2(X_0', \mathcal{O}_{X_0'}),$$

hence X_0' carries three 2-forms coming from $\tilde{\mathcal{X}}$. But then it is clear that also some of the holomorphic 1-forms on $\tilde{\mathcal{X}}$ give non-zero 1-forms on X_0', since the 2-forms are wegdge products of the 1-forms. Hence X_0' is Kummer and so does X_0. □

If problem 8.2 has a positive answer in dimension 3, Theorem 8.3 excludes the existence of simple non-Kummer threefolds.

References

[BS95] Beltrametti, M.; Sommese, A.J.: *The adjunction theory of complex projective varieties.* de Gruyter Exp. in Math. **16** (1995).

[Ca94] Campana, F.: *Remarques sur le revêtement universel des variétés Kählériennes compacts;* Bull. Soc. Math. France **122** (1994) 255–284.

[CP01] Campana, F.; Peternell, Th.: *The Kodaira dimension of Kummer threefolds;* Bull. Soc. Math. France **129** (2001) 357–359.

[De92] Demailly, J.-P.: *Regularization of closed positive currents and intersection theory;* J. Alg. Geom. **1** (1992), 361–409.

[De93a] Demailly, J.-P.: *Monge-Ampère operators, Lelong numbers and intersection theory;* Complex Analysis and Geometry, Univ. Series in Math., edited by V. Ancona and A. Silva, Plenum Press, New-York (1993).

[De93b] Demailly, J.-P.: *A numerical criterion for very ample line bundles;* J. Differential Geom. **37** (1993) 323–374.

[DPS94] Demailly, J.-P.; Peternell, Th.; Schneider, M.: *Compact complex manifolds with numerically effective tangent bundles;* J. Alg. Geom. **3** (1994) 295–345.

[DPS00] Demailly, J.-P.; Peternell,Th.; Schneider,M.; Pseudo-effective line bundles on compact Kähler manifolds; Intern. J. Math. **6** (2001) 689–741.

[Ei95] Eisenbud, D.: it Commutative algebra; Graduate Texts in Math. **150**, Springer (1995).

[En87] Enoki, I.: *Stability and negativity for tangent bundles of minimal Kähler spaces;* Lecture Notes in Math. **1339** (1987) 118–127.

[Fl87] Fletcher, A.R.: *Contributions to Riemann-Roch on projective 3-folds with only canonical singularities and applications;* Proc. Symp. Pure Math. **46** (1987) 221–231.

[Gr62] Grauert, H.: *Über Modifikationen und exzeptionelle analytische Mengen;* Math. Ann. **146** (1962) 331–368.

[GR70] Grauert, H.; Riemenschneider,O.: *Verschwindungssätze für analytische Kohomologiegruppen auf komplexen Räumen;* Invent. Math. **11** (1970) 263–292.

[Ka81] Kawamata, Y.: *Characterization of abelian varieties.* Comp. math. **43** (1981) 275-276.

[Ka85] Kawamata, Y.: *Minimal models and the Kodaira dimension of algebraic fiber spaces;* J. reine u. angew. Math. **363** (1985) 1-46.

[Ka88] Kawamata, Y.: *Crepant blowing ups of threedimensional canonical singularities and applications to degenerations of surfaces;* Ann. Math. **119** (1988) 93–163.

[KMM87] Kawamata, Y.; Matsuki, K.; Matsuda, K.: *Introduction to the minimal model program;* Adv. Stud. Pure Math. **10** (1987) 283–360.

[KM92] Kollár, J.; Mori, S.: *Classification of three dimensional flips;* Journal of the AMS **5** (1992), 533–703.

[Ko87] Kobayashi, S.: *Differential geometry of complex vector bundles;* Princeton Univ. Press (1987).

[Ko92] Kollař, J. et al.: *Flips and abundance for algebraic 3-folds;* Astérisque **211**, Soc. Math. France (1992).

[Mi87] Miyaoka, Y.: *The Chern classes and Kodaira dimension of a minimal variety;* Adv. Stud. Pure Math. **10** (1987) 449–476.

[Mi88] Miyaoka, Y.: *On the Kodaira dimension of minimal threefolds;* Math. Ann. **281** (1988) 325–332.

[Mi88a] Miyaoka, Y.: *Abundance conjecture for 3-folds: case $\nu = 1$;* Comp. Math. **68** (1988) 203–220.

[Pa98] Paun, M.: *Sur l'effectivité numérique des images inverses de fibrés en droites;* Math. Ann. **310** (1998) 411–421.

[Pe01] Peternell, Th.: *Towards a Mori theory on compact Kähler 3-folds, III;* Bull. Soc. Math. France **129** (2001) 339–356.

[Re87] Reid, M.: *Young person's guide to canonical singularities;* Proc. Symp. Pure Math. **46**, 345-414 (1987).

[Siu74] Siu, Y.T.: *Analyticity of sets associated to Lelong numbers and the extension of closed positive currents;* Invent. Math. **27** (1974) 53–156.

[Va84] Varouchas, J.: *Stabilité de la classe des variétés Kählériennes par certain morphismes propres;* Invent. Math. **77**, (1984) 117–127.

* Université de Grenoble I, BP 74
Institut Fourier, UMR 5582 du CNRS
38402 Saint-Martin d'Hères, France

** Universität Bayreuth
Mathematisches Institut
D-95440 Bayreuth, Deutschland

Moment maps in differential geometry

S.K. Donaldson

More than twenty years ago, Atiyah and Bott observed that the curvature can be viewed as a moment map for the action of the gauge group on the space of connections [2]. Since then, this notion of a moment map—applied to infinite-dimensional symmetry groups underlying differential-geometric problems—has proved to be very fruitful. It yields a unified point of view on many different questions, and brings with it a package of standard theory which can either be applied directly or at least, in the deeper aspects, suggests what one ought to try to prove. In this article we will first survey briefly some of the well-established applications of these ideas in the literature. Then, in Section 2, we go on to study a new variant of the theme, for diffeomorphism groups acting on sections of bundles. We prove a general abstract result and then begin the study of a particular example, which leads to a hyperkahler extension of the Weil-Petersson metric on the moduli space of Riemann surfaces.

The author is grateful to Richard Thomas and William Goldman for helpful discussions related to this article.

1. Survey of some examples

We recall the definition of a moment map. Suppose a Lie group G acts on a symplectic manifold (X, ω). Thus the derivative of the action is a map

$$r : \mathfrak{g} \to \mathrm{Vect}(X).$$

A moment map for the action is a map

$$\mu : X \to \mathfrak{g}^*$$

whose derivative

$$d\mu : TX \to \mathfrak{g}^*$$

is the transpose of r, where TX is identified with T^*X using the form ω. We will always require that the moment map be an equivariant map, intertwining the G-actions on X and $\mathfrak{g}^*$. One of the main applications of this idea is to the construction of symplectic quotients. If c is an element of $\mathfrak{g}^*$ which is fixed by the co-adjoint action we form the G-invariant subset $Z_c = \mu^{-1}(c) \subset X$. If G acts freely on Z_c this set is a submanifold, and the quotient Z_c/G is a manifold with a natural induced symplectic form—the symplectic quotient of X. In the case when X is Kahler the quotient has a natural Kahler structure. In addition, there is a circle of ideas relating this symplectic quotient to a quotient of a set of stable points in X by the action of the complexified group, but we will not say much about this side of things in the present article.

1.1. Gauge theory. The original Atiyah and Bott example is this. Let K be a compact Lie group and let Σ be a compact oriented surface. We consider the space $\mathcal{A}$ of connections on a principal K-bundle P over Σ. This is an affine space, with tangent space the sections of $T^*\Sigma \otimes \text{ad } P$ where ad P is the bundle of Lie algebras associated to the adjoint representation of K. Fix an invariant inner product on $\mathfrak{k}$. The tensor product of this inner product and the wedge product on 1-forms gives a skew-symmetric map

$$(T^*\Sigma \otimes \text{ad } P) \otimes (T^*\Sigma \otimes \text{ad } P) \to \Lambda^2 T^*\Sigma,$$

and we define a symplectic form on $\mathcal{A}$ by integrating over the compact surface Σ. To make the notation more transparent, let us suppose that K is the unitary group $U(n)$, so we can work with a Hermitian vector bundle E over Σ and regard ad P as a sub-bundle of End E (the sub-bundle of skew-adjoint endomorphisms). Then the symplectic form is given by the formula

$$\Omega(a, b) = \int_\Sigma \text{Tr}(a \wedge b).$$

The gauge group $\mathcal{G}$ of automorphisms of P acts on $\mathcal{A}$, preserving the symplectic form. We can represent an element of $\mathcal{G}$ as a section of End E and the action is given by

$$g(A) = A - (d_A g) g^{-1},$$

where d_A is the covariant derivative on End E induced by the connection. The Lie algebra of $\mathcal{G}$ can be regarded as the sections of ad P and the infinitesimal action is given by

$$r(u)(A) = -d_A u.$$

The curvature $F(A)$ of a connection A is a 2-form with values in the bundle ad P and the variation of the curvature is given by

$$F(A + a) = F(A) + d_A a + a \wedge a,$$

where d_A is the coupled exterior derivative extending the covariant derivative on ad P. Thus the derivative of the curvature, regarded as a map

$$F : \mathcal{A} \to \Omega^2(\text{ad } P),$$

is just

$$d_A : \Omega^1(\text{ad } P) \to \Omega^2(\text{ad } P).$$

The space $\Omega^2(\text{ad } P)$ of bundle-valued 2-forms is regarded as embedded in the dual of the Lie algebra of $\mathcal{G}$ via the pairing

$$(u, F) \mapsto \int_\Sigma \text{Tr}(uF).$$

The assertion that F is a moment for the action is the statement that, for all u and a,

$$\int_\Sigma \text{Tr } (d_A u \wedge a) = -\int_\Sigma \text{Tr } (u d_A a),$$

which follows from Stokes' Theorem since

$$\text{Tr}(d_A u \wedge a + u d_A a) = d(\text{Tr } (ua)).$$

In this case the symplectic quotient $\mathcal{N}$ is a moduli space of (projectively) flat connections over Σ and one of the fruits of Atiyah and Bott's observation is that one immediately sees that this moduli space has a natural symplectic structure.

1.2. Coupled equations. Among the various generalisations of the set-up of the previous subsection, one of the most useful is to the study of connections combined with additional data. A very general formulation of this idea had been given by Mundet i Riera [12] and Cielibak *et al* [5]. In the set-up above, we consider an additional auxiliary symplectic manifold (X, ω) with a K-action and a moment map $\mu : X \to \mathfrak{k}^*$. Now we form the associated bundle $\underline{X} \to \Sigma$ with fibre X. There is a symplectic form, which we still denote by ω, on the vertical subbundle $T_V\underline{X}$ of the tangent bundle of $\underline{X}$. Likewise the moment map defines a bundle map, which we still denote by μ, from $\underline{X}$ to ad P. We consider the space $\Gamma(\underline{X})$ of sections of $\underline{X}$. A tangent vector in $\Gamma(\underline{X})$ is given by a section of the pull-back of the vertical subbundle. Thus if we have two such tangent vectors ξ, η we can form a function $\omega(\xi, \eta)$ on Σ. Now suppose that we fix an area form ρ on Σ. We define a symplectic pairing by

$$\Omega(\xi, \eta) = \int_\Sigma \omega(\xi, \eta)\rho.$$

It is straightforward to see that this yields a closed 2-form on the infinite-dimensional space $\Gamma(\underline{X})$. The gauge group $\mathcal{G}$ acts as fibrewise automorphisms of $\underline{X}$ and hence acts on Γ, and this action preserves the symplectic form. The moment map for the action is given merely by

$$\phi \mapsto \mu(\phi)\rho,$$

for a section ϕ of $\underline{X}$. Now consider the space $\mathcal{A} \times \Gamma(\underline{X})$ with the product symplectic form and the diagonal $\mathcal{G}$-action. The moment map is given by the equation

$$\underline{\mu}(A, \phi) = F(A) + \mu(\phi)\rho.$$

So we can form a symplectic quotient parametrising solutions (A, ϕ) of the equation

$$F(A) + \mu(\phi)\rho = c\rho \tag{1}$$

for a suitable fixed c. This equation, regarded as a PDE for (A, ϕ), is under-determined and the symplectic quotient is infinite-dimensional. The main interest comes from combining this moment map equation with another equation, a coupled Cauchy-Riemann equation. Here we suppose that X is actually a Kahler manifold, with K acting by isometries, and that Σ is a Riemann surface. Given a connection A on P we can form the covariant derivative $\nabla_A\phi$ of a section ϕ of $\underline{X}$. This is an $\mathbf{R}$-linear map from the tangent space of Σ to the vertical tangent bundle of X and under our assumptions each of these spaces have natural complex structures. Thus we can write the covariant derivative as a sum of complex-linear and anti-linear parts

$$\nabla_A\phi = \partial_A\phi + \overline{\partial}_A\phi.$$

The coupled Cauchy-Riemann equation is the equation

$$\overline{\partial}_A\phi = 0. \tag{2}$$

The combined equations (1) and (2) form an elliptic system (when account is taken of the gauge group action) with a finite-dimensional moduli space of solutions.

The Cauchy-Riemann equation (2) can be thought of as cutting out an infinite-dimensional symplectic (in fact complex) submanifold of $\mathcal{A} \times \Gamma(\underline{X})$ and the general theory yields a symplectic (in fact Kahler) structure on the moduli space.

The examples of this general set-up which have been studied most extensively arise when X is a complex vector space, i.e. from a unitary representation of K. See [3],[4] for surveys of these developments. *Hitchin's equations* [10] arise when one takes the complexified adjoint representation. Reverting again to the case when K is $U(n)$, so we can formulate things in terms of a Hermitian vector bundle E, the auxiliary field ϕ—the "Higgs field"—is now a section of End$E$. Then the finite dimensional moment map is given by

$$\mu(\phi) = [\phi, \phi^*].$$

In fact it is more useful to consider a minor variant of the general set-up in which we take ϕ to be a section of the tensor product

$$T^*\Sigma \otimes_{\mathbf{C}} \text{End } E.$$

The preceding discussion goes through without essential change but the advantage is that the symplectic form on the space of Higgs fields can now be defined without fixing an area form on Σ. Likewise $[\phi, \phi^*]$ is a bundle valued 2-form and (1) is modified to

$$F(A) + [\phi, \phi^*] = c\rho \tag{3}$$

The appropriate constant c is determined by the first Chern class of the bundle E. If the bundle is trivial the constant is zero and the theory is entirely conformally invariant. In sum, Hitchin associates to a compact Riemann surface a moduli space $\mathcal{N}^c$ parametrising solutions to (2),(3). Clearly the Kahler moduli space $\mathcal{N}$ forms a submanifold of $\mathcal{N}^c$.

Hitchin found many remarkable properties of the solutions of these equations and of their moduli space $\mathcal{N}^c$. Many of these stem from the fact that the moduli space can be regarded as a "hyperkahler quotient". To review these ideas, recall that a hyperkahler structure on a manifold X is a Riemannian metric together with three complex structures I_1, I_2, I_3 satisfying the algebraic relations of the quaternions and such that X is Kahler with respect to each structure. Thus there are three symplectic forms $\omega_1, \omega_2, \omega_3$. If a group G acts on X, preserving all this structure, then in favourable situations there will be three moment maps $\mu_1, \mu_2, \mu_3 : X \to \mathfrak{g}^*$. If c_i are fixed by the co-adjoint action we can form the *hyperkahler quotient*

$$\frac{\mu_1^{-1}(c_1) \cap \mu_2^{-1}(c_2) \cap \mu_3^{-1}(c_3)}{G}.$$

It is a general fact that (provided the group acts freely on the indicated set) this quotient space has a natural hyperkahler structure.

Going back to Hitchin's equations: we can regard the space $\mathcal{A} \times \Omega^{1,0}(\text{End } E)$ as the cotangent bundle of the complex manifold $\mathcal{A}$. As such it has a canonical holomorphic symplectic structure, and the real and imaginary parts of this yield two real symplectic forms Ω_2, Ω_3. The first symplectic form Ω_1 is the natural Kahler form. These three forms are associated to an obvious flat hyperkahler structure on $\mathcal{A} \times \Omega^{1,0}(\text{End } E)$. The moment map for the gauge group action with this first symplectic structure is $\underline{\mu}_1 = \underline{\mu}$, as we have seen. The special feature now is that the other two moment maps can be identified with the real and imaginary parts of

$\overline{\partial}_A \phi$, that is

$$\overline{\partial}_A \phi = \underline{\mu}_2 + i\underline{\mu}_3.$$

So $\mathcal{N}^c$ has a natural hyperkahler structure, as the hyperkahler quotient of $\mathcal{A} \times \Omega^{1,0}(\Sigma)$.

The theory alluded to above, relating symplectic and complex quotients, gives different descriptions $\mathcal{N}^c$ and the different complex structures are visible in the different descriptions. This manifold can be described as

1. a moduli space of "stable pairs" which is a completion of the cotangent bundle $T^*\mathcal{N}$. Here the visible complex structure is I_1 and $\mathcal{N} \subset \mathcal{N}^c$ is a complex submanifold for this structure.
2. a moduli space of irreducible representations of $\pi_1(\Sigma)$ in $GL(n, \mathbf{C})$. Here the visible complex structure is I_2 and $\mathcal{N} \subset \mathcal{N}^c$ is a totally real submanifold (the unitary representations) for this structure.

We will not atempt to review any more of this rich theory except to recall that, given a solution (A, ϕ) of Hitchin's equations one constructs a flat $GL(n, \mathbf{C})$ connection

$$A + \Phi + \Phi^*.$$

(Here we are considering the case when the bundle is topologically trivial and the constant c is zero.) Conversely, given such a flat connection the Higgs field ϕ appears as the derivative of a section of the associated flat bundle with fibre the hyperbolic 3-space $H^3 = PGL(2, \mathbf{C})/PU(2)$. From this point of view, the first of Hitchin's equations is the requirement that the section be a *harmonic section.*

The extension of the Kahler manifold $\mathcal{N}$ to the hyperkahler "thickening" $\mathcal{N}^c$ is a general phenomenon. Indeed if N is any real-analytic Kahler manifold then it has been shown by Feix [8] and Kaledin [15] that there is a hyperkahler manifold N^c with a circle action, containing N as the fixed submanifold of the action, such that the first Kahler structure on N^c restricts to the given one on N. A neighbourhood of N in N^c can be viewed either as a neighbourhood of the zero section in the cotangent bundle T^*N or as a complexification of N. Further, any two such extensions are isometric, on small neighbourhoods of N. There is one point which we shall want to refer to in this general set-up. The circle action preserves the first symplectic form and so has a Hamiltonian $H : N^c \to \mathbf{R}$, vanishing along the fixed set $N \subset N^c$. It is easy to show that this function H is a *Kahler potential* for the metric with respect to the second complex structure

$$i\overline{\partial}^{I_2} \partial^{I_2} H = \omega_2. \tag{4}$$

1.3. Maps and diffeomorphisms. In [7], the author developed some analogues of this moment map theory where the relevant group is a diffeomorphism group, rather than a gauge group. We do not wish to repeat much of that discussion here but we will recall one of the main points since similar ideas will appear in Section 2 below.

Consider again a symplectic manifold (X, ω) and a compact manifold M. We work with the infinite dimensional space Maps (M, X) of smooth maps from M to X. If f is such a map, a tangent vector to Maps (M, X) at f is a section of the vector bundle $f^*(TX)$ over M. The symplectic form ω defines a symplectic structure on the fibres of this bundle, also denoted by ω. Thus if ξ and η are two tangent vectors to Maps (M, X) at f we have a function $\omega(\xi, \eta)$ on M. Now suppose that M has a fixed volume form ρ. Then we can define a skew pairing on these tangent vectors by

$$\Omega(\xi,\eta) = \int_M \omega(\xi,\eta)\rho. \tag{5}$$

This is essentially the same as the construction at the beginning of (1.2), in the case when the bundle P is trivial, and it is easy to see that Ω is, at least formally, a symplectic structure on Maps (M,X).

The group $\mathcal{G}$ which we will consider in this case is the group of "exact" volume-preserving diffeomorphisms. Going to the level of Lie algebras, a vector field v on M preserves the volume form ρ if

$$L_v(\rho) = d(i_v(\rho)) = 0.$$

A volume-preserving vector field is called exact if the closed form $i_v(\rho)$ can be written as the exterior derivative of an $(n-2)$-form. Of course, if $H_{n-1}(M;\mathbf{R})$ vanishes this condition is vacuous. The group $\mathcal{G}$ can be defined to be the set of volume-preserving diffeomorphisms which can be generated by integrating time-dependent families of such exact vector fields. It is a normal subgroup of the component of the identity $\mathcal{G}_0^+$ in the full group $\mathcal{G}^+$ of volume-preserving diffeomorphisms and the quotient $\mathcal{G}_0^+/\mathcal{G}$ can be identified with the torus

$$A_M = H^{n-1}(M;\mathbf{R})/H^{n-1}(M;\mathbf{Z}).$$

Now the group $\mathcal{G}$ obviously acts on Maps (M,X) by composition, preserving the symplectic form Ω, and so we can seek a moment map for the action. The moment map should take values in the dual of the Lie algebra of $\mathcal{G}$. This Lie algebra can be identified with the quotient of the $(n-2)$-forms on M by the closed $(n-2)$-forms. Thus, using the pairing by wedge product and integration over M, the dual is the space of closed 2-forms on M. (More precisely, we should take closed currents.) So a moment map should assign to a map $f : M \to X$ a closed 2-form over M. As shown in [7], the moment map is simply

$$\underline{\mu}(f) = f^*(\omega). \tag{6}$$

To give one illustration of how these ideas can be applied; consider the case when M is the 2-sphere and restrict to the $\mathcal{G}$-invariant open subset of Maps (X,M) consisting of embeddings with symplectic image, in a given homotopy class. Define the constant c by

$$\int_M f^*(\omega) = c\int_M \rho.$$

for any map f in this homotopy class. Then the symplectic quotient $\underline{\mu}^{-1}(c\rho)/\mathcal{G}$ can be identified with the set of embedded symplectic spheres in the given homotopy class and the general theory tells that this space has a natural symplectic structure. Thus the theory gives a tidy way of passing between parametrised and unparametrised symplectic submanifolds.

We will not say any more about this development here, except to point out that ideas in a similar spirit have been developed in a very exciting way by Thomas recently [15]; introducing the notion of "stable" Lagrangian submanifolds of Calabi-Yau manifolds.

2. Sections of bundles and diffeomorphisms

We now begin the main part of this article, developing a new variation of these ideas. We start with a general, abstract, discussion and then go on to consider two examples in more detail.

2.1. Abstract theory. Let (X, ω) be a symplectic manifold with an $SL(n, \mathbf{R})$-action and an equivariant moment map

$$\mu : X \to \mathfrak{sl}(n)^*.$$

Let M^n be a compact manifold with a fixed volume form $\rho \in \Omega^n(M)$. Thus there is a principal bundle of frames over M with structure group $SL(n, \mathbf{R})$ and we can form the associated bundle $\underline{X} \to M$, with fibre X. As in Section 1.2, there is a natural symplectic form, which we again denote by ω, on the vertical tangent bundle $T_V\underline{X}$ along the fibres over $\underline{X}$. We consider the infinite-dimensional space $\mathcal{S}$ of sections of $\underline{X} \to M$. Again, as in Section 1.2, there is a symplectic form Ω on $\mathcal{S}$ given at a point $\phi \in \mathcal{S}$ by

$$\Omega(\xi, \eta) = \int_M \omega(\xi, \eta)\rho, \tag{7}$$

where ξ and η are sections of $\phi^*(T_V\underline{X})$.

Now let $\mathcal{G}$ be the group of exact volume-preserving diffeomorphisms of M. This group acts naturally on $\mathcal{S}$, preserving the symplectic form, so we seek a moment map $\underline{\mu}$ for the action. As in Section 1.3, this moment map should assign to a section ϕ a closed 2-form over M. The author does not know of a really satisfactory definition of this moment map. The definition we give goes via the choice of an auxiliary structure; a torsion-free $SL(n, \mathbf{R})$-connection ∇ on the tangent bundle of M. Fixing such a connection, we associate three 2-forms over M to a section ϕ of $\underline{X}$, as follows.

1. Using the connection we can differentiate the section to obtain
$$\nabla\phi \in \Gamma(T^*M \otimes \phi^* T_V\underline{X}).$$
Then we define
$$\omega(\nabla\phi, \nabla\phi) \in \Omega^2(M),$$
by taking the tensor product of the wedge-product on the T^*M component and the form ω on the $T_V\underline{X}$ component.
2. The equivariant moment map μ induces a map from sections of $\underline{X}$ to sections of the vector bundle $TM \otimes T^*M$ with trace zero. (Using the standard identifiction of $\mathfrak{sl}(n, \mathbf{R})^*$.) Thus to a section ϕ we associate $\mu_\phi \in \Gamma(TM \otimes T^*M)$, written in index notation as μ^i_j. The curvature of the connection ∇ is a section $R^i_{j\ kl}$ of $TM \otimes T^*M \otimes \Lambda^2 T^*M$. There is a natural pairing $R.\mu_\phi \in \Omega^2(M)$, written in index notation as $R^i_{j\ kl}\mu^j_i$.
3. The covariant derivative of μ_ϕ is a section of $T^*M \otimes TM \otimes T^*M$:
$$\nabla\mu_\phi = \mu^i_{j;k}.$$
We produce a 1-form $c(\nabla\mu_\phi)$ by contracting two indices
$$c(\nabla\mu_\phi) = \mu^i_{j;i}.$$
Then we have a 2-form $d\,(c(\nabla\mu_\phi)) \in \Omega^2(M)$.

With these three ingredients to hand, we define

$$\underline{\mu}(\phi) = \omega(\nabla\phi, \nabla\phi) + R.\mu_\phi + d\,(c(\nabla\mu_\phi))\,. \tag{8}$$

Thus $\underline{\mu}$ is a map from $\mathcal{S}$ to $\Omega^2(M)$. The main result of this section is as follows.

Theorem 9

1. *$\underline{\mu}$ is independent of the choice of the connection ∇, hence it is a $\mathcal{G}$-equivariant map from $\mathcal{S}$ to $\Omega^2(M)$.*
2. *For each ϕ, $\underline{\mu}(\phi)$ is a closed 2-form over M.*
3. *The map $\underline{\mu}$ is an equivariant moment map for the $\mathcal{G}$-action on $(\mathcal{S}, \Omega)$.*

Of course, all of these assertions are, at bottom, straightforward calculations. We begin with item (1). Suppose we make an infinitesimal change γ in the connection ∇. Thus, in index notation, we have a tensor γ^i_{jk}, symmetric in the indices j, k since we are considering torsion-free connections. The change in the curvature, to first order, is given by $\gamma^i_{jk;l} - \gamma^i_{jl;k}$. So the change in $R.\mu_\phi$ is

$$(\gamma^i_{jk;l} - \gamma^i_{jl;k})\mu^j_i. \tag{10}$$

The change in the covariant derivative $\nabla\mu_\phi$ is

$$\mu^l_j\gamma^i_{lk} - \mu^i_l\gamma^l_{jk}.$$

When we make the contraction one term vanishes, since $\gamma^i_{li} = 0$. Thus the change in $c(\nabla\mu_\phi)$ is the 1-form $-\mu^i_j\gamma^j_{ik}$, and the change in its exterior derivative is

$$\mu^i_{j;k}\gamma^j_{il} + \mu^i_j\gamma^j_{il;k} - \mu^i_{j;l}\gamma^j_{ik} - \mu^i_j\gamma^j_{ik;l}. \tag{11}$$

Let $r : \mathfrak{sl}(n;\mathbf{R}) \to \mathrm{Vect}(X)$ be the derivative of the action of $SL(n;\mathbf{R})$ on X. Any section δ of $TM \otimes T^*M$, with trace zero, defines a vertical vector field $r(\delta)$ over $\underline{X}$. That is, $r(\sigma)$ is a section of $T_V\underline{X}$ over $\underline{X}$. Taking the tensor product with T^*M we can define a section $r(\gamma)$ of $T_v\underline{X} \otimes \pi^*T^*M$. Then the infinitesimal change in $\nabla\phi$ is $\phi^*(r(\gamma))$. This means that, to first order, the variation in the term $\omega(\nabla\phi, \nabla\phi)$ is

$$\omega\,(r(\gamma), \nabla\phi) + \omega\,(\nabla\phi, r(\gamma))\,.$$

Now the defining property of the moment map μ asserts that

$$\omega(r(\delta), \eta) = \delta.D\mu(\eta),$$

where δ is any section of $TM \otimes T^*M$, η is any section of $T_V\underline{X}$ and $D\mu$ denotes the fibrewise derivative of μ—a bundle map from $T_V\underline{X}$ to the lift of $T^*M \otimes TM$. The covariant derivative $\nabla\mu$ is given by applying the bundle map $D\mu$ to the covariant derivative of ϕ. Thus the variation in $\omega(\nabla\phi, \nabla\phi)$ can be written as

$$\gamma^i_{jk}\mu^j_{l;i} - \gamma^i_{jl}\mu^j_{k;i}. \tag{12}$$

Putting together the three contributions (10),(11),(12) we see that the overall change in $\underline{\mu}$ vanishes. This completes the proof of item (1) of Theorem 9.

To prove item (2), that the 2-form $\underline{\mu}(\phi)$ is closed, we observe that this is a local property so by item (1) we can reduce to the case the connection ∇ to be flat, and we regard the section as a map from the base into X. Then the form $\omega(\nabla\phi, \nabla\phi)$ is nothing other than the pull back of the form ω on X to the base and hence is closed. The curvature term vanishes and the term $d\,(c(\nabla\mu_\phi))$ is evidently closed.

Finally we come to the crucial property (3); the moment map condition. To derive this we digress to give another interpretation of the formula (8) which will also be useful for other purposes. Let us assume that there is an $SL(n,\mathbf{R})$-equivariant Hermitian line bundle $L \to X$, with connection, and that the moment map μ defines the lift of the action to L in the standard way. This means that there is a line bundle $\underline{L} \to \underline{X}$, defined as an associated bundle to the frame bundle. The line bundle $\underline{L}$ comes endowed with a connection in the vertical directions in $\underline{X}$; the choice of a conection ∇ on TM gives a natural extension of this to a conection $\nabla_{\underline{L}}$ on $\underline{L} \to \underline{X}$. The curvature of this connection is a closed 2-form $\underline{\omega}$ on the total space $\underline{X}$, restricting to ω in the vertical directions. (This form $\underline{\omega}$ exists without any assumption about the line bundle, but it seems easiest to explain the construction using this device. Another framework is provided by the theory of equivariant cohomology.)

Lemma 13 *For a section ϕ of $\underline{X}$ the pull-back $\phi^*(\underline{\omega})$ is*

$$\omega(\nabla\phi, \nabla\phi) + R.\mu_\phi.$$

The proof is a matter of unravelling the definitions which we mostly leave to the reader. At a given point of $\underline{X}$ we split the tangent bundle into horizontal and vertical parts

$$T\underline{X} \cong TM \oplus TX.$$

The 2-form $\underline{\omega}$ thus has *a priori* three components; in $\Lambda^2 T^*X, \Lambda^2 T^*M$ and $T^*X \otimes T^*M$. The first component is given by the form ω and the second by $R.\mu$. The fact that the connection ∇ is torsion-free implies that the third, mixed, component is identically zero. These two terms in $\underline{\omega}$ go over to the two terms in the statement of the lemma under pull-back by ϕ.

With this discussion in place we prove item (3) of Theorem 9. Let λ be a fixed $(n-2)$-form over M and let $\xi \in \Gamma(\phi^* T_V \underline{X})$ be an infinitesimal variation of a section ϕ. It is convenient to think of ξ as being the pull-back of a section of the vertical bundle over the whole of $\underline{X}$. We want to compute the derivative

$$G_1 = \delta_\xi \int_M \underline{\mu}(\phi) \wedge \lambda,$$

with respect to ϕ in the direction ξ. By Lemma 13, this is

$$G_1 = \delta_\xi \int_M \phi^*(\underline{\omega}) \wedge \lambda + \delta_\xi c(\nabla\mu_\phi) \wedge d\lambda,$$

where we have used integration by parts on the second term. For the first term we have

$$\delta_\xi \int_M \phi^*(\underline{\omega}) \wedge \lambda = \int_M \phi^*(i_\xi d + d i_\xi)(\underline{\omega}) \wedge \lambda,$$

which is

$$\int_M d\phi^*(i_\xi(\underline{\omega})) \wedge \lambda,$$

since $\underline{\omega}$ is closed. Integrating by parts on this term we have in sum

$$G_1 = \int_M (\phi^*(i_\xi(\underline{\omega})) + \delta_\xi c(\nabla\mu_\phi)) \wedge d\lambda.$$

Now write $d\lambda = i_v(\rho)$, where v is a volume-preserving vector field on M. Then we have a pointwise identity

$$\phi^*(i_\xi(\underline{\omega})) \wedge i_v(\rho) = \phi^*(\omega(\nabla_v\phi, \xi))\rho,$$

so

$$G_1 = \int_M \phi^*(\omega(\nabla\phi,\xi))\rho + \delta_\xi c(\nabla\mu_\phi)\wedge i_v(\rho).$$

Now consider the integral

$$G_2 = \int_M \omega(\delta_v\phi,\xi)\rho.$$

Here $\delta_v\phi$ denotes the derivative of the action of the volume-preserving vector fields on $\mathcal{S}$: the Lie derivative of ϕ along v. This is given by the formula

$$\delta_v\phi = \nabla_v\phi + r(\nabla v)(\phi),$$

where $r(\nabla v)$ is the vertical vector field defined by $\nabla v \in \Gamma(TM\otimes T^*M)$, as above. By the definition of the moment map μ we have

$$\omega(\xi, r(\nabla v)(\phi)) = \delta_\xi((\nabla v).\mu_\phi).$$

Thus

$$G_1 - G_2 = \delta_\xi \int_M (c(\nabla\mu_\phi)\wedge i_v + (\nabla v).\mu_\phi)\,\rho. \tag{14}$$

In index notation, the integrand on the right hand side of (14) is

$$\mu^j_{i;j} + v^i_{;j}\mu^j_i$$

which is the divergence $(v^i\mu^j_i)_{;j}$. Thus the integral over M vanishes and so $G_1 = G_2$, which is precisely the desired moment map identity.

2.2. The Weil-Petersson metric. The fundamental example of the set-up considered in the previous subsection arises when the base manifold M is an oriented surface Σ of genus genus $(\Sigma)\geq 2$ and X is the homogeneous space $H^2 = SL(2,\mathbf{R})/SO(2)$—the hyperbolic plane. Points of X can be viewed as complex structures on $\mathbf{R}^2$ and a section of the bundle $\underline{X}\to\Sigma$ is a complex structure on the surface. Thus we denote a section by J. The symplectic form on H^2 is uniquely determined by the $SL(2,\mathbf{R})$-invariance up to an overall scale. We fix this scale by decreeing that the moment map $\mu : H^2 \to \mathfrak{sl}(2,\mathbf{R})$ is just one half the natural inclusion, thinking of points of H^2 as trace-free endomorphisms of $\mathbf{R}^2$. A little calculation shows if we take the model of H^2 as the upper half plane in $\mathbf{C}$ the symplectic form we are using is

$$\frac{dxdy}{2y^2}.$$

To identify the moment map $\underline{\mu}$ in this case we can, given a complex structure J, choose the connection ∇ to be the Levi-Civita connection of the metric defined by J and ρ. Thus the covariant derivatives of J and μ_J vanish and the only term remaining in the formula (8) is that involving the curvature, which in this case is just $R = KJ\otimes\rho$, where K is the Gauss curvature. Since $|J|^2 = 2$ the moment map is $\underline{\mu} = K\rho$. Thus we recover the fact that the Gauss curvature furnishes a moment map for the action of the group of exact area preserving diffeomorphisms of a surface on the complex structures. As far as the author knows, this was first shown by Quillen, in about 1983, in answer to a question of Atiyah. (There is a straightforward variant of the whole theory of (2.1) to the case where X is a symplectic manifold with an action of the symplectic group $Sp(2n,\mathbf{R})$, the base manifold M is a symplectic manifold and we consider the action of the group of exact symplectomorphisms of M on sections of the resulting bundle $\underline{X}\to M$. The fundamental example here is when $X = Sp(2n, \mathbf{R})/U(n)$ so sections are compatible almost-complex structures on M. Then the general theory produces the moment-map calculation of [6] which—at least in the case of integrable structures—had been observed previously by Fujiki in [9].)

Just as for the moduli spaces of flat connections considered in Section 1, one simple application of this moment map calculation is the definition of a canonical Kahler metric on the symplectic quotient $\mathcal{M} = \underline{\mu}^{-1}(c\rho)/\mathcal{G}$. Here the appropriate constant c is fixed by Gauss-Bonnet to be

$$c = \frac{2\pi(2 - 2\text{genus }(\Sigma))}{\text{Area }(\Sigma, \rho)}. \tag{15}$$

A subtlety arises here because $\mathcal{M}$ is not quite the same as the usual moduli space $\mathcal{M}_0$ of complex structures which is the quotient of $\underline{\mu}^{-1}(c\rho)$ by the group $\mathcal{G}^+$ of all area-preserving diffeomorphisms. There are two aspects to this. First, the quotient of the identity component $\mathcal{G}_0^+$ of $\mathcal{G}^+$ by $\mathcal{G}$ is the $2g$-torus $A_\Sigma = H^1(\Sigma, \mathbf{R})/H^1(\Sigma, \mathbf{Z})$. This means that the torus A_Σ acts on $\mathcal{M}$ and the quotient is the Teichmuller space $\mathcal{T}$. Second, the quotient of $\mathcal{G}^+$ by its identity component $\mathcal{G}_0^+$ is the mapping class group Γ. This discrete group acts on $\mathcal{T}$ and $\mathcal{M}_0$ with quotient $\mathcal{M}_0$. It is the first aspect which is of a differential geometric nature. The difficulty is that there is no way to extend the moment $\underline{\mu}$ to an equivariant moment map for the full action of $\mathcal{G}_0^+$. Instead we can proceed as follows. Suppose in general that a torus T acts freely on a symplectic manifold (Y, ω) and that the T-orbits are *symplectic submanifolds* of Y. In this special case the naive quotient Y/T has a natural induced symplectic structure. To see this, we take the field of subspaces $H \subset TY$ defined as the anhilliator under ω of the tangent spaces to the T-orbits. Since the orbits are symplectic the subspaces H furnish complements to the orbit tangent spaces and ω is nondegenerate on H. Then identifying H with the pull-back of the tangent bundle of Y/T we can push the form down to define a nondegenerate 2-form on the quotient which one checks is closed. We can apply this to our situation, with the action of A_Σ on $\mathcal{M}$. The A_Σ-orbits are actually complex submanifolds of $\mathcal{M}$. In fact we can identify $\mathcal{M}$ with the moduli space of pairs consisting of a marked Riemann surface (Σ, J) and a choice of holomorphic line bundle of fixed degree over Σ and this moduli space has a natural complex structure. The A_Σ-orbits just arise from varying the line bundle and are obviously complex submanifolds. Using this device, we get a symplectic (Kahler) form on $\mathcal{T} = \mathcal{M}/A_\Sigma$. But then the whole construction is obviously invariant under the mapping class group so we finally obtain a Kahler metric on $\mathcal{M}$ which is of course nothing but the standard Weil-Petersson metric.

3. The hyperkahler extension of the Weil-Petersson metric

We will now consider a new example, in which the aim is to produce an explicit construction of the Feix-Kaledin hyperkahler thickening of the Weil-Petersson metric. As we outline in 3.1 below, we expct that this will be a hyperkahler metric on the "quasi-Fuchsian moduli space". We should mention here that I. Platis has constructed a complex symplectic structure on this space [13] and has investigated the hyperkahler geometry [14]. It will be interesting to compare the formulae of Platis with those derived from our moment map point of view. Our construction is obviously closely modelled on Hitchin's in the gauge theory case. Variants of the same idea, which we do not explore here, give a uniform framework for discussing other moduli spaces of "pairs" consisting of a Riemann surface with an additional tensor field.

3.1. The hyperkahler extnsion of the hyperbolic plane. We consider again the case when the base manifold is a compact surface Σ with a fixed area form ρ. As we have seen, the $SL(2,\mathbf{R})$-space H^2—the upper half plane— leads essentially to the standard Weil-Petersson metric on the moduli space. Now let X be the unit disc bundle in the cotangent bundle T^*H: this has a natural $SL(2;\mathbf{R})$ action induced by that on H^2 which commutes with action of S^1 given by rotation of the fibres.

Lemma 16.
There is an $SL(2;\mathbf{R}) \times S^1$-invariant hyperkahler metric g_X on X which on a fibre of $X \to H$ is given by

$$\frac{1}{4}\frac{d\sigma d\overline{\sigma}}{\sqrt{1-|\sigma|^2}}.$$

Here, in the formula for the metric on the fibre, we understand that the fibre is identified with standard disc in $\mathbf{C}$. The hyperkähler structure comprises complex structures I_1, I_2, I_3 on X and $SL(2;\mathbf{R})$ preserves each of I_1, I_2, I_3 whereas the circle action preserves I_1 but rotates I_2, I_3. This metric is nothing other than the Feix-Kaledin hyperkahler extension of the constant curvature metric on H^2 and the point of Lemma 16 is to find this extension explicitly. The metric g_X is the analogue of the well-known Calabi-Eguchi-Hanson metric on the cotangent bundle of the round 2-sphere. To find the metric from first principles one can proceed as follows. We consider $\mathbf{C}^2$ with the indefinite Hermitian metric $|z|^2 - |w|^2$ and make the standard identification of H^2 with the unit disc, the complex projectivization of the positive cone for this Hermitian form. In this model the symmetry group $SL(2;\mathbf{R})$ appears as the locally isomorphic group $SU(1,1)$. Then the set

$$\tilde{X} = \{(z,w) \in \mathbf{C}^2 : 0 < |z|^2 - |w|^2 < 1\}$$

is a bundle over H^2 with fibre $\mathbf{C}^*$ and it is in a natural way a double covering of X minus the zero section. Thus we can calculate in the more convenient model $\tilde{X}$. Here we seek a $U(1,1)$-invariant Calabi-Yau Kahler metric. Thus we take the metric to be of the form $i\overline{\partial}\partial F$ where $F = F(r)$ is a function of $r = |z|^2 - |w|^2$. A short calculation shows that

$$\overline{\partial}\partial F = (F' + |z|^2) d\overline{z}dz + (|w|^2 F'' - F') d\overline{w}dw - F''(z\overline{w}d\overline{z}dw + \overline{z}wd\overline{w}dz).$$

The condition we need to satisfy is the Monge-Ampere equation

$$(\overline{\partial}\partial F)^2 = dzd\overline{z}dwd\overline{w}.$$

This reduces to the ODE

$$rF'F'' + (F')^2 = -1.$$

Setting $F' = G$, we get a first order equation for G which we can integrate to give

$$G(r) = \sqrt{\frac{b^2}{r^2} - 1},$$

for a constant of of integration b. The desired soluton is obtained when $b = 1$ (other values just give trivial rescaling). One can check that this induces the metric we fixed before on the zero-section of X. On the fibre of $\tilde{X} \to H$ given by $w = 0$ the metric is just

$$\frac{|z|^2}{\sqrt{1-|z|^4}} d\overline{z}dz.$$

To get the metric on the fibre of X we set $\sigma = z^2$, which yields

$$\frac{1}{4}\frac{d\sigma d\overline{\sigma}}{\sqrt{1-|\,\sigma|^2}}$$

as asserted in the Lemma.

We now fit into the general framework of Section 2, forming the bundle $\underline{X}$ over our surface with fibre X and the space of sections of $\underline{X}$, which we will now denote by $\mathcal{S}^c$. We use the symbol $\mathcal{S}$ to denote the sections of the bundle with fibres H^2 considered in the previous section, so we can obviously regard $\mathcal{S}$ as a subset of $\mathcal{S}^c$. Explicitly, a point ϕ in $\mathcal{S}^c$ is given by a pair (J,σ) where J is a complex structure on Σ and σ is a smooth quadratic differential with respect to this structure, with $|\sigma| < 1$ everywhere. (Here of course the norm $|\sigma|$ is computed using the metric defined by J and the area form ρ.) There is a hyperkahler structure on $\mathcal{S}^c$ induced from that on X, preserved by the group $\mathcal{G}$. Thus we are in the familiar general setting, sketched in (1.2), where we can take a hyperkahler quotient. The three symplectic forms $\omega_1, \omega_2, \omega_3$ on X induce forms $\Omega_1, \Omega_2, \Omega_3$ on $\mathcal{S}^c$ and we have moment maps $\underline{\mu}_1, \underline{\mu}_2, \underline{\mu}_3$.

Proposition 17.

The moment maps are given by

$$\underline{\mu}_1(J,\sigma) = \left(\sqrt{1-|\sigma|^2}K + \tfrac{1}{4\sqrt{1-|\sigma|^2}}(|\partial\sigma|^2 - |\overline{\partial}\sigma|^2) + i\overline{\partial}\partial(\sqrt{1-|\sigma|^2})\right)\rho$$

$$(\underline{\mu}_2 + i\underline{\mu}_3)(J,\sigma) = \overline{\partial}(\iota\overline{\partial}\sigma),$$

where K is the Gauss curvature and ι is the natural isomorphism from $\Omega^{0,1}(T^*\Sigma \otimes T^*\Sigma)$ to $\Omega^{1,0}$ defined by the metric.

To identify $\underline{\mu}_1$ we apply Theorem 9. A complex structure J is, by definition, a section of $T\Sigma \otimes T^*\Sigma$. We claim that the moment map $\mu_1 : X \to \mathfrak{sl}(2, \mathbf{R})$, for the form ω_1 on X, is given by

$$\mu(J,\sigma) = \frac{1}{2}\sqrt{1-|\sigma|^2}.J \tag{18}$$

To see this, observe that symmetry conditions dictate that the moment map has the form $f(|\sigma|)J$ for some function f. To find this function we need only consider the action of the circle subgroup generated by J. That is, we essentially have to find the Hamiltonian H for the rotations acting on the disc with the area form of Lemma 16. In polar co-ordinates this area form is

$$\frac{1}{2\sqrt{1-r^2}} r dr d\theta,$$

and contraction with $\frac{\partial}{\partial\theta}$ yields

$$\frac{r}{2\sqrt{1-r^2}} dr = -dH,$$

where $H(r) = \frac{1}{2}\sqrt{1-r^2}$. Taking account of the fact that the circle generated by J acts with weight 2 on the disc and that $|J|^2 = 2$, we deduce that $f(|\sigma|) = \frac{1}{2}\sqrt{1-|\sigma|^2}$ as required.

We can now identify the three terms in the formula (8). We use the Levi-Civita connection ∇ of the metric defined by J and ρ. The curvature R is thus $KJ \otimes \rho$, where K is the Gauss curvature, and so (since $|J|^2 = 2$):

$$R.\mu_{J,\sigma} = \sqrt{1-|\sigma|^2}K\rho.$$

The term $\omega(\nabla\phi, \nabla\phi)$ only has a contribution from $\nabla\sigma$, since J is parallel. Writing the derivative in holomorphic and anti-holomorphic parts and using the formula for the metric on the fibres of $X \to H$ we get

$$\frac{1}{4\sqrt{1-|\sigma|^2}}(|\partial\sigma|^2 - |\overline{\partial}\sigma|^2).$$

Finally, the term $\nabla\mu$ is just $\frac{1}{2}J \otimes d(\sqrt{1-|\sigma|^2})$ and the contraction $c(\nabla\mu)$ is $\frac{1}{2}J(d(\sqrt{1-|\sigma|^2})$, where J acts on the 1-forms on Σ in the standard way. So the exterior derivative is

$$dc(\nabla\mu) = \frac{1}{2}dJd(\sqrt{1-|\sigma|^2}) = i\overline{\partial}\partial\sqrt{1-|\sigma|^2}.$$

With these identifications we obtain the given formula for $\underline{\mu}_1$.

We could find $\underline{\mu}_2$ and $\underline{\mu}_3$ in the same manner, applying the general discussion of (2.1). However it is much simpler to proceed as follows. Since X is contained in the cotangent bundle of H^2, the space $\mathcal{S}^c$ can be viewed formally as an open set in the cotangent bundle of $\mathcal{S}$ and the form Ω_1 is nothing other than the standard symplectic form on the cotangent bundle. Now suppose, in general, that a group G acts on a manifold Q so we have a linear map $r : \mathfrak{g} \to \Gamma(TQ)$. The tranpose of r gives a map

$$r^T : T^*Q \to \mathfrak{g}^*,$$

and it is a simple fact that this is the moment map for the induced action of G on the cotangent bundle T^*Q. Thus we can apply this procedure to find the moment map $\underline{\mu}_2$. The Lie algebra of $\mathcal{G}$ is identified with the functions on Σ, modulo constants, and the infinitesimal action on $\mathcal{S}_0$ is given by

$$H \mapsto \overline{\partial}(v_H),$$

where v_H is the Hamiltonian vector field on Σ defined by H and ρ and $\overline{\partial}$ denotes the $\overline{\partial}$-operator on the tangent bundle. One readily checks that the transpose of this is given by

$$\sigma \mapsto \mathrm{Re}(\partial\iota\overline{\partial}\sigma),$$

whence the identification of μ_2. The formula for $\underline{\mu}_3$ then follows from the requirement of compatibility with the complex structure.

Our interpretation in Lemma 13 of the moment map formula shows that the integrals of the $\underline{\mu}_i$ over Σ are independent of the point (J, σ) in $\mathcal{S}$. Clearly these integrals vanish for $\underline{\mu}_2$ and $\underline{\mu}_3$ and for $\underline{\mu}_1$ we can find the integral by reducing to the case when $\sigma = 0$. Thus Gauss-Bonnet shows that

$$\int_\Sigma \underline{\mu}_1 = 2\pi(2 - 2\text{genus }(\Sigma)).$$

We have now reached the point where we can apply the general hyperkahler moment map theory to see that we have a manifold $\mathcal{M}^c$, parametrising solutions (J, σ) of the equations $\underline{\mu}_1 = c\rho, \mu_2 = \mu_3 = 0$ (where the constant c is again given by (15)), modulo the action of the group $\mathcal{G}$. The next issue we have to face is that $\mathcal{M}^c$ is larger than the space we really want. This is just the same point that we encountered in (2.2), except now we have to take care because the three symplectic forms behave in different ways. We would really like to take the quotient by the larger group $\mathcal{G}^+$ of all area-preserving diffeomorphism but, just as we have seen in (2.2), there is no way to define an equivariant moment map for the action of this space, extending $\underline{\mu}_1$. For $\underline{\mu}_2$ and $\underline{\mu}_3$ however the picture is different. Indeed it follows from the discussion in the proof of Proposition 17 that the map

$$(J, \sigma) \mapsto \iota(\overline{\partial}\sigma),$$

defines a moment map for the action of $\mathcal{G}^+$ (where we identify $\Omega^{0,1}$ with real 1-forms and pair these with vector fields). Likewise the map $(J, \sigma) \mapsto T(J\overline{\partial}\sigma)$ extends $\underline{\mu}_3$. To handle this, we consider the map

$$m : \mathcal{M}^c \to H^1(\Sigma; \mathbf{R})$$

defined by mapping a pair (J, σ) to the cohomology class of $\mathrm{Re}\,\iota(\overline{\partial}\sigma)$. General principles show that this is the moment map for the action of the torus $A_\Sigma = \mathcal{G}_0^+/\mathcal{G}$ on $\mathcal{M}^c$ for the symplectic form Ω_2. Thus we get an induced symplectic form Ω_2 on the symplectic quotient

$$\mathcal{M}_1^c = m^{-1}(0)/A_\Sigma.$$

Similarly the map $(J, \sigma) \mapsto Jm(J, \sigma)$, using the complex structure on $H^1(\Sigma; \mathbf{R})$ defined by the metric J, is a moment map for the same action with respect to the form Ω_2, so we also get a symplectic form Ω_2 induced on $\mathcal{M}_1^c$. For the first symplectic structure we need to proceed differently. Let V be the complex vector bundle over the moduli space $\mathcal{M}_0$ consisting of isomorphism classes of pairs (J, α), where J is a complex structure and α is a class in the cohomology group $H^{0,1}(\Sigma)$ defined using this complex structure. Thus the total space V is naturally a complex manifold. We have a map

$$\tilde{m} : \mathcal{M}^c \to V$$

defined by $\tilde{m}(J, \sigma) = (J, m(J, \sigma))$, where we use the complex structure J to identify $H^{0,1}$ with $H^1(\Sigma; \mathbf{R})$. One readily checks that this is a holomorphic map, with respect to the first complex structure on $\mathcal{M}^c$. This means that $m^{-1}(0) \subset \mathcal{M}^c$, which is the pre-image of the zero section of V, is a complex submanifold of $\mathcal{M}^c$ with respect to the first complex structure. In particular the form Ω_1 restricts to a nondegenerate form on $m^{-1}(0)$. Further, one checks much as in (2.2) that the orbits of the action of J on $m^{-1}(0)$ are complex submanifolds, thus we can use the same quotient construction as in (2.2) to obtain an induced symplectic form Ω_1 on $\mathcal{M}_1^c$. Again, everything is invariant under the mapping class group so we finally get three algebraically-compatible symplectic forms on $\mathcal{M}_0^c = \mathcal{M}_1^c/\Gamma$ and hence, by a lemma of Hitchin [10], a hyperkahler structure.

Now the points of $\mathcal{M}_0^c$ have a more straightforward geometric interpretation. If $\underline{\mu}_2$ and $\underline{\mu}_3$ and m all vanish on a pair (J, σ) we have $\partial\iota(\overline{\partial}\sigma) = 0$ and $\iota(\overline{\partial}\sigma) = \overline{\partial}f$ for some function f on Σ. But then $\partial\overline{\partial}f = 0$ so f is a constant, thus $\overline{\partial}\sigma = 0$. So the points of $\mathcal{M}_0^c$ correspond to equivalence classes of pairs (g, σ) where g is a Riemannian metric on Σ; σ is a *holomorphic* quadratic differential for the complex stucture defined by g which satisfy the equation $\underline{\mu}_1 = c\rho$. The equivalence classes are now taken under the action of the full group of diffeomorphisms of Σ. When σ is holomorphic we can write the moment map $\underline{\mu}_1$ in a neater way.

Lemma 18.
If σ is a holomorphic quadratic differential we have

$$\underline{\mu}_1(J,\sigma) = \left(K + \frac{1}{2}\Delta \log(1+\sqrt{1-|\sigma|^2})\right)\rho$$

To establish this formula we can work around a point where σ does not vanish. Let h be the smooth function $|\sigma|^2$, with the norm defined by the metric. Thus the curvature form of $T^*\Sigma$ is $\frac{1}{2}\overline{\partial}\partial \log h$ and

$$|\partial\sigma|^2 = h^{-1}|\partial h|^2.$$

The formula we need then follows from the identity, for any function h,

$$\sqrt{1-h}\,\overline{\partial}\partial \log h - \frac{1}{2h\sqrt{1-h}}\overline{\partial}h\partial h - 2\overline{\partial}\partial\sqrt{1-h} = \overline{\partial}\partial \log h - 2\overline{\partial}\partial \log(1+\sqrt{1-h}),$$

which we leave as an exercise for the reader. (The author's solution to this exercise goes via the identity:

$$-2\partial \log(1+\sqrt{1-h}) = \sqrt{1-h}\,\partial \log h + \frac{1}{\sqrt{1-h}}\partial h - \partial \log h.\ ) \tag{19}$$

To sum up then we have

Proposition 20 *The moduli space $\mathcal{M}_0^c$ is the quotient by the diffeomorphism group of the set of pairs (g,σ) where g is a Riemannian metric on the surface Σ and σ is a holomorphic quadratic differential on (Σ, g) such that*

$$K + \frac{1}{2}\Delta \log(1+\sqrt{1-|\sigma|^2}) = c,$$

where the constant c is given by (15). There is a hyperkahler metric on $\mathcal{M}_0^c$ extending the Weil-Petersson metric on the moduli space $\mathcal{M}_0 \subset \mathcal{M}_0^c$.

3.2. Geometric interpretation. The equation we have found in the previous section does not look very natural but we can transform it to a much more familiar shape. We make a start on this story here, although we have not yet been able to explore it in full detail.

Suppose we have a Riemannian surface (Σ, g) with a holomorphic quadratic differential σ and $|\sigma| < 1$ everywhere. We define a function F by

$$F = 1 + \sqrt{1-|\sigma|^2}.$$

Let g_1 be the scaled metric $g_1 = Fg$. Thus the curvature 2-form of (Σ, g_1) is

$$\Theta_1 = \Theta - \overline{\partial}\partial \log F,$$

where Θ is the curvature form of (Σ, g). On the other hand from the definition of F we have

$$(F-1)^2 = 1 - |\sigma|^2$$

which implies that

$$1 = \frac{2}{F} - \frac{|\sigma|^2}{F^2}. \tag{21}$$

If we write, for the original metric

$$K + \frac{1}{2}\Delta \log F = c + \chi,$$

then the Gauss curvature of g_1 is

$$K_1 = \frac{c+\chi}{F}.$$

On the other hand the norm of σ in the rescaled metric is

$$|\sigma|_1^2 = \frac{|\sigma|^2}{F^2}.$$

So

$$K_1 - \frac{c}{2}|\sigma|_1^2 = \frac{\chi}{F} + \frac{c}{2}.$$

using (21). We see then that the original metric satisfies the equation of Proposition 20 if and only if the rescaled metric satisfies the equation

$$K_1 - \frac{c}{2}|\sigma|_1^2 = \frac{c}{2}. \tag{22}$$

Notice that $|\sigma|_1 < 1$ and that the function

$$y = \frac{x}{1+\sqrt{1-x^2}}$$

gives a diffeomorphism from $[0,1)$ to itself, so the construction is invertible. Thus we can equally well regard $\mathcal{M}_0^c$ as a moduli space of pairs (g_1, σ) which satisfy the equation (22). It is now convenient to assume that the area of our original metric was normalised so that $c = -2$, thus the equation is

$$K_1 + |\sigma|_1^2 = -1. \tag{23}$$

The equation (23) is a reduced form of Hitchin's equation. In making this identification however we should be clear that the context is different. In Hitchin's case the complex structure on the underlying Riemann surface is fixed whereas in our case it is allowed to vary. Suppose we have any Riemannian surface (Σ, g_1) with a quadratic differential σ. We choose a square root L of the tangent bundle of Σ, so L is a holomorphic line bundle over Σ. Let a denote the $U(1)$-connection on L induced by the Levi-Civita connection. Now consider the vector bundle $L \oplus L^{-1}$ with the connection

$$A = \begin{pmatrix} a & \sigma \\ \overline{\sigma} & -a \end{pmatrix}.$$

Here $-a$ denotes the connection on L^{-1} and σ is regarded as an element of $\Omega^{1,0}(L^{-2}) = \Omega^{0,1}(\mathrm{Hom}(L, L^{-1}))$. We define the Higgs field $\Phi \in \Omega^{1,0}(\mathrm{End}(E))$ by

$$\Phi = \begin{pmatrix} 0 & 1 \\ 0 & 0 \end{pmatrix}.$$

Here the element "1" is regarded as an element of $\Omega^{1,0}(L^2) = \Omega^{1,0}(\mathrm{Hom}(L^{-1}, L)$ by the natural isomorphism. Then

$$F(A) + [\Phi, \Phi^*] = \begin{pmatrix} K_1 + 1 + |\sigma|_1^2 & 0 \\ 0 & -(K_1 + 1 + |\sigma|_1^2) \end{pmatrix} \frac{i\rho}{2}$$

So the solutions to Hitchins equation of this shape precisely correspond to the solutions of equation (23). All this is very similar to, but not the same as, the special solutions of Hitchin's equation studied in [10] of the form

$$A = \begin{pmatrix} a & 0 \\ 0 & -a \end{pmatrix}, \Phi = \begin{pmatrix} 0 & 1 \\ \sigma & 0 \end{pmatrix}$$

which lead to a parametrisation of Teichmuller space.

Now, given a solution of equation (22) we get a flat $SL(2,\mathbf{C})$-connection $A + \Phi + \Phi^*$ over Σ with a harmonic section of the associated bundle with fibre the hyperbolic 3-space $H^3 = PSL(2,\mathbf{C})/PU(2)$. In other words, we get a $\pi_1(\Sigma)$-equivariant harmonic map from the universal cover $\tilde{\Sigma}$ to H^3. The derivative of this map is represented by the Higgs field Φ, and the special form of this implies that the map is actually an *isometric immersion.* Thus the image is a minimal surface in H^3, by the standard relation between harmonicity and minimality. The quadratic differential σ appears now as the second fundamental form of the surface. Let us assume now that the quotient of H^3 by the action of $\pi_1(\Sigma)$ is a manifold: hence a hyperbolic 3-manifold Y. We then get an isometric minimal immersion of Σ in Y. Conversely, given such an immersion we can go backwards to recover the data (g_1, σ).

The overall picture that can be expected to emerge from this brings in the "quasi-Fuchsian" moduli space of Riemann surface theory. Let Σ_+, Σ_- be two compact Riemann surfaces of the same genus and fix a homotopy class $[f]$ of homeomorphisms between them. The simultaneous uniformisation theorem of Bers [1] asserts that there is a discrete subgroup $\pi \subset SL(2,\mathbf{C})$ and a Jordan curve whose complement has two components Ω_+, Ω_-, such that Ω_+/π is a uniformisation of Σ_+ and Ω_-/π is a uniformisation of $\overline{\Sigma_-}$ (the surface with the opposite complex structure). The group π also acts on the hyperbolic space H^3 and the quotient gives a hyperbolic 3-manifold $Y(\Sigma_+, \Sigma_-, [f])$ homeomorphic to $\Sigma_\pm \times \mathbf{R}$. This is a "hyperbolic cobordism" from Σ_+ to Σ_-, in that the conformal structures naturally induced on the two ends of the 3-manifold are the given ones. The quasifuschian moduli space $\mathcal{QF}$ is the moduli space of this data: it can be regarded either as an open subset of the moduli space of representations of $\pi_1(\Sigma_+)$ in $SL(2,\mathbf{C})$, modulo the mapping class group, or as the quotient of $\mathcal{T} \times \overline{\mathcal{T}}$ by the mapping class group (a bundle over the moduli space $\mathcal{M}_0$ with fibre $\overline{\mathcal{T}}$).

It is reasonable to hope that our hyperkahler manifold $\mathcal{M}_0^c$ can identified with $\mathcal{QF}$ via the correspondences above. What this would mean is that one could find a unique minimal surface, of a suitable kind, in any $Y(\Sigma_+, \Sigma_-, [f])$. This should be related to old work of Uhlenbeck [16]. Thus, in rough analogy with Hitchin's case, we would get three different descriptions of the same manifold:

1. as an open subset of the moduli space of representations of a surface group in $SL(2,\mathbf{C})$ modulo the mapping class group;
2. as an open subset in the cotangent bundle $T^*\mathcal{M}_0$ (viewed as pairs (J,σ));
3. as pairs of Riemann surfaces Σ_+, Σ_- with a given homotopy class of homeomorphisms between them.

Let us finally note that the last description gives a particularly attractive representation of the metric. Recall that, in the general framework of the Feix-Kaledin hyperkahler extensions, the Hamiltionian for the circle action with respect to the first structure gives a Kahler potential for the metric in the second complex structure, (see (4) above). Returning to our original picture in (3.1), it is easy to see that the Hamiltionian is given by

$$H(J,\sigma) = \int_{\Sigma} (\sqrt{1-|\sigma|^2} - 1)\ \rho.$$

But, up to a constant, this is just the *area* of the surface in the rescaled metric g_1. The second complex structure is the apparent structure in the third (conjectural) representation above. So we conjecture that there should be a hyperkahler metric on $\mathcal{T} \times \overline{\mathcal{T}}/\Gamma$ defined by taking the obvious complex structure and a Kahler potential $H(\Sigma_+, \Sigma_-, [f])$ given by the area of the preferred minimal surface in $Y(\Sigma_+, \Sigma_-, [f])$.

References

[1] L.V. Ahlfors, *Lectures on quasiconformal mappings,* Van Nostrand, 1966.

[2] M. F. Atiyah and R. Bott, *The Yang-Mills equations over Riemann surfaces,* Phil. Trans. Roy. Soc. London, **308** (1982), 523-615.

[3] D. Banfield, *Stable pairs and principle bundles,* Quarterly J. Math., **51** (2000), 417-36.

[4] S. Bradlow, G. Daskalopoulos, O. Garcia-Prada and R. Wentworth, *Stable augmented vector bundles,* Vector bundles in algebraic geometry, Cambridge UP, 1995, 15-67.

[5] K. Cielibak, R. Gaio and D. Salamon, *J-holomorphic curves, moment maps and invariants of Hamiltonian group actions,* Int. Math. Res. Notices, **16** (2000), 831-882.

[6] S. Donaldson, *Remarks on gauge theory, complex geometry and 4-manifold topology,* Fields medallists lectures, World Scientific, 1997, 384-403.

[7] S. Donaldson, *Moment maps and diffeomorphisms,* Asian J. Math., **3** (1999), 1-15.

[8] B. Feix, *Hyperkahler metrics on cotangent bundles,* Jour. Reine Ang. Math., **532** (2001), 33-46.

[9] A. Fujiki, *The moduli spaces and kahler metrics of polarised algebraic manifolds,* Sugaku, **42** (1990), 231-243 (English trans. Sugaku Expositions **5** (1992) 173-191)

[10] N.J. Hitchin, *The self-duality equations on a Riemann surface,* Proc. London Math. Soc, **55** (1987), 59-126.

[11] N. J. Hitchin, A. Karhlede, U.Lindstrom and M. Rocek, *Hyperkahler metrics and supersymmetry,* Comm. Math. Phys., **108** (1987), 535-89.

[12] I. Mundet i Riera, *A Hitchin-Kobayashi correspondence for Kahler fibrations,* Jour. Reine Ang. Math, **528** (2000), 41-80.

[13] I. D. Platis, *Complex symplectic geometry of quasi-Fuchsian space,* Geom. Dedicata, **87** (2001), 17-34.

[14] I.D. Platis, *Hyperkahler geometry of quasi-Fuchsian space,* preprint.

[15] R. Thomas, *Moment maps, monodromy and mirror manifolds,* Proc. Seoul 2000, World Scientific 2001, pp. 467-489.

[16] K. Uhlenbeck, *Closed minimal surfaces in hyperbolic 3-manifolds,* Seminar on minimal submanifolds, Princeton UP, 1983, pp. 147-168.

[17] M. Verbitsky and D. Kaledin, Hyperkahler manifolds, International Press, 1999.

Imperial College
London

Local rigidity for cocycles

David Fisher and G. A. Margulis

ABSTRACT. In this paper we study perturbations of constant cocycles for actions of higher rank semi-simple algebraic groups and their lattices. Roughly speaking, for ergodic actions, Zimmer's cocycle superrigidity theorems implies that the perturbed cocycle is measurably conjugate to a constant cocycle modulo a compact valued cocycle. The main point of this article is to see that a cocycle which is a continuous perturbation of a constant cocycle is actually continuously conjugate back to the original constant cocycle modulo a cocycle that is continuous and "small".

We give some applications to perturbations of standard actions of higher rank semisimple Lie groups and their lattices. Some of the results proven here are used in our proof of local rigidity for affine and quasi-affine actions of these groups.

We also improve and extend the statements and proofs of Zimmer's cocycle superrigidity.

1. Introduction

Let G be a connected semisimple Lie group with no compact factors and all simple factors of real rank at least two. Further assume G is simply connected as a Lie group or simply connected as an algebraic group. The latter means that $G = \mathbb{G}(\mathbb{R})$ where $\mathbb{G}$ is a simply connected semisimple $\mathbb{R}$-algebraic group. Let $\Gamma < G$ be a lattice and L be the k points of an algebraic k-group where k is a local field of characteristic zero. Roughly speaking, Zimmer's cocycle superrigidity theorems imply that any cocycle into L over an ergodic action of G or Γ is measurably conjugate to a constant cocycle, modulo some compact noise. (See below for a precise formulation.) This theorem has many consequences for the dynamics of smooth actions of these groups. Even stronger results would follow if one could produce a continuous or smooth conjugacy. The main purpose of this paper is to prove that a perturbation of a constant cocycle is conjugate back to the constant cocycle via a small (and often continuous) conjugacy, modulo "small" noise. We also prove stronger and more general versions of the cocycle superrigidity theorems than had previously been known. In particular, we do not need to pass to a finite ergodic extension of the action and we obtain more general statements when k is non-Archimedean.

First author partially supported by NSF grants DMS-9902411 and DMS-0226121. Second author partially supported by NSF grant DMS-9800607. The authors would also like to thank the FIM at ETHZ for hospitality and support.

Throughout we work with a more general group G. We let I be a finite index set and for each $i \in I$, we let k_i be a local field of characteristic zero and $\mathbb{G}_i$ be a connected simply connected semisimple algebraic k_i-group. We first define groups G_i, and then let $G = \prod_{i \in I} G_i$. If k_i is non-Archimedean, $G_i = \mathbb{G}_i(k_i)$ the k_i-points of $\mathbb{G}_i$. If k_i is Archimedean, then G_i is either $\mathbb{G}_i(k_i)$ or its topological universal cover. (This makes sense, since when $\mathbb{G}_i$ is simply connected and k_i is Archimedean, $\mathbb{G}_i(k_i)$ is topologically connected.) Throughout the introduction, we assume that the k_i-rank of any simple factor of any $\mathbb{G}_i$ is at least two.

We first state a version of our main result for G actions and cocycles.

THEOREM 1.1. *Let G be as above, $L = \mathbb{L}(k)$ where $\mathbb{L}$ is an algebraic k-group and k is a local field of characteristic zero and let $\pi_0 : G \to L$ be a continuous homomorphism. Let (S, μ) be a standard probability measure space, ρ a measure preserving action of G on S, and $\alpha_{\pi_0} : G \times S \to L$ be the constant cocycle over the action ρ given by $\alpha_{\pi_0}(g, x) = \pi_0(g)$. Assume $\alpha : G \times S \to L$ is a Borel cocycle over the action ρ such that α is L^∞ close to α_{π_0}. Then there exists a measurable map $\phi : S \to L$, and a cocycle $z : G \times S \to Z$ where $Z = Z_L(\pi_0(G))$, the centralizer in L of $\pi_0(G)$, such that*

1. *we have $\alpha(g, x) = \phi(gx)^{-1}\pi_0(g)z(g, x)\phi(x)$;*
2. *$\phi : S \to L$ is small in L^∞;*
3. *the cocycle z is L^∞ close to the trivial cocycle*
4. *the cocycle z is measurably conjugate to a cocycle taking values in a compact subgroup C of Z where C is contained in a small neighborhood of the identity.*

Furthermore if S is a locally compact topological space, μ is a Borel measure on S with $\operatorname{supp}(\mu) = S$ and α and ρ are continuous then both ϕ and z can be chosen to be continuous.

Remark: If k is Archimedean, point (4) implies that z is measurably conjugate to the trivial cocycle.

Before stating the analogous theorem for Γ actions and cocycles, we need to recall a consequence of the superrigidity theorems [**M1, M2, M3**]. We will use the notation introduced here in the statements below. If G is as above and $\Gamma < G$ is a lattice, we call a homomorphism $\pi : \Gamma \to L$ *superrigid* if it almost extends to a homomorphism of G. This means that there is a continuous homomorphism $\pi^E : G \to L$ and a homomorphism $\pi^K : \Gamma \to L$ with bounded image such that $\pi(\gamma) = \pi^E(\gamma)\pi^K(\gamma)$ and $\pi^E(\Gamma)$ commutes with $\pi^K(\Gamma)$. The superrigidity theorems imply that any continuous homomorphism of Γ into an algebraic group is superrigid. This can be deduced easily from Lemma VII.5.1 and Theorems VII.5.15 and VII.6.16 of [**M3**].

THEOREM 1.2. *Let Γ be as above, $L = \mathbb{L}(k)$ be as in Theorem 1.1 and $\pi_0 : \Gamma \to L$ be a continuous homomorphism. Let (S, μ) be a standard probability measure space, ρ be a measure preserving action of Γ on S, and let $\alpha_{\pi_0} : \Gamma \times S \to L$ be the constant cocycle over the action ρ given by $\alpha_{\pi_0}(\gamma, x) = \pi_0(\gamma)$. Assume $\alpha : \Gamma \times S \to L$ is a Borel cocycle over the action ρ such that α is L^∞ close to α_{π_0}. Then there exists a measurable map $\phi : S \to L$, and a cocycle $z : \Gamma \times S \to Z$ where $Z = Z_L(\pi_0^E(G))$ such that*

1. *we have $\alpha(\gamma, x) = \phi(\gamma x)^{-1}\pi_0^E(\gamma)z(\gamma, x)\phi(x)$;*
2. *$\phi : S \to L$ is small in L^∞;*
3. *the cocycle z is L^∞ close to the constant cocycle defined by π_0^K*

4. *z is measurably conjugate to a cocycle taking values in a compact subgroup C of Z where C is contained in a small neighborhood of $\pi_0^K(\Gamma)$.*

Furthermore if S is a locally compact topological space, μ is a Borel measure on S with $\mathrm{supp}(\mu) = S$ *and α and ρ are continuous then both ϕ and z can be chosen to be continuous.*

Remark: If k is Archimedean, point (4) implies that z is measurably conjugate to a cocycle taking values in the closure of $\pi_0^K(\Gamma)$.

To prove Theorems 1.1 and 1.2, we prove a very general result about perturbations of cocycles over measure preserving actions of groups with property T. The result shows that any perturbation of a cocycle taking values in a compact group also takes values in a compact group, see Theorem 5.4.

Use of (an extension and modification of) Zimmer's cocycle superrigidity theorems is a key step in the proof of Theorem's 1.1 and 1.2. The cocycle superrigidity theorems are generalizations of the second author's superrigidity theorems. Our strongest results require an integrability condition on the cocycles considered.

DEFINITION 1.3. *Let D be a locally compact group, (S, μ) a standard probability measure space on which D acts preserving μ and L be a normed topological group. We call a cocycle $\alpha : D \times S \rightarrow L$ over the D action* D-integrable *if for any compact subset $M \subset D$, the function $Q_{M,\alpha}(x) = \mathrm{supp}_{m \in M} \ln^+ ||\alpha(m,x)||$ is in $L^1(S)$.*

Any continuous cocycle over a continuous action on a compact topological space is automatically D-integrable. We remark that a cocycle over a cyclic group action is D-integrable if and only if $\ln^+ ||(\alpha(\pm 1, x)||$ is in $L^1(S)$.

THEOREM 1.4. *Let G, S, μ, L be as in Theorem 1.1. Assume G acts ergodically on S preserving μ. Let $\alpha : G \times S \rightarrow L$ be a G-integrable Borel cocycle. Then α is cohomologous to a cocycle β where $\beta(g,x) = \pi(g)c(g,x)$. Here $\pi : G \rightarrow L$ is a continuous homomorphism and $c : G \times S \rightarrow C$ is a cocycle taking values in a compact group centralizing $\pi(G)$.*

THEOREM 1.5. *Let G, Γ, S, L and μ be as Theorem 1.2. Assume Γ acts ergodically on S preserving μ. Assume $\alpha : \Gamma \times S \rightarrow L$ is a Γ-integrable, Borel cocycle. Then α is cohomologous to a cocycle β where $\beta(\gamma, x) = \pi(\gamma)c(\gamma, x)$. Here $\pi : G \rightarrow L$ is a continuous homomorphism of G and $c : \Gamma \times X \rightarrow C$ is a cocycle taking values in a compact group centralizing $\pi(G)$.*

The principal improvements over earlier results is that we do not need to pass to a finite ergodic extension of the action and that we consider the case where k is a non-Archimedean fields of characteristic 0. This builds on work of the second author, Zimmer, Stuck, Lewis, Lifschitz, Venkataramana and others [**L, Li, M1, M2, M3, Z2, Z1, Z4, Stu, V**]. In the case where S is a single point, Theorem 1.5 is equivalent to the fact that all homomorphisms of Γ to algebraic groups are superrigid. Theorem 1.4 is equivalent to the same fact when applied to $S = G/\Gamma$.

Remark: When k is non-Archimedean, it is not always the case that the algebraic hull of the cocycle is reductive unlike the case $k = \mathbb{R}$ treated in [**Z4**].

Remark: We also prove a result showing uniqueness of the homomorphism π occurring in Theorems 1.4 and 1.5. See subsection 3.8 for details.

Remark: Most of the results here should be true for k_i and k of positive characteristic as well, though additional arguments, similar to those in [**V, Li**] are apparently required. Some partial results in this direction are in [**Li**].

The main applications of our results on perturbations of constant cocycles are to studying perturbations of affine actions of G and Γ.

DEFINITION 1.6. **1** *Let A and D be topological groups, and $B < A$ a closed subgroup. Let $\rho : D{\times}A/B{\rightarrow}A/B$ be a continuous action. We call ρ* affine, *if, for every $d{\in}D$ there is a continuous automorphism L_d of A and an element $t_d{\in}A$ such that $\rho(d)[a] = [t_d{\cdot}(L_d(a))]$.*

2 *Let A and B be as above. Let C and D be two commuting groups of affine diffeomorphisms of A/B, with C compact. We call the action of D on $C\backslash A/B$ a* generalized affine action.

3 *Let A, B, D and ρ be as in* 1 *above. Let M be a compact Riemannian manifold and $\iota : D{\times}A/B{\rightarrow}\mathrm{Isom}(M)$ a C^1 cocycle. We call the resulting skew product D action on $A/B{\times}M$ a* quasi-affine action. *If C and D are as in* 2, *and $\alpha : D{\times}C\backslash A/B{\rightarrow}\mathrm{Isom}(M)$ is a C^1 cocycle, then we call the resulting skew product D action on $C\backslash A/B{\times}M$ a* generalized quasi-affine action.

Our notion of *generalized affine action* is from [**F**]. The main application of our results on local rigidity of constant cocycles is as part of our work on local rigidity of volume preserving quasi-affine actions of G and Γ on compact manifolds. We believe that volume preserving generalized quasi-affine actions on compact manifolds are locally rigid as well. As evidence for this, we have the following local entropy rigidity result. For any measure preserving action ρ of D, we denote by $h_\rho(d)$ the entropy of $\rho(d)$. Let $\mathbb{H}$ be an algebraic group defined over $\mathbb{R}$. We will refer to the connected component of the identity in $\mathbb{H}(\mathbb{R})$ as a connected real algebraic group.

COROLLARY 1.7. *Let H be a connected real algebraic group, $\Lambda < H$ a cocompact lattice and $K < H$ a compact subgroup. Let $D = G$ or Γ be as above and let ρ be a C^2 generalized affine action of D on $K\backslash H/\Lambda$. Let ρ' be any C^2 action sufficiently C^1 close to ρ. Then $h_\rho(d) = h_{\rho'}(d)$ for all $d{\in}D$.*

This result generalizes the one in [**QZ**]. Given the description of generalized standard affine actions below, the proof in [**QZ**] actually applies. We will prove Corollary 1.7 as a corollary of (part of) the proof of Theorems 1.1 and 1.2.

We note here that our techniques prove local rigidity results for perturbations of more general cocycles over actions of G and Γ than those in Theorems 1.1 and 1.2. We can prove an analogous theorem for perturbations of cocycles that are products of compact valued cocycles with constant cocycles. More generally, the original cocycle and the perturbed cocycle need not be cocycles over the same action, but only over actions that are "close". For example if S is a topological space, then the actions being C^0 close is sufficient. (Since constant cocycles are cocycles over any action, one need only consider a single action in the formulations of Theorems 1.1 and Theorem 1.2.) The proof of Corollary 1.7 then implies a local entropy rigidity result for generalized quasi-affine actions of G and Γ. The interested reader is welcome to adjust the proofs below to cover these situations, but for the sake of clarity we have restricted to the generality that we need for our next set of applications.

We now state a theorem which is used in our work on local rigidity of quasi-affine actions [**FM1, FM2**]. This theorem shows that any perturbation of any quasi-affine action is continuously semi-conjugate back to the original action, at least "along hyperbolic directions".

Let H be a connected real algebraic group and $\Lambda < H$ a discrete cocompact subgroup and let D be either G or Γ. Let ρ be a quasi-affine action of D on $H/\Lambda \times M$ which lifts to $H \times M$. By the discussion in section 6 there is a unique subgroup Z in H which is the maximal subgroup of H such that the derivative of ρ on Z cosets is an isometry for an appropriate choice of metric on H/Λ. The description given there shows that the lift of ρ to $H \times M$ descends to an action $\bar{\rho}$ on $Z\backslash H$. For example, if $G < H$ acts on H/Λ by left translations, then $Z = Z_H(G)$.

THEOREM 1.8. *Let $H/\Lambda \times M, \rho, D, Z$ and $\bar{\rho}$ be as in the preceding paragraph. Given any action ρ' sufficiently C^1 close to ρ, there is a continuous $D \times \Lambda$ equivariant map $f : (H \times M, \rho') \to (Z\backslash H, \bar{\rho})$, and f is C^0 close to the natural projection map.*

For actions by left translations this follows from Theorems 1.1 and 1.2. To prove Theorem 1.8 as stated here, we need a stronger result which is Theorem 5.1 in section 5. Theorem 1.8 holds more generally for any skew product action of D on H/Λ which is affine on H/Λ and given by a cocycle $\iota : D \times H/\Lambda \to \mathrm{Diff}^1_\omega(M)$ where ω is a volume form on M and M is compact. The version stated here is what is needed in [**FM1**]. We note that, by Theorems 6.4 and 6.5 below any quasi-affine D action on $H/\Lambda \times M$ lifts to $H \times M$ on a finite index subgroup $D' < D$.

Theorem 1.1, Theorem 1.2 and their applications hold in a wider setting than the groups G and Γ discussed above. The proof uses only that the cocycles we are considering satisfy the conclusion of the cocycle superrigidity theorems and that the group G has "few" representations. For example, for $Sp(1,n)$, F_4^{-20} and their lattices, our techniques can be combined with the results of [**CZ**] to obtain local rigidity theorems for certain perturbations of certain cocycles of these groups. If variants of Theorems 1.4 and 1.5 hold for $Sp(1,n)$ and F_4^{-20} and their lattices, then Theorems 1.1, 1.2 and 5.1 hold for these groups as well.

In section 2 we collect various standard definitions used throughout the paper. Section 3 concerns superrigidity for cocycles. Section 4 proves that certain orbits in representation varieties are closed. Section 5 contains the proof of our main results. The final section of the paper contains the proofs of Corollary 1.7 and Theorem 1.8. This section also contains a detailed description of all affine actions of G and Γ as above.

2. Preliminaries

We now collect various definitions that will be used in the course of the paper.

2.1. Algebraic groups. In this paper the words "algebraic group" mean a linear algebraic group defined over a local field k in the sense of [**B2**]. Unless otherwise noted, throughout this paper k will be a local field of characteristic zero. For background on algebraic groups particularly relevant to what follows, we refer the reader to [**M3**, I.1-2].

2.2. Cocycles and ergodic theory. Given a group D, a space X and an action $\rho : D \times X \to X$, we define a *cocycle over the action* as follows. Let L be a group, the cocycle is a map $\alpha : D \times X \to L$ such that $\alpha(g_1 g_2, x) = \alpha(g_1, g_2 x)\alpha(g_2, x)$ for all $g_1, g_2 \in D$ and all $x \in X$. The regularity of the cocycle is the regularity of the map α. If the cocycle is measurable, we only insist on the equation holding almost everywhere in X. Note that the cocycle equation is exactly what is necessary to define a skew product action of D on $X \times L$ or more generally an action of D on $X \times Y$ by $d(x,y) = (dx, \alpha(d,x)y)$ where Y is any space with an L action.

We say two cocycles α and β are *cohomologous* if there is a map $\phi : X \to L$ such that $\alpha(d,x) = \phi(dx)^{-1}\beta(d,x)\phi(x)$. Again we can define the cohomology relation in any category, depending on how much regularity we seek or can obtain on ϕ. A cocycle is called *constant* if it does not depend on x, i.e. $\alpha_\pi(d,x) = \pi(d)$ for all $x \in X$ and $d \in D$. One can easily check from the cocycle equation that this forces the map π to be a homomorphism $\pi : D \to L$. When α is cohomologous to a constant cocycle α_π we will often say that α is cohomologous to the homomorphism π. The cocycle superrigidity theorems imply that many cocycles are cohomologous to constant cocycles, at least in the measurable category.

A measurable cocycle $\alpha : D \times S \to L$ is called *strict* if it is defined for all points in $D \times S$ and the cocycle equation holds everywhere instead of almost everywhere. For a dictionary translating facts about strict cocycles on homogeneous D-spaces to facts about homomorphisms of subgroups of D, see [**Z2**, Section 4.2].

An action of a group D on a topological space X is called *tame* if the quotient space $D\backslash X$ is T_0, i.e. if for any two points in $D\backslash X$, there is an open set around one of them not containing the other.

Given a locally compact group D and a discrete subgroup $\Gamma < D$, there is a particularly important strict cocycle $\beta_X : D \times D/\Gamma \to \Gamma$. We define this by choosing a fundamental domain X for the Γ action on D. By this we mean that there is a unique representation $d = \omega(d)\tau(d)$ where $\omega(d)$ is in X and $\tau(d)$ is in Γ. Identifying D/Γ with $X \subset D$, we define $\beta_X(d,x) = \tau(dx)^{-1}$. This cocycle is of particular interest when $\Gamma < D$ is a lattice. We call β_X the *strict cocycle corresponding to the fundamental domain X*.

Let D be a compactly generated group, with compact generating set K. Let A be a metrizable, locally compact group and fix a distance function $d : A \times A \to \mathbb{R}$. Given two measurable cocycles $\alpha, \beta : D \times S \to A$ into a locally compact group A, we can define a measurable function on S by $d(\alpha(d,x), \beta(d,x))$. We say that α and β are L^∞ *close* if there exists a small $\varepsilon > 0$ such that $\|d(\alpha(k,x), \beta(k,x))\|_\infty < \varepsilon$ for any $k \in K$.

Let $\mathbb{L}$ be an algebraic k-group and $L = \mathbb{L}(k)$. Let a group D act ergodically on a measure space S and let $\alpha : D \times S \to L$ be a cocycle. There is a unique (up to conjugacy), minimal algebraic subgroup $\mathbb{H}$ in $\mathbb{L}$ such that α is cohomologous to a cocycle taking values in $H = \mathbb{H}(k)$. The group H is referred to as the *algebraic hull* for the cocycle. This is a generalization the Zariski closure of a subgroup of an algebraic group. For more details, see chapter 9 of [**Z2**].

We recall that given any group D acting on a compact metric space X preserving a Borel measure μ, there is an *ergodic decomposition* of μ. That is, there are Borel measures μ_i on X, where each μ_i is an invariant ergodic measure for the action of D, and the measure μ is obtained as an integral of the μ_i over a specific measure $\tilde{\mu}$ on the space of measures on X. Furthermore, the measures μ_i are mutually singular.

2.3. The space of actions. In the introduction, some statements are made about actions being C^k close. Let D be a locally compact topological group and X a smooth manifold. Since an action is a map $D \to \mathrm{Diff}^k(X)$ we can topologize the space of actions by taking the compact open topology on $\mathrm{Hom}(D, \mathrm{Diff}^k(X))$. Two actions are C^k close if they are close with respect to this topology. If D is compactly generated with compact generating set K, this means that ρ and ρ' are C^k close if and only if $\rho(d)\circ\rho'(d)^{-1}$ is in a small neighborhood of the identity in $\mathrm{Diff}^k(X)$ for all $d\in K$. Given a manifold or a space X equipped with an action ρ, we often write (X,ρ) to denote the space with the action. Similarly a map written $(X,\rho)\to(X',\rho')$ is a map of D-spaces or a D equivariant map.

3. Superrigidity for Cocycles

In this section we prove Theorems 1.4 and 1.5 as well as some related results. Our integrability condition allows us to use Oseledec' Multiplicative Ergodic Theorem to obtain our general result. Some partial results below do not require the integrability condition. Theorem 1.5 is deduced from Theorem 1.4. The proof of Theorem 1.4 requires that one first argue the case where L is semi-simple and then use the result in that case to prove the more general result.

Theorems 1.4 and 1.5 imply a general result on the algebraic hull of the cocycles considered. In fact, at least for G cocycles, this result is a step in the proof of Theorem 1.4, see Theorem 3.10. It is proved in [**M3**] that for any field k and any homomorphism $\pi:\Gamma\to\mathbb{L}(k)$, the Zariski closure of $\pi(\Gamma)$ is semisimple. This is equivalent to saying that the algebraic hull of the cocycle $\pi\circ\beta : G\times G/\Gamma\to L$ is semisimple. In [**Z4**], it is shown that if $k=\mathbb{R}$, any G-integrable cocycle $\alpha : G\times X\to L$ has algebraic hull reductive with compact center. If k is non-Archimedean, it is no longer the case that the algebraic hull is reductive. The following example shows that our results on the algebraic hull are sharp.

EXAMPLE 3.1. We let J be a finite index set and for each $j\in J$, we let k_j be a local field of characteristic zero and $\mathbb{H}_j$ be a connected simply connected semisimple algebraic k_j-group. We let $H_j=\mathbb{H}_j(k_j)$ the k_j-points of $\mathbb{H}_j$ and $H=\prod_{j\in J}H_j$. We further assume that there is an irreducible lattice $\Lambda<H$. For many examples where irreducible Λ exist, we refer the reader to [**M3**, IX.1.7]. Let $\pi : G\to H$ be a homomorphism and assume that $Z_H(\pi(G))$ contains a non-trivial unipotent subgroup $U<H_l$ for some $l\in J$ where k_l is non-Archimedean. (We leave the easy construction of explicit examples to the reader.) Let $K<U$ be a Zariski dense compact subgroup and consider the G action on $K\backslash H/\Lambda$ and H/Λ. Choosing a measurable trivialization of the K-bundle $H/\Lambda\to K\backslash H/\Lambda$ defines a cocycle $\alpha :$ $G\times K\backslash H/\Lambda\to K$, which we view as $\alpha : G\times K\backslash H/\Lambda\to U$ via the inclusion of $K<U$. Standard arguments using Mautner's Lemma show that the G actions on H/Λ and $K\backslash H/\Lambda$ are ergodic. A simple argument using the fact that the Mackey range of the cocycle α is H/Λ and ergodicity of the G action on H/Λ shows that U is the algebraic hull of α. See [**Z2**, 4.2.24] for definitions and discussion of the Mackey range.

The reader should note the following

1. the above construction yields the same results when applied to the restriction of the actions and cocycles to any lattice $\Gamma<G$;
2. the construction gives non-trivial examples even when $G=\mathbb{G}(\mathbb{R})$;
3. one can take products of cocycles constructed as above with constant cocycles to obtain cocycles whose algebraic hull is neither unipotent nor reductive;
4. the argument above works for more general subgroups $Z<Z_H(\pi(G))\cap H_l$ where $K<Z$ is a Zariski dense compact subgroup. One can construct

examples where $Z = F \ltimes U$ is a Levi decomposition and the F action on U is non-trivial.

Let $\mathbb{L}$ be an algebraic group over k_l and $L = \mathbb{L}(k_l)$ and $D = G$ or Γ. The above outline constructs cocycles $\alpha : D \times S \to L$ of the form $\alpha = \pi \cdot c$ where $\pi : G \to L$ is a continuous homomorphism and $c : D \times S \to C$ is a cocycle taking values in a compact group $C < Z_L(\pi(G))$. We can construct α with algebraic hull L for any $\mathbb{L}$ provided we choose π so that $\pi(G)$ commutes with the unipotent radical of $\mathbb{L}$.

We now briefly indicate the plan of this section. Subsection 3.1 fixes notation for all of section 3 and contains some technical lemmas used throughout. In subsection 3.2 we prove a key technical result which shows that certain cocycles are cohomologous to constant cocycles. Subsection 3.3 applies the results of subsection 3.2 to prove a variant of Theorem 1.4 where the algebraic hull of the cocycle is assumed to be semisimple. Subsection 3.4 proves some conditional results on G-integrable cocycles, again using the results from subsection 3.2. We show how to use property T to control cocycles into amenable and reductive groups in subsection 3.5 and then prove Theorem 1.4 in subsection 3.6. Theorem 1.5 is also proven in subsection 3.6 modulo some facts concerning G-integrability of certain induced cocycles. These facts are then proven in subsection 3.7. Subsection 3.8 concerns the uniqueness of the homomorphism π in Theorems 1.4 and 1.5. These results are used in subsection 3.9 to prove some results on cocycles with constrained projections. The result on cocycles with constrained projections is required to prove Theorem 5.1 which is used in the proof of Theorem 1.8.

3.1. Notations and reductions. In this subsection, we fix notations and definitions for all of section 3. We also prove some technical lemmas that are used throughout this section.

The group G will be as specified in the introduction, but we both weaken the rank assumption and make some preliminary reductions. Let $\mathcal{S}$ be the union of primes of $\mathbb{Z}$ and $\{\infty\}$ and let $\mathbb{Q}_p$ be the p-adic completion of $\mathbb{Q}$, where as usual, $\mathbb{Q}_\infty = \mathbb{R}$. By application of restriction of scalars, we can assume that each $k_i = \mathbb{Q}_{p_\alpha}$, where the p_i are distinct elements of the set $\mathcal{S}$. As before, for the Archimedean factor, we can replace $\mathbb{G}_i(\mathbb{R})$ by it's topological universal cover. Actually this can be done or not done for each simple factor independently, though we simplify exposition by ignoring this nuance. Instead of assuming that each simple factor of $\mathbb{G}_i(k_i)$ has k_i-rank at least two, we let $r_i = k_i\text{-rank}(\mathbb{G}_i(k_i))$ and define the *rank of* G as $\sum_{i \in I} r_i$ and assume that the rank of G is at least two and that G has no non-trivial compact factors (or, equivalently, that every simple factor of $\mathbb{G}_i$ has k_i-rank at least one).

We specify a certain compact homogeneous G space, often called a *boundary* for G. Let $\mathbb{P}_i < \mathbb{G}_i$ be a minimal parabolic subgroup. we define P_i to be $\mathbb{P}_i(k_i)$ if $G_i = \mathbb{G}_i(k_i)$. If G_i is the topological universal cover of $\mathbb{G}_i(k_i)$, we define P_i to be the pre-image of $\mathbb{P}_i(k_i)$ under the covering map from G_i to $\mathbb{G}_i(k_i)$. We let $P = \prod_{i \in I} P_i$ and the homogeneous space we consider is $P \backslash G$. We note that the G action on $P \backslash G$ factors through the projection to $\prod_{i \in I} \mathbb{G}_i(k_i)$ and the space $P \backslash G$ can be identified with $\prod_{i \in I} \mathbb{P}_i(k_i) \backslash \mathbb{G}_i(k_i)$ which can be identified as a variety with $\prod_{i \in I} (\mathbb{P}_i \backslash \mathbb{G}_i)(k_i)$.

We fix (S, μ) to be a standard probability measure space. Also $\mathbb{L}$ will denote an algebraic k-group and $L = \mathbb{L}(k)$. We denote by $\mathbb{L}^0$ the connected component of

$\mathbb{L}$ and let $L^0 = \mathbb{L}^0(k)$. As above, we apply restriction of scalars and assume that $k = \mathbb{Q}_p$ for some $p \in \mathcal{S}$.

By a simple factor of G, we mean a subgroup $F < G$ which is either $\mathbb{F}(k_i)$ or its topological universal cover, where $\mathbb{F}$ is almost simple. We note that under our hypothesis, G is the direct product of all of its simple factors. We say a simple factor F_i has rank one if the k_i rank of $\mathbb{F}_i$ is one. If F_i is a simple factor of G then there is a group $F_i^c < G$ such that $G = F_i \times F_i^c$. We call F_i^c the complement of F_i.

DEFINITION 3.2. *Let (S, μ) be a finite measure space. Given a group G acting ergodically on S preserving μ, we call the action* weakly irreducible *if for any rank one simple factor $F < G$, the complement F^c acts ergodically on S.*

If no simple factor of G has rank 1, this is equivalent to the ergodicity of the G action. This is weaker than the standard definition of irreducibility where it is assumed that all simple factors act ergodically [**Z2**]. The definition of an irreducible action is motivated by properties of irreducible lattices. We call a lattice $\Gamma < G$ *weakly irreducible* if the projection of Γ to any rank 1 factor of G is dense. Standard arguments using the generalized Mautner phenomenon, see [**M3**, II.3.3], show that a lattice is weakly irreducible if and only if the action of G on G/Γ is weakly irreducible.

We will use the following elementary lemmas repeatedly. The first is obvious.

LEMMA 3.3. *Let A be a group and let $\alpha : D \times S \to A$ and $\beta : D \times S \to A$ be cocycles over the action of a group D on a set S. Assume that $\beta(D \times X)$ is contained in a subgroup $B < A$ and let $Z = Z_A(B)$. Let $\eta : D \times S \to Z$ be a map. If $\alpha(d, x) = \beta(d, x)\eta(d, x)$, then η is a cocycle over the D action.*

We let $\tau_i : G \to \mathbb{G}_i(k_i)$ be the natural projection.

LEMMA 3.4. *Given a non-trivial continuous homomorphism $\pi : G \to L$ there is $i \in I$ such that $k = k_i$ and a k-rational homomorphism $\pi_i : \mathbb{G}_i \to \mathbb{L}$ such that $\pi = \pi_i \circ \tau_i$. From this we can deduce:*

1. *the Zariski closure of $\pi(G)$ is semisimple and connected and;*
2. *if $\mathbb{L}' \to \mathbb{L}$ is a k-isogeny, then π lifts to a continuous homomorphism $\pi' : G \to \mathbb{L}'(k)$*

PROOF. We first give the proof where all G_i are k_i-points of algebraic k_i-groups and then describe the modifications necessary when G_i is the topological universal of such a group.

Let the projection from G to G_i be τ_i. Since $k = \mathbb{Q}_p$, by [**M3**, I.2.6] any continuous homomorphism of any G_i into L is the restriction of rational map from $\mathbb{G}_i$ to $\mathbb{L}$. This implies there is an i and a rational homomorphism $\bar{\pi} : \mathbb{G}_i \to \mathbb{L}$ such that π is the restriction of $\tau_i \circ \bar{\pi}$.

Since $\mathbb{G}_i$ is connected and semisimple and the characteristic of k is zero, it follows that the Zariski closure of $\bar{\pi}(\mathbb{G}_i)$ is connected and semisimple. If $\mathbb{L}' \to \mathbb{L}$ is an isogeny, then $\bar{\pi}$ lifts to a map $\bar{\pi}' : \mathbb{G} \to \mathbb{L}'$ since $\mathbb{G}_i$ is simply connected.

Now assume that G_i is the topological universal cover of $\mathbb{G}_i(\mathbb{R})$. If $k \neq \mathbb{R}$ then any continuous homomorphism from G_i to L is trivial, so we are done by the discussion above. If $k = \mathbb{R}$ then π factors through a continuous homomorphism $\bar{\pi} : G_i \to L$. The image of $\bar{\pi}$ is a closed subgroup of L and so is the real points of a real algebraic subgroup. This implies that $\bar{\pi}$ factors through the covering map $G_i \to \mathbb{G}_i(\mathbb{R})$. The conclusions of the lemma now follow as before. □

3.2. α-invariant maps into algebraic varieties. Given two G-spaces S and Y, an L space R and a cocycle $\alpha : G\times S\to L$, we call a map $f : Y\times S\to R$ *α-invariant* if $f(gy, gs) = \alpha(g, s)f(y, s)$ for all g and almost every (y, s). Note that this definition differs slightly from the one in [**Z2**], where this map would be called $\tilde{\alpha}$-invariant where $\tilde{\alpha}$ is the pullback of α to $G\times Y\times S$.

The following theorem will play a key role in all proofs in this section. The assumption on the rank of G is only used to be able to apply this theorem.

THEOREM 3.5. *Assume G acts weakly irreducibly on S preserving μ. Let M be the k points of an algebraic variety $\mathbb{M}$ defined over k on which $\mathbb{L}$ acts k rationally. Assume that $\alpha : G\times S\to L$ is a Borel cocycle whose algebraic hull is L and that there exists a measurable α-invariant map $\phi : P\backslash G\times S\to M$ such that the essential image of ϕ is not contained in the set of L-fixed points of M. Then there is a normal k-subgroup $\mathbb{H} < \mathbb{L}$ of positive codimension such that:*

1. *$p_H\circ\alpha$ is cohomologous to a continuous homomorphism $\pi_H : G\to L/H$, where $p_H : L\to L/H$ and $H = \mathbb{H}(k)$;*
2. *$\mathbb{L}/\mathbb{H}$ is semisimple and connected.*

PROOF. Let $\phi_s(x) = \phi(x, s)$ where $x\in P\backslash G$ and $s\in X$. First one shows that either ϕ_s is rational for almost every s or ϕ_s is constant for almost every s. By rational we mean that there is $i\in I$ such that $k = k_i$, the map ϕ factors through the projection $p_i : P\backslash G\to\mathbb{P}_i(k_i)\backslash\mathbb{G}_i(k_i)$ which means that $\phi = p_i\circ\bar{\phi}$ where that $\bar{\phi}$ is a k rational map $\mathbb{P}_i\backslash\mathbb{G}_i\to\mathbb{M}$. Rationality of ϕ was shown by Zimmer in [**Z1**] for irreducible actions with each $G_i = \mathbb{G}_i(k_i)$ using an adaptation of an argument due to the second author [**M1, M2**]. The proof goes through almost verbatim for weakly irreducible actions, as well as for the case where one G_i is the universal cover of $\mathbb{G}_i(\mathbb{R})$. See also pages 104-5 of [**Z2**] or [**Fu3**] for accessible presentations of special cases. Our definition of weak irreducibility is motivated by the ergodicity needed at this step of the proof. We now assume that ϕ_s is rational and proceed in this case, the case of ϕ_s constant is discussed at the end of the proof.

Secondly, one sees that the map $\Phi : S\to \mathrm{Rat}(P\backslash G, M)$ defined by $\Phi(s) = \phi_s$ takes values in a single orbit. This follows from tameness of the $G\times L$ action on $\mathrm{Rat}(P\backslash G, M)$ and the ergodicity of the G action on S, see the "Proof of Step 3" on pages 105-6 and also Proposition 3.3.2 of [**Z2**].

One now picks a rational map ψ in this orbit and defines a map $l : S\to L$ such that $\phi_s = l(s)\psi$. Letting $\beta(g, s) = l(gs)^{-1}\alpha(g, s)l(s)$ we have that $\beta(g, s)\psi(x) = \psi(gx)$. Let H denote the point-wise stabilizer of $\psi(P\backslash G)$ in M. Since $M = \mathbb{M}(k)$ and $\mathbb{L}$ acts rationally on $\mathbb{M}$, $H = \mathbb{H}(k)$ where $\mathbb{H} < \mathbb{L}$ is an algebraic subgroup defined over k. Since $\beta(g, s)\psi(x) = \psi(gx)$, the Zariski closure of $\psi(P\backslash G)$ is invariant under $\beta(G\times S)$ and since the algebraic hull of β is L, the Zariski closure of $\psi(P\backslash G)$ is L-invariant. Therefore H is normal in L, and $\mathbb{H}$ is normal in $\mathbb{L}$. Fixing (almost any) s, and writing $\beta_s(g) = \beta(g, s)$, we have that $\beta_s(g_1g_2)\psi(x) = \psi(g_1g_2x) = \beta_s(g_1)\psi(g_2x) = \beta_s(g_1)\beta_s(g_2)\psi(x)$. Therefore $\beta_s(g_1g_2)\beta(g_2)^{-1}\beta(g_1)^{-1}$ fixes $\psi(P\backslash G)$ pointwise. It follows that $p_H\circ\beta_s : G\to L/H$ is a homomorphism. That $\pi = p_H\circ\beta_s$ is continuous follows from a result of Mackey, see [**Z2**, B.3]. The remaining conclusions of the theorem follow from Lemma 3.4.

If ϕ is constant for almost every $s\in S$, we have an α-invariant map $\Phi : S\to M$. The image of this map is contained in a single H orbit since the L action on M is tame and the G action on S is ergodic. Since the L action on M is defined by an algebraic action of $\mathbb{L}$ on $\mathbb{M}$, the stabilizer of this orbit is $H = \mathbb{H}(k)$ where $\mathbb{H} < \mathbb{L}$ is an algebraic subgroup. This means that we have an α invariant map $\phi : S \to L/H$. By [**Z2**, Lemma 5.2.11], this implies that the cocycle α is equivalent to one taking values in H, which contradicts our assumption on the algebraic hull of the cocycle unless $H = L$ in which case we contradict our assumption that the essential image of ϕ is not contained in the set of L-fixed points. □

The reader should note that essentially the same result can be proven by considering equivariant measurable maps from $G \times X$ to vector spaces, as in [**M3**, VII.1-4]. The argument there requires some modification since, in the language of that text, one needs to consider maps that are not strictly effective.

3.3. Algebraic hull semisimple. We now prove Theorems 1.4 and Theorem 1.5 in the case where the algebraic hull of the cocycle is semisimple.

THEOREM 3.6. *Let G act weakly irreducibly on S preserving μ and let $\alpha : G \times S \to L$ be a Borel cocycle with algebraic hull L. Further assume that $\mathbb{L}$ is semisimple. Then α is cohomologous to a cocycle $\beta = \pi \cdot c$. Here $\pi : G \to L$ is a continuous homomorphism and $c : G \times S \to C$ is a cocycle taking values in a compact group C centralizing $\pi(G)$.*

THEOREM 3.7. *Let $\Gamma < G$ be a weakly irreducible lattice. Assume Γ acts ergodically on S preserving μ. Let $\alpha : \Gamma \times S \to L$ be a Borel cocycle with algebraic hull L. Further assume $\mathbb{L}$ is semisimple. Then α is cohomologous to a cocycle β where $\beta(\gamma, x) = \pi(\gamma)c(\gamma, x)$. Here $\pi : G \to L$ is a continuous homomorphism and $c : \Gamma \times S \to C$ is a cocycle taking values in a compact group centralizing $\pi(G)$.*

To reduce Theorem 3.6 to Theorem 3.5 we need to find a k variety M on which H acts k-rationally and an α-invariant map $f : P \backslash G \times S \to M$. To produce α invariant maps, one uses the following modification of a lemma of Furstenberg from [**Fu2**] which can be deduced from Propositions 4.3.2, 4.3.4 and 4.3.9 of [**Z2**]. The lemma holds under more general circumstances than those needed here. For the lemma, G can be a locally compact, σ-compact group, P a closed amenable subgroup and L any topological group.

LEMMA 3.8. *Assume G acts on S preserving μ. Let $\alpha : G \times S \to L$ be a Borel cocycle. Let B be any compact metrizable space on which L acts continuously and $\mathcal{P}(B)$ the space of Borel regular probability measures on B. Then there is an α-invariant map $f : P \backslash G \times S \to \mathcal{P}(B)$.*

We note here that we give a proof using only amenability of P, without reference to the notion of an amenable actions, though we do rely on ideas of Zimmer's to construct a convex compact space on which P acts affinely and continuously.

PROOF. Let B be any compact L-space. Via α we can define a skew product action of G on $S \times B$. We consider the diagonal G action on $G \times S \times B$ given by the right G action on G and the skew product action on $S \times B$, which we note commutes with the left G action on G. Let μ_G be Haar measure on G and $\mathcal{M}(G \times S \times B)$ be the space of regular Borel measures on $G \times S \times B$ which are invariant under the diagonal action and project to $\mu_G \times \mu$ on $G \times S$. We want to topologize $\mathcal{M}(G \times S \times B)$ so the left G action is continuous and the space is a compact convex affine G-space. Using disintegration of measures, we can identify $\mathcal{M}(G \times S \times B)$ with $F(G \times S, \mathcal{P}(B))$ the space of measurable maps from $G \times S$ to $\mathcal{P}(B)$. Let $C(B)$ be the Banach space of continuous functions on B. We identify $F(G \times S, \mathcal{P}(B))$ as a subset of

$L^\infty(G\times S, C(B)^*)$ and give $L^\infty(G\times S, C(B)^*)$ the weak topology coming from the identification $L^\infty(G\times S, C(B)^*) = L^1(G\times S, C(Y))^*$. In this topology the action of G is continuous and $F(G\times S, \mathcal{P}(B))$ is the unit ball in $L^\infty(G\times S, C(B)^*)$. See [**Z2**, Section 4.3] for more discussion of this and related constructions. It follows that there is a fixed point $\mu^P \in \mathcal{M}(G\times S\times B)$ for the left P action. By applying disintegration of measures, this is left P-invariant, α-invariant map $\tilde{f} : G\times S \to \mathcal{P}(B)$ or equivalently an α-invariant map $f : P\backslash G\times S \to \mathcal{P}(B)$ □

We will also need the following lemma essentially due to Furstenberg.

LEMMA 3.9. *Let $J < GL(V)$, where $V = k^n$ and k is a local field of characteristic zero. Let J act on the projective space $P(V)$ preserving a measure μ. Then either J is projectively compact (i.e. the image of J in $PGL(V)$ is compact) or there is a proper subspace $W < V$ with $\mu(W) > 0$.*

For a proof, we refer the reader to [**Z2**, Lemma 3.2.2] or the original article of Furstenberg [**Fu1**] in the case where $k = \mathbb{R}$.

PROOF OF THEOREM 3.6. We call a representation of an algebraic group *almost faithful* if the kernel of the representation is finite. We choose an almost faithful irreducible k-rational representation σ of $\mathbb{L}$ on V such that the restriction of σ to any $\mathbb{L}^0$ invariant subspace is almost faithful, where as usual $\mathbb{L}^0$ denotes the connected component. (This can be done by inducing an almost faithful irreducible $\mathbb{L}^0$ representation.) Let $B = \mathbb{P}(V)$ be the corresponding projective space.

Since $P < G$ is an amenable subgroup Lemma 3.8 provides an α equivariant map $f : P\backslash G\times X \to \mathcal{M}(B)$. In fact this map takes values in a single H orbit $\mathcal{O}$ in $\mathcal{P}(B)$. This is deduced from ergodicity of the G action on S and the tameness of the action of L on $\mathcal{P}(B)$ which is [**Z2**, Corollary 3.2.12].

Let J be the stabilizer of a point μ for the L action on $\mathcal{O}$. We prove that either J is compact or J is contained in an algebraic subgroup of positive codimension in $\mathbb{L}$. If $\mathbb{L}$ is connected, this is Proposition 3.2.15 of [**Z2**]. By Lemma 3.9, if J is not projectively compact, then there is a proper subspace $W < V$ such that $\mu([W]) > 0$. Since $\mathbb{L}$ is semisimple and the representation on V is almost faithful, the map from $\mathbb{L}$ to $PGL(V)$ has finite kernel and only compact subgroups of L are projectively compact. Assuming J is non-compact, we choose W of minimal dimension among subspaces with positive μ measure. Since the measure of W is positive, $J{\cdot}W$ must be a finite union of disjoint subspaces $\cup_{l=1}^n W_l$, and we let $\mathbb{F}$ be the stabilizer of the J orbit $J{\cdot}W$. Let $F = \mathbb{F}(k)$. The stabilizer J^W in J of W is of finite index in J and, by minimality of W and Lemma 3.9, acts on $P(W)$ via a homomorphism to a compact subgroup of $PGL(W)$. If $\dim(\mathbb{F}) = \dim(\mathbb{L})$, then the connected component of $\mathbb{L}$ preserves $\cup_{l=1}^n W_l$ and by connectedness preserves W. Since $J^W < L$ and acts compactly on W, we then have that $J^W \cap \mathbb{L}^0(k)$ is projectively compact. But, since we have that $\mathbb{L}^0$ is semisimple and the representation of $\mathbb{L}^0$ on W is almost faithful, the map from $\mathbb{L}^0$ to $PGL(W)$ has finite kernel. This implies that $J^W \cap \mathbb{L}^0(k)$ is compact. The group $J^W \cap \mathbb{L}^0(k)$ has finite index in J^W which has finite index in J, so in this case, J is compact. Therefore either J is compact or $\mathbb{F}$ is of positive codimension in $\mathbb{L}$.

If J is compact, then Lemma 5.2.10 of [**Z2**] applies and shows that the cocycle α is cohomologous to one with bounded image. If $J < \mathbb{F}$ an algebraic subgroup of positive codimension, then we compose $\phi : P\backslash G\times X \to \mathcal{O}$ with the projection $p : \mathcal{O} \to L/F \subset (\mathbb{L}/\mathbb{F})(k)$. We note that the set of L fixed points in L/F is empty so we can apply Theorem 3.5. This theorem produces a normal k-subgroup of positive codimension $\mathbb{H} < \mathbb{L}$, such that the projection of α to $\mathbb{L}(k)/\mathbb{H}(k)$ is cohomologous to a continuous homomorphism π of G. That theorem also implies that $\mathbb{L}/\mathbb{H}$ is semisimple and connected. Since $\mathbb{L}$ is semisimple, there is a connected normal subgroup $\mathbb{H}^c < \mathbb{L}$ such that the map $\mathbb{H}^c \to \mathbb{H}/\mathbb{L}$ is an isogeny. Now $\mathbb{L} = \mathbb{H}\cdot\mathbb{H}^c$ where $\mathbb{H}\cap\mathbb{H}^c$ is finite. Because $[\mathbb{H}, \mathbb{H}^c] \subset \mathbb{H}\cap\mathbb{H}^c$ which is finite and $\mathbb{H}^c$ is connected, $\mathbb{H}$ and $\mathbb{H}^c$ commute. By Lemma 3.4 we can lift π to a homomorphism $\pi' : G \to \mathbb{H}^c$. It then follows that α is cohomologous to $\pi'\cdot\alpha'$ where α' takes values in $\mathbb{H}(k)$ and is a cocycle by Lemma 3.3. One can now replace α by α' and complete the proof of the theorem by induction on the dimension of L. □

PROOF OF THEOREM 3.7. This is proved by inducing actions and cocycles, exactly as in [**Z2**, Theorem 9.4.14]. We briefly outline the argument. Given a Γ action on S and a Γ cocycle $\alpha : \Gamma\times S \to L$, we consider the induced G action on $(G\times S)/\Gamma$ and a cocycle $\tilde{\alpha} : G\times G/\Gamma\times S \to L$. We define $\tilde{\alpha}$ by taking a fundamental domain X for Γ in G and the strict cocycle $\beta_X : G\times G/\Gamma \to \Gamma$ corresponding to X and letting $\tilde{\alpha}(g, [g_0, x]) = \alpha(\beta(g, [g_0]), x)$. It is straightforward to verify that weak irreducibility of Γ and ergodicity of the Γ action imply that the induced G action is weakly irreducible. One then shows that the algebraic hull of $\tilde{\alpha}$ is L and applies Theorem 3.6 to $\tilde{\alpha}$. Straightforward manipulation allows one to deduce the desired conclusions concerning α. □

3.4. Conditional results using G-integrability of α. In this section we prove a conditional result concerning the algebraic hull of G-integrable cocycles. The assumption of G-integrability is only used here.

Before stating our result, we fix some notation and assumptions. We assume that G has property T and that G acts weakly irreducible on (S, μ). Let $\alpha : G\times S \to L$ be a G-integrable Borel cocycle and assume that L is the algebraic hull of the cocycle. We can write $\mathbb{L} = \mathbb{F} \ltimes \mathbb{U}$ where $\mathbb{F}$ and $\mathbb{U}$ are k-subgroups, $\mathbb{U}$ is the unipotent radical of $\mathbb{L}$ and $\mathbb{F}$ is reductive. Let $p_F : \mathbb{L} \to \mathbb{F}$ and be the natural projection. We assume that the cocycle $p_F \circ \alpha$ is cohomologous to a cocycle of the form $\pi\cdot c$ where $\pi : G \to F$ is a continuous homomorphism and c is a cocycle taking values in a compact subgroup $C < Z_F(\pi(G))$. We note that π can be viewed as defining a homomorphism of G into L, and we let α' be the cocycle cohomologous to α that projects to $\pi\cdot c$.

THEOREM 3.10. *Under the hypotheses discussed in the preceding paragraph, U commutes with $\pi(G)$.*

We prove the theorem by contradiction. The general scheme is as follows. If U does not commute with $\pi(G)$ there exists a k-rational action of $\mathbb{L}$ on a variety $\mathbb{M}$ and an α-invariant map ϕ into $\mathbb{M}(k)$ such that the pointwise stabilizer $H = \mathbb{H}(k)$ of the image does not contain all of U. Applying Theorem 3.5 we obtain a contradiction, since number 2 of that theorem implies that $\mathbb{L}/\mathbb{H}$ is semisimple and this implies that $\mathbb{U} < \mathbb{H}$.

We will construct a measurable map ϕ that satisfies the hypotheses of Theorem 3.5 by using Oseledec' multiplicative ergodic theorem. We will give an argument that is close to the one in [**M3**, Section V.3-4], but also refer the reader to [**Z4**] for a somewhat different approach.

Let $\mathfrak{l}$ be the Lie algebra of L. Let $Gr_j(\mathfrak{l})$ be the Grassmann variety of j planes in $\mathfrak{l}$. We have an action of L on $\mathfrak{l}$ by the adjoint representation which also defines an action of L on $Gr_j(\mathfrak{l})$.

We look at the representation $\mathrm{Ad}_{\mathfrak{l}} \circ \pi$.

THEOREM 3.11. *Assume $\pi(G)$ does not commute with U. Then there is an integer $0 < m < \dim(\mathfrak{l})$ and an α-equivariant measurable map $\phi : P\backslash G \times S \to Gr_m(\mathfrak{l})$ such that the pointwise stabilizer of the image does not contain all of U.*

Before proving Theorem 3.11, we show how that result implies Theorem 3.10.

PROOF OF THEOREM 3.10. We now apply Theorem 3.5 to the map ϕ from Theorem 3.11. This is possible since the stabilizer of the essential image does not contain U and so the essential image is not contained in L fixed points. If H is the stabilizer of the essential image of ϕ this implies that $\mathbb{L}/\mathbb{H}$ is semisimple and therefore that $U < H$. But this implies that U is contained in the stabilizer of the essential image of ϕ', a contradiction. □

Before proving Theorem 3.11 we recall several facts from [**M3**]. The following, which is [**M3**, I.4.6.2], is a simple corollary of the Poincaré recurrence theorem.

LEMMA 3.12. *Let A be an automorphism of (S, μ) and f a non-negative measurable function on X. Then for almost all $x \in X$*

$$\liminf_{m \to \infty} \frac{1}{m} f(A^m(x)) = 0$$

and

$$\liminf_{m \to -\infty} \frac{-1}{m} f(A^m(x)) = 0.$$

Let A be an ergodic automorphism of (S, μ) and W be a k vector space. Let $u : \mathbb{Z} \times S \to GL(W)$ be a $\mathbb{Z}$-integrable cocycle over the $\mathbb{Z}$-action generated by A. If we define

$$\chi_+(u, x, w) = \lim_{m \to \infty} \frac{1}{m} \ln \|u(A^m, x)w)\|$$

and

$$\chi_-(u, x, w) = \lim_{m \to -\infty} \frac{1}{m} \ln \|u(A^m, x)w)\|$$

it follows from Oseledec multiplicative ergodic theorem that both $\chi_+(u, x, w)$ and $\chi_-(u, xw)$ exist for almost all $x \in X$ and all $w \in W$. Furthermore, that theorem shows that there exists a finite set J and real numbers $\chi_j(u)$ and maps $\omega_j(u, x) : S \to Gr_{l(j)}(W)$ such that

1. for almost all $x \in S$, the space W is the direct sum $\oplus_J \omega_j(u, x)$;
2. for almost all $x \in S$ the sequence $\{\frac{1}{m} \ln \|u(m, x)\|/\|w\|\}_{m \in \mathbb{N}^+}$ converges to $\chi_j(u)$ uniformly in $w \in \omega_j(u, x) - \{0\}$.

Furthermore

$$\{0\} \cup \{w \in \mathfrak{h} - \{0\} | \chi_+(u, x, w) \leq a\} = \oplus_{\chi_j \leq a} \omega_j(u, x)$$

and

$$\{0\} \cup \{w \in \mathfrak{h} - \{0\} | \chi_-(u, x, w) \geq a\} = \oplus_{\chi_j \geq a} \omega_j(u, x).$$

All of this is contained in [**M3**, V.2.1]. We refer to $\omega_j(u, x)$ as the *characteristic subspace* for u with *characteristic number* χ_j.

We call a cocycle $v : G{\times}X{\to}H$ *G-quasi-integrable* if v is cohomologous to a G-integrable cocycle u. If v is a $\mathbb{Z}$-quasi-integrable cocycle then we have $v(g,x) = \psi(gx)u(g,x)\psi(x)^{-1}$ for some $\mathbb{Z}$-integrable cocycle u. We then define $\omega_j(v,x) = \psi(x)\omega_j(u,x)$. It is easy to verify, using Lemma 3.12, that if v is in fact $\mathbb{Z}$-integrable, our two definitions of $\omega_j(v,x)$ agree and so $\omega_j(v,x)$ is well-defined and independent of the choice of u. Though $\omega_j(v,x)$ does not satisfy the dynamical condition 2 above, it can be shown to satisfy weaker dynamical conditions, see Definition 2.4 and Remark 2.4 in [**Z4**].

In the proof below we will need some functorial properties of characteristic subspaces. For $\mathbb{Z}$-integrable cocycles these are [**M3**, V.2.3] and it follows easily from the definition that they hold for $\mathbb{Z}$-quasi-integrable cocycles as well. Let $u : \mathbb{Z}{\times}S{\to}GL(W)$ be a $\mathbb{Z}$-quasi-integrable cocycle and let $Q < W$ be a subspace such that $u(m,x)Q = Q$. Let V be the quotient W/Q and $p : W{\to}V$ the projection. We have two additional cocycles $u^Q(m,x) = u(m,x)|_Q$ and $u^V(m,x) = p{\circ}u(m,x)$ both of which can easily be seen to be quasi-integrable. Then for any characteristic subspace $\omega_l(u^V,x)$ (respectively $\omega_l(u^Q,x)$) there is a characteristic subspace $\omega_{l(j)}(u,x)$ such that $\omega_l(u^V,x) = p(\omega_{l(j)}(u,x))$ (respectively $\omega_l(u^Q,x) = \omega_{l(j)}(u,x){\cap}Q$).

For certain cocycles it is easy to compute characteristic subspaces and numbers. Let $\sigma : \mathbb{Z}{\to}GL(W)$ be a homomorphism, let $M = \sigma(1)$ and let $c : \mathbb{Z}{\times}S{\to}GL(W)$ be a cocycle taking values in a compact subgroup of $GL(W)$. We let $u(m,x) = \sigma(m)c(m,x)$. We let $\Omega(M)$ be the set of all eigenvalues of M, $W_\lambda(M)$ the eigenspace corresponding to $\lambda{\in}\Omega(M)$ and $W_d(M) = [\oplus_{\ln|\lambda|=d}W_\lambda(M)]_k$. We also let $W_+(M) = \oplus_{d>0}W_d(M)$ and $W_{\leq 0}(M) = \oplus_{d\leq 0}W_d(M)$. We will call d a *characteristic number* of M and $W_d(M)$ a *characteristic subspace* of M. Then the characteristic numbers of u are the characteristic numbers of M and the characteristic subspaces for u are the characteristic numbers for M. Furthermore the space $\oplus_{\chi_j\leq 0}\omega_j(u,x) = W_{\leq 0}(M)$ and $\oplus_{\chi_j>0}\omega_j(u,x) = W_+(M)$

PROOF OF THEOREM 3.11. As the proof is very involved, we divide it into several steps. The basic idea is to choose an element t of G and use Oseledec theorem to construct characteristic maps from $S{\to}Gr_m(\mathfrak{l})$ for α and each $g^{-1}tg$. This gives an α-invariant map $\phi : G{\times}S{\to}Gr_m(\mathfrak{l})$, which we show descends to an α-invariant map $\phi : P\backslash G{\times}S{\to}Gr_m(\mathfrak{l})$. We then pass to characteristic subspaces for the cocycle α' which is cohomologous to α and where $p_F{\circ}\alpha' = \pi{\cdot}c'$. We use the functoriality of characteristic subspaces and the form of α' to compute the characteristic subspaces quite explicitly. Finally using the assumption that $\pi(G)$ does not commute with U, we show that the stabilizer of the essential image does not contain U.

Step One: Choosing t.

We call a subgroup diagonalizable if it can be conjugated to a subgroup of the group of diagonal matrices. Recall that a subgroup $\mathbb{S}_i < \mathbb{G}_i$ is called a *maximal torus* if it is maximal diagonalizable subgroup of $\mathbb{G}_i$. We fix a torus $\mathbb{S}_i$ in each $\mathbb{G}_i$. We let $\mathbb{T}_i < \mathbb{S}_i$ be the *maximal split torus*, i.e. the maximal subgroup of $\mathbb{S}_i$ that is diagonalizable over k_i. We let $X(\mathbb{T}_i)$ be the group of k_i characters of $\mathbb{T}_i$, and $T_i = \mathbb{T}_i(k_i)$

LEMMA 3.13. *There exists an element $t{\in}\prod_{i\in I}T_i$ such that*

1. *the group generated by t is not contained in any proper normal subgroup of G*
2. *for any $\chi{\in}X(\mathbb{T}_i)$ where $\chi(\tau_i(t))$ has modulus one it follows that $\chi(\tau_i(t)) = 1$.*

PROOF. To satisfy 1, it suffices to choose t such that it projects to an element which generates an infinite subgroup in each simple factor of each $\mathbb{G}_i$.

For $k_i = \mathbb{R}$ it suffices to assume that $\chi(\tau_i(t))$ is positive for every $\chi \in X(\mathbb{T}_i)$. If k_i is non-Archimedean, we identify T_i with $(k_i^*)^{l(i)}$ where $l(i)$ is the k_i-rank of $\mathbb{G}_i$. We choose π a uniformizer of k_i. We assume that the projection of $\tau_i(t)$ to each copy of k_i^* in $(k_{\alpha_0}^*)^{i(\alpha_0)}$ is the product of a unit of k_i^* with a non-zero power of π. □

Remark: Let π be finite dimensional representation of G on a vector space V. It follows from our choice of t that if $\pi(t)$ has all eigenvalues of modulus one then π is trivial.

Step two: Oseledec theorem and characteristic maps.

We will construct the map $\phi : G \times S \to Gr_k(\mathfrak{h})$ by applying Oseledec theorem to certain cocycles over the action of $g^{-1}tg$ on S.

Since G acts ergodically on S, it follows from the Mautner phenomenon that t and therefore $g^{-1}tg$ does as well [**M3**, II.3.3]. We define a map $\phi' : G \times S \to Gr_k(\mathfrak{l})$. The element $g^{-1}tg$ generates a $\mathbb{Z}$ action on S. We define a cocycle $u_g : \mathbb{Z} \times S \to GL(\mathfrak{l})$ over this $\mathbb{Z}$ action by $u_g(m,x) = \mathrm{Ad}_{\mathfrak{l}}\, \alpha(g^{-1}t^m g, x)$ and apply Oseledec theorem to each cocycle u_g. Since different choices of $g \in G$ define cohomologous cocycles over conjugate actions, it follows that the characteristic numbers $\chi_j(u_g)$ do not depend on g nor do the dimensions $l(j)$ of the subspaces $\omega_j(u_g, x)$. We $\chi_j = \chi_j(u_g)$ and $\omega_j(g,x) = \omega(u_g, x)$. We now have maps $\omega_j : G \times S \to Gr_{l(j)}(\mathfrak{l})$. If we let G on $G \times X$ by $h(g,x) = (gh^{-1}, hx)$, the map $\omega_j : G \times X \to H$ is α-invariant. To show this one uses the cocycle identity to see that

$$\alpha(hg^{-1}s^m gh^{-1}, hx) = \alpha(h, g^{-1}s^m gx)\alpha(g^{-1}s^m g, x)\alpha(h,x)^{-1}$$

and notes that

$$\liminf_{m \to \infty} \frac{1}{m} \ln^+ \|\alpha(h, g^{-1}s^m gx)\| = 0$$

for almost every x by Lemma 3.12.

Step 3: P-invariance of characteristic maps.

We now show that there is a minimal parabolic P such that the map $\omega_{\leq \chi}(g,x) = \oplus_{\chi_j \leq \chi} \omega_i(g,x)$ descends to a map from $P\backslash G \times S \to Gr_k(\mathfrak{h})$. This follows as in [**M3**, Theorem V.3.3]. If we let P be the set of elements in G such that the set $M = \{s^m p s^{-m} | m \in \mathbb{N}^+\}$ is relatively compact in G then P is a minimal parabolic in G as discussed in [**M3**, VI.4.8 and Lemma II.3.1(b)]. One then can compute that

$$\alpha((pg)^{-1}s^m pg, x) = \alpha((pg)^{-1}s^m ps^{-m} gg^{-1}s^m g, x)$$
$$= \alpha((pg)^{-1}, s^m pgx)\alpha(s^m ps^{-m}, s^m gx)\alpha(g, g^{-1}s^m gx)\alpha(g^{-1}s^m g, x).$$

Because α is G-integrable and M is precompact, it then follows that $Q_{M,\alpha}(s^m x) = \mathrm{supp}_{m \in \mathbb{N}^+} \ln^+ \|\alpha(s^m ps^{-m}, y)\| < \infty$ for almost all $y \in X$ and therefore that

$$\liminf_{m \to \infty} \frac{1}{m} \ln^+ \|\alpha(s^m ps^{-m}, s^m gx)\| = 0$$

almost everywhere by Lemma 3.12. Similarly both $\ln^+ \|\alpha((pg)^{-1}, y)\| < \infty$ and $\ln^+ \|\alpha(g,y)\| < \infty$ for almost all $y \in X$ and therefore

$$\liminf_{m \to \infty} \frac{1}{m} \ln^+ \|\alpha((pg)^{-1}, s^m pgx)\| = 0$$

and

$$\liminf_{m\to\infty} \frac{1}{m} \ln^+ \|\alpha(g, g^{-1}s^m gx)\| = 0$$

almost everywhere by Lemma 3.12. It follows that $\chi^+(pg, x, w) \leq \chi^+(g, x, w)$ and the reverse inequality follows by replacing p by p^{-1}. See [**M3**, V.3.3] for more detailed computations for certain choices of cocycle.

Let $\phi' = \oplus_{\chi_j \leq 0} \omega_i(g, x) : P\backslash G \times S \to Gr_k(\mathfrak{h})$.

Step 4: Modifying the map.

Let α' be the cocycle cohomologous to α which projects via p_F to $\pi{\cdot}c$, and let ψ be the map implementing the cohomology between α and α'. We will let $\phi([g], x) = \psi(x)\phi'([g], x)$. Since α' is G-quasi-integrable, it follows from the definitions that $\phi([g], x)$ is $\oplus_{\chi_j \leq 0} \omega'_j([g], x)$ where $\omega'_j([g], x) = \psi(x)\omega_j([g], x)$ are the characteristic subspaces for the cocycle $u'(m, x) = \alpha'(g^{-1}t^m g, x)$ over the $\mathbb{Z}$ action on S generated by $g^{-1}tg$.

Step 5: Application of functoriality.

Since L is an algebraic group with unipotent radical U and Levi complement F, we have that $\mathfrak{l} = \mathfrak{f} \oplus \mathfrak{u}$. Now $\mathfrak{f}$ and $\mathfrak{u}$ are invariant under $\mathrm{Ad}_{\mathfrak{l}}$ and so are invariant under the cocycle $\alpha' : G \times S \to GL(\mathfrak{l})$. By the functoriality of characteristic subspaces discussed above, we have that $\phi([g], x) \cap \mathfrak{f}$ is $\oplus_{\chi_j \leq 0} \omega^{\mathfrak{f}}_j([g], x)$ where $\omega^{\mathfrak{f}}_j([g], x)$ are the characteristic subspaces for the cocycle $u'^{\mathfrak{f}}_g(m, x) = \alpha'(g^{-1}t^m g, x)|_{\mathfrak{f}}$. Since $\mathrm{Ad}_{\mathfrak{l}}|_{\mathfrak{f}}$ factors through the map $p_F : L \to F$, and $p_F \circ \alpha'(g, x) = \pi(g)c(g, x)$ where c is a cocycle taking values in a compact group, it follows that $\phi([g], x) \cap \mathfrak{f} = \mathrm{Ad}_{\mathfrak{h}}(\pi(g)^{-1}) W_{\leq 0}(\mathrm{Ad}_{\mathfrak{f}} \circ \pi(t))$.

The intersection of $\phi([g], x)$ with $\mathfrak{u}$ is more complicated to describe. To do this, we let $\mathfrak{u}_0 = \mathfrak{u}$ and $\mathfrak{u}_i = [\mathfrak{u}, \mathfrak{u}_{i-1}]$. Since $\mathfrak{u}$ is unipotent, there is a number k such that $\mathfrak{u}_k$ is the center of $\mathfrak{u}$ and $\mathfrak{u}_l = 0$ for all $l > k$. Furthermore, $\mathfrak{u}_{i+1}$ is an ideal in $\mathfrak{u}_i$ and we have a sequence of quotients $\mathfrak{u}_i/\mathfrak{u}_{i+1}$. A key fact for what follows is that $\mathrm{Ad}(u)(\mathfrak{u}_i) \subset \mathfrak{u}_{i+1}$ for any $u \in U$. Since $\mathfrak{u}_i$ is $\mathrm{Ad}_{\mathfrak{l}}$ invariant, it follows that the cocycle $\mathrm{Ad}_{\mathfrak{l}} \circ \alpha'$ leaves each $\mathfrak{u}_i$ invariant. Let $p_i : \mathfrak{u}_i \to \mathfrak{u}_{i+1}$ be the projection. It follows that

$$p_i((\mathrm{Ad}_{\mathfrak{l}} \circ \alpha'(g, x)|_{\mathfrak{u}_i})v) = p_i(\mathrm{Ad}_{\mathfrak{l}}|_{\mathfrak{u}_i}(\pi(g)c(g, x))v) \tag{1}$$

for any $v \in \mathfrak{u}_i$.

Since the representation of G on $\mathfrak{h}$ is defined by $\mathrm{Ad}_{\mathfrak{h}} \circ \pi$, and each $\mathfrak{u}_i$ is $\mathrm{Ad}_{\mathfrak{h}}$ invariant, the G action leaves each $\mathfrak{u}_i$ invariant . Therefore,since G is semisimple, there are G invariant subspaces $\mathfrak{v}_i$ in $\mathfrak{u}_i$ such that $\mathfrak{u}_i = \mathfrak{v}_i \oplus \mathfrak{u}_{i+1}$. Since c takes values in a compact subgroup $C < F$, we can assume that $\mathfrak{v}_i$ is also c invariant. We identify $\mathfrak{v}_i$ and $\mathfrak{u}_i/\mathfrak{u}_{i+1}$ as G modules in the following paragraphs.

Now for any $v \in \mathfrak{v}_i$, we can rewrite equation 1 to

$$p_i(\mathrm{Ad}_{\mathfrak{l}} \circ \alpha'(g, x)|_{\mathfrak{u}_i})v) = \mathrm{Ad}_{\mathfrak{l}}((\pi(g)c(g, x)))v). \tag{2}$$

It follows from the definition of $\phi([g], x)$ in terms of characteristic subspaces, the functoriality of characteristic subspaces and equation 2 above, that $p_i(\phi([g], x) \cap \mathfrak{u}_i) = p_i(\mathrm{Ad}_{\mathfrak{l}}(\pi(g)^{-1}) W_{\leq 0}((\mathrm{Ad}_{\mathfrak{f}} \circ \pi)|_{\mathfrak{v}_i}(t)))$.

Step 6: U does not stabilize the essential image of ϕ.

Since $\pi(G)$ does not commute with U by assumption, the representation of G on $\mathfrak{v}_i$ is non-trivial for some i. We fix one such i for what follows. Since G is semisimple $\mathrm{Ad}_{\mathfrak{l}} \circ \pi(G)|_{\mathfrak{v}_i} < SL(\mathfrak{v}_i)$. By our choice of t and i this implies that the decomposition

$$\mathfrak{v}_i = \big(\mathfrak{v}_i \cap W_+((\mathrm{Ad}_{\mathfrak{l}} \circ \pi)_{\mathfrak{v}_i}(t))\big) \oplus \big(\mathfrak{v}_i \cap W_{\leq 0}((\mathrm{Ad}_{\mathfrak{l}} \circ \pi)_{\mathfrak{v}_i}(t))\big)$$

is non-trivial. We let

$$V_+ = \mathfrak{v}_i \cap W_+((\mathrm{Ad}_{\mathfrak{l}} \circ \pi)|_{\mathfrak{v}_i}(t))$$

and

$$V_- = \mathfrak{v}_i \cap W_{\leq 0}((\mathrm{Ad}_{\mathfrak{l}} \circ \pi)|_{\mathfrak{v}_i}(t)).$$

For any choice of $g \in G$ we write V_+^g for $(\mathrm{Ad}_{\mathfrak{l}} \circ \pi(g)^{-1})V_+$ and V_-^g for $(\mathrm{Ad}_{\mathfrak{l}} \circ \pi(g)^{-1})V_-$. Note that $\mathfrak{v}_i = V_+^g \oplus V_-^g$ is non-trivial for all choices of g. Letting $\bar{V}_+^g$ (resp. $\bar{V}_-^g$) be the projection of $p_i(V_+^g)$ (resp. $p_i(V_-^g)$) to and $\bar{\phi}([g], x) = p_i(\phi([g], x) \cap \mathfrak{u}_i)$ we have that $\bar{V}_-^g = \bar{\phi}([g], x)$ so $\bar{V}_+^g \cap \bar{\phi}([g], x) = 0$.

Let $\mathfrak{t}$ be the Lie algebra of $\pi(T)$ and write ${}^g\mathfrak{t}$ for $\mathrm{Ad}_{\mathfrak{l}}(\pi(g)^{-1}(\mathfrak{t})$. Then for almost every $([g], x)$ we have ${}^g\mathfrak{t} \subset \phi'([g], x) \cap \mathfrak{f}$. Since $\mathrm{Ad}_{\mathfrak{h}}(\pi(g)^{-1}\pi(t)\pi(g))V_+^g = V_+^g$ we have $V_+^g \subset [{}^g\mathfrak{t}, V_+^g]$. We let U_g be the collection of elements of U of the form $\mathrm{Exp}(V_+^g)$ where $\mathrm{Exp} : \mathfrak{u} \to U$ is the Lie group exponential map. That $V_+^g \subset [{}^g\mathfrak{t}, V_+^g]$ implies that, for some $u_0 \in U_g$ the projection of $(\mathrm{Ad}_{\mathfrak{l}}(u_0)(\phi([g], x) \cap \mathfrak{f})) \cap \mathfrak{u}_i$ contains non-zero vectors in $\bar{V}_+^g$ and $p_i((\mathrm{Ad}_{\mathfrak{l}}(u_0)(\phi([g], x) \cap \mathfrak{f})) \cap \mathfrak{u}_i)$ is not contained in $\bar{\phi}([g], x) = \bar{V}_-^g$. This implies that, for $g = g_0$ fixed, the subgroup of U generated by u_0 does not stabilize the essential image of $\phi([g_0], x)$. We want the same conclusion for a set of g of positive measure. We note that the spaces $(\phi([g], x) \cap \mathfrak{f}) = \mathrm{Ad}_{\mathfrak{l}}(\pi(g^{-1}))W_{\leq 0}((\mathrm{Ad}_{\mathfrak{l}} \circ \pi)_{\mathfrak{f}}(t))$ and $\bar{V}_-^g$ depend continuously on g, and that the action of u_0 via $\mathrm{Ad}_{\mathfrak{l}}$ is continuous. So there exist a small $\epsilon > 0$ such that

$$p_i((\mathrm{Ad}_{\mathfrak{l}}(u_0)(\phi([g_0], x) \cap \mathfrak{f})) \cap \mathfrak{u}_i) \not\subseteq \bar{V}_-^{g_0}$$

implies

$$p_i((\mathrm{Ad}_{\mathfrak{l}}(u_0)(\phi([g], x) \cap \mathfrak{f})) \cap \mathfrak{u}_i) \not\subseteq \bar{V}_-^g$$

for all g in $B(g_0, \epsilon)$. This immediately implies that $\phi([B(g_0, \epsilon)] \times S)$ is not stabilized by u_0 which suffices to see that u_0 is not in the stabilizer of the essential image of ϕ. □

3.5. Property T and cocycles into amenable and reductive groups. In this section we note some results for cocycles for groups with property T of Kazhdan .

THEOREM 3.14. *Let D be a group with property T of Kazhdan and A be an amenable group. Assume D acts ergodically on S preserving μ. Let $\alpha : D \times S \to A$ be a cocycle. Then α is cohomologous to a cocycle taking values in a compact subgroup of A.*

This is [**Z2**, Theorem 9.1.1]. Our first application of this result is to cocycles with reductive target. We will also require the following algebraic lemma.

LEMMA 3.15. *Let $\mathbb{L}$ be a reductive group and $p_Z : \mathbb{L} \to \mathbb{L}/Z(\mathbb{L}^0)$ the natural projection. Let $\mathbb{F} < [\mathbb{L}^0, \mathbb{L}^0]$ be a connected subgroup. Let $g \in \mathbb{L}$, and assume that $[p_Z(g), p_Z(\mathbb{F})]$ is trivial. Then $[g, \mathbb{F}]$ is also trivial.*

PROOF. The commutator $[g, \mathbb{F}]$ is contained in $Z(\mathbb{L}^0)$ by assumption. Since $[\mathbb{L}^0, \mathbb{L}^0]$ is normal in $\mathbb{L}$ it follows that $[g, \mathbb{F}] < [\mathbb{L}^0, \mathbb{L}^0]$. Therefore $[g, \mathbb{F}]$ is a connected subgroup of the finite group $Z(\mathbb{L}^0) \cap [\mathbb{L}^0, \mathbb{L}^0]$. □

THEOREM 3.16. *Let G act weakly irreducibly on S preserving μ. Let $\alpha : G \times S \to L$ be a cocycle with algebraic hull L. Assume in addition that G has property T of Kazhdan and that $\mathbb{L}$ is reductive. Then α is cohomologous to a cocycle $\beta = \pi \cdot c$. Here $\pi : G \to L$ is a continuous homomorphism and $c : G \times S \to C$ is a cocycle taking values in a compact group C centralizing $\pi(G)$.*

PROOF. Since $\mathbb{L}$ is reductive, the connected component $\mathbb{L}^0$ is reductive. This implies that $\mathbb{L}^0 = [\mathbb{L}^0, \mathbb{L}^0]Z(\mathbb{L}^0)$ where $[\mathbb{L}^0, \mathbb{L}^0]$ is semisimple, $Z(\mathbb{L}^0)$ is the center of $\mathbb{L}^0$ and $[\mathbb{L}^0, \mathbb{L}^0] \cap Z(\mathbb{L}^0)$ is finite. Since $Z(\mathbb{L}^0)$ is characteristic in $\mathbb{L}^0$, it is normal in $\mathbb{L}$ and the quotient $\mathbb{J} = \mathbb{L}/Z(\mathbb{L}^0)$ is semisimple. We let $p_J : \mathbb{L} \to \mathbb{J}$ and look at the cocycle $p_J \circ \alpha$. By Theorem 3.6, we have that $p_J \circ \alpha$ is cohomologous to a $\pi' \cdot c'$ where $\pi' : G \to J$ is a continuous homomorphism and c' is a cocycle taking values in a compact subgroup C' of J which commutes with $\pi'(G)$. By Lemma 3.4, the Zariski closure $\mathbb{J}_E$ of $\pi'(G)$ is semisimple and connected. Letting $\mathbb{J}_K$ be the Zariski closure of $c'(G \times S)$, it is clear that $\mathbb{J}$ is the almost direct product $\mathbb{J}_E \times \mathbb{J}_K$. The map $[\mathbb{L}, \mathbb{L}] \to \mathbb{J}$ is an isogeny, so again by Lemma 3.4, we can lift π' to a homomorphism π from G to $[\mathbb{L}, \mathbb{L}] < \mathbb{L}$. Then the Zariski closure of $\pi(G)$ is a connected semisimple subgroup of $\mathbb{L}$ whose projection to $\mathbb{J}$ is $\mathbb{J}_E$. Let $\mathbb{L}_K < \mathbb{L}$ be the pre-image of $\mathbb{J}_K$. Then by Lemma 3.15, the Zariski closure of $\pi(G)$ commutes with $\mathbb{L}_K$. We now see that α is cohomologous to $\pi \circ \alpha'$ where α' takes values in a subgroup H of $\mathbb{L}_K(k) < L$ and is a cocycle by Lemma 3.3. In fact H is the pre-image in $\mathbb{L}_K(k)$ of C' and so is a compact extension of $Z(\mathbb{L}^0)(k)$ and is therefore an amenable group. Since G has property T of Kazhdan, it follows from 3.14 that α' is cohomologous to a cocycle taking values in a compact group. □

Remark: In Theorem 3.16 G can be replaced by Γ. This can be proven in two ways. It follows from Theorem 3.16 exactly as Theorem 3.7 follows from Theorem 3.6. Or, using the Borel density theorem and Theorem 3.7, one can modify the proof of Theorem 3.16 to prove the same result for Γ.

3.6. Proofs of Theorems 1.4 and 1.5. We note here that Theorem 1.4 holds under weaker hypotheses. Namely, it holds for G-integrable cocycles over weakly irreducible actions for G as in Theorem 3.16. However, the existence of ergodic decompositions of measures makes the formulation of Theorem 1.4 more useful in applications.

PROOF OF THEOREM 1.4. We first verify that we can use Theorems 3.10 and 3.16 under the assumptions of Theorem 1.4.

For a proof that G has property T, we refer the reader to [**M1**, III.5] for the case where all factors of G are algebraic. When we replace some factors by their topological universal covers, the resulting G is a central extension of a T group, and the extension does not split with respect to any non-trivial subgroup of the center. It follows that G has T by an argument due to Serre presented in [**HV**, Proposition.3.d.17]. The G action is weakly irreducible since it is ergodic and G contains no rank one simple factors.

Let $L = J \ltimes U$ be a Levi decomposition as above, and $p_J : L \to J$ be the natural projection. Combining Theorem 3.10 with Theorem 3.16, we have the following:

1. a cocycle β cohomologous to α;
2. a homomorphism $\pi : G \to J < L$ such that $\pi(G)$ commutes with U
3. a cocycle $c : G \times S \to C$ where $C < Z_J(\pi(G))$ is compact
4. $p_J \circ \beta(g,x) = \pi(g)c(g,x)$

We can define a map $\tilde{\alpha}(g,x) = \pi(g)^{-1}\beta(g,x)$. That $\tilde{\alpha}$ is a cocycle follows by Lemma 3.3 since U commutes with $\pi(G)$. It suffices to show that $\tilde{\alpha}$ is cohomologous to a cocycle taking values in a compact group. But $p_J \circ \tilde{\alpha} = c$ so $\tilde{\alpha}$ takes values in $K \ltimes U$, an amenable group. Since G has property T of Kazhdan, we are done by 3.14. □

PROOF OF THEOREM 1.5. We prove Theorem 1.5 from Theorem 1.4 by inducing cocycles and actions exactly as in the proof of Theorem 3.7. It is easy to see that the induced action is ergodic. The only additional difficulty is verifying that the induced cocycle is G-integrable. If Γ is cocompact in G, the argument is straightforward. Since, G/Γ is compact, the strict cocycle $\beta : G{\times}G/\Gamma{\rightarrow}\Gamma$ can be defined using a precompact fundamental domain. This forces $\beta(g, S)$ to be finite for any $g{\in}G$. In this case it is easy to verify that $\tilde{\alpha}$ is G-integrable.

For Γ non-uniform, we need to prove that the induced cocycle is G-integrable. Since these arguments take us rather far afield, we relegate the proof to the next subsection, see Proposition 3.17. □

3.7. Integrability of Cocycles. When Γ is non-uniform, we need to be careful to verify that the induced cocycle is G-integrable before we can apply Theorem 1.4 to the induced action and cocycle. For G algebraic, the solution is the close to the content of Lewis' note [**L**]. Since Lewis' note continues to circulate in draft form and not appear and since our integrability assumption for the Γ cocycle is weaker than his and our groups G and Γ are more general than his, we give an argument here.

PROPOSITION 3.17. *Let G be as in the introduction and $\Gamma < G$ a lattice. Assume Γ acts on a standard measure space (S, μ) preserving μ. Let $\alpha : \Gamma{\times}S{\rightarrow}L$ be a Γ-integrable Borel cocycle. Then there is a choice of fundamental domain $X{\subset}G$ for Γ such that the induced cocycle $\tilde{\alpha} : G{\times}(G{\times}S)/\Gamma{\rightarrow}L$ is G-integrable.*

As noted above, the proposition is only non-trivial when Γ is non-uniform. For now we restrict our attention to the case where $G = \prod_i \mathbb{G}_i(k_i)$ and consider the case where a factor of G is the topological universal cover of $\mathbb{G}_i(\mathbb{R})$ only at the end of the proof. We will work with a fundamental domain X that is a contained in a finite union of (generalized) Siegel domains.

Recall that the choice of a fundamental domain X allows us to write any element of G as $\omega(g)\tau(g)$ where $\omega(g){\in}X$ and $\tau(g){\in}\Gamma$. We fix an embedding of each $\mathbb{G}_i(k_i)$ in $GL(n, k_i)$ and use this to define a norm. We then take the supremum norm over factors to define a norm on G. In the case where $\mathbb{G}_i(\mathbb{R})$ is replaced by it's topological universal cover, we define a norm as in [**F**], section 7.2.

PROPOSITION 3.18. *There exists a fundamental domain X for Γ in G such that:*

1. $\int_X \ln^+ \|x\| d\mu_G(x) < \infty$;
2. *for any compact set $M{\subset}G$, there is a constant C_M such that*

$$\sup_{g\in M} \|w(gx)\|{\leq}C_M\|x\|$$

for all $x{\in}X$.

Remark: Proposition 3.18 is true with no assumption on the rank of $\mathbb{G}_\alpha$. For uniform lattices, the proposition is trivial. It can easily be reduced to the case where Γ is non-uniform and irreducible. For irreducible Γ there are two cases, one where the rank of G is one and the other where the rank of G is at least two. Though we do not use the first case, the fundamental domain constructed by Garland and Ragunathan for such Γ and G can easily be seen to satisfy the proposition [**GR**]. In the second case, by the second authors arithmeticity theorems, Γ is arithmetic in G. We give a proof for the case $G = \mathbb{G}(\mathbb{R})$ and Γ a non-uniform irreducible

arithmetic lattice, in which case Γ is commensurable to $\mathbb{G}(\mathbb{Z})$. The case of non-uniform irreducible arithmetic Γ in more general G is analogous though the notation becomes more involved. We note that for irreducible Γ, Proposition 3.17 is true as long as $G{\neq}SL_2(\mathbb{R})$. This requires different estimates from the ones given below.
Remark: The careful reader may have noted that Proposition 3.18 follows from [**M3**, VIII.1.2]. Though there may exist a fundamental domain for which that proposition is correct, the proof indicated there, using a fundamental domain X contained in a finite union of Siegel sets is not. More precisely, for a fundamental domain X of this type, part a of the conclusions there is true as stated, but part b is true only if X is contained in a single Siegel domain. All applications of [**M3**, VIII.1.2.b] both in that text and in the articles [**F, L**] can be replaced by the estimate in part 2 of Proposition 3.18.

Proof of proposition 3.18. Recall that we are proving the proposition in the case where $G = \mathbb{G}(\mathbb{R})$ and Γ commensurable to $\mathbb{G}(\mathbb{Z})$. Let T be a maximal $\mathbb{Q}$ split torus in G. Let F be a root system for G with respect to A, and let Δ be a set of simple roots. Let $A = \{a{\in}A|\alpha(a){\leq}1\forall\alpha{\in}\Delta\}$ be a Weyl chamber. By standard reduction theory, we can assume there is a finite set $F{\in}\mathbb{G}(\mathbb{Q})$ and a bounded set $L{\subset}G$ such that $X{\subset}LAF$. The first conclusion of the proposition is standard, and proofs can be found in [**B1**] or [**PlRa**]. The proof of the second conclusion depends on standard facts from reduction theory. Our principle reference for these facts is [**B1**], particularly Section 14.4. Though the discussion there is restricted to real groups, analogous statements are known for more general G as above.

Let $l = \dim(A)$. As in [**B1**, Section 14], we can find a finite collection $\rho_1, \cdots, \rho_l$ of $\mathbb{Q}$ representations from $\mathbb{G}$ into $GL(W_i)$ and vectors $w_i{\in}W_i(\mathbb{Z})$ such that

$$\mathbb{P} = \{g{\in}\mathbb{G}|\rho_i(g)w_i = \chi_i(g)w_i\}$$

where $\chi_i : \mathbb{P}{\to}\mathbb{R}^*_{\geq 0}$ restricted to the split torus $\mathbb{A}$ is the highest weight of ρ_i and $\chi_i|_{\mathbb{A}} = d_i\alpha_i$ for some simple root α_i and an integer $d > 0$. This implies that for any other weight $\chi{\neq}\chi_i$ of ρ_i we have $|\chi(s)| > |\alpha_i(s)|^{-d}|\chi_i(s)|$ where $d > 0$ is an integer.

Given two real valued functions f and g on x, we write $f \prec g$ if $f(x){\leq}Cg(x)$ and $f \asymp g$ if $f \prec g$ and $g \prec f$.

Given $x{\in}X$, we write $x = laf$. We take a compact set $M{\subset}G$ and $x{\in}LA_1F$. We write $gx = glaf = l'a'f'\gamma$ for any $g{\in}M$ where $\gamma = \tau(gx)^{-1}$. It follows that $l'a'f' = \omega(gx)$. Since L is compact and F is finite, to prove the proposition it suffices to show that $\|a\| \asymp \|a'\|$. Since χ_i for $1{\leq}i{\leq}l$ form a basis for $X(T)\bigotimes\mathbb{R}$, it is enough to show that $|\chi_i(a)| \asymp |\chi_i(a')|$ for all i.

We apply gx to $f^{-1}w_i$ for each ρ_i. Then $\|(glaf)(f^{-1}w_i)\| = \|glaw_i\|{\asymp}|\chi_i(a)|$. We also have $\|(glaf)(f^{-1}w_i)\| = \|(l'a'f'\gamma)(f^{-1}w_i)\|$. If $f'\gamma f^{-1}w_i$ is proportional to w_i we have $\|(l'a'f'\gamma)(f^{-1}w_i)\|{\asymp}|\chi_i(a')|$ and $\|(l'a'f'\gamma)(f^{-1}w_i)\|{\succ}|\alpha_i(a')|^{-d}|\chi_i(a')|$ otherwise. Therefore $|\chi_i(a')| \prec |\chi_i(a)|$. Replacing $f^{-1}w_i$ with $(f'\gamma)^{-1}w_i$ and arguing in the same manner yields $|\chi_i(a)| \prec |\chi_i(a')|$. Therefore we have $|\chi_i(a)| \asymp |\chi_i(a')|$ which suffices to prove the proposition. (We note that the constants implicit in the signs $\asymp$ and $\prec$ used here depend on the compact set M.) □

Corollary 3.19. *There exists a fundamental domain X for Γ in G such that for any compact set $M{\subset}G$ there is a constant C_M such that*

$$\ln^+\|\beta(g,x)\|{\leq}C_1(M)\ln^+\|x\| + C_2(M)$$

for all x in X.

PROOF. We write $gx = \omega(gx)\tau(gx)$ and recall that $\beta(g,x) = \tau(gx)^{-1}$ is the strict cocycle $\beta : G \times G/\Gamma \to \Gamma$ corresponding to X. This implies that $\beta(g,x) = x^{-1}g^{-1}\omega(gx)$. Therefore

$$\ln^+ \|\beta(g,x)\| \leq \ln^+ \|x^{-1}\| + \ln^+ \|g^{-1}\| + \ln^+ \|\omega(gx)\|.$$

Since G is algebraic there exist positive constants c and d such that $\|x^{-1}\| < c\|x\|^d$. Combined with Proposition 3.18 this implies that the previous equation can be rewritten

$$\ln^+ \|\beta(g,x)\| \leq dC_M \ln^+ \|x\| + \ln^+(c) + \ln^+ \|g\|.$$

Letting $C_1(M) = dC_M$ and $C_2(M) = \text{supp}_{g \in M} \ln^+ \|g\| + \ln^+(c)$, we are done. □

PROOF OF PROPOSITION 3.17. We have assumed that the cocycle α is Γ-integrable. If we let K be a finite generating set for Γ, and K^j the set of all words in S of length less than j, then Γ-integrability implies that $\int_S \ln^+ \|\alpha(\gamma, x)\| \leq Cj$ almost everywhere for some constant C (not depending on j) and all $\gamma \in K^j$. Recall that the cocycle $\tilde{\alpha}$ is defined by $\tilde{\alpha}(g, [g_0, s]) = \alpha(\beta(g, [g_0]), s)$. To show that $\tilde{\alpha}$ is quasi-integrable, it therefore suffice to show that the word length of $\beta(g, [g_0])$ is an L^1 function on G/Γ. We choose a fundamental domain to define β as in Proposition 3.18. For such a domain it follows that $\ln^+ \|x\|$ is in $L^1(X)$. In what follows, we identify G/Γ with the fundamental domain X and consider $\beta : G \times X \to \Gamma$ written as $\beta(g,x)$. To finish the argument, one then uses a theorem of Lubotzky, Mozes and Ragunathan. Define a distance function on G by choosing a right G and left K invariant Riemannian metric on G. Then Lubotzky, Mozes, and Ragunathan show that the word length metric on Γ is bilipschitz equivalent to the induced metric as a subset of G [**LMR**]. This result, combined with a simple computation [**F**, Proof of Proposition 7.9], shows that the word length of $\beta(g,x)$ is bounded by a multiple of $\ln^+ \|\beta(g,x)\|$ plus a constant. One then applies Corollary 3.19 to see that $\ln^+ \|\beta(g,x)\| < C_1(M) \ln^+ \|x\| + C_2(M)$ for any $g \in M$ where $M \subset G$ is pre-compact.

Letting $M \subset G$ be any pre-compact set, writing $\|\gamma\|_\Gamma$ for the word length of γ and collecting inequalities, we have:

$$\begin{aligned}
\int_{S \times X} Q_{M,\tilde{\alpha}}(s,x) &= \int_{S \times X} \text{supp}_{g \in M} \tilde{\alpha}(g,s,x) \\
&\leq C \int_X \text{supp}_{g \in M} \|\beta(g,x)\|_\Gamma \\
&\leq C' \int_X \text{supp}_{g \in M} \ln^+ \|\beta(g,x)\| + B \\
&\leq C_1(M)C' \int_X \ln^+ \|x\| + B + C_2(M).
\end{aligned}$$

This shows that $\tilde{\alpha}$ is G-integrable whenever G is an algebraic group.

When G or a simple factor of G is not algebraic, we need to extend the reduction theory arguments to fundamental domains for G/Γ in this setting. This is done at the end of section 7.2. of [**F**]. A key step in the argument there is showing that $\Gamma \cap Z(G) < Z(G)$ is a subgroup of finite index. This allows one to choose a fundamental domain for Γ in G which is contained in a finite union of connected components of pre-images of Siegel sets. In this context there is a choice of the norm on G, since G is not a linear group. For the choice made in [**F**], the extension of the results in [**LMR**] is obvious. We note that some statements in [**F**] inherit the inaccuracy of [**M3**, VIII.1.2]. All of these inaccuracies can be fixed easily using Proposition 3.18 above. □

3.8. Uniqueness of the superrigidity homomorphism. In this section, we show that the homomorphism π appearing in the formulation of Theorems 1.4 and 1.5 is unique up to conjugacy. In fact, we prove a more general fact that requires no assumption on the rank of G. In this subsection, G will be as in subsection 3.1, but with no assumption on the rank of G, but still assuming G has no compact factors. As usual, $\Gamma < G$ is a lattice.

THEOREM 3.20. *Let $D = G$ or Γ. Assume D acts on S preserving μ. For $j = 1, 2$, let $\pi_j : G \to L$ be continuous homomorphism, let $Z_k = Z_L(\pi_j(G))$ and let $c_j : D \times S \to C_j$ be a cocycle over the D action taking values in a compact subgroup $C_j < Z_j$. Let $\alpha_j : D \times S \to L$ be the cocycle over the D action defined by $\alpha_j(d, x) = \pi_j(d)c_j(d, x)$. Then if α_1 is cohomologous to α_2, the homomorphism π_1 and π_2 are conjugate.*

Before proving the theorem, we recall some terminology and notation. For any element g of $GL_n(\mathbb{R})$, there is a unique decomposition of $g = us = su$ where u is unipotent and s is semisimple. Further, we have a unique decomposition $s = cp = pc$ where all eigenvalues of p are positive and all eigenvalues of c have modulus one. We refer to p as the *polar part* of g and denote it by $\mathrm{pol}(g)$. For any subset $\Omega \subset GL_n(\mathbb{R})$ we define

$$P(\Omega) = \{\,\mathrm{pol}(h) : h \in \Omega\}.$$

In general $P(\Omega)$ is not a subset of Ω, but if Ω is a semisimple subgroup without compact factors, then the Zariski closure of $P(\Omega)$ is Ω.

For non-Archimedean fields k, the situation is more complicated. We fix a uniformizer π for the field k and define a *polar element* of $GL_n(k)$ to be an element p all of whose eigenvalues are powers of π. We call an element c in $GL_n(k)$ *compact* if c generates a bounded subgroup in $GL_n(k)$. Each element of $GL_n(k)$ can be written uniquely as su where u is unipotent and s is semisimple. We call an element g *quasi-polar* if $g = su$ as above and s can be written as $s = cp = pc$ where p is polar and c is compact. Once π is fixed, this decomposition is unique. We note that if s is semisimple then $s^{n!}$ is quasi-polar. As above for h quasi-polar, we denote the polar part by $\mathrm{pol}(h)$. For a subset Ω in $GL_n(k)$, we define

$$P(\Omega) = \{\mathrm{pol}(h) : h \in \Omega \text{ and } h \text{ quasi-polar}\}.$$

As before, in general $P(\Omega)$ is not a subset of Ω, but if Ω is a semisimple subgroup without compact factors, then the Zariski closure of $P(\Omega)$ is Ω.

Recall from subsection 3.4 that if M is a linear transformation and we let $\Omega(M)$ be the set of all eigenvalues of M, we call the numbers $d = |\lambda|$ for $\lambda \in \Omega(M)$ characteristic numbers of M. If $W_\lambda(M)$ is the eigenspace corresponding to $\lambda \in \Omega(M)$, we let $W_d(M) = [\oplus_{\ln|\lambda| = d} W_\lambda(M)]_k$ be the characteristic subspace of M with characteristic number d.

Remark: The key fact about polar elements is that a polar element is completely determined by it's characteristic numbers and subspaces. We note that under any rational homomorphism $\pi : \mathbb{G}(k) \to GL_n(k)$ the image of a polar element is a polar element. In fact, for g quasi-polar, $\mathrm{pol}(\pi(g)) = \pi(\mathrm{pol}(g))$. This implies that we can define polar and quasi-polar elements of a linear algebraic group $\mathbb{G}$ and that the definition is independent of the realization $\mathbb{G}$. However, the set of polar elements of $\mathbb{G}$ does depend on the choice of uniformizer for k.

For each $i \in I$ we fix an almost faithful representation of $\mathbb{G}_i(k_i)$ in $GL_n(k)$. We will call an element g of G polar if $\tau_i(g)$ is polar whenever it is non-trivial. We call a subgroup $F < G$ *Zariski dense* if $\tau_i(F)$ is Zariski dense in $\mathbb{G}_i$ for each i.

LEMMA 3.21. *There exists a finite collection of quasi-polar element $g_1, \dots, g_l \in \Gamma$ such that the group F generated by $pol(g_1), \dots, pol(g_l)$ is Zariski dense in G.*

PROOF. This is similar to [**MQ**, Lemma 4.5]. Let Z be the Zariski closure of $< \mathrm{pol}(\gamma) | \gamma \in \Gamma >$. The proof follows from the fact Z is invariant under conjugation by elements of Γ and so also by elements of G by the Borel-Wang density theorem [**M3**, Theorem II.4.4]. Hence $G \cap Z$ is a normal subgroup of G. That it is all of G follows from results of Mostow and Prasad-Ragunathan [**M, PR**] which show that there is a maximal split torus $A < G$ such that $A \cap \Gamma$ is a lattice in A. This implies that $A \cap \Gamma$ projects to a non-compact subgroup of semisimple elements of each simple factor of G.

Since $< \mathrm{pol}(\gamma) | \gamma \in \Gamma >$ is Zariski dense in G and algebraic groups satisfy an ascending chain condition, it follows that there is a finite collection $\gamma_1, \dots, \gamma_l$ such that $< \gamma_1, \dots, \gamma_l >$ is Zariski dense in G.

Though the results in [**M, PR**] are only stated for the case of real algebraic G, the interested reader may generalize the proof of [**PR**] to the more general G considered in the statement of our theorems. □

PROOF OF THEOREM 3.20. If $D = \Gamma$, we apply Lemma 3.21 to obtain the group F and $g_1, \dots, g_l$. If $D = G$, there exists a Zariski dense finitely generated subgroup F generated by polar elements $g_1, \dots, g_l$. In either case, we fix F and $g_1, \dots, g_l$ for the remainder of the proof.

We define associated actions of D on $S \times \mathfrak{l}$ by $\rho_j(d)(x, v) = (dx, \mathrm{Ad}_{\mathfrak{l}} \circ \alpha_j(d, x))$. Oseledec' multiplicative ergodic theorem implies that there are characteristic exponents and subspaces for any $\mathbb{Z}$ action defined by powers of an element d in D. Let $\psi : S \to L$ be the measurable function such that $\alpha_1(d, x) = \psi(dx)^{-1} \alpha_2(d, x) \psi(x)$. It follows easily from Lemma 3.12 that for any $d \in D$ the characteristic numbers for $\rho_2(d)$ and $\rho_1(d)$ are equal. In fact, that lemma shows that if λ is a characteristic number for $\rho_1(d)$ with characteristic subspace ${}^1W^d_\lambda(x)$ then $(\psi(x)^{-1})\,{}^1W^d_\lambda(x)$ is a characteristic subspace for $\rho_2(d)$ with characteristic number λ. Since each α_j is the product of a constant cocycle and a compact valued cocycle, it follows from the discussion immediately proceeding the proof of Theorem 3.11 that the characteristic numbers and subspaces for $\rho_i(d)$ are just the characteristic numbers and subspaces for the linear representation $\mathrm{Ad}_{\mathfrak{a}} \circ \pi_j$. This implies that ${}^jW^d_\lambda(x) = W_\lambda(\pi_j(d))$. Combining these two facts, we see that $\psi(x) W_\lambda(\pi_1(d)) = W_\lambda(\pi_2(d))$ for every d, every λ, and almost every x.

We let $\{{}^j\lambda^m_k\}$ be the characteristic numbers of $\pi_j(g_m)$ for each $1 \le m \le l$. Then by the argument above, we see that ${}^1\lambda^m_k = {}^2\lambda^m_k$ and also that $\psi(x) W_{{}^1\lambda^m_k}(\pi_1(g_m)) = W_{{}^2\lambda^m_k}(\pi_2(g_m))$ for almost every x and $1 \le m \le l$. Fixing x for which the equation holds for all m, and letting $l = \psi(x)$, the definition of the polar part of an element implies that $\mathrm{pol}(\pi_1(g_m)) = l^{-1} \mathrm{pol}(\pi_2(g_m)) l$. This implies that $(\pi_1(\mathrm{pol}(g_m))) = l^{-1}(\pi_2(\mathrm{pol}(g_m))l$. Since the group F generated by $\mathrm{pol}(g_1), \dots, \mathrm{pol}(g_l)$ is Zariski dense in G and since Lemma 3.4 implies each π_j factors through a rational homomorphism of some $\mathbb{G}_i$ this implies that $\pi_1 = l^{-1} \pi_2 l$. □

Remark: If one of the cocycles α_j is simply a homomorphism, it is possible to give a simpler proof based on the Borel density theorem.

3.9. Cocycles with prescribed projections. In this subsection we state and prove some variants on the cocycle superrigidity theorems. Though these variants hold in many settings, we only state variants of Theorems 1.4 and 1.5. Therefore, throughout this subsection G will be as in the introduction, i.e. with the assumption that G has no rank 1 simple factors. The variants stated below are needed for our applications to local rigidity of affine action and are used in the proof of Theorem 5.1, which in turn is used to prove Theorem 1.8.

Throughout this subsection $\mathbb{A}$ and $\mathbb{H}$ will be algebraic k-subgroups of $\mathbb{L}$ such that $\mathbb{L} = \mathbb{A}\ltimes\mathbb{H}$. We further assume that $\mathbb{A}$ is a connected semisimple k-group. We will denote the k points $\mathbb{H}(k) = H$ and $\mathbb{A}(k) = A$. We fix homomorphisms $\pi_A^G : G \to A$ and $\pi_A^\Gamma : \Gamma \to A$ with Zariski dense image. We let $p_A : L \to A$ denote the standard projection.

THEOREM 3.22. *Assume G acts ergodically on S preserving μ. Let $\alpha : G\times S \to L$ be a G-integrable Borel cocycle such that $p_A \circ \alpha = \pi_A^G$. Further assume that L is the algebraic hull of the cocycle. Then there is a measurable map $\phi : S \to H$ such that $\beta = \phi(gx)^{-1}\alpha(g,x)\phi(x) = \pi(g)c(g,x)$ where $\pi : G \to L$ is a continuous homomorphism and $c : G\times S \to C$ is a cocycle taking values in a compact subgroup $C < Z_L(\pi(\Gamma))$. The fact that $\phi(S) \subset H$ implies that $p_A \circ \beta = p_A^\Gamma$.*

THEOREM 3.23. *Assume Γ acts ergodically on S preserving μ. Let $\alpha : \Gamma\times S \to L$ be a Γ-integrable, Borel cocycle such that $p_A \circ \alpha = \pi_A^\Gamma$. Further assume L is the algebraic hull of the cocycle. Then there is a measurable map $\phi : S \to H$ such that $\beta = \phi(\gamma x)^{-1}\alpha(\gamma,x)\phi(x) = \pi(\gamma)c(\gamma,x)$ where $\pi : G \to L$ is a continuous homomorphism and $c : \Gamma\times S \to C$ is a cocycle taking values in a compact subgroup $C < Z_L(\pi(\Gamma))$. The fact that $\phi(S) \subset H$ implies that $p_A \circ \beta = p_A^\Gamma$.*

These variants are proven from Theorems 1.4 and 1.5 using Theorem 3.20, the following lemma and some facts about the structure of algebraic groups.

LEMMA 3.24. *Let D, R, F and A be groups. Assume $A\times F$ acts on R by a (possibly trivial) homomorphism into $Aut(R)$. We will write an element $g \in (A\times F)\ltimes R$ as (g_A, g_F, g_R) where $g_A \in A, g_F \in F$ and $g_R \in R$. Let $\pi_A : D \to A$ be a homomorphism and p_A the projection of from $(A\times F)\ltimes R$ to A. Let D act on a set X. Let $\alpha : D\times X \to (A\times F)\ltimes R$ be a cocycle over the D action on X. Assume:*

1. *$p_A \circ \alpha(d,x) = \pi_A(d)$;*
2. *Let $C < R$ be a subgroup such that A and F normalize C;*
3. *α is cohomologous to a cocycle β taking values in $(A\times F)\ltimes C$.*

Assume $\lambda(x) = (\lambda_A(x), \lambda_F(x), \lambda_R(x))$ is the function such that

$$\lambda(dx)^{-1}\alpha(d,x)\lambda(x) = \beta(d,x)$$

Let $\lambda'(x) = (1_A, \lambda_F(x), \lambda_R(x))$. Then

$$\lambda'(gx)^{-1}\alpha(g,x)\lambda'(x) = (\pi(d), \beta_F(d,x), \beta'_R(d,x)$$

where $\beta'_R(d,x) = {}^{\lambda_A(dx)}\beta_R(d,x)$ and $\beta'_R(D\times X) \subset C$.

PROOF. This is checked directly by multiplication. We write the multiplication in $(A\times F)\ltimes R$ as $(g_A, g_F, g_R)(h_A, h_F, h_R) = (g_A h_A, g_F h_F, g_R{}^{(g_A, g_F)}(h_R))$ where

${}^{(g_A,g_F)}(h_R)$ is the image of h_R under the automorphism of R given by (g_A, g_F). Then $\lambda'(dx)^{-1}\alpha(d,x)\lambda'(x) = \lambda_A(dx)\beta(d,x)\lambda_A(x)^{-1}$. Then

$$(\lambda_A(dx), 1_F, 1_R)(\beta_A(d,x), \beta_F(d,x), \beta_R(d,x))(\lambda_A(x)^{-1}, 1_F, 1_R)$$
$$= (\lambda_A(dx)\beta_A(d,x)\lambda_A(x)^{-1}, \beta_F(d,x), {}^{\lambda_A(dx)}\beta_R(d,x)).$$

Defining $\beta'(d,x) = {}^{\lambda_A(dx)}\beta_R(d,x)$, since $\beta_R(D\times X)\subset C$ and A normalizes C, we have $\beta'(D\times X)\subset C$. □

PROOF OF THEOREM 3.22. We can write $\mathbb{H} = \mathbb{F}\ltimes\mathbb{U}$ where $\mathbb{F}$ is semisimple and $\mathbb{U}$ is solvable. We first show that we can change $\mathbb{A}$ so that it commutes with $\mathbb{F}$. If not, then since $\mathbb{A}$ is connected and semisimple $\mathbb{A}$ acts on $\mathbb{F}$ by inner automorphisms and therefore $\mathbb{A}$ is an almost direct product $\mathbb{A}_1\mathbb{A}_2$ where $\mathbb{A}_1$ commutes with $\mathbb{F}$ and $\mathbb{A}_2$ is (virtually) a subgroup of $\mathbb{F}$. Let $\Delta^{-1}(\mathbb{A}_2)$ be the antidiagonal embedding of $\mathbb{A}_2$ in $\mathbb{A}\ltimes\mathbb{F}$. We replace $\mathbb{A}$ by $\mathbb{A}' = \mathbb{A}_1\Delta^{-1}(\mathbb{A}_2)$. A simple computation shows that $\mathbb{F}$ and $\mathbb{A}'$ commute in the semidirect product $\mathbb{A}'\ltimes\mathbb{H}$. In what follows we replace $\mathbb{A}$ by $\mathbb{A}'$.

If we apply Theorem 1.4 to α we see that α is cohomologous to a cocycle β where β is of the form $\pi\cdot c$ for a homomorphism $\pi : G\to L$ and a cocycle $c : G\times S\to K$ where $K < L$ is compact. We let $C = K\cap\mathbb{U}(k)$. It follows from Theorem 3.20 that $p_A\circ\beta$ is cohomologous to a homomorphism conjugate to π_A^G and that A is the algebraic hull of $p_A\circ\alpha$. Since the algebraic hull of α contains the Zariski closure of C, it follows from Theorem 3.10 that $\mathbb{A}$ and C commute.

We now apply Lemma 3.24 with $A = A$, $F = \mathbb{F}(k), R = \mathbb{U}(k)$ and $C = C$. □

PROOF OF THEOREM 3.23. In the case when π_A^Γ is the restriction to Γ of a continuous homomorphism of G, the proof above applies verbatim. Though Theorem 3.10 is only stated for G actions and cocycles, the analogous statement for Γ actions and cocycles is easily proven by inducing cocycles and actions.

When $k = \mathbb{R}$ and π_A^Γ does not extend, the argument is even simpler. We let $\mathbb{H} = \mathbb{F}\ltimes\mathbb{U}$ where $\mathbb{F}$ is reductive and $\mathbb{U}$ is unipotent. As above we modify $\mathbb{A}$ so that $\mathbb{A}$ commutes with $\mathbb{F}$. This is possible with $\mathbb{F}$ reductive since $\mathbb{A}\ltimes\mathbb{F}$ is reductive and therefore $\mathbb{A}$ commutes with the torus in $\mathbb{A}\ltimes\mathbb{F}$ which contains the torus in $\mathbb{F}$. We let $F = \mathbb{F}(\mathbb{R})$ and $U = \mathbb{U}(\mathbb{R})$. It follows from Theorem 1.5 and the fact that U contains no compact subgroups that $\beta(g,x)$ takes values in $A\times F$. The theorem now follows from Lemma 3.24 applied to $A, F, R = U$ and $C = 1_U$.

The case of k non-Archimedean and π_A^Γ not extendable is the most complicated and the only place where we need the full strength of both Theorem 3.20 and Lemma 3.24. Let $\mathbb{H} = \mathbb{F}\ltimes\mathbb{U}$ and $\mathbb{L} = (\mathbb{A}\times\mathbb{F})\ltimes\mathbb{U}$ as in the proceeding paragraph. Then $L = (A\times F)\ltimes U$ where $U = \mathbb{U}(k)$ and $F = \mathbb{F}(k)$. We can write $\mathbb{A}$ and A as almost direct products $\mathbb{A} = \mathbb{A}_1\mathbb{A}_2$ and $A = A_1A_2$ where $A_1 = \mathbb{A}_1(k), A_2 = \mathbb{A}_2(k)$ and $p_{A_1}\circ\pi_A$ extends to a continuous homomorphism of G and $p_{A_2}\circ\pi_A$ has bounded image. By Theorem 3.10 applied as above, $\mathbb{A}_1$ commutes with $\mathbb{U}$.

Note that $\mathbb{U}(k)$ is the injective limit of it's compact subgroups. This follows from the same fact for the additive group $\mathbb{Q}_p$, where the compact subgroups are subgroups with bounded denominators. Apply Theorem 1.5 to α to obtain a cocycle β cohomologous to α, where $\beta(\gamma,x) = \pi(\gamma)c(\gamma,x)$ where π is a continuous homomorphism of G and c is a cocycle taking values in a compact subgroup $C < Z_L(\pi(G))$. Write $\beta(\gamma,x) = (\beta_A(\gamma,x), \beta_F(\gamma,x), \beta_U(\gamma,x))$ using coordinates as in Lemma 3.24. Let C_1 be the smallest compact subgroup of U containing $\beta_U(\Gamma\times S)$. Note that A_1 commutes with C_1.

Let $p_A \circ \beta = \beta_A$. This is cohomologous to $(p_A \circ \pi) \cdot (p_A \circ c)$. By Theorem 3.20, $p_A \circ \pi$ is conjugate to π_A^E by an element $a \in A$. Let β' be the conjugate of β by a. Writing $\beta'(\gamma, x) = (\beta'_A(\gamma, x), \beta'_F(\gamma, x), \beta'_U(\gamma, x))$ as above, it is clear that $\beta'_F = \beta_F$ and that $\beta'_U(\Gamma \times X) \subset C_2 = {}^a C_1$ and that C_2 is a compact subgroup of U. Note that A_1 commutes with C_2. It also follows that $\beta'_A(\gamma, x) = \pi_A^E(\gamma) c_A(\gamma, x)$ where $c_A(\Gamma \times X) \subset C_A$ where $C_A < A_2$ is a compact subgroup and $\pi_A^E(\Gamma)$ is contained in A_1. We note that C_A acts on U by automorphisms, and let K be the set of all images of C_2 under the action of C_A. It is clear that K is compact and is the union of subgroups of U. Since U is the projective limit of it's compact subgroups, there is a compact subgroup $C < U$ with $K \subset C$. We now apply Lemma 3.24 with $A = A, F = F, R = U$ and $C = C$. □

4. Orbits in the space of representations

In this section we prove an independent result that is used in the proof of our results on local rigidity of constant cocycles. This result appears to be known, but we include a proof for completeness. In this section D will be any finitely presented group and $H = \mathbb{H}(k)$ will be the k points of a algebraic k-group $\mathbb{H}$ where k is a local field of characteristic 0. We will fix a realization $\mathbb{H} < GL(W)$. We will call a homomorphism ρ from a group D to $\mathbb{H}$ completely reducible if the representation on W given by $\rho(D) < \mathbb{H} < GL(W)$ is completely reducible. We let $\mathrm{Hom}(D, \mathbb{H})$ be the space of homomorphisms of D into $\mathbb{H}$ which has a natural structure as an algebraic subvariety of $\mathbb{H}^m$ where m is the number of generators of D. We note that the structure of $\mathrm{Hom}(D, \mathbb{H})$ as a variety is independent of the presentation of D and that $\mathrm{Hom}(D, H)$ is the set of k-points of the variety. Note that $\mathbb{H}$ (resp. H) acts on the space $\mathrm{Hom}(D, \mathbb{H})$ (resp. $\mathrm{Hom}(D, H)$) by conjugation.

THEOREM 4.1. *Let D and H be as above. Let $\pi : D \to H$ be any completely reducible homomorphism. Then*

1. *the $\mathbb{H}$ orbit of π in* $\mathrm{Hom}(D, \mathbb{H})$ *is Zariski closed and*
2. *the H orbit of π in* $\mathrm{Hom}(D, H)$ *is Hausdorff closed.*

Point (2) of Theorem 4.1 follows from point (1) and a result of Bernstein and Zelevinsky [**BZ**]. The result of Bernstein and Zelevinsky says that, given an action of a k group G on a k variety V, the k points of any orbit are locally closed in the Hausdorff topology. An examination of the proof shows that for Zariski closed orbits, the orbit is also Hausdorff closed. For an accessible proof in characteristic zero see [**AB**, proof of Theorem 6.1]. Let K be the algebraic closure of k and consider $\mathbb{H} < GL(W)$ where $W = K^n$. We prove part 1 of Theorem 4.1 from the following result.

THEOREM 4.2. *Let D be as above and let $\pi : D \to GL(W)$ be any completely reducible representation. Then the $GL(W)$ orbit of π in* $\mathrm{Hom}(D, GL(W))$ *is Zariski closed.*

PROOF OF THEOREM 4.2. We let $K[D]$ be the group ring of D. The representation π defines a representation $\tilde{\pi}$ of $K[D]$. This representation factors through a finite dimensional quotient $A = K[D]/\ker(\tilde{\pi})$ and since π is completely reducible A is a semisimple algebra. (See for example [**La**, XVII.6], particularly Theorem 6.1.)

If $\tilde{\pi}'$ is in the closure of the $GL(W)$ orbit of π then $\tilde{\pi}'$ also vanishes on $\ker(\tilde{\pi})$ and so $\tilde{\pi}'$ is also a representation of A. We recall that two representations of a semisimple algebra A are conjugate if and only if they have the same character, see for example [**La**, Theorem XVII.3.8]. This implies that the $GL(W)$ orbit of π is closed. □

The proof of Theorem 4.1 is modelled on the proof that conjugacy classes of semisimple elements in algebraic groups are closed.

PROOF OF (1) IN THEOREM 4.1. The space $\mathrm{Hom}(D,\mathbb{H})$ is an algebraic variety over k. Assume that the $\mathbb{H}$ orbit of π is not closed. Then the orbit closure $\mathrm{Zar}(\mathbb{H}\cdot\pi)$ is the union of $\mathbb{H}\cdot\pi$ with a collection of subvarieties of strictly smaller dimension.

Given any homomorphism $\sigma : D \to G$ where G is any group, the orbit of σ in $\mathrm{Hom}(D,G)$ is naturally identified with $G/Z_G(\sigma(D))$. So given a homomorphism $\bar{\pi} \in \mathrm{Zar}(\mathbb{H}\cdot\pi)\backslash(\mathbb{H}\cdot\pi)$, we have that $\dim(Z_{\mathbb{H}}(\bar{\pi}(D))) > \dim(Z_{\mathbb{H}}(\pi(D)))$. We now show that $\dim(Z_{\mathbb{H}}(\bar{\pi}(D))) = \dim(Z_{\mathbb{H}}(\pi(D)))$ for any completely reducible homomorphisms π and $\bar{\pi}$ with $\bar{\pi} \in \mathrm{Zar}(\mathbb{H}\cdot\pi)$.

Let $\mathfrak{h}$ be the Lie algebra of $\mathbb{H}$ and $\mathrm{Ad}_{\mathfrak{h}}$ the adjoint representation of $\mathbb{H}$ on $\mathfrak{h}$. Let $\mathfrak{z}(\pi)$ (resp. $\mathfrak{z}(\bar{\pi})$) be the $\mathrm{Ad}_{\mathfrak{h}} \circ \pi(D)$ (resp. $\mathrm{Ad}_{\mathfrak{h}} \circ \bar{\pi}(D)$) invariant vectors in $\mathfrak{h}$. Since the characteristic of K is zero, we have that $\mathfrak{z}(\pi)$ is the Lie algebra of $Z_{\mathbb{H}}(\pi(D))$ and $\mathfrak{z}(\bar{\pi})$ is the Lie algebra of $Z_{\mathbb{H}}(\bar{\pi}(D))$ (see [**B2**], II.7). By construction $\mathrm{Ad}_{\mathfrak{h}} \circ \bar{\pi}$ is in the closure of the $\mathbb{H}$ orbit of $\mathrm{Ad}_{\mathfrak{h}} \circ \pi$ in $\mathrm{Hom}(D, GL(\mathfrak{h}))$. By Theorem 4.2 $\mathrm{Ad}_{\mathfrak{h}} \circ \bar{\pi}$ is conjugate to $\mathrm{Ad}_{\mathfrak{h}} \circ \pi$ by an element of $GL(\mathfrak{h})$ and therefore $\dim(\mathfrak{z}(\pi)) = \dim(\mathfrak{z}(\bar{\pi}))$. This implies that the $\mathbb{H}$ orbit of π is closed in the Zariski topology on $\mathrm{Hom}(D,\mathbb{H})$. □

Theorem 4.1 would suffice to prove Theorems 1.1 and 1.2. However, to prove Theorem 5.1, we need the following.

COROLLARY 4.3. *Let $\mathbb{L} = \mathbb{A} \ltimes \mathbb{H}$ where all groups are k-algebraic. Let D be a finitely generated group. If $H = \mathbb{H}(k)$ and $L = \mathbb{L}(k)$, then H orbits of completely reducible homomorphisms in* $\mathrm{Hom}(D, L)$ *are Hausdorff closed.*

PROOF. It follows from part 1 of Theorem 4.1 that $\mathbb{L}$ orbits in $\mathrm{Hom}(D,\mathbb{L})$ are Zariski closed. We let U be an $\mathbb{L}$ orbit in $\mathrm{Hom}(D,\mathbb{L})$. Then for any $u \in U$, then $\mathbb{L}u = \cup_{a \in \mathbb{A}} a\mathbb{H}u$. Since $\mathbb{H}$ is normal in $\mathbb{L}$, $a\mathbb{H}u = \mathbb{H}au$. The $\mathbb{H}$ action on U is algebraic, so the closure of an orbit must consist of the orbit plus sets of strictly lower dimension. Since all the sets $a\mathbb{H}u$ have the same dimension, this means that the closure of each $\mathbb{H}au$ in $\mathbb{L}u$ is $\mathbb{H}au$. Since $\mathbb{L}u$ is Zariski closed, so is $\mathbb{H}au$ and in particular $\mathbb{H}u$. That Hu is Hausdorff closed follows from the (proof of) the result of Bernstein-Zelevinsky as in the proof of Theorem 4.1. □

5. Proof of Local Rigidity for Cocycles

In this section we prove Theorems 1.1 and 1.2. Actually we prove Theorem 5.1 below which implies Theorems 1.1 and 1.2. As mentioned in the introduction Theorem 5.1 is need to prove Theorem 1.8. We first fix some notations for the entire section. In the first subsection we formulate Theorem 5.1 and prove the theorem modulo a result on perturbations of cocycles taking values in compact groups. This last result Theorem 5.4, which holds for any compact valued cocycle over an action of a group with property T, is proven in the second subsection.

5.1. Theorem 5.1 and proof. We will let k be a local field of characteristic zero, $\mathbb{L}$ an algebraic k-group and $\mathbb{H}, \mathbb{A} < \mathbb{L}$ k-algebraic subgroups such that $\mathbb{L} = \mathbb{A} \ltimes \mathbb{H}$. We further assume that $\mathbb{A}$ is a connected semisimple k-group. We let L, A and H denote the k points of $\mathbb{L}, \mathbb{A}$ and $\mathbb{H}$ respectively. We will denote by D a group that is either G or Γ where G and Γ are as in the introduction. If π_1 and π_2 are two homomorphisms of a group B into a group C whose images commute, we will denote by (π_1, π_2) the homomorphism defined by $(\pi_1, \pi_2)(b) = (\pi_1(b), \pi_2(b))$ for all $b \in B$. We also fix a continuous homomorphism $\pi_0 : D \to L$ such that $\pi_0 = (\pi_A, \pi_H)$ where $\pi_A : D \to A$ and $\pi_H : D \to H$ are continuous homomorphisms such that $\pi_A(D)$ and $\pi_H(D)$ commute in L. If $D = \Gamma$, recall that π_0 is superrigid, which means that $\pi_0 = (\pi_0^E, \pi_0^K)$ where π_0^E is the restriction to Γ of a continuous homomorphism from G to H, and π_0^K is homomorphism from Γ to H with bounded image, and $\pi_0^E(\Gamma)$ and $\pi_0^K(\Gamma)$ commute. Similar statements are true for π_H and π_A. If $D = G$, we abuse notation by writing π_0^E for π_0 and π_0^K for the trivial homomorphism and similarly for π_A and π_H. We note that our assumption that $\pi_0 = (\pi_A, \pi_H)$ actually follows from the structure theory of algebraic group when D is G or the superrigidity theorems when D is Γ, though in both cases possibly only after replacing $\mathbb{A}$ by an isomorphic subgroup of $\mathbb{L}$.

THEOREM 5.1. *Let (S, μ) be a standard probability measure space, ρ a measure preserving action of D on S, and let $\alpha_{\pi_0} : D \times S \to L$ be the constant cocycle over the action ρ given by $\alpha_{\pi_0}(d, x) = \pi_0(d)$. Furthermore, let $\alpha : D \times S \to L$ be a Borel cocycle over the action ρ such that:*

1. *the cocycles α_{π_0} and α are L^∞ close*
2. *the projection of α to A is π_A.*

Then there exist measurable maps $\phi : S \to H$ and $z : D \times S \to Z$ where we define $Z = Z_L(\pi_0^E(D)) \cap H$ such that

1. *we have $\alpha(d, x) = \phi(dx)^{-1}(\pi_A(d), \pi_H^E(d))z(d, x)\phi(x)$;*
2. *$\phi : S \to H$ is small in L^∞;*
3. *the map $(\pi_A^K(d), z(d, x))$ is a cocycle and is L^∞ close to the constant cocycle defined by π_0^K.*
4. *the cocycle (π_A^K, z) is measurably conjugate to a cocycle taking values in a compact subgroup K of Z and K is contained in a small neighborhood of $\pi_0^K(D)$.*

Furthermore if S is a topological space, $\operatorname{supp}(\mu) = S$ *and α and ρ are continuous then both ϕ and z can be chosen to be continuous.*

Remark: Theorem 5.1 is an immediate consequence of Theorem 5.4 if π_0 has bounded image, so we assume throughout that π_0 has unbounded image.
Remark: If $D = G$ or k is Archimedean or π_A^K is trivial, the map z is a cocycle. More generally, it is a *twisted cocycle*, as discussed in [**F**, Section 7.1].

As in the proof of Theorem 3.1 of [**MQ**], we deduce our result from the stability of partially hyperbolic vector bundle maps, though the details of the argument are quite different. To make the line of the argument clear, we outline it briefly for G actions assuming that the group A above is trivial and that the G action on S is ergodic. (In other words we sketch a proof of Theorem 1.1 from the introduction, with the additional assumption that the G action is ergodic.) It follows from Theorem 1.4 that $\alpha(g, x) = \phi(gx)^{-1}\pi'(g)c(g, x)\phi(x)$ where $\pi' : G \to L$ is a continuous homomorphism and $c : G \times S \to C$ is a cocycle taking values in a compact group

$C < Z_L(\pi'(G))$ and $\phi : X \to L$ is a measurable map. Since the set of polar elements of G is Zariski dense in G (see subsection 3.9) we can find a finitely generated Zariski dense subgroup $< g_1, \ldots, g_l >= F < G$ where each generator g_i has the property that it's image under any rational homomorphism $\pi : G \to L$ is uniquely determined by the characteristic numbers and subspaces of $\pi(g_i)$. We realize L as a subgroup of $GL(n,k)$ and study the dynamics of the skew product actions $\rho_{\alpha_{\pi_0}}$ and ρ_α on $S \times k^n$ determined by α_{π_0} and α. It follows from standard arguments on stability of partially hyperbolic vector bundle maps, see Lemma 5.2 below, that for any compact set $K < G$, if we pick α close enough to α_{π_0}, the characteristic numbers and subspaces for $\rho_{\alpha_{\pi_0}}(g)$ can be made arbitrarily close to those of $\rho_\alpha(g)$. This allows us to show that for almost any $x \in S$ and any i, the image of ${}^{\phi(x)}\pi'(g_i)$ is close to the image of $\pi_0(g_i)$. It then follows from Theorem 4.1 that π_0 and π' are conjugate as homomorphisms of F, and therefore as homomorphisms of G since $F < G$ is Zariski dense. Most of the remaining conclusions of the theorem are deduced by a more careful analysis of the data coming from Lemma 5.2 and Theorem 4.1. That z is measurably conjugate to a cocycle taking values in a compact group contained in a neighborhood of the identity uses Theorem 5.4. The general case follows more or less the same outline, using Theorems 3.22 and 3.23 in place of Theorem 1.4 and requiring somewhat more care due to the presence of many ergodic components. For Γ actions and cocycles there is an additional nuance since we cannot choose $F < \Gamma$ and here we use Lemma 3.21.

If $D = \Gamma$ then by Lemma 3.21 of subsection 3.8 there are elements $g_1, \ldots, g_l \in D$ and such that the group F generated by their polar parts is Zariski dense in G. If $D = G$, we pick a collection $g_1, \ldots, g_l$ of polar elements in G such that the group F generated by $g_1, \ldots, g_l$ is Zariski dense in G. In either case, we fix F and $g_1, \ldots, g_l$ for the remainder of the section. The reader is referred to subsection 3.8 for a discussion of polar elements.

Remark: Under any rational homomorphism $\pi : \mathbb{G}(k) \to GL_n(k)$ the image of a polar element is a polar element and $\mathrm{pol}(\pi(g)) = \pi(\mathrm{pol}(g))$.

As stated above, we will use a dynamical argument to show that $\mathrm{pol}(\tilde{\pi}_i(g_j))$ is close to $\mathrm{pol}(\pi_0(g_j))$ for a finite collection of g_j in Γ and then use this to conclude that $\tilde{\pi}_i^E$ and π_0^E are close in the compact open topology on homomorphisms.

We fix an almost faithful representation $\sigma : L \to GL(n,k)$. We can associate to any action ρ of D on a space X and any L valued cocycle α over ρ, an action $d_\alpha \rho$ of D on the trivial bundle $X \times k^n$ via $g(x,v) = (\rho(g)x, \sigma(\alpha(g,x))v)$. We use this to define two actions $d_{\alpha_{\pi_0}}\rho$ and $d_\alpha \rho$, both of which are linear extensions of ρ.

For any linear map $A : k^n \to k^n$, there is a finite field extension k' of k and a decomposition $\mathbb{R}^n = \bigoplus_{j=1}^l W_j$ such that the eigenvalues of $A|_{W_j \otimes k'}$ have the same absolute value λ_j for each j, and $\lambda_1 > \lambda_2 > \ldots \lambda_l$. We call W_j the *characteristic subspace* of A with *characteristic number* λ_j.

Let S, μ be a finite measure space, T a measure preserving transformation of S, and $\alpha : \mathbb{Z} \times S \to GL(n,k)$ a cocycle. We let

$$W_{\alpha,\varepsilon,\lambda}(x) = \{w \in W| \quad \limsup_{m \to \infty} |\frac{1}{m} \log \|\alpha(m,x)w\| - \lambda| < \varepsilon\} \cup \{0\}.$$

LEMMA 5.2. *Let $A \in GL(n,k)$ be linear transformation, (S,μ) a finite measure space, T a measure preserving transformation of (S,μ) and α_0 be the constant cocycle over the T action defined by A. Then there exists ε_0 depending only on A, such that for any $\varepsilon < \varepsilon_0$ and any cocycle $\alpha : \mathbb{Z} \times S \to GL(n,k)$ over the T action that* *is sufficiently L^∞ close (depending on ε) to A the spaces W_{A,λ_j} and $W_{\alpha,\varepsilon,\lambda_j}(x)$ are L^∞ close. Furthermore, the subspaces $W_{\alpha,\varepsilon,\lambda_j}(x)$ are measurable functions of $x \in S$ and if T and α are continuous, then $W_{\alpha,\varepsilon,\lambda_i}(x)$ is defined for all x and depends continuously on x, T and α.*

Remark: The number ε_0 is explicitly known and can be taken to be one third of $\min_{1 \leq j < l}(\lambda_j - \lambda_{j-1})$.
This lemma follows from standard arguments on the stability of invariant distributions for partially hyperbolic vector bundle maps. There are many possible sources for such arguments, which go back at least as far as Anosov [**An**]. For a proof that is easily adapted to this setting see the proofs of Theorem 1 and Lemma 1 in [**P**].

Let $g_1, \ldots, g_k$ be as above. For π_0, we let λ_j^l be the characteristic numbers of $\pi_0(g_l)$ and let W_j^l be the corresponding subspaces. For α as in Theorem 5.1, we apply Lemma 5.2 to $g_1, \ldots, g_k$. We denote the resulting subspaces of W by $W_{\alpha,\varepsilon,g_l,\lambda_j^l}(x)$.

Before proceeding with the proof of Theorem 5.1, we record the information given by applying the results of section 3 to α.

PROPOSITION 5.3. *Let D, S, ρ, μ, α and π be as above. On each ergodic component μ_i of the measure μ the cocycle $\alpha : D \times S \to L$ is cohomologous to the product of a constant cocycle with a compact valued cocycle. More precisely there exist μ_i-measurable maps $\phi_i : S \to L$ such that*

$$\alpha(d, x) = \phi_i(dx)^{-1}(\pi_A, \pi_i)(d)c_i(d, x)\phi_i(x)$$

for all $d \in D$ and μ_i almost every x. Here π_i is the restriction to D of a continuous homomorphism $\pi_i : G \to H$ and $c_i : D \times S \to C_i$ are measurable maps taking values in compact subgroups $C_i < L$ commuting with $(\pi_A^E, \pi_i)(D)$. Furthermore, we can assume that there is a finite set Π of homomorphisms of G in H containing all π_i and that each ϕ_i takes values in H.

PROOF OF PROPOSITION 5.3. The proposition follows by applying Theorems 3.22 and 3.23 to α. To apply those theorems, we need to see that α is D-integrable. Since α is L^∞ close to a constant cocycle, it follows that $\ln^+ Q_{M,\alpha}(x)$ is essentially bounded for any precompact M and so in L^1. Therefore the cocycle is D-integrable for almost every ergodic component μ_i of μ.

Theorems 3.22 and 3.23 show that α is cohomologous to β where $p_A \circ \beta = \pi_A$ and $\beta = \bar{\pi}_i c_i$ where $\bar{\pi}_i : G \to L$ is a continuous homomorphism and $c_i : D \times S \to C$ is a cocycle taking values in a compact group $C_i < Z_L(\bar{\pi}_i(D))$. Furthermore, we have that $p_A \circ \beta = \pi_A$ and that the cohomology ϕ_i takes values in H. To prove the proposition, we need to see that we can write $\bar{\pi}_i = (\pi_A^E, \pi_i)$.

Let $\mathbb{H} = \mathbb{F} \ltimes \mathbb{U}$ be Levi decomposition, with $\mathbb{F}$ reductive and $\mathbb{U}$ unipotent. Since $\mathbb{A}$ is connected, it follows that $\mathbb{F} = \mathbb{F}_1 \mathbb{F}_2$ an almost direct product, where $\mathbb{A}$ and $\mathbb{F}_2$ commute as subgroups of $\mathbb{L}$ and $\mathbb{A}$ acts on $\mathbb{F}_1$ by automorphisms via a homomorphism $\mathbb{A} \to \mathbb{F}_1$ composed with the adjoint action of $\mathbb{F}_1$ on itself. We write $\mathbb{L} = ((\mathbb{A} \ltimes \mathbb{F}_1)\mathbb{F}_2) \ltimes \mathbb{U}$ and as usual denote $F_1 = \mathbb{F}_1(k), F_2 = \mathbb{F}_2(k)$ and $U = \mathbb{U}(k)$. The map $\Delta(a, f) = (a, a^{-1}f)$ is an isomorphism between $\mathbb{A} \ltimes \mathbb{F}_1$ and $\mathbb{A} \times \mathbb{F}_1$. By replacing $\mathbb{A} < \mathbb{L}$ by $\Delta(\mathbb{A})$, we have that $\mathbb{A}$ and $\mathbb{F}$ commute. This does not effect π_0, since we have assumed that π_A and π_H commute which forces π_H to take values in F_2. After conjugating by some element of U, we may assume $\bar{\pi}_i$ takes values in $\mathbb{A} \times \mathbb{F}$. Writing $\bar{\pi}_i(d) = \pi_A^E(d)\pi_i(d)$ where π_i takes values in $\mathbb{F}$, it follows from the fact that $\mathbb{F}$ and $\mathbb{A}$ commute that π_i is a homomorphism.

That all π_i are contained in a finite collection Π follows from the fact that there are only finitely many (conjugacy classes of) homomorphisms of G into L. $\square$

We denote by $\tilde{\pi}_i$ the homomorphism (π_A, π_i) and call Π' the collection of all $\tilde{\pi}_i$. We fix the collection Π' of homomorphisms of G into L (or equivalently π_A and the collection Π of homomorphisms of G into H) for the remainder of the section.

Let $\tilde{\pi}_i(g_l)$ have characteristic numbers ${}^i a^l_1, \ldots, {}^i a^l_{s_i}$ and ${}^i W^l_j$ be the characteristic subspaces corresponding to ${}^i a^l_j$. Then for any $w \in W - \{0\}$ and almost every $x \in \operatorname{supp} \mu_i$, there is a j such that

$$\limsup_{m\to\infty} \frac{1}{m} \log \|\alpha(g_l^m, x)w\| = {}^i a^l_j.$$

Since the set $\{\lambda^l_m, {}^i a^l_j\}_{i,l,j,m}$ is finite, if we choose α close enough to π_0 and ε small enough, after re-indexing we have

$$\phi_i(x)\, {}^i W^l_j = W_{\alpha,\varepsilon,g_l,\lambda^l_m}(x) \tag{3}$$

for μ_i almost all x and all i. Furthermore, we have that for each j there is an m such that ${}^i a^l_j = \lambda^l_m$ and that $\dim({}^i W^l_j) = \dim(W^l_m)$. This proof of equation 3 is essentially contained in [**MQ**] discussion preceding Lemma 3.4 or [**QZ**] proof of Theorem A.

PROOF OF THEOREM 5.1. First we show that $\tilde{\pi}_i = (\pi_A, \pi^E_H)$ for all i. Recall that $\pi_0 = (\pi^E_0, \pi^K_0)$ and that $\tilde{\pi}_i = (\tilde{\pi}^E_i, \tilde{\pi}^K_i)$ where π^E_0 and π^K_0 (resp. $\tilde{\pi}^E_i$ and $\tilde{\pi}^K_i$) commute and $\pi^E_0, \tilde{\pi}^E_i$ are restrictions of continuous homomorphisms of G and $\tilde{\pi}^K_i, \pi^K_0$ have bounded image. Note that for any $g \in \Gamma$ we have $\operatorname{pol}(\tilde{\pi}_i(g)) = \operatorname{pol}(\tilde{\pi}^E_i(g))$ and that the same is true for π_0. Since $\pi_0 = (\pi_A, \pi_H)$ and $\tilde{\pi}_i = (\pi_A, \pi_i)$ and π_i is the restriction of a rational homomorphism from $\mathbb{G}$ to $\mathbb{L}$, it suffices to show that $\tilde{\pi}^E_i = \pi^E_0$.

By Lemma 5.2,by choosing ε small enough and α close enough to π_0, the space $W_{\alpha,\varepsilon,g_l,\lambda^l_m}(x)$ can be made arbitrarily close to the space W^l_m for almost every $x \in S$. By equation 3, we have that $\phi_i(x)\, {}^i W^l_j = W_{\alpha,\varepsilon,g_l,\lambda^l_m}(x)$ for μ_i almost every $x \in S$, so W^l_m is close to $\phi(x)\, {}^i W^j_l$ for almost every x. Furthermore, by the remark following equation 3, for each j there is an m such that the action of the polar part of ${}^{\phi(x)}\tilde{\pi}_i(\gamma_l)$ on $\phi(x)\, {}^i W^l_j$ is the same as the action of the polar part of $\pi_0(\gamma_l)$ on W^l_m. This implies that, for α close enough to π_0, we can make $\operatorname{pol}({}^{\phi_i(x)}\tilde{\pi}_i(\gamma_l)) = {}^{\phi_i(x)}\tilde{\pi}_i(\operatorname{pol}(\gamma_l))$ arbitrarily close to $\operatorname{pol}(\pi_0(\gamma_l)) = \pi_0(\operatorname{pol}(\gamma_l)$ for every l and i and μ_i almost every x. Therefore, for almost every $x \in S$, the homomorphisms ${}^{\phi(x)}\tilde{\pi}^E_i$ and π^E_0 can be made arbitrarily close as homomorphisms of F by choosing α close enough to π_0. Since there are only finitely many homomorphisms in Π' and H orbits in $\operatorname{Hom}(F, L)$ are closed by Corollary 4.3, this implies that $\tilde{\pi}^E_i$ and π^E_0 are in the same H orbit in $\operatorname{Hom}(F, L)$ and so are conjugate by an element of H. Since F is Zariski dense in G and $\tilde{\pi}^E_i$ and π^E_0 are restrictions of rational homomorphisms, it follows that $\tilde{\pi}^E_i$ is conjugate to π^E_0 as homomorphisms of G.

By relabelling, we now have

$$\alpha(g, x) = \phi_i(gx)^{-1}(\pi_A, \pi^E_H)(g)c_i(g, x)\phi_i(x)$$

for each i and μ_i almost every x where each c_i is a cocycle taking values in a compact subgroup of $Z = Z_L((\pi^E_0)(D)) \cap H$.

We define a map $\bar{\phi}_h : S \to \mathrm{Hom}(F, L)$ by taking x to ${}^{\phi_i(x)}\tilde{\pi}_i|_F$. By Lemma 5.2 and the discussion above, this map is measurable and L^∞ small and has image contained in a single H orbit in $\mathrm{Hom}(F, L)$. Since F and D are Zariski dense in G, and π_0^E is rational, it follows that $Z_L(\pi_0^E(F)) = Z_L(\pi_0^E(G)) = Z_L(\pi_0^E(D))$ and so we may identify the H orbit in $\mathrm{Hom}(F, L)$ with H/Z where $Z = Z_L(\pi_0^E(D)) \cap H$ as before. Therefore, choosing a point w in the image, we can define a measurable map $\phi_h : S \to H$ such that $\bar{\phi}_h(x) = \phi_h(x)w$. Furthermore, we can choose ϕ_h to be L^∞ small. This uses that we can choose a Borel section of $H \to H/Z$ which is continuous in a small neighborhood of $[1_H]$ and does not increase norms.

It follows from the definitions of ϕ_i and ϕ_h that for each i, we have that $\phi_i(x) = \phi_h(x)\phi_i^z(x)$ where ϕ_i^z takes values in Z. Then a simple computation shows that:

$$\alpha(g,x) = \phi_h(gx)\phi_i^z(gx)(\pi_A, \pi_H^E)(g)c_i(g,x)\phi_i^z(x)^{-1}\phi_h(x)^{-1}$$

$$= \phi_h(gx)(\pi_A, \pi_H^E)(g)z(g,x)\phi_h(x)^{-1}$$

where $z(g,x) = \pi_A^K(g)^{-1}\phi_i^z(gx)\pi_A^K(g)c_i(g,x)\phi_i^z(x)^{-1}$ takes values in Z.

Since α is measurable and L^∞ close to π_0 and ϕ_h is measurable and L^∞ small, it follows that $z(g,x) = \phi_h(gx)\alpha(g,x)\phi_h(x)^{-1}(\pi_A, \pi_i^E)(g)^{-1}$ is measurable and L^∞ close to π_H^K. It is clear that $(\pi_A^K(g), z(g,x))$ is a cocycle taking values in $Z_L(\pi_0^E)$, which implies that z is a cocycle when π_A^K is trivial. Theorem 5.4 in the next section shows that $(\pi_A^K, z(g,x))$ is measurably conjugate to a cocycle taking values in a group that is contained in a small open neighborhood of $(\pi_A^K, \pi_H^K)(D)$. When k is Archimedean, this implies that $(\pi_A^K, z(g,x))$ is measurably conjugate into the closure of $(\pi_A^K, \pi_H^K)(D)$ by Lemma 5.6. This implies that π_A^K and z take values in groups which commute, which by Lemma 3.3 implies that z is a cocycle.

When ρ and α are continuous, we deduce continuity of ϕ_h from the continuity of $W_{\alpha,\varepsilon,g_l,\lambda_m^l}(x)$ which follows from Lemma 5.2. Since we know that $\pi_i^E = \pi_0^E$, we can rewrite equation 3 as $\phi(x)W_m^l = W_{\alpha,x,g_l,\lambda_m^l}$ for each l. This implies that the polar part of ${}^{\phi(x)}\pi_0^E(g_l)$ depends continuously on x for each l. Combined with Lemma 3.21 this implies that ${}^{\phi(x)}\pi_0^E$ depends continuously on x. Therefore $\phi(x)$ is continuous modulo the stabilizer of π_0^E in H, or ϕ_h is continuous. That z is then continuous as well follows from the formula defining z two paragraphs above. □

5.2. Local rigidity of compact valued cocycles. In this subsection D will denote a locally compact group with property T of Kazhdan, A will be a locally compact group and $C < A$ will be a compact subgroup. As before (S, μ) will be a standard measure space and we will assume D acts on S preserving μ. We will say that a group C' is close to C if there is a small neighborhood U of C such that $C' \subset U$. Throughout ν will denote right Haar measure on A. We will say that two cocycles $\alpha, \alpha' : D \times S \to A$ are close if there is a compact generating set K for D and a small neighborhood U of 1 in A such that $\alpha(d,s)\alpha'(d,s)^{-1}$ is in U for all $d \in K$.

THEOREM 5.4. *Let $\alpha_0 : D \times S \to C$ be a cocycle over the D action. Any cocycle $\alpha : D \times S \to A$ which is close to α_0 is conjugate to a cocycle into a compact group C' that is close to C.*

Remark: The proof of Theorem 5.4 is simpler if one assumes the D action on S is ergodic.

We will need the following proposition in order to find C' satisfying the conclusions of the theorem.

PROPOSITION 5.5. *Let $C < A$ be a compact subgroup. Then given any small enough neighborhood U of C, there exists a compact group $C' < A$, contained in U, such that any subgroup $C'' < A$ contained in U is conjugate to a subgroup of C' by a small element of A.*

LEMMA 5.6. *Let $C < A$ be a compact subgroup such that A/C is a manifold. Then any compact subgroup C' sufficiently close to C is conjugate to a subgroup of C by a small element of A.*

PROOF. This follows from a barycenter argument that is similar to the one that shows that any two maximal compact subgroups in a semisimple group are conjugate. We let $X = A/C$ and take the C' orbit $\mathcal{O}$ of the coset of the identity $[1_A]$. Since C' is close to C, it follows that $\mathcal{O}$ is contained in a small neighborhood of $[1_A]$. We can then take the barycenter or center of gravity for $\mathcal{O}$. This is defined as the unique minimum of the function $d_{\mathcal{O}}(x) = \int_{\mathcal{O}} d(x,y)^2 d\mu(y)$ where μ is the push-forward of Haar measure on C' to $\mathcal{O}$. That a barycenter exists and is unique can be proven from convexity of the distance function on a small enough neighborhood $U_{[1_A]}$. Convexity of the distance function on this neighborhood can be proven by comparison with the sphere whose sectional curvature is the maximum of the sectional curvatures of two planes in $T(A/C)$, using the fact that for small enough neighborhoods on the sphere, the distance function is convex, see for example [**BH**, Exercises 2.3(1),p.176]. The barycenter is then a fixed point for the C' action and is close to $[1_A]$ since $c'[1_A]$ is close to $[1_A]$ for all $c' \in C'$. This implies that $C' < aCa^{-1}$ for $a \in A$ small. □

PROOF OF PROPOSITION 5.5. We let A^0 be the connected component of the identity in A, and $p : A \to A/A^0$ be the projection. Then, since A/A^0 is totally disconnected, there is an open subgroup $\bar{C}$ containing $p(C)$. Since $\bar{C}$ is open, if U is a sufficiently small neighborhood of C, then $p(U)$ is contained in $\bar{C}$. We will find $C' < p^{-1}(\bar{C})$ and so replace A by $A' = p^{-1}(\bar{C})$.

Given any open set U containing the identity in A' there is a compact normal subgroup $N \subset U$ such that A'/N has no small subgroups, and is therefore a manifold. This is an extension by Glushkov [**Gl**] of results due to Gleason, Montgomery and Zippin, see [**K**] for further discussion, particularly Theorem 18 and the remark following on page 137. We let $C' = CN$. Then $A/C' = (A/N)/(C/(C \cap N))$ and so is a manifold. It then follows from Lemma 5.6 that if U is small enough, any subgroup contained $C'' \subset U$ is conjugate to a subgroup of $C' = CN$. □

Given a non-negative, integrable function h on A and a unitary representation ρ of A on a Hilbert space $\mathcal{H}$, we define $\rho(h) = \int_A h(a)\rho(a)d\nu(a)$. If $\int_A h(a)d\nu(a) = 1$, then $\|\rho(h)\| \leq 1$ as verified in [**M3**, III.1.0].

We recall that a locally compact group D has property T of Kazhdan if the trivial representation is isolated in the unitary dual. This has the following consequence, which can be seen as an effective version of the standard statement "any (ε, K)-invariant vector is close to a D invariant vector".

LEMMA 5.7. *Let D be a locally compact group with property T. Then for any compact generating set K for A there is a non-negative continuous function h with support contained in K^2 and $\int_A f d\mu_A = 1$ and a constant $B = B(K,h)$, such that for any unitary representation ρ of D on a Hilbert space $\mathcal{H}$ and for any vector $v \in \mathcal{H}$, we have*

1. $\lim_{n\to\infty} \rho(h)^n v = v_F$ *exists;*
2. v_F *is fixed by* D;
3. $d(v_F, v)\leq B \operatorname{supp}_{k\in K} d(kv, v)$.

PROOF. This follows from the definition of property T and from [**M3**, III.1.3], which shows that $\rho(h)$ is a contraction on the orthogonal complement of the D fixed vectors in $\mathcal{H}$. That h can be chosen with support in K^2 follows from the construction of h in the proof of [**M3**, III.1.1]. □

PROOF OF THEOREM 5.4. We let $\mathcal{H} = L^2(A)$ and write the natural action of A coming from the right regular representation as $v \to va^{-1}$. Fix a vector $v_0 \in \mathcal{H}$ whose stabilizer is C. We define two representations ρ_{α_0} and ρ_α of D on $L^2(S, \mathcal{H}, \mu)$ by $(\rho_{\alpha_0}(d)f)(x) = f(d^{-1}x)\alpha_0(d, x)^{-1}$ and $(\rho_\alpha(d)f)(x) = f(d^{-1}x)\alpha(d, x)^{-1}$. Then the function $f_0 : S \to \mathcal{H}$ defined by $f_0(x) = v_0$ for all x is ρ_{α_0} invariant. It is easy to see that $\operatorname{supp}_{k\in K} d(\rho_\alpha(k)f_0, f_0) < \varepsilon$ where ε only depends on how close α is to α_0.

First assume that the action of D on S is ergodic. By 2 and 3 of Lemma 5.7, there is a function $f \in L^2(S, \mathcal{H})$ such that f is $\rho_\alpha(D)$ invariant and $\|f - f_0\|_2$ is small. By the proof of [**Z2**, Lemma 9.1.2] one sees that the A action on $\mathcal{H}$ is tame and so ergodicity of the D action on S implies that f takes values in a single A orbit $\mathcal{O}$ in $\mathcal{H}$. This then implies by [**Z2**, Lemma 5.2.11] that α is equivalent to a representation into the stabilizer A_v of some vector $v \in \mathcal{O}$. It is easy to verify that A_v is compact. Since f is L^2 close to f_0, we can choose v to be close to v_0. This immediately implies that the stabilizer of A_v is Hausdorff close to the stabilizer of v_0.

If the action is not ergodic, we cannot conclude that f takes values in a single orbit. If one traces through the above argument, one sees that, on each ergodic component μ_i of μ, f takes values in a single A orbit we can view f as a μ measurable map $S \to S \times H/C_i$ where $C_i < A$ is a compact subgroup depending on μ_i. That we can find a μ measurable function conjugating α to a cocycle α' taking values in C_i for μ_i almost every x follows from the existence of a Borel section for the map $S \times H \to S \times H/C_i$. The existence of such a section can be deduced from [**Z2**, Theorem A.5]. However, since we only know that f is close to f_0 in $L^2(S, \mathcal{H}, \mu)$, it only follows from this argument that C_i is close to C on "most" ergodic components. We want to see that f is actually close to f_0 in $L^2(S, \mathcal{H}, \mu_i)$ for almost every ergodic component. This is deduced from 1 of Lemma 5.7, since $f = \lim_{n\to\infty} \rho(h)^n f$ and this equation holds in both $L^2(S, \mathcal{H}, \mu)$ and $L^2(S, \mathcal{H}, \mu_i)$. That f is close to f_0 in $L^2(S, \mathcal{H}, \mu_i)$ for almost every ergodic component then follows from 3 of Lemma 5.7. This implies that each C_i is contained in a small neighborhood of C, and so is conjugate into a subgroup $C' < A$ contained in a small neighborhood of C by an element a_i. Conjugating by a map $\phi : S \to A$ such that $\phi(s) = a_i$ for μ_i almost every s, we have that α is conjugate to a cocycle taking values in C'. □

6. Affine actions, perturbations and cocycles

In this section we prove Corollary 1.7 and Theorem 1.8. In order to do so, we need a more detailed description of the actions in Definition 1.6 when the group acting is G or Γ as above. The section is divided into two subsections, the first giving an algebraic description of affine actions, the second proving Corollary 1.7 and Theorem 1.8.

6.1. Description of affine actions. Let H be a connected Lie group and $\Lambda < H$ a discrete cocompact subgroup. We will let $\mathrm{Aff}(H/\Lambda)$ be the group of affine transformation of H/Λ. Any affine transformation f of H/Λ has a lift of the form $h_f \circ L_f$ where $h_f \in H$ and $L_f \in \mathrm{Aut}(H)$ where $\mathrm{Aut}(H)$ denotes the group of continuous automorphisms of H. Let $N_{\mathrm{Aut}(H)}(\Lambda)$ be the group of elements $L \in \mathrm{Aut}(H)$ such that $L{\cdot}\Lambda \subset \Lambda$. Then $L_f \in N_{\mathrm{Aut}(H)}(\Lambda)$ and we have a map $\phi : N_{\mathrm{Aut}(H)}(\Lambda) \ltimes H \rightarrow \mathrm{Aff}(H/\Lambda)$. There is a map $\Delta^{-1} : \Lambda \rightarrow N_{\mathrm{Aut}(H)}(\Lambda) \ltimes \Lambda$ given by $(\lambda) \rightarrow \mathrm{Ad}((\lambda)^{-1}), (\lambda))$. We denote the image of this map by $\Delta^{-1}(\Lambda)$. (If H has trivial center, then the image is in fact an anti-diagonal embedding of Λ.)

PROPOSITION 6.1. *The kernel of the map* $\phi : N_{\mathrm{Aut}(H)}(\Lambda) \ltimes H \rightarrow \mathrm{Aff}(H/\Lambda)$ *is* $\Delta^{-1}(\Lambda)$.

PROOF. First note that any diffeomorphism f of H/Λ gives rise to an element f_* of $\mathrm{Out}(\pi_1(H/\Lambda))$. If f is trivial, then f_* must be trivial as well. If $f = \phi((a, h_0))$ for $a \in N_\Lambda(\mathrm{Aut}(H))$ and $h_0 \in H$ and f_* is trivial, then a must be an inner automorphism of H preserving Λ. This implies that $a = z\,\mathrm{Ad}(\lambda)$ for some $\lambda \in \Lambda$ and $z \in Z_{\mathrm{Aut}(H)}(\Lambda)$, the centralizer in $\mathrm{Aut}(H)$ of Λ. But then $(a, h_0)[h] = [z{\cdot}(\lambda h_0 h \lambda^{-1})] = [z{\cdot}(\lambda(h_0)h)]$ for all $h \in H$. So $(a, h_0)[h] = [h]$ for all $h \in H$ if and only if $z{\cdot}(\lambda h_0)^{-1} z{\cdot}h = h$ for all $h \in H$. Since z centralizes λ this is equivalent to $\lambda(z{\cdot}h_0)(z{\cdot}h) = h$ for all $h \in H$. Picking $h = 1$ this forces $z{\cdot}h_0 = \lambda^{-1}$. This implies that $z{\cdot}h = h$ for all $h \in H$ which implies that $z = 1$ and so $h_0 = \lambda^{-1}$. □

Most of the difficulty in proving the theorems we need describing affine actions derive from tori in the reductive component of H. To deal with this difficulty we replace H and Λ by groups H' and Λ' such that the respective quotients are diffeomorphic and the affine groups are the same. First we note the a simple fact about covers.

LEMMA 6.2. *Let H be a real Lie group and $\Lambda < H$ a cocompact lattice. Let $p : H' \rightarrow H$ be a covering map and $\Lambda' = p^{-1}(\Lambda)$. Then*

1. *H/Λ is diffeomorphic to H'/Λ'*
2. $\mathrm{Aff}(H/\Lambda) < \mathrm{Aff}(H'/\Lambda')$

PROOF. The first claim of the lemma is immediate. To see the second, we note that any continuous automorphism A of H lifts to a continuous automorphism A' of H'. This uses the fact that the fundamental group of H is abelian and so any cover is a normal cover. If $A{\cdot}\Lambda = \Lambda$ then $A'{\cdot}\Lambda' = \Lambda'$. Also, given any element of h, we can choose an element h' in H' projecting to h. It is easy to verify that hA and $h'A'$ induce the same diffeomorphism of $H/\Lambda = H'/\Lambda'$. □

We now show how to replace H by a cover H', though we need to use an algebraic structure on H' so that the cover is not a rational map.

PROPOSITION 6.3. *Given a real algebraic group H and a cocompact lattice Λ there is a cover $p : H' \rightarrow H$ and a realization of H' as $\mathbb{H}'(\mathbb{R})$ for a connected $\mathbb{R}$ algebraic group $\mathbb{H}'$, such that*

1. *there is a finite index subgroup* $\mathrm{Aut}^A(H') < \mathrm{Aut}(H')$ *such that all elements of* $\mathrm{Aut}^A(H')$ *are rational automorphisms of H' and*
2. $\mathrm{Aut}^A(H') \ltimes H'$ *is the real points of a real algebraic group which we denote by* $\mathrm{Aut}^A(\mathbb{H}') \ltimes \mathbb{H}'$.

PROOF. We first define the group H'. By definition $H = \mathbb{H}(\mathbb{R})$ where $\mathbb{H}$ is an algebraic $\mathbb{R}$-group. We take a Levi decomposition $\mathbb{H} = \mathbb{L} \ltimes \mathbb{U}$ where $\mathbb{L}$ is reductive and $\mathbb{U}$ is unipotent. We first pass to a finite central extension $\tilde{\mathbb{H}}$ so as to be able to assume that $\mathbb{L}$ is a direct product of a torus $\mathbb{T}$ and a simply connected semisimple group $\mathbb{J}$. We let $\sigma : \mathbb{L} \to \mathrm{Aut}(\mathbb{U})$ be the representation defining the semidirect product. We let $\mathbb{T}_1 = \ker(\sigma) \cap \mathbb{T}$. This is a finite extension of a connected group $\mathbb{T}_1^0$, and $\mathbb{T}_1^0 < Z(\tilde{\mathbb{H}})$ and so $\tilde{\mathbb{H}} = \mathbb{T}_1^0 \times \mathbb{H}^*$. The universal cover of $\mathbb{T}_1^0(\mathbb{R})$ is isomorphic to $\mathbb{R}^n$ for $n = \dim(\mathbb{T}_1^0)$, and we can realize $\mathbb{R}^n$ as the real points of a unipotent algebraic group which we denote by $\mathbb{U}^*$. We replace $\mathbb{H}$ by $\mathbb{H}' = \mathbb{H}^* \times \mathbb{U}^*$ and H by $H' = \mathbb{H}'(\mathbb{R})$. There is a covering map $p_1^* : \mathbb{U}^*(\mathbb{R}) \to \mathbb{T}_1^0(\mathbb{R})$ which defines a covering map $p_1 : \mathbb{H}'(\mathbb{R}) \to \tilde{\mathbb{H}}(\mathbb{R})$ which we compose with the covering map $p_2 : \tilde{\mathbb{H}}(\mathbb{R}) \to \mathbb{H}(\mathbb{R})$ to define a covering map $p : \mathbb{H}'(\mathbb{R}) \to \mathbb{H}(\mathbb{R})$. We let $\Lambda' = p^{-1}(\Lambda)$.

To continue the proof, we will need a Levi decomposition of $\mathbb{H}' = \mathbb{L}' \ltimes \mathbb{U}'$. By the discussion above, we can write $\mathbb{L}' = \mathbb{J}' \times \mathbb{T}'$ where $\mathbb{J}'$ is semisimple and $\mathbb{T}'$ is a torus. We have constructed $\mathbb{H}'$ such that the homomorphism $\mathbb{T}' \to \mathrm{Aut}(\mathbb{U}')$ has finite kernel. (The attentive reader will note that $\mathbb{J}'$ is isomorphic to $\mathbb{J}$ above and that $\mathbb{T}$ above is $\mathbb{T}_1 \times \mathbb{T}'$, but we will not need this in the discussion that follows.) As usual, $L' = \mathbb{L}'(\mathbb{R}), U' = \mathbb{U}'(\mathbb{R}), J' = \mathbb{J}'(\mathbb{R})$ and $T' = \mathbb{T}'(\mathbb{R})$.

The group $\mathrm{Aut}^A(H')$ will consist of those automorphisms of H' which project to inner automorphisms of L. We first show this has finite index in $\mathrm{Aut}(H)$. The group of outer automorphisms of J is finite, so it suffices to show that the group of automorphisms of T' that extend to automorphisms of H' is finite. In fact, the group Ξ of automorphisms of T' which extend to $T' \ltimes U'$ is finite. Any such automorphism must induce a permutation of the finite collection Δ of weights defining the representation σ of $\mathbb{T}$ on $\mathbb{U}$, so by passing to a subgroup $\Xi' < \Xi$ of finite index, we may assume that Ξ' fixes Δ pointwise. Since the kernel of σ is finite, Δ forms a basis for the group of characters of $\mathbb{T}'$ which vanish on $\ker \sigma$. Therefore Ξ' acts trivially on a subgroup of finite index in the group of characters of $\mathbb{T}'$ and, since $\mathbb{T}'$ is connected, acts trivially on $\mathbb{T}'$.

We can write any element of $\phi \in \mathrm{Aut}^A(H')$ as a composition of three elements. First we translate by an element u of U' so that $u \circ \phi(L) = L$. Then we conjugate by an element l of L so that $\mathrm{Ad}(l) \circ u \circ \phi$ is trivial on L. The automorphism $\mathrm{Ad}(l) \circ u \circ \phi = a$ is clearly an automorphism of U which commutes with the action of L on U. We write $\phi = alu$. Viewing a as belonging to $Z_{\mathrm{Aut}(U)}(L)$, l as an element of the adjoint group of $\bar{L}$ of L, and u as an element of $U/Z_U(L)$, this decomposition is unique, and clearly makes $\mathrm{Aut}^A(H')$ the set of $\mathbb{R}$ point of an $\mathbb{R}$-variety. Writing the multiplication on $\mathrm{Aut}^A(H') \cong Z_{\mathrm{Aut}(U)}(L) \times \bar{L} \times (U/Z_U(L))$ it is clear that all factors commute pairwise except the last two. The product $\bar{L} \ltimes (U/Z_U(L))$ is clearly a quotient of the adjoint group of $J \ltimes U$ by the image of $Z_U(L)$ in the adjoint group of $J \ltimes U$, and so is the real points of an algebraic group defined over $\mathbb{R}$, and so $\mathrm{Aut}^A(H')$ is the real points of an algebraic group defined over $\mathbb{R}$.

It also follow easily from our description of $\mathrm{Aut}^A(H')$ that every element of $\mathrm{Aut}^A(H')$ is the restriction of a rational automorphism of $\mathbb{H}'$. This follows from the fact that $\mathrm{Aut}(\mathbb{U})$ acts rationally on $\mathbb{U}$ which follows from the fact that exp and ln are rational diffeomorphisms between U and $\mathfrak{u}$ by the Baker-Campbell-Hausdorff formula, and that any automorphism of $\mathfrak{u}$ is linear and therefore rational.

To show that $\mathrm{Aut}^A(H') \ltimes H'$ is the real points of an algebraic group defined over $\mathbb{R}$ only requires that we show that $\mathrm{Aut}^A(H') \ltimes H' \to H'$ is the restriction of a rational map. Using the coordinates on $\mathrm{Aut}^A(H')$ described above this reduces to showing that the map $Z_{\mathrm{Aut}(\mathbb{U})}(\mathbb{L})\ltimes\mathbb{U}\to\mathbb{U}$ is rational. This follows from the fact that $\mathrm{Aut}(\mathbb{U})\ltimes\mathbb{U}\to\mathbb{U}$ is rational which follows from the fact that $\mathrm{Aut}(\mathbb{U})$ is defined as an algebraic subgroup of $GL(\mathfrak{u})$ and from rationality of automorphisms of U discussed above. $\square$

THEOREM 6.4. *Let G be as in section 3.1, but with no assumption on the rank of G. Assume H is a connected real algebraic group and Λ a cocompact discrete subgroup. Let ρ be an affine action of G on H/Λ. Then the action ρ is given by $\rho(g)[h]=[\pi_0(g)h]$ where $\pi_0:G\to H$ is a continuous homomorphism.*

THEOREM 6.5. *Let G be as in section 3.1 and let $\Gamma<G$ be a weakly irreducible lattice. Let H and Λ be as above. Let ρ be an affine action of Γ on H/Λ. Then there is a finite index subgroup $\Gamma'<\Gamma$ such that, possibly after replacing H by H' as in Proposition 6.3, the Γ' action on H/Λ is given by $\rho(\gamma)[h]=[\pi_H(\gamma)\cdot\pi_A(\gamma)h]$. Here $\pi_H:\Gamma'\to H'$ and $\pi_A:\Gamma'\to\mathrm{Aut}(H')$ are homomorphisms whose images commute as subgroups of $\mathrm{Aut}(H')\ltimes H'$. Furthermore, we can assume that $(\pi_A,\pi_H)(\Gamma')$ is contained in $\mathrm{Aut}^A(H')\ltimes H'$, an algebraic group.*

Remark: Using the results of [**C, GS, Rg1, Rg2, St**] in combination with the arguments in [**M3**], one can assume only that Γ projects to a dense subgroup of a rank one simple factors not locally isomorphic to F_4^{-20} or $Sp(1,n)$.

It is obvious from this description that the action of G or Γ on H/Λ lifts to H on a subgroup of finite index.

PROOF OF THEOREM 6.4. The action ρ defines a continuous homomorphism

$$\pi:G\to(N_{\mathrm{Aut}(H)}(\Lambda)\ltimes H)/(Z(H)\cap\Lambda).$$

Since the target is a Lie group, any simple factor $\mathbb{F}(k)$ of G which is defined over a non-Archimedean field k has trivial image, since is totally disconnected and topologically almost simple. Therefore it suffices to consider the case where G is a connected Lie group. We replace H by a group H' and Λ by Λ' as in Proposition 6.3. Let $\mathrm{Aut}^A(H')<\mathrm{Aut}(H')$ be the subgroup of finite index such that every element of $\mathrm{Aut}^A(H')$ is rational. Then since $\pi(G)$ is connected, $\pi(G)$ must be contained in the image of $(\mathrm{Aut}^A(H')\ltimes H')\cap(N_{\mathrm{Aut}(H')}(\Lambda)\ltimes H')$.

It follows from generalizations of Borel's density theorem that Λ' is Zariski dense in a cocompact normal subgroup of H', see for example [**D**] or [**Sh**, Theorem 1.1]. Therefore by Lemma 6.3 any automorphism of H' that fixes Λ' pointwise factors through an automorphism of a compact quotient $\bar{H}$ of H. It is easy to see that $\mathrm{Aut}(\bar{H})$ is a discrete extension of a compact group, since $\bar{H}$ is an almost direct product of compact simple groups and compact torii. Since Λ' is discrete $N_{\mathrm{Aut}(H')}(\Lambda')/Z_{\mathrm{Aut}(H')}(\Lambda')$ is discrete and therefore $N_{\mathrm{Aut}(H')}(\Lambda')$ is a discrete extension of a compact group. As remarked above $\pi(G)$ is contained in the connected component of $\mathrm{Aff}(H/\Lambda')$. Letting Z^0 be the connected component of $Z_{\mathrm{Aut}(H')}(\Lambda')$, the connected component of $\mathrm{Aff}(H/\Lambda)$ is $\phi(Z^0\ltimes H')$. Now $Z^0\ltimes H'\cap\ker(\phi)=Z(H')\cap\Lambda'$ so $\phi(Z^0\ltimes H')\cong(Z^0\ltimes H')/(Z(H')\cap\Lambda')$. Since G has no compact factors and Z^0 is compact, the map $\pi:G\to(Z^0\ltimes H')/(Z(H')\cap\Lambda')$ takes values in $H/(Z(H)\cap\Lambda')$.

Since G is either simply connected or simply connected as an algebraic group, we can lift π to a homomorphism $\tilde{\pi}:G\to H'$. This also define a homomorphism

$\pi : G \to H$, and it is easy to verify that $\tilde{\pi}$ and π define the same affine action on $H/\Lambda \cong H'/\Lambda'$. □

The proof for Γ actions is more complicated and requires the use of the superrigidity theorems. In addition to a direct application, we will also use the following consequence of the superrigidity theorems.

LEMMA 6.6. *Let G and Γ be as above. Let $\pi : \Gamma \to D$ be any Zariski dense homomorphism into a real algebraic group D. Let $\tilde{D}$ be a real algebraic group and $\tilde{D} \to D$ an isogeny. Then there is a finite index subgroup $\Gamma' < \Gamma$ and a homomorphism $\tilde{\pi} : \Gamma' \to \tilde{D}$ such that $p \circ \tilde{\pi} = \pi$ where $p : \tilde{D} \to D$ is the natural covering map.*

PROOF OF LEMMA 6.6. Since by [**M3**, IX.5.8], the image of any homomorphism from Γ into a real algebraic group has semisimple algebraic closure, it suffices to consider the case where D is semisimple. Since it also suffices to consider the case where $\mathbb{D}$ is simply connected as an algebraic group and simply connected semisimple algebraic groups are direct products of simple groups by [**M3**, I.1.4.10], it suffices to consider the case where $\mathbb{D}$ is simple and simply connected as an algebraic group.

We first assume $G = \prod_I G_i$ where each G_i is algebraic and $G_1 = \mathbb{G}_1(\mathbb{R})$. We have a homomorphism $\pi : \Gamma \to \mathbb{D}(\mathbb{R})$. It is clear that $\pi(\Gamma) < \mathbb{D}(k)$ where k is a finite extension of $\mathbb{Q}$, and we let $\bar{k}$ be the algebraic closure of k. Since Γ is Zariski dense in D, it follows that D is defined over k. A corollary of the superrigidity theorems, see [**M3**, Theorems VII.6.5 and VII.6.6], shows that, after passing to a subgroup of finite index, there is an embedding σ of $\bar{k}$ in $\mathbb{C}$, and a $\mathbb{C}$-rational map $\eta : \mathbb{G}_1 \to {}^{\sigma}\mathbb{D}$ such that $\pi(\gamma) = \sigma^{-1}(\eta(\gamma))$. Here ${}^{\sigma}\mathbb{D}$ is the $\sigma(k)$ algebraic group defined by the image under σ of the equations defining $\mathbb{D}$. Since $\mathbb{G}_1$ is simply connected as an algebraic group, it follows that we can lift η to a map to $\tilde{\mathbb{D}}$, and then define the lift of π by the same equation.

For G_1 a topological cover of a real algebraic group $\bar{G}_1$, the argument above gives the same conclusion concerning π, η and σ where η is a continuous homomorphism of G_1 which factors through $\bar{G}_1$. □

PROOF OF THEOREM 6.5. The action ρ is described by a homomorphism $\pi : \Gamma \to (N_{\mathrm{Aut}(H)}(\Lambda) \ltimes H)/\Delta^{-1}(\Lambda)$. We observe that a finite index subgroup in the group $(N_{\mathrm{Aut}(H)}(\Lambda) \ltimes H)/\Delta^{-1}(\Lambda)$ maps into $(\mathrm{Aut}^A(H) \ltimes H)/\Delta^{-1}(\Lambda)$ which maps onto $(\mathrm{Aut}^A(H) \ltimes H)/\Delta^{-1}(H)$. After passing to a subgroup of finite index in Γ and composing π with this inclusion and surjection, we obtain a homomorphism $\bar{\pi} : \Gamma \to (\mathrm{Aut}^A(H) \ltimes H)/\Delta^{-1}(H)$. Recall that $(\mathrm{Aut}^A(H) \ltimes H)$ is an algebraic group, and note that $\Delta^{-1}(H)$ is an algebraic subgroup, so the quotient $(\mathrm{Aut}^A(H) \ltimes H)/\Delta^{-1}(H)$ is an algebraic group. Applying the superrigidity theorems to π, we see that $\bar{\pi} = (\bar{\pi}_E, \bar{\pi}_K)$ where $\bar{\pi}_E$ is the restriction of a continuous homomorphism of G and $\bar{\pi}_K$ has bounded image. By [**M3**, IX.5.8], we know that $\bar{\pi}(\Gamma)$ has semisimple Zariski closure $\mathbb{J}$. Here $\mathbb{J} = \mathbb{J}_1\mathbb{J}_2$ where $\mathbb{J}_1$ is isotropic over $\mathbb{R}$ and $\mathbb{J}_2$ is anisotropic over $\mathbb{R}$.

We let $\mathrm{Aut}^A(H) \ltimes H = \mathbb{L}_1(\mathbb{R}) \ltimes \mathbb{U}_1(\mathbb{R})$ where $\mathbb{L}_1$ is reductive algebraic group and $\mathbb{U}_1$ is the unipotent radical. Similarly, we let $\Delta^{-1}(H) = \mathbb{L}_2(\mathbb{R}) \ltimes \mathbb{U}_2(\mathbb{R})$ and $(\mathrm{Aut}^A(H) \ltimes H)/\Delta^{-1}(H) = \mathbb{L}_3(\mathbb{R}) \ltimes \mathbb{U}_3(\mathbb{R})$ where $\mathbb{L}_2$ and $\mathbb{L}_3$ are reductive and $\mathbb{U}_2$ and $\mathbb{U}_3$ are unipotent. Then $\mathbb{L}_3 = \mathbb{L}_1/\mathbb{L}_2$ and these are all reductive, we can find a subgroup $\tilde{\mathbb{L}}_3$ such that $\mathbb{L} = \mathbb{L}_2\tilde{\mathbb{L}}_3$ is an almost direct product and the map $\tilde{\mathbb{L}}_3 \to \mathbb{L}_3$ is an isogeny. By Lemma 6.6, there is a finite index subgroup $\Gamma' < \Gamma$ on which we can lift $\tilde{\pi}$ to a representation into $\tilde{\pi}$ into $\mathbb{L}_3(\mathbb{R})$ and therefore into $\mathrm{Aut}^A(H)\ltimes H$. It is easy to verify that $\tilde{\pi}$ and π define the same affine action of Γ'.

That π can be chosen to be of the form (π_A, π_H) requires a supplementary argument. Let $\mathbb{H} = \mathbb{L}\ltimes\mathbb{U}$ be Levi decomposition. It follows from the proof of Proposition 6.3 that the Levi complement of $\mathrm{Aut}^A(H)$ a direct product of the adjoint group $\bar{\mathbb{L}}$ of $\mathbb{L}$ and a reductive subgroup $\mathbb{L}'$ of $\mathrm{Aut}(\mathbb{U})$ that commutes with $\bar{\mathbb{L}}$. In the description above, one can take $\mathbb{L}_2 = \Delta^{-1}(\mathbb{L})$ and $\mathbb{L}_1 = \mathbb{L}'\times(\bar{\mathbb{L}}\ltimes\mathbb{L})$. Therefore $\mathbb{L}_3$ is isomorphic to $\mathbb{L}'\times\bar{\mathbb{L}}$, and if one chooses $\tilde{\mathbb{L}}_3$ to be $\mathbb{L}'\times\mathbb{L} < \mathbb{L}'\times(\bar{\mathbb{L}}\ltimes\mathbb{L})$ one has the desired conclusion. □

6.2. Applications. We now proceed to prove the applications listed in the introduction. For the remainder of this section G is as defined in the second paragraph of the introduction and Γ is a lattice in G. For any affine action of G or Γ on H/Λ, or any associated quasi-affine or generalized affine action, we assume that H satisfies the conclusions of Proposition 6.3 and so we can describe the action by Theorem 6.4 or 6.5. We call the homomorphism defining the action $\pi_0 = (\pi_A, \pi_H)$ and let $\mathbb{A}$ be the Zariski closure of $\pi_A(\Gamma')$ in $\mathrm{Aut}^A(\mathbb{H})$. If we are concerned with Γ actions, π_0 only defines the action on a subgroup of finite index Γ'. For the proof of Theorem 1.8, we replace Γ' by a subgroup of finite index to assure that $\mathbb{A}$ is connected.

Proof of Corollary 1.7. Let $D = G$ or Γ and ρ be a generalized standard affine action of D on a manifold M. Since the entropy of an element is determined by the entropy of it's kth power, it suffices to prove the corollary for a subgroup D' of finite index.

We first prove the corollary for ρ affine and $M = H/\Lambda$ and then describe the modifications for the general case. For an affine action we have $TM = H/\Lambda\times\mathfrak{h}$, and by Theorems 6.4 and 6.5 there is a subgroup of finite index $D' < D$ and a homomorphism $\pi_0 : D\to Aut(H)\ltimes H$ such that the derivative cocycle of the D' action is $\tilde{\pi}_0 = (\mathrm{Ad}_{\mathrm{Aut}(H)\ltimes H}\circ\pi_0)|_{\mathfrak{h}}$.

Let ρ' be a C^2 action C^1 close to ρ. By a result of Seydoux, ρ' preserves a measure that is in the same measure class as Lebesgue measure [**S**]. Since the derivative cocycle $\alpha_{\rho'}$ of ρ' is C^0 close to the constant cocycle given by $\tilde{\pi}_0$ (which is the derivative cocycle for ρ), it follows from Theorem 5.1 that $\alpha_{\rho'}$ is cohomologous to $\tilde{\pi}_0^E\cdot c$ where $\tilde{\pi}_0^E$ is the extendable part of $\tilde{\pi}_0$, and c is cocycle over the D action taking values in a compact group that commutes with $\tilde{\pi}_0^E(D')$. The corollary now follows from the fact that the entropies $h_\rho(d)$ and $h_{\rho'}(d)$ can both be computed in the same manner from the eigenvalues of $\tilde{\pi}_0^E(d)$. This follows from Pesin's formula relating entropy to Lyapunov exponents as observed by Furstenberg in [**Fu3**], see also [**Z2**, Chapter 9].

We now pass to the case of a generalized affine action ρ of D on $K\backslash H/\Lambda$. By definition ρ is the quotient of an affine action D on H/Λ, so on a subgroup of finite index, ρ is given by a homomorphism $\pi_0 : D\to \mathrm{Aut}(H)\ltimes H$. Since $K < H$ commutes with D, and since K is compact and the Zariski closure of $\pi_0(D)$ in $\mathrm{Aut}(H)\ltimes H$ is semisimple, there is a splitting $\mathfrak{h} = \mathfrak{k}\oplus\mathfrak{m}$ invariant under $\mathrm{Ad}_{\mathrm{Aut}(H)\ltimes H}$ restricted to both K and $\pi_0(D)$. The tangent bundle to $K\backslash H/\Lambda$ can be identified with $K\backslash H/\Lambda\times\mathfrak{m}$ and, on a subgroup $D' < D$ of finite index, the derivative cocycle is $(\mathrm{Ad}_{\mathrm{Aut}(H)\ltimes H}\circ\pi_0)|_{\mathfrak{m}}$. The remainder of the proof follows as before. □

We remark that the proof of the corollary only uses part of Theorems 1.1 and 1.2, namely the conclusion that the derivative cocycle for the perturbed action is cohomologous to $\pi_0^E{\cdot}c$ where c takes values in a compact group.

We now prove Theorem 1.8 from the introduction. Actually, we prove a slightly stronger statement. Let H be a connected real algebraic group and $\Lambda < H$ a discrete cocompact subgroup. Let ρ be a standard affine action of $D = G$ or Γ on $H/\Lambda{\times}M$. By Theorems 6.4 and 6.5 above, there is a finite index subgroup $D' < D$ such that the action of D' on $H/\Lambda{\times}M$ is given by $\rho(d)([h], m) = (\pi_0(d)h, \iota(d,h)m)$ where $\pi_0 : D{\to}\mathrm{Aut}(H){\ltimes}H$ is a homomorphism and $\iota : D{\times}H/\Lambda{\to}\,\mathrm{Isom}(M)$ is a cocycle. It is clear from this description that the action of D' lifts to $H{\times}M$. For G actions, we let $Z = Z_H(\pi_0(G))$. For Γ actions the description of Z is more complicated. Let Γ' be the subgroup of finite index given by Theorem 6.5. Recall that $\pi_0 = (\pi_A, \pi_H)$ and let A be the set of real points of the Zariski closure of $\pi_A(\Gamma')$, and define $L = A{\ltimes}H$. We view π_0 as taking values in L. We let $Z = Z_L(\pi_0^E(\Gamma)){\cap}H$.

THEOREM 6.7. *Let ρ be a standard affine action of $D = G$ or Γ on $H/\Lambda{\times}M$ as above. Let D' and Z be as above. Given any action ρ' sufficiently C^1 close to ρ, there is a cocycle $z : D{\times}H/\Lambda{\times}M{\to}Z$ and a continuous map $f : H/\Lambda{\times}M{\to}H/\Lambda$, such that*

1. *f is C^0 close to the natural projection*
2. *for any $d{\in}D'$ and any $([h], m){\in}H/\Lambda{\times}M$ we have*

$$f(\rho'(d)([h], m)) = (\pi_A(d), \pi_H^E(d)z(d, ([h], m)))f([h], m).$$

To prove Theorem 6.7, we recall, and generalize slightly, the construction, from [**MQ**] of a cocycle from a perturbation of an affine action. We will prove Theorem 6.7 by applying Theorem 5.1 to this cocycle.

Once again, we let D denote our acting group. For the construction of the cocycle, D can be more general than G or Γ above, as long as the D action is as in the conclusions of Theorems 6.4 or 6.5. First we define the cocycle for actions by left translations. Let the D action ρ on H/Λ be defined via a homomorphism $\pi_0 : D{\to}H$. Let ρ' be a perturbation of ρ. If D is connected it is clear that the action lifts to $\tilde{H}$ and therefore to H. If D is discrete, this lifting still occurs, since the obstacle to lifting is a cohomology class in $H^2(D, \pi_1(H/\Lambda))$ which does not change under a small perturbation of the action. (A direct justification without reference to group cohomology can be found in [**MQ**] section 2.3.) Write the lifted actions of D on H by $\tilde{\rho}$ and $\tilde{\rho}'$ respectively. We can now define a cocycle $\alpha : D{\times}H{\to}H$ by

$$\tilde{\rho}'(g)x = \alpha(g, x)x$$

for any g in D and any x in H. It is easy to check that this is a cocycle and that it is right Λ invariant, and so defines a cocycle $\alpha : D{\times}H/\Lambda{\to}H$. See [**MQ**] section 2 for more discussion.

Similarly if D acts on H/Λ affinely we can obtain a cocycle as well. Note this only occurs if D is discrete. For a finite index subgroup $D' < D$ the action is described by a homomorphism π_0 into a group $A{\ltimes}H = L$ as described before the statement of Theorem 6.7. This action D' lifts to H as noted above. We define the cocycle only after replacing D with D'. As before the D action defined by ρ' also lifts to H. We let π_A be the projection of π_0 on A. Recall that $\pi_A(D)$ must normalize Λ and that $\pi_A(D)$ is Zariski dense in A. Here we consider H/Λ as the space $\pi_A(D){\ltimes}H/\pi_A(D){\ltimes}\Lambda$, and then define the cocycle by the same equation as before, noting that we get a cocycle $\alpha : D{\times}H/\Lambda{\to}\pi(D){\ltimes}H$. Observe that there is a map $\alpha' : D\times H/\Lambda \to H$ such that $\alpha(d,x) = (\pi_A(d), \alpha'(d,x))$. It is possible to define α' directly by the same equation as α, but it is not a cocycle. The map α' is called a twisted cocycle in [**F**], further discussion can be found there and in [**MQ**] section 2, example 2.2.

Whenever we have the cocycle α defined as taking values in $\pi_A(D)\ltimes H$, we extend this to a cocycle into $L = A\ltimes H$ where A is as above.

If D acts on $H/\Lambda\times M$ we can also define a cocycle as above. For simplicity we discuss the definition of this cocycle when the D action on H/Λ is given by a homomorphism $\pi_0 : D\to H$. Let $p : H/\Lambda\times M\to H/\Lambda$ be the projection. In this case we define $\alpha : D\times H/\Lambda\times X\to H$ via the formula

$$\alpha(g,x)p(x) = p(\tilde{\rho}'(g)x).$$

The verification that this is a cocycle and defined on $D\times H/\Lambda\times M$ follows exactly as in [**FW**]. The case of more general π_0 follows exactly as above and again results in a cocycle into $A\ltimes H$. Note that to define this cocycle, we do not need to know that the skew product action is isometric on M.

In all three cases, we can define a cocycle α_{π_0} corresponding to the action ρ it is clear from the definition that $\alpha_{\pi_0}(g,x) = \pi_0(g)$ in all cases.

PROOF OF THEOREM 6.7. Given ρ' a perturbation of ρ, we construct a cocycle $\alpha : D\times H/\Lambda\to A\ltimes H$ as described above. It is clear from the construction that α is continuous and close to the constant cocycle defined by π_0. Applying the result of Seydoux, we see that ρ' preserves a measure μ that is in the same measure class as Lebesgue measure. Therefore μ has full support. Applying Theorem 5.1 to α, we have that there is a map $\phi_h : H/\Lambda\times M\to H$ and a cocycle $z : H/\Lambda\times M\to H$ such that

1. $\alpha(g,x) = \phi_h(gx)^{-1}(\pi_A(g), \pi_H^E(g)z(g,x))\phi_h(x)$
2. ϕ_h is continuous, C^0 small and depends continuously on $\rho'\in \mathrm{Diff}^1(H/\Lambda\times M)$
3. z is continuous and C^0 close to the constant cocycle defined by π_H^K

We define $f : H/\Lambda\times M\to Z\backslash H/\Lambda$ by $f(x) = p(\phi_h(x)x)$ where we write $p : H/\Lambda\times M\to H/\Lambda$ for the natural projection. It then follows from the definition of α and our conclusions about ϕ that:

$$\begin{aligned} f(\rho'(g)x) &= p(\phi_h(gx)\rho'(g)x) = p(\phi_h(gx)\alpha(g,x)x) \\ &= p(\phi_h(gx)\phi_h(gx)^{-1}(\pi_A(g), \pi_H^E(g)z(g,x))\phi_h(x)x) \\ &= (\pi_A(g), \pi_H^E(g)z(g,x))p(\phi_h(x)x) \\ &= (\pi_A(g), \pi_H^E(g)z(g,x))f(x). \end{aligned}$$

□

Theorem 1.8 is an immediate consequence of Theorem 6.7 and the definition of Z.

References

[AB] N. A'Campo and M.Burger, Réseaux arithmétiques et commensurator d'après G.A.Margulis, *Invent. Math.* 116 (1994) 1-25.

[An] D.V.Anosov, Geodesic flows on compact manifolds of negative curvature, *Trudy Mat. Inst. Steklov* 90 (1967) .

[BZ] I.N.Bernstein and A.V.Zelevinsky, Representations of the group $GL(n,F)$ where F is a non-archimedean local field, Russ. Math. Surveys, 31.3 (1976), 1-68.

[B1] A.Borel, *Introduction aux groupes arithmtiques.* (French) Publications de l'Institut de Mathmatique de l'Universit de Strasbourg, XV. Actualits Scientifiques et Industrielles, No. 1341 Hermann, Paris 1969 125 pp.

[B2] A.Borel, *Linear Algebraic Groups*, Springer-Verlag, New York, 1991.

[BH] M.Bridson, A.Haefliger, *Metric spaces of non-positive curvature.* Grundlehren der Mathematischen Wissenschaften, 319. Springer-Verlag, Berlin, 1999.

[C] K.Corlette, Archimedean superrigidity and hyperbolic geometry. *Ann. of Math.* 135 (1992), no. 1, 165–182.

[CZ] K.Corlette and R.J.Zimmer, Superrigidity for cocycles and hyperbolic geometry. *Internat. J. Math.* 5 (1994) 273–290.

[D] S.G.Dani, A simple proof of Borel's density theorem, *Math. Z.* 174 (1980) 81-94.

[F] D. Fisher, On the arithmetic structure of lattice actions on compact spaces, *Erg. Th. Dyn. Sys.* (2002) 22, 1141-1168.

[FW] D. Fisher and K.Whyte, Continuous quotients for lattice actions on compact spaces, *Geom. Ded.* 2001 (87) 181-189.

[FM1] D.Fisher and G.A.Margulis, Local rigidity of standard actions of higher rank groups and their lattices, preprint.

[FM2] D.Fisher and G.A.Margulis, Almost isometric actions, Property T, and local rigidity, preprint.

[Fu1] H.Furstenberg, A Poisson formula for semisimple Lie groups, *Annals of Math.* 77(1963), 335-383.

[Fu2] H.Furstenberg, Boundary theory and stochastic processes on homogeneous spaces in: Harmonic analysis on homogeneous spaces. *Symposia Pure and Appl. Math.* 26 (1973) 193-229.

[Fu3] H.Furstenberg, Rigidity and cocycles for ergodic actions of semisimple Lie groups (after G.A.Margulis and R.Zimmer), *Seminaire Bourbaki*, 1979/80, 272-292.

[GR] H.Garland and M.S.Ragunathan, Fundamental domains for lattices in (R-)rank 1 semisimple Lie groups. *Ann. of Math.* (2) 92 1970 279–326.

[Gl] V.M.Glushkov, The structure of locally compact groups and Hilbert's Fifth problem, *Uspehi Mat. Naut* 12 (1957) 3-41.

[GS] Gromov, Mikhail; Schoen, Richard. Harmonic maps into singular spaces and p-adic superrigidity for lattices in groups of rank one. Inst. Hautes tudes Sci. Publ. Math. No. 76 (1992), 165–246.

[HV] P. de la Harpe, A.Valette, *La propriete (T) de Kazhdan pour les groupes localement compacts*, Asterisque 175, Soc. Math. de France, Paris, 1989.

[K] I.Kaplansky, *Lie algebras and locally compact groups.* Reprint of the 1974 edition. Chicago Lectures in Mathematics. University of Chicago Press, Chicago, IL, 1995.

[La] S. Lang, *Algebra, Third Revised Edition*, Springer-Verlag, New York, 2002.

[L] J. Lewis, The algebraic hull of a measurable cocycle, to appear *Geom. Ded.*

[Li] L. Lifschitz, Superrigidity thoerems in positive characteristic, *J. Algebra* 229 (2000) 375-404.

[LMR] A. Lubotzky, S. Mozes, M.S. Raghunathan, The word and Riemannian metrics on lattices of semisimple groups, *Publ IHES* 91 (2000) 5-53.

[M1] G.A. Margulis, Discrete groups of motions of manifolds of non-positive curvature, *Proc. Int. Cong. Math.* Vancouver 1974. *AMS Translations* 109 (1977) 33-45.

[M2] G.A. Margulis, Arithmeticiy of irreducible lattices in semisimple gropus of rank greater than 1, appendix to the Russian translation of M.Ragunathan, *Discrete subgroups of Lie groups*, Mir, Moscow, 1977 (in Russian). English translation in *Invent. Math.* 76 (1984) 93-120.

[M3] G.A. Margulis, *Discrete subgroups of semisimple Lie groups*, Springer-Verlag, New York, 1991.

[MQ] G. Margulis and N. Qian, Local rigidity of weakly hyperbolic actions of higher real rank semisimple Lie groups and their lattices, *Ergodic Theory and Dynamical Systems* 21 (2001) 121-164.

[M] G.D.Mostow, *Strong rigidity of locally symmetric spaces*, Annals of Math. Studies 78, 1973, Princeton University Press, Princeton, NJ.

[P] Ya. B. Pesin, On the existence of invariant fiberings for a diffeomorphism of a smooth manifold, *Math. USSR Sbornik.*, 20 (1973), 213-222.

[PlRa] V. Platonov and A. Rapinchuk, *Algebraic Groups and Number Theory*, Academic Press, New York, 1994.

[PR] G.Prasad and M.S.Ragunathan, Cartan subgroups and lattices in semisimple groups, *Ann. Math.*, 96 (1972), 296-317.

[QZ] N.Qian and R.J. Zimmer, Entropy rigidity for semisimple group actions, *Israel J. Math.*, 99 (1997) 55-67.

[Rg1] M.S. Raghunathan, Cohomology of arithmetic subgroups of algebraic groups. I, II. *Ann. of Math.* 86 (1967), 409-424; ibid. (2) 87 (1967) 279–304.

[Rg2] M.S. Raghunathan, On the first cohomology of discrete subgroups of semisimple Lie groups. *Amer. J. Math.* 87 (1965) 103–139.

[S] Seydoux, G., Rigidity of ergodic volume-preserving actoins of semisimple groups of higher rank on compact manifolds, *Trans. Am.* 345 (1994) 753-776.

[Sh] Y.Shalom, Invariant measures for algebraic actions,Zariski dense subgroups and Kazhdan's property (T), *Trans. Am.* 351 (1999) 3387-3412.

[St] A.N.Starkov, Vanishing of the first cohomologies for lattices in Lie groups. *J. Lie Theory* 12 (2002), no. 2, 449-460.

[Stu] G.Stuck Cocycles of ergodic group actions and vanishing of first cohomology for S-arithmetic groups. *Amer. J. Math.* 113 (1991), no. 1, 1–23.

[V] T.N. Venkataramana, On superrigidity and arithmeticity of lattices in semsimple groups over local fields of arbitrary characteristic, *Invent. Math.* 92 (1988) 255-306.

[Z1] R.J, Zimmer Strong rigidity for ergodic actions of semisimple Lie groups. *Ann. of Math.* (2) 112 (1980), no. 3, 511–529.

[Z2] R.J. Zimmer,*Ergodic Theory and semisimple Groups*, Birkhäuser, Boston, 1984.

[Z3] R.J. Zimmer, Actions of semisimple groups and discrete subgroups, *Proc. Internat. Cong. Math.*, Berkeley, 1986, 1247-1258.

[Z4] R.J. Zimmer, On the algebraic hull of an automorphism group of a principal bundle, *Comment. Math. Helv.*, 65 (1995) 375-387.

David Fisher
Department of Mathematics and Computer Science
Lehman College - CUNY
250 Bedford Park Boulevard W
Bronx, NY 10468
dfisher@lehman.cuny.edu

G.A. Margulis
Department of Mathematics
Yale University
P.O. Box 208283
New Haven, CT 06520
margulis@math.yale.edu

Einstein Metrics, Four-Manifolds, and Differential Topology

Claude LeBrun

A Riemannian metric g on a smooth manifold M is said to be *Einstein* if it has constant Ricci curvature, or in other words if

$$r = \lambda g, \tag{1}$$

where r is the Ricci tensor of g and λ is some real constant [**7**]. We still do not know if there are any obstructions to the existence of Einstein metrics on high-dimensional manifolds, but it has now been known for three decades that not every 4-manifold admits such metrics [**20, 37**]. Only recently, however, has it emerged that there are also obstructions to the existence of Einstein metrics which depend on the differentiable structure rather than just on the homotopy type of a 4-manifold [**27, 23, 28, 22**]. This article will attempt to give a concise explanation of this state of affairs.

Our current understanding of the problem rests largely on certain curvature estimates which are deduced from the Seiberg-Witten equations. The strongest of these [**28**] was originally proved indirectly, by invoking the solution to a generalized version of the Yamabe problem. One main purpose of the present article is to give (§4) a new and simpler proof of this estimate, using conformal rescaling only in order to introduce a generalized form of the Seiberg-Witten equations. But this article will also attempt to clarify the nature of the resulting obstructions, by systematically reformulating them in terms of a new diffeomorphism invariant, called $\alpha(M)$, which is introduced in §3.

As this article will make abundantly clear, Blaine Lawson's work on spin geometry and scalar curvature has had a deep and lasting impact on my own research. On a more personal level, Blaine has also been tremendous source of inspiration and encouragement throughout my many years at Stony Brook. I am lucky indeed to be able to call him a friend and colleague, and it is a very great pleasure for me to be able to contribute an article to this volume.

1. Differential Geometry on 4-Manifolds

The curvature and topology of 4-manifolds are interrelated in a number of ways that have no adequate analogs in other dimensions. Many of these phenomena are intimately related to the fact that the bundle Λ^2 of 2-forms over an oriented

---

Supported in part by NSF grant DMS-0305865.

Riemannian 4-manifold (M, g) invariantly decomposes as the direct sum

$$\Lambda^2 = \Lambda^+ \oplus \Lambda^-, \tag{2}$$

of the eigenspaces of the Hodge star operator

$$\star : \Lambda^2 \to \Lambda^2.$$

The sections of Λ^+ are characterized by $\star\varphi = \varphi$, and so are called *self-dual 2-forms*, whereas the sections of Λ^- satisfy $\star\varphi = -\varphi$, and are called *anti-self-dual 2-forms.* Writing an arbitrary 2-from uniquely as

$$\varphi = \varphi^+ + \varphi^-,$$

where $\varphi^\pm \in \Lambda^\pm$, we then have

$$\varphi \wedge \varphi = \left(|\varphi^+|^2 - |\varphi^-|^2\right) d\mu_g,$$

where $d\mu_g$ denotes the metric volume form associated with the fixed orientation. The real importance of all this stems from the fact that the curvature of any connection on a vector bundle over M is a bundle-valued 2-form, and (2) therefore always gives rise to a decomposition of any curvature tensor into simpler pieces.

The Riemann curvature tensor of g is a good case in point. By raising an index, we can turn the Riemannian curvature into a self-adjoint linear map

$$\mathcal{R} : \Lambda^2 \to \Lambda^2$$

called the *curvature operator.* Decomposing the 2-forms as in (2), this linear endomorphism of Λ^2 can then be decomposed into irreducible pieces

$$\mathcal{R} = \left(\begin{array}{c|c} W^+ + \frac{s}{12} & \overset{\circ}{r} \\ \hline \overset{\circ}{r} & W^- + \frac{s}{12} \end{array} \right). \tag{3}$$

Here s denotes the scalar curvature, whereas $\overset{\circ}{r} = r - \frac{s}{4}g$ denotes the trace-free part of the Ricci curvature. The tensors $W_\pm$ are the trace-free pieces of the appropriate blocks, and are respectively called the *self-dual* and *anti-self-dual Weyl curvature* tensors. The corresponding pieces of the Riemann tensor enjoy the remarkable property of being *conformally invariant*, in the sense that they remain unaltered when the metric is multiplied by an arbitrary smooth positive function. Notice that Einstein condition (1) is satisfies iff $\mathcal{R}$ commutes with the star operator; in more elementary terms, this amounts to the statement that a Riemannian 4-manifold is Einstein iff the sectional curvatures coincide for every orthogonal pair of 2-planes $P, P^\perp \subset T_xM$, $x \in M$.

Now let us suppose that (M, g) is a *compact* oriented Riemannian 4-manifold. The Hodge theorem then tells us that every de Rham class on M has a unique harmonic representative, so that we have a canonical identification

$$H^2(M, \mathbb{R}) = \{\varphi \in \Gamma(\Lambda^2) \mid d\varphi = 0,\ d\star\varphi = 0\}.$$

However, the Hodge star operator $\star$ defines an involution of the right-hand side. We therefore have a direct-sum decomposition

$$H^2(M, \mathbb{R}) = \mathcal{H}_g^+ \oplus \mathcal{H}_g^-, \tag{4}$$

where

$$\mathcal{H}_g^{\pm} = \{\varphi \in \Gamma(\Lambda^{\pm}) \mid d\varphi = 0\}$$

are the spaces of self-dual and anti-self-dual harmonic forms.

The intersection form

$$\begin{array}{rcl} Q : H^2(M,\mathbb{R}) \times H^2(M,\mathbb{R}) & \longrightarrow & \mathbb{R} \\ ([\varphi] , [\psi]) & \mapsto & \int_M \varphi \wedge \psi \end{array}$$

becomes positive-definite when restricted to $\mathcal{H}_g^+$, and negative-definite when restricted to $\mathcal{H}_g^-$. Moreover, these two subspaces are mutually orthogonal with respect to Q. Indeed, combining an L^2-orthonormal basis for $\mathcal{H}_g^+$ with an L^2-orthonormal basis for $\mathcal{H}_g^-$ gives one a basis for $H^2(\mathbb{R})$ in which the intersection form is represented by the diagonal matrix

$$\left[\begin{array}{ccc} \underbrace{\begin{matrix} 1 & & \\ & \ddots & \\ & & 1 \end{matrix}}_{b_+(M)} & & \\ & b_-(M)\left\{\begin{matrix} -1 & & \\ & \ddots & \\ & & -1 \end{matrix}\right. \end{array}\right].$$

The numbers $b_{\pm}(M) = \dim \mathcal{H}_g^{\pm}$ are thus oriented homotopy invariants of M; namely, b_+ (respectively, b_-) is the maximal dimension possible for a linear subspace of $H^2(M,\mathbb{R})$ on which the restriction of Q is positive (respectively, negative) definite.

The assignment $g \mapsto \mathcal{H}_g^+$ gives one an important canonical map

$$\{\text{Riemannian metrics on } M\} \longrightarrow Gr_{b_+}^+[H^2(M,\mathbb{R})]$$

from the infinite-dimensional space of all metrics to the finite-dimensional Grassmannian of $b_+(M)$-dimensional subspaces of $H^2(M,\mathbb{R})$ on which the intersection form is positive definite. This map is called the *period map* of M. It is obviously invariant under the action of the identity component $\mathit{Diff}_0(M)$ of the diffeomorphism group, and can also be shown to be invariant under the action of the smooth functions $M \to \mathbb{R}^+$ by conformal rescaling. A more subtle fact is that the period map is smooth, and has no critical points [**13**].

The difference $\tau(M) = b_+(M) - b_-(M)$ is called the *signature* of M. Like the Euler characteristic $\chi(M) = 2 - 2b_1(M) + b_2(M)$, it is the index of a geometric elliptic operator, and so may be expressed as a curvature integral. Indeed, to be explicit, one has

$$\begin{aligned} \chi(M) &= \frac{1}{8\pi^2}\int_M \left[\frac{s^2}{24} + |W^+|^2 + |W^-|^2 - \frac{|\mathring{r}|^2}{2}\right] d\mu \\ \tau(M) &= \frac{1}{12\pi^2}\int_M \left[|W^+|^2 - |W^-|^2\right] d\mu \end{aligned}$$

for absolutely any Riemannian metric g on M. In particular, we may combine these to obtain the Gauss-Bonnet-like formulæ

$$(2\chi \pm 3\tau)(M) = \frac{1}{4\pi^2}\int_M \left[\frac{s^2}{24} + 2|W_\pm|^2 - \frac{|\mathring{r}|^2}{2}\right] d\mu. \tag{5}$$

Notice, however, that the integrand in this last expression is non-negative for any Einstein metric. This gives us an important constraint, discovered independently by Thorpe [**37**] and Hitchin [**20**], on the topology of 4-dimensional Einstein manifolds.

THEOREM 1 (Hitchin-Thorpe Inequality). *If the smooth compact oriented 4-manifold M admits an Einstein metric g, then*

$$(2\chi + 3\tau)(M) \geq 0,$$

with equality iff (M, g) is finitely covered by a flat 4-torus T^4 or by $K3$ with a hyper-Kähler metric.

Moreover, (M, g) must also satisfy

$$(2\chi - 3\tau)(M) \geq 0,$$

and this inequality is strict unless (M, g) is finitely covered by a flat 4-torus T^4 or by the orientation-reversed version of $K3$ with a hyper-Kähler metric.

Here a smooth Riemannian metric g on an oriented 4-manifold M is called *hyper-Kähler* if the induced connection on Λ^+ is flat and trivial. Up to diffeomorphism, there is exactly one simply connected compact 4-manifold admits such a metric, called $K3$, in honor of Kummer, Kähler, and Kodaira. This 4-manifold is spin (meaning that its tangent bundle has $w_2 = 0$), and has $b_+ = 3$, $b_- = 19$. One model of $K3$ is the quartic hypersurface

$$z_0^4 + z_1^4 + z_2^4 + z_3^4 = 0$$

in $\mathbb{CP}_3$. The important point is that $K3$ is the underlying smooth oriented 4-manifold of a simply connected compact complex surface with $c_1 = 0$, because a truly remarkable theorem of Kodaira asserts that all the complex surfaces satisfying these conditions are deformation equivalent, and hence mutually diffeomorphic [**4**].

In order to put the Hitchin-Thorpe inequality to work, let us consider the following important way of constructing new manifolds.

DEFINITION 1. *Let M_1 and M_2 be smooth connected compact oriented n-manifolds.*

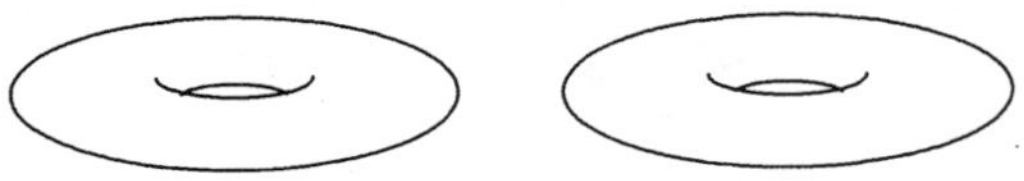

Their connected sum $M_1 \# M_2$ is then the smooth connected oriented n-manifold obtained by deleting a small ball from each manifold

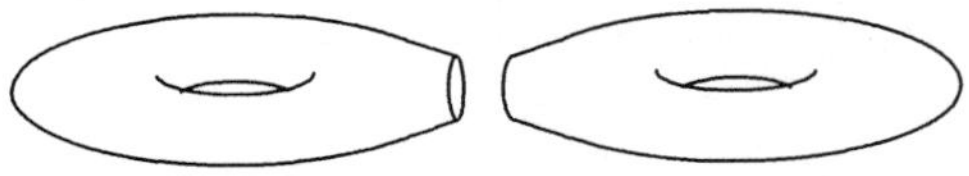

and identifying the resulting S^{n-1} boundaries

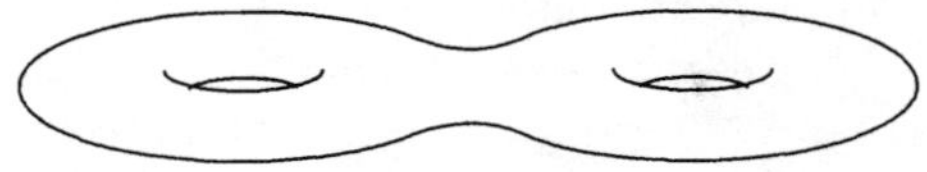

via a reflection.

If M_1 and M_2 are simply connected 4-manifolds, then $M = M_1\#M_2$ is also simply connected, and has $b_\pm(M) = b_\pm(M_1) + b_\pm(M_2)$.

Now let us use $\mathbb{CP}_2$ denote the complex projective plane with its *standard* orientation, and $\overline{\mathbb{CP}}_2$ denote the same smooth 4-manifold with the *opposite* orientation. Then the iterated connected sum

$$k\mathbb{CP}_2\#\ell\overline{\mathbb{CP}}_2 = \underbrace{\mathbb{CP}_2\#\cdots\#\mathbb{CP}_2}_{k}\#\underbrace{\overline{\mathbb{CP}}_2\#\cdots\#\overline{\mathbb{CP}}_2}_{\ell}$$

is a simply connected 4-manifold with $b_+ = k$ and $b_- = \ell$. In particular,

$$(2\chi + 3\tau)(k\mathbb{CP}_2\#\ell\overline{\mathbb{CP}}_2) = 4 + 5k - \ell,$$

so this construction gives us lots of simply connected 4-manifolds which do not admit Einstein metrics, by taking ℓ to be sufficiently large with respect to k.

We have now seen that Euler characteristic χ and signature τ have a vital rôle to play in the theory of 4-dimensional Einstein manifolds. However, knowing the Euler characteristic and signature of a *simply connected* 4-manifold is equivalent to knowing the invariants $b_\pm$, or in other words knowing the intersection form Q up to isomorphism as a quadratic form over $\mathbb{R}$. Now a celebrated result of Michael Freedman [**14**] asserts that simply connected smooth 4-manifolds are determined up to homeomorphism by their intersection forms over $\mathbb{Z}$. However, the integer coefficient version of the intersection form certainly contains more information than just $b_\pm$; indeed, the *parity* of the form (even or odd) determines whether or not the 4-manifold is spin. Indefinite quadratic forms over $\mathbb{Z}$ turn out to classified [**16**] by parity and $b_\pm$. On the other hand, a gauge-theoretic argument of Donaldson [**12**] shows that only the simplest definite forms can arise as intersection forms of 4-manifolds. One thus obtains the following remarkable classification result:

THEOREM 2 (Freedman). *Two smooth compact simply connected oriented 4-manifolds are orientedly homeomorphic if and only if*

- *they have the same Euler characteristic χ;*
- *they have the same signature τ; and*
- *both are spin, or both are non-spin.*

As a consequence, any smooth compact simply connected non-spin 4-manifold is homeomorphic to a connected sum $k\mathbb{CP}_2\#\ell\overline{\mathbb{CP}}_2$. For spin manifolds, the situation is a bit more unsettled, but the connected sums $mK3\#n(S^2 \times S^2)$ and their orientation-reversed versions, together with S^4, at least exhaust all the simply connected homeotypes for which the additional constraint $\chi \geq \frac{11}{8}|\tau| + 2$ is satisfied. The so-called 11/8 conjecture asserts that this last inequality is in fact automatically satisfied, or in other words that the above list of spin homeotypes is complete. A strong partial result in this direction has been proved by Furuta [**15**].

For the purposes of Riemannian geometry, however, this beautiful classification is somewhat aside from the point, since it is concerned with classification up to *homeomorphism*, not *diffeomorphism.* In order to do differentiable geometry, we need a differentiable structure. However, many of the above topological manifolds turn out to have *infinitely many* inequivalent differentiable structures, and the difference between these smooth structures is often detectable by asking questions about the curvature of Riemannian metrics. This article will attempt to explain some of the ramifications that this interplay between Riemannian geometry and differential topology is now known to have for the existence and uniqueness of Einstein metrics on 4-manifolds.

2. Examples of Einstein Manifolds

Most of the currently available examples of compact 4-dimensional Einstein manifolds are Kähler. Recall that a Riemannian manifold (M, g) is called *Kähler* if it admits an almost-complex structure $J : TM \to TM$, $J^2 = -1$, which is invariant under parallel transport with respect to g. Such an almost-complex structure is automatically integrable, and (M, J) may therefore be viewed as a complex manifold.

One may ask when a given compact complex manifold admits a compatible Kähler metric which is also Einstein. For Einstein metrics of negative Ricci curvature, the definitive solution to this problem was found independently by Aubin [**3**] and Yau [**40**]:

THEOREM 3 (Aubin/Yau). *A compact complex manifold (M^{2m}, J) admits a compatible Kähler-Einstein metric with $s < 0$ iff its canonical line bundle $K = \Lambda^{m,0}$ is ample. When such a metric exists, it is unique, up to an overall multiplicative constant.*

Here a holomorphic line bundle $L \to M$ is called *ample* if it it has a positive power $L^{\otimes k}$ with enough holomorphic sections to yield an embedding $M \hookrightarrow \mathbb{CP}_n$. As it turns out, a compact complex manifold (M, J) of real dimension 4 has ample canonical line bundle K iff it is a minimal complex surface of general type without (-2)-curves [**4**]. These exist in great profusion.

Yau [**40**] also gave a definitive solution to the analogous problem for Ricci-flat metrics:

THEOREM 4 (Yau). *A compact complex manifold (M, J) admits a compatible Kähler-Einstein metric with $s = 0$ iff (M, J) admits a Kähler metric and $K^{\otimes \ell}$ is trivial for some positive integer ℓ. When this happens, there is exactly one such metric in each Kähler class.*

In real dimension 4, there are exactly two diffeotypes of compact Kähler manifolds for which K is trivial, namely $K3$ and T^4. The Ricci-flat Kähler metrics on these manifolds are exactly the hyper-Kähler metrics alluded to in our previous discussion. Any other compact Kähler surface with K^{ℓ} trivial for some positive ℓ is the quotient of $K3$ or T^4 by the free action of a finite group of isometries of one of these hyper-Kähler metrics.

The existence of Kähler-Einstein metrics is much more delicate in the case of positive Ricci curvature. In real dimension 4, however, a complete solution to the problem was given by Tian [**38**]:

THEOREM 5 (Tian). *A compact complex surface (M^4, J) admits a compatible Kähler-Einstein metric with $s > 0$ iff its anti-canonical line bundle K^{-1} is ample and its Lie algebra of holomorphic vector fields is reductive.*

The 4-manifolds which carry Einstein metrics by virtue of this last result are $\mathbb{CP}_2$, $S^2 \times S^2$, and $\mathbb{CP}_2\#k\overline{\mathbb{CP}}_2$, $3 \leq k \leq 8$.

Until quite recently, only sporadic examples of non-Kähler compact 4-dimensional Einstein manifolds were known. An interesting case in point is the Page metric [**7, 31**] on $\mathbb{CP}_2\#\overline{\mathbb{CP}}_2$, which is beautiful, but has yet to lead to the construction of other compact examples. Of course, we have long known [**9**] that there is an infinite class of compact hyperbolic manifolds $\mathcal{H}^4/\Gamma$, but it is easy to dismiss these constant-sectional-curvature spaces as a bit boring in the small. A recent construction of Michael Anderson [**2**], however, puts these hyperbolic manifolds in a new, non-trivial context. Anderson's construction begins with a complete, non-compact hyperbolic manifold of finite volume, replaces the cusps with Schwarzschild-anti-deSitter metrics, and then, under mild additional technical hypotheses, perturbs the resulting metric so as to make it Einstein. Of course, these new Einstein manifolds still bear a family resemblance to their hyperbolic cousins, as they always have infinite fundamental group and vanishing signature, and always contain large regions where the sectional curvature is nearly constant. Nevertheless, Anderson's construction does seem to represent the first systematic method of constructing such a large class of compact non-locally-symmetric Einstein manifolds of general holonomy.

3. The Seiberg-Witten Equations

If M is a smooth oriented 4-manifold, $w_2(TM) \in H^2(M, \mathbb{Z}_2)$ is always in the image of the natural homomorphism $H^2(M, \mathbb{Z}) \to H^2(M, \mathbb{Z}_2)$ induced by $\mathbb{Z} \to \mathbb{Z}_2$. Consequently, we can always find Hermitian line bundles $L \to M$ such that $c_1(L) \equiv w_2(TM)$ mod 2. For any such L, and for any Riemannian metric g on M, one can then find rank-2 Hermitian vector bundles $\mathbb{V}_\pm$ which formally satisfy

$$\mathbb{V}_\pm = \mathbb{S}_\pm \otimes L^{1/2},$$

where $\mathbb{S}_\pm$ are the locally defined left- and right-handed spinor bundles of (M, g). Such a choice of $\mathbb{V}_\pm$, up to isomorphism, is called a spin^c structure $\mathfrak{c}$ on M. Moreover, $\mathfrak{c}$ is completely determined by the first Chern class $c_1(L) = c_1(\mathbb{V}_\pm) \in H^2(M, \mathbb{Z})$ if we assume that $H_1(M, \mathbb{Z})$ does not contain any elements of order 2.

Every unitary connection A on L induces a connection

$$\nabla_A : \Gamma(\mathbb{V}_+) \to \Gamma(\Lambda^1 \otimes \mathbb{V}_+),$$

and composition of this with the natural *Clifford multiplication* homomorphism

$$\Lambda^1 \otimes \mathbb{V}_+ \to \mathbb{V}_-$$

gives one a spin^c version

$$D_A : \Gamma(\mathbb{V}_+) \to \Gamma(\mathbb{V}_-)$$

of the Dirac operator [**19, 25**]. This is an elliptic first-order differential operator, and in many respects closely resembles the usual Dirac operator of spin geometry. In particular, one has the Weitzenböck formula

$$(6) \qquad \langle \Phi, D_A^* D_A \Phi \rangle = \frac{1}{2}\Delta|\Phi|^2 + |\nabla_A \Phi|^2 + \frac{s}{4}|\Phi|^2 + 2\langle -iF_A^+, \sigma(\Phi)\rangle$$

for any $\Phi \in \Gamma(\mathbb{V}_+)$, where F_A^+ is the self-dual part of the curvature of A, and where $\sigma : \mathbb{V}_+ \to \Lambda^+$ is a natural real-quadratic map satisfying

$$|\sigma(\Phi)| = \frac{1}{2\sqrt{2}}|\Phi|^2.$$

This, of course, generalizes the Weitzenböck formula used by Lichnerowicz [**29**] to prove that metrics with $s > 0$ cannot exist when M is spin and $\tau(M) = -8\hat{A}(M) \neq 0$. However, one cannot hope to derive interesting geometric information about the Riemannian metric g by just using (6) for an arbitrary connection A, since one would have no control at all over the F_A^+ term. Witten [**39**], however, had the brilliant insight that one could remedy this by considering both Φ and A as unknowns, subject to the *Seiberg-Witten equations*

$$D_A\Phi = 0 \tag{7}$$

$$F_A^+ = i\sigma(\Phi). \tag{8}$$

These equations are non-linear, but they become an elliptic first-order system once one imposes the 'gauge-fixing' condition

$$d^*(A - A_0) = 0$$

to eliminate the natural action of the 'gauge group' of automorphisms of the Hermitian line bundle $L \to M$.

Because the Seiberg-Witten equations are non-linear, one cannot use something like an index formula to predict that they must have solutions. Nonetheless, there exist spin^c structures on many 4-manifolds for which there is at least one solution for every metric g. This situation is conveniently described by the following terminology [**24**]:

DEFINITION 2. *Let M be a smooth compact oriented 4-manifold with $b_+ \geq 2$. An element $a \in H^2(M, \mathbb{Z})/\text{torsion}$ will be called a* **monopole class** *of M iff there exists a spin^c structure $\mathfrak{c}$ on M with first Chern class*

$$c_1(L) \equiv a \mod \text{torsion}$$

which has the property that the corresponding Seiberg-Witten equations (7–8) have a solution for every Riemannian metric g on M.

For example, if (M, ω) is a symplectic 4-manifold with $b_+ \geq 2$, and if J is any almost-complex structure which is compatible with ω, then $\pm c_1(M, J)$ are both monopole classes [**35**]. Usually, one detects the presence of a monopole class by thinking of the moduli space of the Seiberg-Witten equations (that is, solutions modulo gauge equivalence) as a cycle which represents an element of the homology of a certain configuration space [**39, 36, 30**]. The resulting homology class is then metric-independent, and one may, for example, then check that it is non-zero by analyzing the moduli space for some particular metric. A sophisticated recent refinment of this idea, however, instead detects the presence of a monopole class by means of an element of a stable cohomotopy group [**5, 6, 21**].

Because the Seiberg-Witten equations imply the Weitzenböck formula

$$0 = 2\Delta|\Phi|^2 + 4|\nabla\Phi|^2 + s|\Phi|^2 + |\Phi|^4, \tag{9}$$

where $\nabla = \nabla_A$, one immediately sees that they admit no solution with $\Phi \not\equiv 0$ relative to a metric g with $s > 0$. This allows one to prove, in particular, that there are lots of simply connected non-spin 4-manifolds which do not admit positive-scalar-curvature metrics, in complete contrast to the situation in higher dimensions [**18**]. But indeed, this Weitzenböck formula makes an even more remarkable prediction concerning the behavior of the scalar curvature [**39, 26**]:

PROPOSITION 6. *Let (M, g) be a smooth compact oriented Riemannian manifold, let $\mathfrak{c}$ be a spin^c structure on M, and let c_1^+ denote the self-dual part of the harmonic 2-form representing the first Chern class $c_1(L)$ of $\mathfrak{c}$. If there is a solution of the Seiberg-Witten equations (7–8) on M for g and $\mathfrak{c}$, then the scalar curvature s_g of g satisfies*

$$\int_M s_g^2 d\mu_g \geq 32\pi^2 [c_1^+]^2.$$

When $[c_1^+] \neq 0$, moreover, equality can only occur if g is a Kähler metric of constant scalar curvature.

PROOF. Integrating (9), we have

$$0 = \int [4|\nabla\Phi|^2 + s|\Phi|^2 + |\Phi|^4] d\mu,$$

and it follows that

$$\int (-s)|\Phi|^2 d\mu \geq \int |\Phi|^4 d\mu.$$

Applying the Cauchy-Schwarz inequality to the left-hand side therefore gives us

$$\left(\int s^2 d\mu\right)^{1/2} \left(\int |\Phi|^4 d\mu\right)^{1/2} \geq \int |\Phi|^4 d\mu,$$

so that

$$\int s^2 d\mu \geq \int |\Phi|^4 d\mu = 8 \int |F_A^+|^2 d\mu,$$

and the inequality is strict unless $\nabla\Phi \equiv 0$ and s is constant. However, $F_A^+ - 2\pi c_1^+$ is an exact form plus a co-exact form, and so is L^2-orthogonal to the harmonic forms. This gives us the inequality

$$\int |F_A^+|^2 d\mu \geq 4\pi^2 \int |c_1^+|^2 d\mu = 4\pi^2 \int c_1^+ \wedge c_1^+,$$

and the last expression may be re-interpreted as the intersection pairing $[c_1^+]^2$ of the de Rham class of c_1^+ with itself. This gives us the desired inequality

$$\int s^2 d\mu \geq 32\pi^2 [c_1^+]^2,$$

and, when the right-hand side is non-zero, equality can only happen if g has special holonomy and constant scalar curvature. □

One important application of this estimate is the following fundamental fact:

PROPOSITION 7. *Let M be any smooth compact oriented 4-manifold with $b_+(M) \geq 2$. Then $\mathcal{C} = \{$monopole classes of $M\}$ is a finite set.*

PROOF. Let $g_1 = g$ be any Riemannian metric on M, and let $\mathbf{e}_1 = [\omega_1]$ be the cohomology class of a harmonic self-dual form with respect to g, normalized so that $\mathbf{e}_1^2 := Q(\mathbf{e}_1, \mathbf{e}_1) = 1$. Because every metric is a regular point of the period map [**13**, Prop. 4.3.14], it follows that $[\omega_1]$ has an open neighborhood in $H^2(M, \mathbb{R})$ in which every element can be represented by a self-dual harmonic form relative to some perturbation of g. However, any open set in a finite-dimensional vector space spans the entire space. Hence we can find a basis $\{\mathbf{e}_j\}$ for $H^2(M,\mathbb{R})$, together with a collection of Riemannian metrics $\{g_j \mid j = 1,\ldots,b_2\}$, such that the g_j-harmonic 2-form ω_j representing the de Rham class $\mathbf{e}_j$ is self-dual with respect to g_j.

For convenience, let us now normalize these basis elements so that $\mathbf{e}_j^2 = 1$ for each j, and let $L_j : H^2(M,\mathbb{R}) \to \mathbb{R}$ be the linear functionals $L_j(x) = \mathbf{e}_j \cdot x := Q(\mathbf{e}_j, x)$. Then Proposition 6, together with the Cauchy-Schwarz inequality, implies that any monopole class $a \in H^2(M,\mathbb{Z})/\text{torsion}$ satisfies

$$|L_j(a)| = |\mathbf{e}_j \cdot a^+_{g_j}| \leq \sqrt{(a^+_{g_j})^2} \leq \left(\frac{1}{32\pi^2}\int_M s^2_{g_j}\, d\mu_{g_j}\right)^{1/2} = \kappa_j,$$

where the constant κ_j is independent of a. This shows that $\mathcal{C} \subset H^2(M,\mathbb{R})$ is contained in the $b_2(M)$-dimensional parallelepiped

$$\left\{x \in H^2(M,\mathbb{R}) \;\middle|\; |L_j(x)| \leq \kappa_j \;\forall j = 1,\ldots,b_2(M)\right\},$$

which is a compact set. Since $\mathcal{C} \subset H^2(M,\mathbb{Z})/\text{torsion}$ is also discrete, it follows that $\mathcal{C}$ is finite. □

We now want to extract a numerical invariant from the set of monopole classes which captures those features of Seiberg-Witten theory which are of the greatest relevance to problems in Riemannian geometry. To this end, let us once again consider the open Grassmannian $\mathbb{G} = Gr^+_{b_+}[H^2(M,\mathbb{R})]$ of all maximal linear subspaces $\mathcal{H}$ of the second cohomology for which the restriction $Q|_{\mathcal{H}}$ of the intersection pairing is positive definite. Each element $\mathcal{H} \in \mathbb{G}$ then determines an orthogonal decomposition

$$H^2(M,\mathbb{R}) = \mathcal{H} \oplus \mathcal{H}^\perp$$

with respect to Q. Let

$$\mathfrak{C} := \mathcal{C} - \{0\} \subset H^2(M,\mathbb{Z})/\text{torsion} \subset H^2(M,\mathbb{R})$$

denote the set of all the non-zero monopole classes $a \neq 0$ of M. Given a monopole class $a \in \mathfrak{C}$ and a positive subspace $\mathcal{H} \in \mathbb{G}$, we may then define a^+ to be the orthogonal projection of a into $\mathcal{H}$ with respect to Q. Using this, we now define an oriented-diffeomorphism invariant $\alpha(M) \in [0,\infty)$.

DEFINITION 3. *Let M be a smooth compact oriented manifold with $b_+ \geq 2$, and let $\mathfrak{C} \subset H^2(M,\mathbb{Z})/\text{torsion}$ be the set of the non-zero monopole classes of M. If $\mathfrak{C} = \emptyset$, we declare that $\alpha(M) = 0$. Otherwise, we set*

$$\alpha(M) = \inf_{\mathcal{H}\in\mathbb{G}} \left[\max_{a\in\mathfrak{C}} \sqrt{(a^+)^2}\right].$$

As a matter of notational convenience, let us also set $\alpha^2(M) := [\alpha(M)]^2$.

Needless to say, this is not the simplest invariant that one can cook up using the Seiberg-Witten equations. However, it precisely captures those aspects of Seiberg-Witten theory which are of the greatest relevance to many problems in Riemannian geometry. In particular, Proposition 6 has the following important consequence:

PROPOSITION 8. *Let M be any smooth compact oriented 4-manifold with $b_2 \geq 2$. Then every Riemannian metric g on M satisfies*

$$\int_M s_g^2\, d\mu_g \geq 32\pi^2\alpha^2(M). \tag{10}$$

Moreover, if $\alpha(M) \neq 0$, then equality can hold only if g is a Kähler metric of constant negative scalar curvature.

In particular, any metric, or sequence of metrics, on M gives one an explicit upper bound for $\alpha(M)$. On the other hand, one can obtain a lower bound for $\alpha(M)$ by replacing $\mathfrak{C}$ with any known set of non-zero monopole classes. Remarkably, the upper and lower bounds obtained in this manner actually coincide for many 4-manifolds, and in these circumstances one can therefore determine the invariant α even without necessarily knowing all the monopole classes on M. For example, the results of [**27, 21**] allow one to read off the following:

PROPOSITION 9. *Let M be the underlying smooth oriented 4-manifold of any compact complex surface (M, J) with $b_+ > 1$. Let X be the minimal model of (M, J). Then*

$$\alpha^2(M) = c_1^2(X).$$

PROPOSITION 10. *Let X_1, X_2, X_3 be minimal, simply connected complex surfaces with $b_+ \equiv 3 \bmod 4$, and let $M = X_1 \# X_2 \# X_3$. Then*

$$\alpha^2(M) = c_1^2(X_1) + c_1^2(X_2) + c_1^2(X_3).$$

Recall that a complex surface is said to be *minimal* if it is not obtained from another complex surface by blowing up. Any complex surface M can be obtained from some minimal surface X, called its minimal model, by blowing up a finite number of times [**4**], and this minimal model X is unique when $b_+ > 1$. As smooth manifolds, one then has

$$M \approx X \# k\overline{\mathbb{CP}}_2$$

for some integer $k \geq 0$, and the statement that X is minimal amounts to saying that this value of k is maximal.

To see that all this is relevant to the study of Einstein metrics, notice that (10) can be rewritten as

$$\frac{1}{4\pi^2}\int_M \frac{s^2}{24} d\mu \geq \frac{1}{3}\alpha^2(M).$$

Comparing the left-hand side with our Gauss-Bonnet formula (5) for $(2\chi - 3\tau)(M)$, we thus immediately obtain the following improvement of one of the Hitchin-Thorpe inequalities:

THEOREM 11. *If a smooth compact oriented 4-manifold M with $b_+ > 1$ admits an Einstein metric g, then*

$$(2\chi - 3\tau)(M) \geq \frac{1}{3}\alpha^2(M).$$

Moreover, if $\alpha(M) \neq 0$, equality occurs if and only if (M, g) is a compact quotient $\mathbb{C}\mathcal{H}_2/\Gamma$ of the complex hyperbolic plane, equipped with a constant multiple of its standard Kähler-Einstein metric.

In particular, this tells us that the Einstein metric on a complex-hyperbolic manifold $\mathbb{C}\mathcal{H}_2/\Gamma$ is *unique*, modulo diffeomorphisms and rescaling [**27**].

Now one can obviously imitate this argument by using $2\chi + 3\tau$ instead of $2\chi - 3\tau$. However, one would expect for the results obtained in this way [**27**] to be quite far from optimal, since the Kähler metrics which saturate (10) when $\alpha \neq 0$ always have $W_+ \neq 0$. In the next section, we will remedy this, by replacing Proposition 6 with an estimate which involves the self-dual Weyl curvature W_+ as well as the scalar curvature s.

4. The Main Curvature Estimate

The Dirac operator is conformally invariant, provided that the relevant spinors are viewed as having appropriate conformal weights [**25, 32**]. Thus, if $\hat{\Phi}$ solves the spin^c Dirac equation $\hat{D}_A\hat{\Phi} = 0$ with respect to the conformally rescaled metric $f^{-2}g$, there is a corresponding solution Φ of the equation $D_A\Phi = 0$ with respect to g, with $|\Phi|_g = f^{-3/2}|\hat{\Phi}|_{\hat{g}}$ and $\sigma(\Phi) = f\hat{\sigma}(\hat{\Phi})$. If $(\hat{\Phi}, A)$ is a solution of the Seiberg-Witten equations with respect to $f^{-2}g$, it thus follows that (Φ, A) solves the *rescaled Seiberg-Witten equations*

$$D_A\Phi = 0 \tag{11}$$
$$F_A^+ = if\sigma(\Phi). \tag{12}$$

with respect to g. We will now show that directly studying equations (11–12) for an appropriate family of choices of f will yield an efficient avenue for deducing curvature estimates for the fixed metric g.

Indeed, plugging (12) into the Weitzenböck formula for (11) gives us

$$\begin{aligned} 0 &= 2\Delta|\Phi|^2 + 4|\nabla_A\Phi|^2 + s|\Phi|^2 + 8\langle -iF_A^+, \sigma(\Phi)\rangle, \\ &= 2\Delta|\Phi|^2 + 4|\nabla_A\Phi|^2 + s|\Phi|^2 + f|\Phi|^4, \end{aligned}$$

so that multiplying by $|\Phi|^2$ gives us

$$0 = 2|\Phi|^2\Delta|\Phi|^2 + 4|\Phi|^2|\nabla_A\Phi|^2 + s|\Phi|^4 + f|\Phi|^6.$$

Integration therefore yields the inequality

$$0 \geq \int_M \left[4|\Phi|^2|\nabla_A\Phi|^2 + s|\Phi|^4 + f|\Phi|^6\right] d\mu. \tag{13}$$

We will now use this to obtain an estimate involving f and curvature of g.

LEMMA 12. *Let (M, g) be a compact oriented Riemannian 4-manifold, and let $f > 0$ be a smooth positive function on M. Suppose, for some spin^c structure with first Chern class $c_1(L)$, that there is a a solution of the Seiberg-Witten equations with respect to the conformally related metric $f^{-2}g$. Let c_1^+ denote the self-dual part of the harmonic 2-form representing $c_1(L)$, relative to the metric g. Then*

$$\left(\int_M f^4 d\mu_g\right)^{1/3}\left(\int_M |s_g + 3w_g|^3 f^{-2} d\mu_g\right)^{2/3} \geq 72\pi^2[c_1^+]^2, \tag{14}$$

where $s_g : M \to \mathbb{R}$ denotes the scalar curvature of g and $w_g : M \to \mathbb{R}$ is the lowest eigenvalue of $W_+ : \Lambda^+ \to \Lambda^+$ at each $x \in M$.

PROOF. Any self-dual 2-form ψ on any oriented 4-manifold satisfies the Weitzenböck formula [**10**]

$$(d + d^*)^2\psi = \nabla^*\nabla\psi - 2W_+(\psi, \cdot) + \frac{s}{3}\psi, \tag{15}$$

where W_+ is the self-dual Weyl tensor. It follows that

$$\int_M \left(|\nabla\psi|^2 - 2W_+(\psi, \psi) + \frac{s}{3}|\psi|^2\right) d\mu \geq 0,$$

and hence that

$$\int_M |\nabla\psi|^2 d\mu \geq \int_M \left(2w - \frac{s}{3}\right)|\psi|^2 d\mu.$$

However, the particular self-dual 2-form $\sigma(\Phi)$ satisfies

$$\begin{aligned} |\sigma(\Phi)|^2 &= \frac{1}{8}|\Phi|^4, \\ |\nabla\sigma(\Phi)|^2 &\leq \frac{1}{2}|\Phi|^2|\nabla\Phi|^2. \end{aligned}$$

Setting $\psi = \sigma(\Phi)$, we thus have

$$\int_M 4|\Phi|^2|\nabla\Phi|^2\, d\mu \geq \int_M \left(2w - \frac{s}{3}\right)|\Phi|^4 d\mu.$$

But (13) tells us that

$$0 \geq \int_M \left[4|\Phi|^2|\nabla_A\Phi|^2 + s|\Phi|^4 + f|\Phi|^6\right] d\mu,$$

so we obtain

$$0 \geq \int_M \left[\left(\frac{2}{3}s + 2w\right)|\Phi|^4 + f|\Phi|^6\right] d\mu,$$

which we may rewrite as

$$\int_M \left[-\frac{2}{3}(s+3w)\,f^{-2/3}\right]\left(f^{2/3}|\Phi|^4\right) d\mu \geq \int_M f|\Phi|^6 d\mu.$$

Applying the Hölder inequality to the left-hand side now gives

$$\left[\int_M \left|\frac{2}{3}(s+3w)\right|^3 f^{-2}d\mu\right]^{1/3}\left[\int_M f|\Phi|^6 d\mu\right]^{2/3} d\mu \geq \int_M f|\Phi|^6 d\mu,$$

and we therefore deduce that

$$\int_M \left|\frac{2}{3}(s+3w)\right|^3 f^{-2}d\mu \;\geq\; \int_M f|\Phi|^6 d\mu.$$

But the Hölder inequality also tells us that

$$\left(\int_M f^4 d\mu\right)^{1/3}\left(\int_M f|\Phi|^6 d\mu\right)^{2/3} \geq \int_M f^{4/3}(f^{2/3}|\Phi|^4)d\mu\ ,$$

so it follows that

$$\left(\int_M f^4 d\mu\right)^{1/3}\left(\int_M \left|\frac{2}{3}(s+3w)\right|^3 f^{-2}d\mu\right)^{2/3} \geq \int_M f^2|\Phi|^4 d\mu\ .$$

However, since $-iF_A^+ = f\sigma(\Phi)$, we also have

$$\int_M f^2|\Phi|^4 d\mu = 8\int_M |F_A^+|^2 d\mu \geq 8\int_M |2\pi c_1^+|^2 d\mu = 32\pi^2[c_1^+]^2$$

because $iF_A^+ = 2\pi c_1^+ + d\theta - d^*(\star\theta)$ for some 1-form θ. Thus

$$\left(\int_M f^4 d\mu\right)^{1/3}\left(\int_M \left|\frac{2}{3}(s+3w)\right|^3 f^{-2}d\mu\right)^{2/3} \geq 32\pi^2[c_1^+]^2.$$

Multiplying both sides by $9/4$ now yields the promised inequality. □

LEMMA 13. *Let M be a smooth compact oriented 4-manifold with $b_+(M) \geq 2$, and let $a \in H^2(M, \mathbb{Z})/torsion$ be a monopole class. Let g be any Riemannian metric on M, and let $u > 0$ be any smooth positive function on M for which*

$$u \geq |s + 3w|$$

at each point $x \in M$. Then

$$\int u^2 d\mu_g \geq 72\pi^2 (a^+)^2.$$

PROOF. For any such function u, set $f = u^{1/2}$. and notice that we have

$$\left(\int u^2 d\mu\right)^{1/3} = \left(\int f^4 d\mu\right)^{1/3}$$

and

$$\left(\int u^2 d\mu\right)^{2/3} = \left(\int u^3 f^{-2} d\mu\right)^{2/3} \geq \left(\int |s + 3w|^3 f^{-2} d\mu\right)^{2/3}.$$

Setting $c_1(L) = a$ and invoking Lemma 12, we therefore have

$$\int_M u^2 d\mu_g \geq \left(\int_M f^4 d\mu_g\right)^{1/3} \left(\int_M |s + 3w|_g^3 f^{-2} d\mu_g\right)^{2/3} \geq 72\pi^2 (a^+)^2,$$

as claimed. □

LEMMA 14. *Let M be a smooth compact oriented 4-manifold with $b_+(M) \geq 2$, and let $a \in H^2(M, \mathbb{Z})/torsion$ be a monopole class. Then every Riemannian metric g on M satisfies*

$$\int (s + 3w)^2 \, d\mu_g \geq 72\pi^2 (a^+)^2.$$

PROOF. Let u_j be a sequence of smooth positive functions satisfying $u_j > |s + 3w|$ and with $u_j \to |s + 3w|$ in the C^0 topology. (Since the smooth functions are dense in C^0, such a sequence may of course be constructed by setting $u_j = \frac{1}{j} + v_j$, where v_j is a smooth function whose sup-norm distance from the continuous function $|s + 3w|$ is less than $1/j$.) Then Lemma 13 tells us that

$$\int_M (s + 3w)^2 d\mu = \inf_j \int_M u_j^2 d\mu \geq 72\pi^2 (a^+)^2,$$

as claimed. □

Note that Lemma 14 essentially reproves [**28**, Theorem 2.3], but does so in a much more elementary manner. What has been lost here, however, is a precise understanding of what happens when the inequality is saturated. For the purpose of finding obstructions to the existence of Einstein metrics, however, this shortcoming will in practice turn out to be irrelevant.

PROPOSITION 15. *Let M be a smooth compact oriented 4-manifold with $b_+(M) \geq 2$. Then every Riemannian metric g on M satisfies*

$$||s|| + \sqrt{6}||W_+|| \geq 6\sqrt{2}\pi\alpha(M),$$

where $||\cdot||$ denotes the L^2 norm with respect to g. If equality occurs, moreover, then the two largest eigenvalues of $W_+ : \Lambda^+ \to \Lambda^+$ are equal at each $x \in M$, and $|W_+|$ is a constant multiple of the scalar curvature s.

PROOF. By Lemma 14, we have

$$||s + 3w|| \geq 6\pi\sqrt{2(a^+)^2},$$

so the triangle inequality tells us that

$$||s|| + 3||w|| \geq ||s + 3w|| \geq 6\pi\sqrt{2(a^+)^2},$$

and that equality can only hold if w and s are proportional, as vectors in L^2.

Now if $\alpha(M) = 0$, the claim is of course trivial. Otherwise, the definition of $\alpha(M)$ guarantees that for every metric g there is a monopole class a with $(a^+)^2 \geq \alpha^2(M)$. Now invoking our calculations for this choice of a, we then have

$$||s|| + 3||w|| \geq 6\pi\sqrt{2}\alpha(M).$$

Moreover, since a 4-manifold M with $\alpha(M) \neq 0$ cannot admit metrics with $s \equiv 0$, equality can only occur if w is a constant times s.

On the other hand, because W_+ is a trace-free endomorphism of Λ^+, we have the point-wise inequality

$$\sqrt{\frac{2}{3}}|W_+| \geq |w|,$$

with equality only when the two largest eigenvalues of W_+ are equal. Hence

$$||s|| + \sqrt{6}||W_+|| \geq ||s|| + 3||w|| \geq 6\pi\sqrt{2}\alpha(M).$$

The resulting inequality is now strict unless the largest eigenvalue of W_+ has multiplicity ≥ 2 everywhere, and unless the functions $|W_+|$ is a constant multiple of the scalar curvature s. □

LEMMA 16. *Let M be a smooth compact oriented 4-manifold with $b_+(M) \geq 2$. Then every Riemannian metric g on M satisfies*

$$\frac{1}{4\pi^2}\int_M \left(\frac{s^2}{24} + 2|W_+|^2\right) d\mu \geq \frac{2}{3}\alpha^2(M).$$

Moreover, equality could only possibly hold for metrics for which $s \equiv -8\sqrt{6}|W_+|$ and for which the largest eigenvalue of W_+ has multiplicity two at each point of M.

PROOF. By Proposition 15, we have

$$(1, \frac{1}{\sqrt{8}}) \cdot \left(||s||, \sqrt{48}||W_+||\right) \geq 6\sqrt{2}\pi\alpha(M),$$

where the left-hand side is to be read as a dot product in $\mathbb{R}^2$. Applying the Cauchy-Schwarz equality to this dot product now gives us

$$\left(1 + \frac{1}{8}\right)^{1/2} \left(||s||^2 + 48||W_+||^2\right)^{1/2} \geq (72\pi^2\alpha^2(M))^{1/2},$$

so that

$$\frac{1}{4\pi^2}\int_M \left(\frac{s^2}{24} + 2|W_+|^2\right) d\mu \geq \frac{2}{3}\alpha^2(M), \tag{16}$$

as claimed. If equality holds, we must have $\left(||s||, \sqrt{48}||W_+||\right) \propto (1, \frac{1}{\sqrt{8}})$ in addition to the previously deduced conditions, and the value of the ratio of $|W_+|$ and s then follows from this, together with the fact that there cannot be any metrics with $s \geq 0$ when $\alpha(M) \neq 0$. □

LEMMA 17. *Let M be a smooth compact oriented 4-manifold with $b_+(M) \geq 2$, and suppose that g is a Riemannian metric on M with constant scalar curvature and harmonic self-dual Weyl curvature:*

$$(\delta W_+)_{bcd} = -\nabla^a (W_+)_{abcd} = 0.$$

Then either g satisfies the strict inequality

$$\frac{1}{4\pi^2}\int_M \left(\frac{s^2}{24} + 2|W_+|^2\right) d\mu > \frac{2}{3}\alpha^2(M),$$

or else $\alpha(M) = 0$, g is a hyper-Kähler metric, and M is diffeomorphic to either $K3$ or T^4.

PROOF. Suppose that g is a metric with constant s which saturates inequality (16). Then $|W_+| \equiv -s/8\sqrt{6}$ is constant, and the largest two eigenvalues of W_+ coincide at each point of M. We now want to ask what this tells us if we also assume that $\delta W_+ = 0$.

One way this may certainly happen is for s and W_+ to both vanish identically. In this case, however, the Weitzenböck formula (15) then implies that every self-dual harmonic form on M is parallel, and, since $b_+(M) \geq 2$ by assumption, it then follows that the orientable rank-3 vector bundle Λ^+ is trivialized by parallel sections, so that g is hyper-Kähler. The classification of complex surfaces then tells us that M must be diffeomorphic to $K3$ or T^4.

On the other hand, the $s \neq 0$ case is ruled out by an observation due to Derdziński [**11**]. Indeed, in this case we would know that W_+ was a non-zero constant times $\omega \otimes \omega - \frac{1}{3}\mathbf{1}_{\Lambda^+}$, where ω is a (sign-ambiguous) self-dual 2-form of norm $\sqrt{2}$ spanning the negative eigenspace of W_+. The harmonicity of W_+ thus implies that

$$\omega_{ab}\nabla^a\omega_{cd} + \omega_{cd}\nabla^a\omega_{ab} = 0.$$

But since ω has constant length, contraction with ω^{cd} tells us that $\delta\omega = 0$, and plugging this back into the original equation then tells us that $\nabla\omega = 0$. This shows that (M, g) is locally Kähler. However, any Kähler manifold of real dimension 4 satisfies $|s| \equiv 2\sqrt{6}|W_+|$, whereas in our case we can already take it as given that $|s| \equiv 8\sqrt{6}|W_+|$. This contradiction eliminates the $s \neq 0$ case, and we are done. □

This now allows us to recover [**28**, Theorem 3.3]:

THEOREM 18. *Let M be a smooth compact oriented 4-manifold with $b_+(M) \geq 2$. If M admits an Einstein metric g, then*

$$(2\chi + 3\tau)(M) \geq \frac{2}{3}\alpha^2(M),$$

with equality only if both sides vanish, in which case M must be diffeomorphic to $K3$ or T^4, and g must be a hyper-Kähler metric.

PROOF. If (M, g) is a compact oriented 4-dimensional Einstein manifold, the generalized Gauss-Bonnet formula (5) tells us that the left-hand side of (16) is given by $(2\chi + 3\tau)(M)$, and the desired inequality therefore follows. However, the Bianchi identities also tell us that a 4-dimensional Einstein manifold has constant s and harmonic W_+, so Lemma 17 tells us that this inequality can only be saturated when g is hyper-Kähler. □

5. An Illustration

Theorem 18 gives us an obstruction to the existence of Einstein metrics involving the differential topology, rather than just the homeomorphism type, of the smooth manifold in question. The following example should help to clarify this point.

Let X be a triple cyclic cover of of $\mathbb{CP}_2$, ramified at a non-singular complex curve B of degree 6.

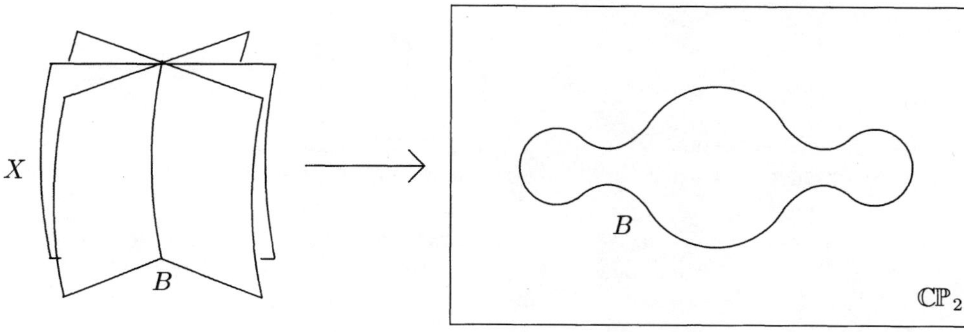

To be more explicit, we could, for example, take X to consist of those elements η of the $\mathcal{O}(2)$ line bundle over $\mathbb{CP}_2 = \{[x : y : z]\}$ which satisfy the equation

$$\eta^3 = x^6 + y^6 + z^6,$$

where the right-hand side is of course interpreted as a section of $\mathcal{O}(6)$. The canonical line bundle of the compact complex surface X is then exactly the pull-back of the $\mathcal{O}(1)$ line bundle on $\mathbb{CP}_2$, and from this we read off that $c_1^2(X) = 3 \cdot 1^2 = 3$, and $h^{2,0}(X) = h^0(\mathbb{CP}_2, \mathcal{O}(1)) = 3$. Moreover, an easy application of the Lefschetz hyperplane-section theorem tells us that X is simply connected.

Now let M be obtained from X by blowing up a point, so that

$$M \approx X \# \overline{\mathbb{CP}}_2.$$

Then $\alpha^2(M) = c_1^2(X) = 3$, whereas $(2\chi + 3\tau)(M) = c_1^2(M) = c_1^2(X) - 1 = 2$. Since $(2\chi + 3\tau)(M) = \frac{2}{3}\alpha^2(M) \neq 0$, Theorem 18 therefore tells us that the smooth compact 4-manifold M cannot admit an Einstein metric.

Next, for the sake of comparison, let N be the double branched cover of $\mathbb{CP}_2$, ramified at a non-singular complex curve B' of degree 8. To be more explicit, one could take N to consist of those elements ζ of the $\mathcal{O}(4)$ line bundle over $\mathbb{CP}_2$ which satisfy the equation

$$\zeta^2 = x^8 + y^8 + z^8.$$

Once again, this branched cover of $\mathbb{CP}_2$ is simply connected. Careful inspection also reveals that, in this example, the canonical line bundle is once again exactly the pull-back of the $\mathcal{O}(1)$ line bundle on $\mathbb{CP}_2$; thus we read off that $c_1^2(N) = 2 \cdot 1^2 = 2$, and $h^{2,0}(N) = h^0(\mathbb{CP}_2, \mathcal{O}(1)) = 3$, which is to say that these two numerical invariants are exactly the same for M and N.

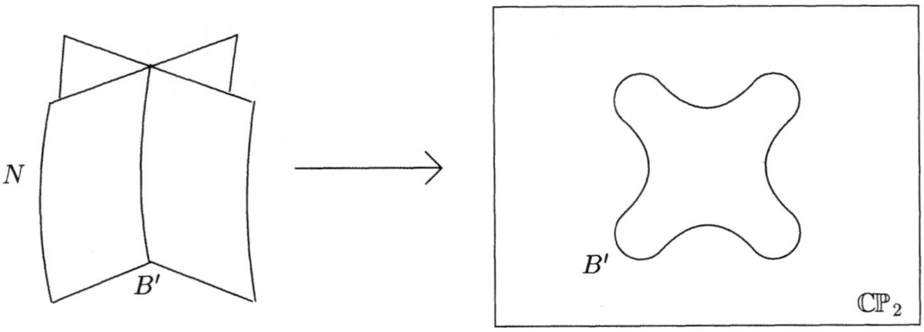

But in another respect, N is wildly unlike from M. Indeed, because it is a compact complex surface with $c_1 < 0$. Theorem 3 tells us that N, unlike M, admits an Einstein metric.

However, each of these compact complex surfaces is simply connected and non-spin. Moreover, both of them have $c_1^2 = 2$ and $h^{2,0} = 3$, and from this it can be deduced that both have $b_+ = 7$ and $b_- = 37$. Thus Theorem 2 tells us that both M and N are homeomorphic to

$$7\mathbb{CP}_2 \# 37\overline{\mathbb{CP}}_2.$$

This topological 4-manifold therefore admits one smooth structure for which there is an Einstein metric, and yet another smooth structure for which no such metric can exist.

6. The Big Picture

The problem of understanding Einstein metrics on compact 4-manifolds has been an extremely active area of research in recent years, and the Seiberg-Witten techniques which have been highlighted in this article do not by any means represent the only strand of thought which has led to significant progress. It is therefore appropriate that this article conclude with a rough indication of some of these other developments, without pretending to offer a definitive survey of the subject.

One such important strand of thought grew out of the work of Gromov [**17**], who, for example, first introduced minimal volume invariants of a smooth manifold. If M is a smooth compact n-manifold, one such invariant is

$$\mathrm{Vol}_r(M) = \inf\{\mathrm{Vol}(M,g) \mid r_g \geq -(n-1)g\},$$

and is sometimes called the *minimal volume with respect to Ricci curvature.* Gromov showed that $\mathrm{Vol}_r(M)$ is positive for certain compact manifolds with infinite fundamental group by giving a lower bound for it in terms of a homotopy invariant which he called the *simplicial volume.* He then went on to observe that this implies that there are obstructions to the existence of Einstein metrics in dimension 4 which are not detected by the Hitchin-Thorpe inequality.

While Gromov's bounds are actually quite weak in practice, they nonetheless represented a breakthrough in the subject, and eventually led to a hunt for sharper estimates of the same flavor. This culminated in the work of Besson, Courtois, and Gallot [**8**], who related Gromov's work to the entropy of the geodesics flow. One of their most remarkable results goes as follows:

THEOREM 19 (Besson-Courtois-Gallot). *Let (X, g_0) be a compact hyperbolic n-manifold, $n > 2$, and let M be a compact manifold of the same dimension. Let $f : M \to X$ be any smooth map. Then*

$$\mathrm{Vol}_r(M) \geq \deg(f)\ \mathrm{Vol}(X, g_0), \tag{17}$$

where $\deg(f)$ *denotes the degree of f. Moreover, if equality holds, and if the infimum (17) is achieved by some metric g, then (M, g) is an isometric Riemannian covering of (X, g_0), with covering map $M \to X$ homotopic to f.*

When $n = 4$ and f is the identity map $X \to X$, this implies that g_0 is the only Einstein metric on X, up to rescalings and diffeomorphisms. Moreover, this same result also gives rise to new obstructions to the existence of Einstein metrics. Indeed, Sambusetti [**34**] used this method to prove the following:

THEOREM 20 (Sambusetti). *Any integer pair (χ, τ) with $\chi \equiv \tau \bmod 2$ can be realized as the Euler characteristic and signature of a smooth compact oriented 4-manifold M which does not admit any Einstein metrics.*

This nicely highlights how much there is to be said about the subject beyond the Hitchin-Thorpe inequality. Note, however, that Sambusetti's examples all have huge fundamental group, and are never even *homotopy equivalent* to an Einstein manifold.

Anderson's work [**1**] on Gromov-Hausdorff limits with Ricci-curvature bounds represents another major area of progress in understanding 4-dimensional Einstein manifolds. Anderson shows that any sequence of unit-volume Einstein manifolds of bounded Euler characteristic and *bounded diameter* must necessarily have a subsequence which converges in the Gromov-Hausdorff sense to a compact Einstein orbifold. In particular, this implies that a topological 4-manifold can only admit finitely many smooth structures for which there exists an Einstein metric of unit volume and diameter $< D$.

It is interesting to reconsider the Kähler-Einstein case in this light. Mumford's finiteness theorem [**4**] implies that a 4-manifold can only admit finitely many smooth structures for which there exist *Kähler*-Einstein metrics, although this finite number can be arbitrarily large [**23, 33**]. Notice that this conclusion holds without the imposition of an extraneous diameter bound. On the other hand, inspection of the Kähler-Einstein case also reveals that the diameter can tend to infinity even for sequences of unit-volume Einstein metrics on a fixed smooth 4-manifold, so a Mumford-type finiteness statement certainly cannot be deduced from a compactness result like Anderson's.

While we do not know at present whether there can only be finitely many smooth structures admitting Einstein metrics on a fixed topological 4-manifold, we *do* at least know that there are often infinitely many smooth structures for which Einstein metrics *don't* exist, even for simply connected manifolds for which the Hitchin-Thorpe inequality is strict [**22**]; cf. [**23**]. In the non-spin case, such an assertion can actually be made for a large sector of choices for (χ, τ). The spin case is much more delicate, however, and the range of homeotypes for which one can make such an assertion is, at present, much more tightly constrained.

References

[1] M.T. Anderson. The L^2 structure of moduli spaces of Einstein metrics on 4-manifolds. *Geom. Func. An.*, 2:29–89, 1992.

[2] M.T. Anderson. Dehn filling and Einstein metrics in higher dimensions. e-print math.DG/0303260, 2003.

[3] T. Aubin. Equations du type Monge-Ampère sur les variétés Kählériennes compactes. *C. R. Acad. Sci. Paris*, 283A:119–121, 1976.

[4] W. Barth, C. Peters, and A. Van de Ven. *Compact Complex Surfaces*. Springer-Verlag, 1984.

[5] S. Bauer and M. Furuta. A stable cohomotopy refinement of Seiberg-Witten invariants: I. Inv. Math. to appear.

[6] S. Bauer. A stable cohomotopy refinement of Seiberg-Witten invariants: II. Inv. Math. to appear.

[7] A. Besse. *Einstein Manifolds*. Springer-Verlag, 1987.

[8] G. Besson, G. Courtois, and S. Gallot. Entropies et rigidités des espaces localement symétriques de courbure strictement négative. *Geom. and Func. An.*, 5:731–799, 1995.

[9] A. Borel. Compact Clifford-Klein forms of symmetric spaces. *Topology*, 2:111–122, 1963.

[10] J.-P. Bourguignon. Les variétés de dimension 4 à signature non nulle dont la courbure est harmonique sont d'Einstein. *Invent. Math.*, 63(2):263–286, 1981.

[11] A. Derdziński. Self-dual Kähler manifolds and Einstein manifolds of dimension four. *Compositio Math.*, 49(3):405–433, 1983.

[12] S. K. Donaldson. An application of gauge theory to four-dimensional topology. *J. Differential Geom.*, 18:279–315, 1983.

[13] S. K. Donaldson and P. B. Kronheimer. *The Geometry of Four-Manifolds*. Oxford University Press, Oxford, 1990.

[14] M. Freedman. On the topology of 4-manifolds. *J. Differential Geom.*, 17:357–454, 1982.

[15] M. Furuta. Monopole equation and the $\frac{11}{8}$-conjecture. *Math. Res. Lett.*, 8(3):279–291, 2001.

[16] R.E. Gompf and A.I. Stipsicz. *4-manifolds and Kirby calculus*. American Mathematical Society, Providence, RI, 1999.

[17] M. Gromov. Volume and bounded cohomology. *Publ. Math. IHES*, 56:5–99, 1982.

[18] M. Gromov and H.B. Lawson. The classification of simply connected manifolds of positive scalar curvature. *Ann. Math.*, 111:423–434, 1980.

[19] N.J. Hitchin. Harmonic spinors. *Advances in Mathematics*, 14:1–55, 1974.

[20] N.J. Hitchin. On compact four-dimensional Einstein manifolds. *J. Differential Geom.*, 9:435–442, 1974.

[21] M. Ishida and C. LeBrun. Curvature, connected sums, and Seiberg-Witten theory. to appear in Comm. Anal. Geom.

[22] M. Ishida and C. LeBrun. Spin manifolds, Einstein metrics, and differential topology. *Math. Res. Lett.*, 9:229–240, 2002.

[23] D. Kotschick. Einstein metrics and smooth structures. *Geom. Topol.*, 2:1–10, 1998.

[24] P.B. Kronheimer. Minimal genus in $S^1 \times M^3$. *Invent. Math.*, 135(1):45–61, 1999.

[25] H.B. Lawson and M.L. Michelsohn. *Spin Geometry*. Princeton University Press, 1989.

[26] C. LeBrun. Polarized 4-manifolds, extremal Kähler metrics, and Seiberg-Witten theory. *Math. Res. Lett.*, 2:653–662, 1995.

[27] C. LeBrun. Four-manifolds without Einstein metrics. *Math. Res. Lett.*, 3:133–147, 1996.

[28] C. LeBrun. Ricci curvature, minimal volumes, and Seiberg-Witten theory. *Inv. Math.*, 145:279–316, 2001.

[29] A. Lichnerowicz. Spineurs harmoniques. *C.R. Acad. Sci. Paris*, 257:7–9, 1963.

[30] P. Ozsváth and Z. Szabó. Higher type adjunction inequalities in Seiberg-Witten theory. *J. Differential Geom.*, 55(3):385–440, 2000.

[31] D. Page. A compact rotating gravitational instanton. *Phys. Lett.*, 79B:235–238, 1979.

[32] R. Penrose and W. Rindler. *Spinors and space-time. Vol. 2*. Cambridge University Press, Cambridge, 1986. Spinor and twistor methods in space-time geometry.

[33] M. Salvetti. On the number of nonequivalent differentiable structures on 4-manifolds. *Manuscripta Math.*, 63(2):157–171, 1989.

[34] A. Sambusetti. An obstruction to the existence of Einstein metrics on 4-manifolds. *C. R. Acad. Sci. Paris*, 322:1213–1218, 1996.

[35] C.H. Taubes. The Seiberg-Witten invariants and symplectic forms. *Math. Res. Lett.*, 1:809–822, 1994.

[36] C.H. Taubes. The Seiberg-Witten and Gromov invariants. *Math. Res. Lett.*, 2:221–238, 1995.

[37] J.A. Thorpe. Some remarks on the Gauss-Bonnet formula. *J. Math. Mech.*, 18:779–786, 1969.

[38] G. Tian. On Calabi's conjecture for complex surfaces with positive first chern class. ***Inv. Math.***, 101:101–172, 1990.
[39] E. Witten. Monopoles and four-manifolds. *Math. Res. Lett.*, 1:809–822, 1994.
[40] S.-T. Yau. Calabi's conjecture and some new results in algebraic geometry. *Proc. Nat. Acad. USA*, 74:1789–1799, 1977.

DEPARTMENT OF MATHEMATICS, SUNY AT STONY BROOK, STONY BROOK, NY 11794-3651
E-mail address: `claude@@math.sunysb.edu`

Topological Quantum Field Theory for Calabi-Yau threefolds and G_2-manifolds

Naichung Conan Leung

1. Introduction

In the past two decades we witness many fruitful interactions between mathematics and physics. One example is the Donaldson-Floer theory for oriented four manifolds. Physical considerations leads to the discovery of the Seiberg-Witten theory which has profound impact to our understandings of four manifolds. Another example is the Mirror Symmetry for Calabi-Yau manifolds. This duality transformation in the string theory leads to many surprising predictions in the enumerative geometry.

String theory in physics studies a ten dimensional space-time $X \times \mathbb{R}^{3,1}$ with X a six dimensional Riemannian manifold with its holonomy group inside $SU(3)$, the so-called *Calabi-Yau threefold.* Certain parts of the Mirror Symmetry conjecture, as studied by Vafa's group, are specific for Calabi-Yau manifolds of complex dimension *three.* They include the Gopakumar-Vafa conjecture on Gromov-Witten invariants of *arbitrary* genus for Calabi-Yau threefolds, the Ooguri-Vafa conjecture on the relationships between knot invariants and enumerations of holomorphic disks and so on. The key reason is they belong to dualities for G_2-manifolds. G_2-manifolds can be naturally interpreted as special Octonion manifolds [**23**]. For any Calabi-Yau threefold X, the seven dimensional manifold $X \times S^1$ is automatically a G_2-manifold because of the natural inclusion $SU(3) \subset G_2$.

In recent years, there are many studies of G_2-manifolds in M-theory including works of Archaya, Atiyah, Gukov, Vafa, Witten, Yau, Zaslow and many others (e.g. [**1**], [**5**], [**13**], [**2**]).

In the studies of the symplectic geometry of a Calabi-Yau threefold X, we consider unitary flat bundles over three dimensional (special) Lagrangian submanifolds L in X. The corresponding theory for a G_2-manifold M is called the *special $\mathbb{H}$-Lagrangian geometry* (or *C-geometry* in [**19**]). where we consider Anti-Self-Dual (abbrev. ASD) bundles over four dimensional coassociative submanifolds, or equivalently *special $\mathbb{H}$-Lagrangian submanifolds of type II* [**23**], (abbrev. $\mathbb{H}$-SLag) C in M.

Counting ASD bundles over a fixed four manifold C is the well-known theory of Donaldson differentiable invariants, $Don(C)$. Similarly, counting unitary flat bundles over a fixed three manifold L is Floer's Chern-Simons homology theory $HF_{CS}(L)$. When C is a connected sum $C_1 \#_L C_2$ along a homology three sphere, the relative Donaldson invariants $Don(C_i)$'s take values in $HF_{CS}(L)$ and

$Don\,(C)$ can be recovered from individual pieces by a gluing theorem, $Don\,(C) = \langle Don\,(C_1), Don\,(C_2)\rangle_{HF_{CS}(L)}$ (see e.g. [**7**]). Similarly when L has a handlebody decomposition $L = L_1 \#_{\Sigma} L_2$, each L_i determines a Lagrangian subspace $\mathcal{L}_i$ in the moduli space $\mathcal{M}^{flat}(\Sigma)$ of unitary flat bundles over the Riemann surface Σ and Atiyah conjectures that we can recover $HF_{CS}(L)$ from the Floer's Lagrangian intersection homology group of $\mathcal{L}_1$ and $\mathcal{L}_2$ in $\mathcal{M}^{flat}(\Sigma)$, $HF_{CS}(L) = HF_{Lag}^{\mathcal{M}^{flat}(\Sigma)}(\mathcal{L}_1, \mathcal{L}_2)$. Such algebraic structures in the Donaldson-Floer theory can be formulated as a Topological Quantum Field Theory (abbrev. TQFT), as defined by Segal and Atiyah [**3**].

In this paper we propose a TQFT by counting ASD bundles over four dimensional $\mathbb{H}$-SLags C in any closed (almost) G_2-manifold M, called *$\mathbb{H}$-SLags cycles.* They can be identified as zeros of a naturally defined closed one form on the configuration space of topological cycles. We expect to obtain a homology theory $H_C(M)$ by applying the Witten's Morse theory. When M is non-compact with an asymptotically cylindrical end $X \times [0, \infty)$, then the set of boundary data of relative $\mathbb{H}$-SLag cycles determines a Lagrangian submanifold $\mathcal{L}_M$ in the moduli space $\mathcal{M}^{SLag}(X)$ of special Lagrangian cycles in the Calabi-Yau threefold X.

When we decompose $M = M_1 \#_X M_2$ along an infinite asymptotically cylindrical neck, it is reasonable to expect to have a gluing formula,

$$H_C(M) = HF_{Lag}^{\mathcal{M}^{SLag}(X)}(\mathcal{L}_{M_1}, \mathcal{L}_{M_2}).$$

The main technical difficulty in defining this TQFT rigorously is the *compactness* issue for the moduli space of $\mathbb{H}$-SLags in M. We do not know how to resolve this problem and our homology groups are only defined in the *formal* sense (and physical sense?).

2. G_2-manifolds and $\mathbb{H}$-SLag geometry

We first review some basic definitions and properties of G_2-geometry, see [**19**] for more details.

DEFINITION 1. *A seven dimensional Riemannian manifold M is called a G_2-manifold if the holonomy group of its Levi-Civita connection is inside $G_2 \subset SO(7)$.*

The simple Lie group G_2 can be identified as the subgroup of $SO(7)$ consisting of isomorphism $g : \mathbb{R}^7 \to \mathbb{R}^7$ preserving the linear three form Ω,

$$\Omega = f^1 f^2 f^3 - f^1\left(e^1 e^0 + e^2 e^3\right) - f^2\left(e^2 e^0 + e^3 e^1\right) - f^3\left(e^3 e^0 + e^1 e^2\right),$$

where $e^0, e^1, e^2, e^3, f^1, f^2, f^3$ is any given orthonormal frame of $\mathbb{R}^7$. Such a three form, or up to conjugation by elements in $GL(7, \mathbb{R})$, is called *positive*, and it determines a unique compatible inner product on $\mathbb{R}^7$ [**6**].

Gray [**12**] shows that G_2-holonomy of M can be characterized by the existence of a positive harmonic three form Ω.

DEFINITION 2. *A seven dimensional manifold M equipped with a positive closed three form Ω is called an almost G_2-manifold.*

Remark: The relationship between G_2-manifolds and almost G_2-manifolds is completely analogous to the relationship between Kahler manifolds and symplectic manifolds.

For example, suppose that X is a complex three dimensional Kähler manifold with a trivial canonical line bundle, i.e. there exists a nonvanishing holomorphic three form Ω_X. Yau's celebrated theorem says that there is a Kähler form ω_X on X with holonomy in $SU(3)$, i.e. a Calabi-Yau threefold. In particular both Ω_X and ω_X are parallel forms. Then the product $M = X \times S^1$ is a G_2-manifold with

$$\Omega = \operatorname{Re}\Omega_X + \omega_X \wedge d\theta.$$

Conversely, one can prove, using Bochner arguments, every G_2-metric on $X \times S^1$ must be of this form. More generally, if ω_X is a general Kähler form on X, then $(X \times S^1, \Omega)$ is an *almost* G_2-manifold and the converse is also true.

Next we quickly review the geometry of $\mathbb{H}$-SLag cycles in an almost G_2-manifold (see [**19**]).

DEFINITION 3. *An orientable four dimensional submanifold C in an almost G_2-manifold (M, Ω) is called a coassociative submanifold, or simply a $\mathbb{H}$-SLag, if the restriction of Ω to C is identically zero,*

$$\Omega|_C = 0.$$

If M is a G_2-manifold, then such a C is calibrated by $*\Omega$ in the sense of Harvey and Lawson [**14**], in particular, it is an absolute minimal submanifold in M. The normal bundle of any $\mathbb{H}$-SLag C can be naturally identified with the bundle of self-dual two forms on C. McLean [**27**] shows that infinitesimal deformations of any $\mathbb{H}$-SLag are unobstructed and they are parametrized by the space of harmonic self-dual two forms on C, i.e. $H^2_+(C, \mathbb{R})$.

For example, if S is a complex surface in a Calabi-Yau threefold X, then $S \times \{t\}$ is a $\mathbb{H}$-SLag in $M = X \times S^1$ for any $t \in S^1$. Similarly, if L is a three dimensional special Lagrangian submanifold in X with phase $\pi/2$, i.e. $\omega|_L = \operatorname{Re}\Omega_X|_L = 0$, then $L \times S^1$ is also a $\mathbb{H}$-SLag in $M = X \times S^1$.

DEFINITION 4. *A $\mathbb{H}$-SLag cycle in an almost G_2-manifold (M, Ω) is a pair (C, D_E) with C a $\mathbb{H}$-SLag in M and D_E an ASD connection over C.*

Remark: $\mathbb{H}$-SLag cycles are supersymmetric cycles in physics as studied in [**26**]. Their moduli space admits a natural three form and a cubic tensor [**19**], which play the roles of correlation function and Yukawa coupling in physics.

We assume that the ASD connection D_E over C has rank one, i.e. a $U(1)$ connection. This avoids the occurrence of reducible connections, thus $\mathcal{M}^{\mathbb{H}-SLag}(M)$ is a smooth manifold. It has a natural orientation and its expected dimension equals $b^1(C)$, the first Betti number of C. This is because the moduli space of $\mathbb{H}$-SLags has dimension equals $b^2_+(C)$ [**27**] and the existence of an ASD $U(1)$-connection over C is equivalent to $H^2_-(C, \mathbb{R}) \cap H^2(C, \mathbb{Z}) \neq \phi$. The number $b^1(C)$ is responsible for twisting by a flat $U(1)$-connection.

For simplicity, we assume that $b^1(C) = 0$, otherwise, one can cut down the dimension of $\mathcal{M}^{\mathbb{H}-SLag}(M)$ to zero by requiring the ASD connections over C to have trivial holonomy around loops $\gamma_1, ..., \gamma_{b^1(C)}$ in C representing an integral basis of $H_1(C, \mathbb{Z})$. We plan to count the algebraic number of points in this moduli space $\#\mathcal{M}^{\mathbb{H}-SLag}(M)$.

This number, in the case of $X \times S^1$, can be identified with a proposed invariant of Joyce [**17**] defined by counting rigid special Lagrangian submanifolds in any Calabi-Yau threefold. To explain this, we need the following proposition on the strong rigidity of product $\mathbb{H}$-SLags.

Proposition 5. *If $L \times S^1$ is a $\mathbb{H}$-SLag in $M = X \times S^1$ with X a Calabi-Yau threefold, then any $\mathbb{H}$-SLag representing the same homology class must also be a product.*

Proof: For simplicity we assume that the volume of the S^1 factor is unity, $Vol\left(S^1\right) = 1$. If $L \times S^1$ is a $\mathbb{H}$-SLag in M then L is special Lagrangian submanifold in X with phase $\pi/2$, i.e. $\operatorname{Re}\Omega_X|_L = \omega|_L = 0$. Suppose C is another $\mathbb{H}$-SLag in M representing the same homology class, we have $Vol\,(C) = Vol\,(L)$. If we write $C_\theta = C \cap (X \times \{\theta\})$ for any $\theta \in S^1$, then $Vol\,(C_\theta) \geq Vol\,(L)$, as L is a calibrated submanifold in X. Furthermore the equality sign holds only if C_θ is also calibrated. In general we have

$$Vol\,(C) \geq \int_{S^1} Vol\,(C_\theta)\, d\theta,$$

with the equality sign holds if and only if C is a product with S^1. Combining these, we have

$$Vol\,(L) = Vol\,(C) \geq \int_{S^1} Vol\,(C_\theta)\, d\theta \geq \int_{S^1} Vol\,(L)\, d\theta = Vol\,(L)\,.$$

Thus both inequalities are indeed equal. Hence $C = L' \times S^1$ for some special Lagrangian submanifold L' in X. ■

Suppose $M = X \times S^1$ is a product G_2-manifold and we consider product $\mathbb{H}$-SLag $C = L \times S^1$ in M. From the above proposition, every $\mathbb{H}$-SLag representing $[C]$ must also be a product. Since $b_+^2(C) = b^1(L)$, the rigidity of the $\mathbb{H}$-SLag C in M is equivalent to the rigidity of the special Lagrangian submanifold L in X. When this happens, i.e. L is a rational homology three sphere, we have $b^2(C) = 0$ and

$$\text{No. of ASD U(1)-bdl}/C = \#H^2(C, \mathbb{Z}) = \#H^2(L, \mathbb{Z}) = \#H_1(L, \mathbb{Z})\,.$$

Here we have used the fact that the first cohomology group is always torsion free. Thus the number of such $\mathbb{H}$-SLag cycles in $X \times S^1$ equals the number of special Lagrangian rational homology three spheres in a Calabi-Yau threefold X, weighted by $\#H_1(L, \mathbb{Z})$. Joyce **[17]** shows that with this particular weight, the numbers of special Lagrangians in any Calabi-Yau threefold behave well under various surgeries on X, and expects them to be invariants. Thus in this case, we have

$$\#\mathcal{M}^{\mathbb{H}-SLag}\left(X \times S^1\right) = \text{Joyce's proposed invariant for \#SLag. in } X.$$

In the next section, we will propose a homology theory, whose Euler characteristic gives $\#\mathcal{M}^{\mathbb{H}-SLag}(M)$.

3. Witten's Morse theory for $\mathbb{H}$-SLag cycles

We are going to use the parametrized version of $\mathbb{H}$-SLag cycles in any almost G_2-manifold M. We fix an oriented smooth four dimensional manifold C and a rank r Hermitian vector bundle E over C. We consider the *configuration space*

$$\mathcal{C} = Map\,(C, M) \times \mathcal{A}\,(E)\,.$$

DEFINITION 6. *An element* (f, D_E) *in* $\mathcal{C}$ *is called a parametrized* $\mathbb{H}$*-SLag cycles in* M *if*

$$f^*\Omega = F_E^+ = 0,$$

where the self-duality is defined using the pullback metric from M.

Remark: From the positivity of Ω, $f^*\Omega = 0$ implies f is an immersion.

Instead of $Aut\,(E)$, the symmetry group $\mathcal{G}$ in our situation consists of gauge transformations of E which cover *arbitrary* diffeomorphisms on M,

$$\begin{array}{ccc} E & \overset{g}{\to} & E \\ \downarrow & & \downarrow \\ M & \overset{g_M}{\to} & M. \end{array}$$

It fits into the following exact sequence,

$$1 \to Aut\,(E) \to \mathcal{G} \to Diff\,(C) \to 1.$$

The natural action of $\mathcal{G}$ on $\mathcal{C}$ is given by

$$g \cdot (f, D_E) = (f \circ g_M, g^* D_E),$$

for any $(f, D_E) \in \mathcal{C} = Map\,(C, M) \times \mathcal{A}\,(E)$. Notice that $\mathcal{G}$ preserves the set of parametrized $\mathbb{H}$-SLag cycles in M.

The configuration space $\mathcal{C}$ has a natural one form Φ_0: At any $(f, D_E) \in \mathcal{C}$ we can identify the tangent space of $\mathcal{C}$ as

$$T_{(f,D_E)}\mathcal{C} = \Gamma\,(C, f^*T_M) \times \Omega^1\,(C, ad\,(E)).$$

We define

$$\Phi_0\,(f, D_E)\,(v, B) = \int_C Tr\,[f^*\,(\iota_v\Omega) \wedge F_E + f^*\Omega \wedge B],$$

for any $(v, B) \in T_{(f,D_E)}\mathcal{C}$.

PROPOSITION 7. *The one form* Φ_0 *on* $\mathcal{C}$ *is closed and invariant under the action by* $\mathcal{G}$.

Proof: Recall that there is a universal connection $\mathbb{D}_E$ over $C \times \mathcal{A}\,(E)$ whose curvature $\mathbb{F}_E$ at a point (x, D_E) equals,

$$\begin{aligned} \mathbb{F}_E|_{(x,D_E)} &= \left(\mathbb{F}_E^{2,0}, \mathbb{F}_E^{1,1}, \mathbb{F}_E^{0,2}\right) \\ &\in \Omega^2\,(C) \otimes \Omega^0\,(\mathcal{A}) + \Omega^1\,(C) \otimes \Omega^1\,(\mathcal{A}) + \Omega^0\,(C) \otimes \Omega^2\,(\mathcal{A}) \end{aligned}$$

with

$$\mathbb{F}_E^{2,0} = F_E,\ \mathbb{F}_E^{1,1}\,(v, B) = B\,(v),\ \mathbb{F}_E^{0,2} = 0,$$

where $v \in T_xC$ and $B \in \Omega^1\,(C, ad\,(E)) = T_{D_E}\mathcal{A}\,(E)$ (see e.g. [**20**]). The Bianchi identity implies that $Tr\mathbb{F}_E$ is a closed form on $C \times \mathcal{A}\,(E)$. We also consider the evaluation map,

$$\begin{aligned} ev &: C \times Map\,(C, M) \to M \\ ev\,(x, f) &= f\,(x). \end{aligned}$$

It is not difficult to see that the pushforward of the differential form $ev^*(\Omega) \wedge Tr\mathbb{F}_E$ on $C \times Map(C, M) \times \mathcal{A}(E)$ to $Map(C, M) \times \mathcal{A}(E)$ equals Φ_0, i.e.

$$\Phi_0 = \int_C ev^*(\Omega) \wedge Tr\mathbb{F}_E.$$

Therefore the closedness of Φ_0 follows from the closedness of Ω. It is also clear from this description of Φ_0 that it is $\mathcal{G}$-invariant. ■

From now on, we assume that E is a rank one bundle.

LEMMA 8. *The zeros of Φ_0 are the same as parametrized $\mathbb{H}$-SLag cycles in M.*

Proof: Suppose (f, D_E) is a zero of Φ_0. By evaluating it on various $(0, B)$, we have $f^*\Omega = 0$, i.e. $f : C \to M$ is a parametrized $\mathbb{H}$-SLag. This implies that the map

$$\lrcorner\Omega : T_{f(x)}M \to \Lambda^2 T_x^* C$$

has image equals $\Lambda_+^2 T_x^* C$, for any $x \in C$. By evaluating Φ_0 on various $(v, 0)$, we have $F_E^+ = 0$, i.e. (f, D_E) is a parametrized $\mathbb{H}$-SLag cycle in M. The converse is obvious. ■

From above results, Φ_0 descends to a closed one form on $\mathcal{C}/\mathcal{G}$, called Φ. Locally we can write $\Phi = d\mathcal{F}$ for some function $\mathcal{F}$ whose critical points are precisely (unparametrized) $\mathbb{H}$-SLag cycles in M. Using the gradient flow lines of $\mathcal{F}$, we could formally define a Witten's Morse homology group, as in the famous Floer's theory. Roughly speaking one defines a complex $(\mathbf{C}_*, \partial)$, where $\mathbf{C}_*$ is the free Abelian group generated by critical points of $\mathcal{F}$ and ∂ is defined by counting the number of gradient flow lines between two critical points of relative index one.

Remark: The equations for the gradient flow are given by

$$\frac{\partial f}{\partial t} = *(f^*\xi \wedge F_E), \ \frac{\partial D_E}{\partial t} = *(f^*\Omega),$$

where $\xi \in \Omega^2(M, T_M)$ is defined by $\langle \xi(u, v), w \rangle = \Omega(u, v, w)$.

The equation

$$\partial^2 = 0$$

requires a good compactification of the moduli space of $\mathbb{H}$-SLag cycles in M, which we are lacking at this moment. We denote this proposed homology group as $H_C(M)$, or $H_C(M, \alpha)$ when $f_*[C] = \alpha \in H_4(M, \mathbb{Z})$.

This homology group should be invariant under deformations of the almost G_2-metric on M and its Euler characteristic equals,

$$\chi(H_C(M)) = \#\mathcal{M}^{\mathbb{H}-SLag}(M).$$

Like Floer homology groups, they measure the *middle dimensional* topology of the configuration space $\mathcal{C}$ divided by $\mathcal{G}$.

4. TQFT of $\mathbb{H}$-SLag cycles

In this section we study complete almost G_2-manifold M_i with asymptotically cylindrical ends and the behavior of $H_C(M)$ when a closed almost G_2-manifold M decomposes into connected sum of two pieces, each with an asymptotically cylindrical end,

$$M = M_1 \underset{X}{\#} M_2.$$

Nontrivial examples of compact G_2-manifolds are constructed by Kovalev [**18**] using such connected sum approach. The boundary manifold X is necessary a Calabi-Yau threefold. We plan to discuss analytic aspects of M_i's in a future paper [**24**].

Each M_i's will define a Lagrangian subspace $\mathcal{L}_{M_i}$ in the moduli space of special Lagrangian cycles in X. Furthermore we expect to have a gluing formula expressing the above homology group for M in terms of the Floer Lagrangian intersection homology group for the two Lagrangian subspaces $\mathcal{L}_{M_1}$ and $\mathcal{L}_{M_2}$,

$$H_C(M) = HF_{Lag}^{\mathcal{M}^{SLag}(X)}\left(\mathcal{L}_{M_1}, \mathcal{L}_{M_2}\right).$$

These properties can be reformulated to give us a topological quantum field theory. To begin we have the following definition.

DEFINITION 9. *An almost G_2-manifold M is called cylindrical if $M = X \times \mathbb{R}^1$ and its positive three form respect such product structure, i.e.*

$$\Omega_0 = \operatorname{Re}\Omega_X + \omega_X \wedge dt.$$

A complete almost G_2-manifold M with one end $X \times [0, \infty)$ is called asymptotically cylindrical if the restriction of its positive three form equals to the above one for large t, up to a possible error of order $O(e^{-t})$. More precisely the positive three form Ω of M restricted to its end equals,

$$\Omega = \Omega_0 + d\zeta$$

for some two form ζ satisfying $|\zeta| + |\nabla\zeta| + |\nabla^2\zeta| + |\nabla^3\zeta| \leq Ce^{-t}$.

Remark: If M is an almost G_2-manifold with an asymptotically cylindrical end $X \times [0, \infty)$, then (X, ω_X, Ω_X) is a complex threefold with a trivial canonical line bundle, but the Kähler form ω_X might not be Einstein. This is so, i.e. a Calabi-Yau threefold, provided that M is a G_2-manifold. We will simply write $\partial M = X$.

We consider $\mathbb{H}$-SLags C in M which satisfy a *Neumann condition* at infinity. That is, away from some compact set in M, the immersion $f : C \to M$ can be written as

$$f : L \times [0, \infty) \to X \times [0, \infty)$$

with $\partial f / \partial t$ vanishes at infinite [**24**]. A relative $\mathbb{H}$-SLag itself has asymptotically cylindrical end $L \times [0, \infty)$ with L a special Lagrangian submanifold in X. A *relative $\mathbb{H}$-SLag cycle* in M is a pair (C, D_E) with C a relative $\mathbb{H}$-SLag in M and D_E a unitary connection over C with finite energy,

$$\int_C |F_E|^2 \, dv < \infty.$$

Any finite energy connection D_E on C induces a unitary flat connection $D_{E'}$ on L [**7**].

Such a pair $(L, D_{E'})$ of a unitary flat connection $D_{E'}$ over a special Lagrangian submanifold L in a Calabi-Yau threefold X is called a *special Lagrangian cycle* in X. Their moduli space $\mathcal{M}^{SLag}(X)$ plays an important role in the Strominger-Yau-Zaslow Mirror Conjecture [**28**] or [**22**]. The tangent space to $\mathcal{M}^{SLag}(X)$ is naturally identified with $H^2(L, \mathbb{R}) \times H^1(L, ad(E'))$. For line bundles over L, the cup product

$$\cup : H^2(L, \mathbb{R}) \times H^1(L, \mathbb{R}) \to \mathbb{R},$$

induces a symplectic structure on $\mathcal{M}^{SLag}(X)$ [**15**]. Using analytic results from [**24**] about asymptotically cylindrical manifolds, we can prove the following theorem.

Claim 10. *Suppose M is an asymptotically cylindrical (almost) G_2-manifold with $\partial M = X$. Let $\mathcal{M}^{\mathbb{H}-SLag}(M)$ be the moduli space of rank one relative $\mathbb{H}$-SLag cycles in M. Then the map defined by the boundary values,*

$$b : \mathcal{M}^{\mathbb{H}-SLag}(M) \to \mathcal{M}^{SLag}(X),$$

is a Lagrangian immersion.

Sketch of the proof ([**24**]): For any closed Calabi-Yau threefold X (resp. G_2-manifold M), the moduli space of rank one special Lagrangian submanifolds L (resp. $\mathbb{H}$-SLags C) is smooth [**27**] and has dimension $b^2(L)$ (resp. $b^2_+(C)$). The same holds true for complete manifold M with a asymptotically cylindrical end $X \times [0, \infty)$, where $b^2_+(C)_{L^2}$ denote the dimension of L^2-harmonic self-dual two forms on a relative $\mathbb{H}$-SLag C in M.

The linearization of the boundary value map $\mathcal{M}^{\mathbb{H}-SLag}(M) \to \mathcal{M}^{SLag}(X)$ is given by $H^2_+(C)_{L^2} \overset{\alpha}{\to} H^2(L)$. Similar for the connection part, where the boundary value map is given by $H^1(C)_{L^2} \overset{\beta}{\to} H^1(L)$. We consider the following diagram where (i) each row is a long exact sequence of L^2-cohomology groups for the pair (C, L) and (ii) each column in a perfect pairing.

$$\begin{array}{ccccccccc}
0 & \to & H^2_+(C,L) & \to & H^2_+(C) & \overset{\alpha}{\to} & H^2(L) & \to & H^3(C,L) & \to & \cdots \\
 & & \otimes & & \otimes & & \otimes & & \otimes & & \\
0 & \leftarrow & H^2_+(C) & \leftarrow & H^2_+(C,L) & \leftarrow & H^1(L) & \overset{\beta}{\leftarrow} & H^1(C) & \leftarrow & \cdots \\
 & & \downarrow & & \downarrow & & \downarrow & & \downarrow & & \\
 & & \mathbb{R} & & \mathbb{R} & & \mathbb{R} & & \mathbb{R} & &
\end{array}$$

Notice that $H^2_+(C, L)$, $H^2_+(C)$ and $H^2(L)$ parametrize infinitesimal deformation of C with fixed ∂C, deformation of C alone and deformation of L respectively.

By simply homological algebra, it is not difficult to see that $\operatorname{Im}\alpha \oplus \operatorname{Im}\beta$ is a Lagrangian subspace of $H^2(L) \oplus H^1(L)$ with the canonical symplectic structure. Hence the result. ■

We denote the immersed Lagrangian submanifold $b\left(\mathcal{M}^{\mathbb{H}-SLag}(M)\right)$ in $\mathcal{M}^{SLag}(X)$ by $\mathcal{L}_M$. When M decompose as a connected sum $M_1 \#_X M_2$ along a long neck, as in Atiyah's conjecture on Floer Chern-Simons homology group [**3**], we expect to have an isomorphism,

$$H_C(M) \cong HF^{\mathcal{M}^{SLag}(X)}_{Lag}(\mathcal{L}_{M_1}, \mathcal{L}_{M_2}).$$

More precisely, suppose Ω_t with $t \in [0, \infty)$, is a family of G_2-structure on $M_t = M$ such that as t goes to infinite, M decomposes into two components M_1 and

M_2, each has an aymptotically cylindrical end $X \times [0,\infty)$. Then we expect that $\lim_{t\to\infty} H_C(M_t) \cong HF_{Lag}^{\mathcal{M}^{SLag}(X)}(\mathcal{L}_{M_1}, \mathcal{L}_{M_2})$. We summarize these structures in the following table:

Manifold:	(almost) G_2-manifold, M^7	(almost) CY threefold, X^6
SUSY Cycles:	$\mathbb{H}$-SLag. submfds.+ ASD bundles	SLag submfds.+ flat bundles
Invariant:	Homology group, $H_C(M)$	Fukaya category, $Fuk(\mathcal{M}^{SLag}(X))$.

These associations can be formalized to form a TQFT [**4**]. Namely we associate an additive category $F(X) = Fuk(\mathcal{M}^{SLag}(X))$ to a closed almost Calabi-Yau threefold X, a functor $F(M): F(X_0) \to F(X_1)$ to an almost G_2-manifold M with asymptotically cylindrical ends $X_1 - X_0 = X_1 \cup \bar{X}_0$. They satisfy

(i) $F(\phi) =$ the additive tensor category of vector spaces $((Vec))$,
(ii) $F(X_1 \amalg X_2) = F(X_1) \otimes F(X_2)$.

For example, when M is a closed G_2-manifold, that is a cobordism between empty manifolds, then we have $F(M): ((Vec)) \to ((Vec))$ and the image of the trivial bundle is our homology group $H_C(M)$.

5. More TQFTs

There are other TQFTs naturally associated to Calabi-Yau threefolds and G_2-manifolds but (1) they do not involve nontrivial coupling between submanifolds and bundles and (2) new difficulties arise because of corresponding moduli spaces for Calabi-Yau threefolds have virtual dimension zero and could be singular. They are essentially in the paper by Donaldson and Thomas [**9**].

TQFT of associative cycles

We assume that M is a G_2-manifold, i.e. Ω is parallel rather than closed. Three dimensional submanifolds A in M calibrated by Ω is called *associative submanifolds* and they can be characterized by $\chi|_A = 0$ ([**14**]) where $\chi \in \Omega^3(M, T_M)$ is defined by $\langle w, \chi(x,y,z)\rangle = *\Omega(w,x,y,z)$. We define a *parametrized A-cycle* to be a pair $(f, D_E) \in \mathcal{C}_A = Map(A,M) \times \mathcal{A}(E)$, with $f: A \to M$ a parametrized A-submanifold and D_E is a unitary flat connection on a Hermitian vector bundle E over A. There is also a natural $\mathcal{G}$-invariant closed one form Φ_A on $\mathcal{C}_A$ given by

$$\Phi_A(f, D_E)(v, B) = \int_A TrF_E \wedge B + \langle f^*\chi, v\rangle_{T_M},$$

for any $(v, B) \in \Gamma(A, f^*T_M) \times \Omega^1(A, ad(E)) = T_{(f,D_E)}\mathcal{C}_A$. Its zero set is the moduli space of A-cycles in M. As before, we could formally apply arguments in Witten's Morse theory to Φ_A and define a homology group $H_A(M)$.

The corresponding category associated to a Calabi-Yau threefold X would be the Fukaya-Floer category of the moduli space of unitary flat bundles over holomorphic curves in X, denote $\mathcal{M}^{curve}(X)$. We summarize these in the following table:

Manifold:	G_2-manifold, M^7	CY threefold, X^6
SUSY Cycles:	A-submfds.+ flat bundles	Holomorphic curves+ flat bundles
Invariant:	Homology group, $H_A(M)$	Fukaya category, $Fuk(\mathcal{M}^{curve}(X))$.

TQFT of Donaldson-Thomas bundles

We assume that M is a seven manifold with a G_2-structure such that its positive three form Ω is co-closed, rather than closed, i.e. $d\Theta = 0$ with $\Theta = *\Omega$. In [**9**] Donaldson and Thomas introduce a first order Yang-Mills equation for G_2-manifolds,

$$F_E \wedge \Theta = 0.$$

Their solutions are the zeros of the following gauge invariant one form Φ_{DT} on $\mathcal{A}(E)$,

$$\Phi_{DT}(D_E)(B) = \int_M Tr[F_E \wedge B] \wedge \Theta,$$

for any $B \in \Omega^1(M, ad(E)) = T_{D_E}\mathcal{A}(E)$. This one form Φ_{DT} is closed because of $d\Theta = 0$. As before, we can formally define a homology group $H_{DT}(M)$. The corresponding category associated to a Calabi-Yau threefold X should be the Fukaya-Floer category of the moduli space of Hermitian Yang-Mills connections over X, denote $\mathcal{M}^{curve}(X)$. Again we summarize these in a table:

Manifold:	G_2-manifold, M^7	CY threefold, X^6
SUSY Cycles:	DT-bundles	Hermitian YM-bundles
Invariant:	Homology group, $H_{DT}(M)$	Fukaya category, $Fuk(\mathcal{M}^{HYM}(X))$.

It is an interesting problem to understand the transformations of these TQFTs under dualities in M-theory.

Acknowledgments: This paper is partially supported by NSF/DMS-0103355. The author expresses his gratitude to J.H. Lee, R. Thomas, A. Voronov and X.W. Wang for useful discussions.

References

[1] B. S. Acharya, B. Spence, *Supersymmetry and M theory on 7-manifolds,* [hep-th/0007213].

[2] M. Aganagic, C. Vafa, *Mirror Symmetry and a G_2 Flop,* [hep-th/0105225].

[3] M. Atiyah, *New invariants of three and four dimensional manifolds,* in The Mathematical Heritage of Herman Weyl, Proc. Symp. Pure Math., **48**, A.M.S. (1988), 285-299.

[4] M. Atiyah, *Topological quantum field theories.* Inst. Hautes Études Sci. Publ. Math. No. 68 (1988), 175–186 (1989).

[5] M. Atiyah, E. Witten, *M-theory dynamics on a manifold of G_2 holonomy,* [hep-th/0107177].

[6] R. Bryant, *Metrics with exceptional holonomy,* Ann. of Math. 126 (1987) 525-576.

[7] S. Donaldson, *Floer homology group in Yang-Mills theory,* Cambridge Univ. Press (2002).

[8] S. Donaldson, P. Kronheimer, *The geometry of four-manifols,* Oxford University Press, (1990).

[9] S. Donaldson, R. Thomas, *Gauge theory in higher dimension,* The Geometric Universe: Science, Geometry and the work of Roger Penrose, S.A. Huggett et al edited, Oxford Univ. Press (1988).

[10] K. Fukaya, Y.G. Oh, H. Ohta, K. Ono, *Lagrangian intersection Floer theory - anomoly and obstruction*, to appear in International Press.

[11] R. Gopakumar, C. Vafa, *M-theory and topological strings - II,* [hep-th/9812127].

[12] A. Gray, *Vector cross products on manifolds*, Trans. Amer. Math. Soc. 141 (1969) 465-504.

[13] S. Gukov, S.-T. Yau, E. Zaslow, *Duality and Fibrations on G_2 Manifolds,* [hep-th/0203217].

[14] R. Harvey, B. Lawson, *Calibrated geometries*, Acta Math. 148 (1982), 47-157.

[15] N. Hitchin, *The moduli space of special Lagrangian submanifolds*. Dedicated to Ennio De-Giorgi. Ann. Scuola Norm. Sup. Pisa Cl. Sci. (4) 25 (1997), no. 3-4, 503–515 (1998). [dg-ga/9711002].

[16] N. Hitchin, *The geometry of three forms in 6 and 7 dimensions, J. Differential Geom. 55 (2000), no. 3,* 547–576. [math.DG/0010054].

[17] D. Joyce, *On counting special Lagrangian homology 3-spheres,* [hep-th/9907013].

[18] A. Kovalev, *Twisted connected sums and special Riemannian holonomy,* [math.DG/0012189].

[19] J.H. Lee, N.C. Leung, *Geometric structures on G_2 and Spin(7)-manifolds,* [math.DG/0202045].

[20] N.C. Leung, *Symplectic structures on gauge theory,* Comm. Math. Phys., 193 (1998) 47-67.

[21] N.C. Leung, *Mirror symmetry without corrections,* [math.DG/0009235].

[22] N.C. Leung, *Geometric aspects of mirror symmetry,* to appear in the proceeding of ICCM 2001, [math.DG/0204168].

[23] N.C. Leung, *Riemannian geometry over different normed division algebras*, preprint 2002.

[24] N.C. Leung, in preparation.

[25] N.C. Leung, S.Y. Yau, E. Zaslow, *From special Lagrangian to Hermitian-Yang-Mills via Fourier-Mukai transform,* to appear in Adv. Thero. Math. Phys.. [math.DG/0005118].

[26] M. Marino, R. Minasian, G. Moore, and A. Strominger, *Nonlinear Instantons from supersymmetric p-Branes,* [hep-th/9911206].

[27] R. McLean, *Deformations of calibrated submanifolds,* Comm. Analy. Geom., **6** (1998) 705-747.

[28] A. Strominger, S.-T. Yau, and E. Zaslow, *Mirror Symmetry is T-Duality,* Nuclear Physics **B479** (1996) 243-259; [hep-th/9606040].

Address: School of Mathematics, University of Minnesota, Minneapolis, MN 55454, USA.

Email: LEUNG@MATH.UMN.EDU

Geometric results in classical minimal surface theory

William H. Meeks III

1. Introduction.

Although classical minimal surface theory dates back to the work of Euler [**23**] and Lagrange [**52**] in the 18th century, most advances in the global theory have been obtained in the past twenty-five years. The primary goal of this survey is to present the recent theorems in minimal surface theory together with a sufficient amount of background information so that these theorems can be understood, appreciated and applied. The presentation here of both the old and the new results in this classical subject is more from a geometric rather than an analytic point of view. The interested reader can also find more detailed history and further results in the following surveys, reports and popular science articles [**4, 18, 19, 37, 38, 39, 40, 46, 63, 65, 87**].

In the next two sections we introduce the concept of minimal surface from several different equivalent points of view and include many of the basic definitions, notation and results. In Section 4 we give a description of eight well-known classical examples of minimal surfaces. These examples motivate many of the theoretical results in later sections and so, the reader should make an effort to understand their geometry and special properties before proceeding. Section 5 introduces the important notion of stable or locally-least-area minimal surface and includes some of the basic theorems on stable minimal surfaces. Part of the importance of stable minimal surfaces is that they are an essential tool for studying many of the difficult global problems in the classical theory.

Section 6 deals with the questions of existence and of regularity of solutions to the classical Plateau problem; this problem asks whether a simple closed curve in $\mathbb{R}^3$ is the boundary of a least-area surface. This section includes several different formulations of this least-area problem, including a short discussion of the barrier construction of Meeks and Yau [**83**] which we need in some later discussions.

The remainder of the survey deals with advances made in the past decade. Section 7 explains the solution of the generalized Nitsche Conjecture given by Collin [**15**] and the recent theorem of Meeks and Rosenberg [**74**] on the uniqueness of the helicoid. Together, the theorem of Collin and the theorem of Meeks and Rosenberg give a satisfactory theory for describing all properly embedded minimal surfaces of finite topology in $\mathbb{R}^3$ in terms of meromorphic data on their conformal compactifications; these conformal compactifications are closed Riemann surfaces. This analytic representation of finite topology examples leads to real analytic structures on the associated moduli spaces of examples of a fixed topology, to a description of the asymptotic behavior of the examples and, in certain cases, to the classification of all examples of a fixed topological type.

In Section 8 we enter the realm of active research on the local geometry of embedded minimal surfaces in Riemannian three-manifolds near points of large Gaussian curvature, under the additional hypothesis of having a fixed bound on the genus of the surface in a small neighborhood of such a point of large curvature. This description follows the pioneering work of Colding and Minicozzi [**9, 10, 11, 12, 13**] and subsequent geometric extensions by Meeks [**61**], Meeks and Rosenberg [**74**] and Meeks, Perez and Ros [**70, 66, 67, 69, 68**].

Section 9 presents all known topological obstructions for properly minimally embedding a noncompact orientable surface into $\mathbb{R}^3$. Here we include a discussion of the classical results of Schoen [**100**] and Lopez-Ros [**56**], which together with Collin's theorem [**15**] and the Meeks-Rosenberg theorem [**74**], give a complete classification of all properly embedded minimal surfaces of finite topology in $\mathbb{R}^3$ which have either genus-zero or two ends. Recently Meeks, Perez and Ros [**69**] have shown that a properly embedded minimal surface in $\mathbb{R}^3$ of finite genus g and a finite number of ends has a bound on the number of its ends that only depends on g. These topological obstructions and classification theorems represent the most important theoretical results that one might hope to obtain in this subject.

Much of present day research in classical minimal surface theory is focussed on describing the asymptotic geometry of properly embedded minimal surfaces with finite genus and infinitely generated fundamental group; this means that we are considering surfaces of finite genus with an infinite number of ends (see Section 3 for a description of the space of ends of a noncompact surface). In this regard, we include in Section 9 a discussion of the fundamental result of Collin, Kusner, Meeks and Rosenberg [**16**] on the structure of the space of ends of a properly embedded minimal surface with an infinite number of ends. This structure theorem, together with related topological obstructions, plays a fundamental role in restricting the geometry and topology of properly embedded minimal surfaces in $\mathbb{R}^3$ which have infinite topology. This structure theorem is essential in proving that a properly embedded minimal surface in $\mathbb{R}^3$ with finite genus and an infinite number of ends must have exactly two limit ends [**67**].

In Section 10 we describe the recent proof by Frohman and Meeks [**29**] of the "Topological Classification Theorem for Minimal Surfaces". This theorem gives a complete cook-book type description of how a minimal surface is embedded in $\mathbb{R}^3$ in terms of calculable algebraic-topological invariants; it gives necessary and sufficient conditions for two different properly embedded minimal surfaces to be properly ambiently isotopic in $\mathbb{R}^3$.

For many of the global questions we would like to answer for some class of minimal surfaces in $\mathbb{R}^3$, it is essential to know the underlying conformal structure. For example, near the end of the proof of the uniqueness of the helicoid in [**74**] (see Sections 7 and 8), one needs to know that certain properly embedded simply-connected minimal surfaces are conformally the complex plane $\mathbb{C}$. This result on conformal structure depends on a general theorem in [**16**] that asserts that any component of the intersection of a properly immersed minimal surface with a closed halfspace in $\mathbb{R}^3$ is parabolic, in the sense that bounded harmonic functions on the component are determined by their boundary values. In Section 11 we prove this theorem and discuss the related question of recurrence for Brownian motion in minimal surfaces.

In Section 12 we describe the theory of periodic minimal surfaces in $\mathbb{R}^3$ as developed by Meeks and Rosenberg [**75**], [**77**], especially in the case of properly embedded surfaces with quotient having finite topology. Here we also discuss the uniqueness of some of the classical periodic minimal surfaces described in Section 4.

In Section 13 we briefly leave the classical setting in order to describe several surprising and deep results that are likely to have important applications. This section covers the theoretical results that Meeks and Rosenberg [**72, 73**] have obtained for properly embedded minimal surfaces in a Riemannian product $M \times \mathbb{R}$, where M is a compact Riemannian surface. The applications of these results that we have in mind pertain to the case where M is the two-sphere S^2 of constant Gaussian curvature 1. These applications are related to the existence and classification of harmonic maps of Riemann surfaces to S^2. Also, we hope to apply these theoretical results to better understand constant mean curvature surfaces in $\mathbb{R}^3$ and minimal and constant mean curvature surfaces in the three-sphere S^3.

In Section 14 we present a brief discussion of 16 fundamental conjectures of the author and others. Since these conjectures are discussed in detail there, we just briefly list them by name here: Convex Curve Conjecture, 4π Conjecture, Finite Topology Conjecture, Properness Conjecture, Liouville Conjecture, Removal Singularities Conjecture, Isometry Conjecture, Genus-Zero Conjecture, Geometric Flux Conjecture, Scherk Uniqueness Conjecture, Uniqueness of Limit Tangent Cone Conjecture, Graph Connectedness Conjecture, Quadratic Area Growth Conjecture, One-ended Conjecture, Infinite Topology Conjecture, Singular Curve Conjecture and the Uniqueness of the $\mathcal{A}$-family Conjecture.

2. The definition of minimal surface.

By a *surface* in $\mathbb{R}^3$ we mean a subset Σ of $\mathbb{R}^3$ that is locally parametrized by the open unit disk D in $\mathbb{R}^2$. In other words, for each $p \in \Sigma$, there is a neighborhood $U_p \subset \Sigma$ together with a map $f\colon D \to \mathbb{R}^3$ which is an injective smooth immersion with $f(D) = U_p$. If Σ is a smooth abstract Riemannian surface, then we say that $f\colon \Sigma \to \mathbb{R}^3$ is a *smooth embedding* or an *isometric injective immersion* if f is an injective immersion and the flat Riemannian metric in $\mathbb{R}^3$ induces the Riemannian metric on the surface Σ. We say that a surface $\Sigma \subset \mathbb{R}^3$ is *complete* if it is a complete metric space with respect to the natural distance function obtained from taking the infimum of the lengths of curves which join pairs of points. By the Hopf-Rinow theorem, Σ is complete if and only if every geodesic segment on Σ can be continued indefinitely. If Σ is allowed to have boundary, then we take the same definition for completeness, except now the Hopf-Rinow theorem states that a geodesic segment can be continued indefinitely or continued until it arrives at the boundary of Σ. At times it is convenient to consider a surface $\Sigma \subset \mathbb{R}^3$ as an embedding under inclusion: $f\colon \Sigma \to \mathbb{R}^3$ with respect to its underlying induced Riemannian structure. It is a standard fact for two-dimensional Riemannian surfaces that the Riemannian metric of a disk $F \subset \Sigma$ multiplied by some positive function is a new metric on F for which F is isometric to the unit disk D in $\mathbb{R}^2$. It follows that when Σ is orientable (Σ has a well-defined unit normal vector field called an *orientation*), then it has a system of local coordinates $\phi_\alpha\colon D \to U_\alpha \subset \Sigma$ such ϕ_α is conformal or angle preserving; such coordinates are called *isothermal* or *conformal* coordinates. With these elementary concepts in mind, we now define the concept of minimal surface by way of a list of equivalent properties.

THEOREM 2.1. *$f\colon \Sigma \to \mathbb{R}^3$ is an oriented* **minimal** *surface if Σ is oriented and any of the following equivalent properties hold:*

1. *Σ has zero mean curvature;*
2. *Small disks in Σ have least-area relative to their boundaries;*
3. *Small disks in Σ have least-energy relative to their boundaries;*
4. *Small disks in Σ are equal to the unique idealized soap film surfaces with the same boundary;*
5. *The coordinate functions f_1, f_2, f_3 of f are harmonic functions;*
6. *The Gauss or unit normal map $G\colon \Sigma \to S^2$ is conformal(we mean here that the derivative map is angle preserving wherever it is nonzero) and its stereographic projection $g\colon M \to \mathbb{C} \cup \{\infty\}$ is a meromorphic function.*

PROOF. I have just a few comments to make on the equivalences in the above Theorem. If $G\colon \Sigma \to S^2$ is the unit normal map, then the tangent space $T_p\Sigma$ of Σ at $p \in \Sigma$ is parallel in $\mathbb{R}^3$ to the tangent space $T_{G(p)}S^2$ to S^2 at the point $G(p) \in S^2$, and so after identifying these spaces under translation, the derivative map can be thought of as a linear map $S_p\colon T_p\Sigma \to T_p\Sigma$ called the *shape operator*. S_p is a symmetric linear transformation whose orthogonal eigenvectors are called the *principal directions* of Σ and the corresponding eigenvalues are called the *principal curvatures* of Σ. The *mean curvature* function H of Σ is the pointwise trace of the shape operator or, equivalently, $H(p)$ is equal to the sum of the principal curvatures at p.

In reference to Statement 2 of Theorem 2.1, one has the following more general result:

THEOREM 2.2. *(First Variation of Area Formula). If Σ is a compact, not necessarily minimal, surface with unit normal vector field N, and $\Sigma(t)$, $-\varepsilon < t < \varepsilon$, is a smooth deformation of Σ with $\partial\Sigma(t) = \partial\Sigma$, then the first derivative of area of this variation at $t = 0$, can be calculated as:*

$$A'(0) = \frac{dA}{dt}\Big|_{t=0} = \int_\Sigma H < N,\, V > dA,$$

where H is the mean curvature function on Σ and V is the variational vector field of $\Sigma(t)$ at $t = 0$.

It follows from the first variation of area formula that a compact minimal surface is a critical point to the area functional. The fact that Statements 1 and 2 in Theorem 2.1 are equivalent follows from this interpretation of the first variation of area formula, together with the fact that critical points of area are always local minima in small neighborhoods of every point, which can be derived for instance from the minimizing area property of minimal graphs.

The fact that Statement 2 is equivalent to Statement 3 follows from the standard inequality between energy and area. The energy we refer to here is the Dirichlet energy $\int_{\tilde\Sigma} |\nabla F|^2\, d\tilde A$ of a parametrization $F\colon \tilde\Sigma \to \Sigma \subset \mathbb{R}^3$ of the surface Σ by a Riemannian surface $\tilde\Sigma$, where $d\tilde A$ is the area form of $\tilde\Sigma$. If dA denotes the area form for Σ and $F^*(dA)$ denotes the pull-back form, then this inequality states that pointwise $|\nabla F|^2\, d\tilde A \geq 2F^*(dA)$ with equality if and only if the map is conformal. Since the inclusion mapping f is conformal (f is an isometry), the energy form of f is just twice its area form. Thus, Σ locally minimizes area precisely when considered as a map, it locally minimizes its energy.

Statements 3 and 4 are equivalent since idealized soap films are just surfaces which, by surface tension, locally minimize their energy. Surface tension on a minimal surface creates a static force which separates at each point on the surface into two forces that act oppositely and orthogonally to the surface along orthogonal directions on the surface (principal directions) and these forces are proportional pointwise to the principal curvatures at the point.

Statements 1 and 5 are equivalent by the easy to derive formula: If $f\colon M \to \mathbb{R}^3$ is an immersed oriented surface, then

$$\Delta f = (\Delta f_1, \Delta f_2, \Delta f_3) = H \cdot N,$$

where Δ is the Laplace operator on Σ, H is the mean curvature function and N is the unit normal field $N\colon \Sigma \to S^2 \subset \mathbb{R}^3$ of Σ.

Statements 1 and 6 are equivalent since the derivative of the Gauss map at a point $p \in \Sigma$ can be identified with the shape operator $S_p\colon T_p\Sigma \to T_p\Sigma = T_{G(p)}S^2$ which is a symmetric transformation with trace equal to H. If one takes the orientation of $T_{G(p)}S^2$ to be given by the orientation of S^2 coming from stereographic projection, which is opposite to the orientation on $T_{G(p)}S^2$ induced by parallel translation to $T_p\Sigma$, we see that the derivative map $G'_p\colon T_p\Sigma \to T_{G(p)}S^2$ is angle preserving wherever the derivative is not zero. This completes our proof of Theorem 2.1.

□

For later purposes note that the Gaussian curvature $K(p)$ of a point p on a minimal surface Σ is nonpositive and equal to the determinant of $S_p\colon T_p\Sigma \to T_p\Sigma$ which is equal to the product of the principal curvatures at p. Thus, on a compact minimal surface, it follows that the total Gaussian curvature of Σ is equal to

$$C(\Sigma) = \int_\Sigma K dA = -\text{Area}(G\colon \Sigma \to S^2),$$

where the area is counted with multiplicity.

3. Basic definitions and results.

An important analytic result is the classical Weierstrass representation of a minimal surface. Basically it gives a cook-book type recipe for analytically defining a minimal surface $f\colon \Sigma \to \mathbb{R}^3$. The approach we take is a variant of the Weierstrass representation given by Osserman in [**90**].

THEOREM 3.1. *(Weierstrass Representation) Suppose Σ is a Riemann surface and $f\colon \Sigma \to \mathbb{R}^3$ is a conformal harmonic map (i.e., f is a branched minimal surface) with $f(p_0) = (0,0,0)$. Let $\eta = dx_3 + i dx_3^*$ be the holomorphic one-form where x_3 is the third coordinate of Σ and x_3^* is the locally defined harmonic conjugate function of x_3. Let $g\colon \Sigma \to \mathbb{C} \cup \{\infty\}$ be the meromorphic Gauss map for Σ. Then:*

$$f(p) = \text{Re} \int_{p_0}^{p} (\frac{1}{2}(\frac{1}{g} - g)\eta, \frac{i}{2}(\frac{1}{g} + g)\eta, \eta).$$

Conversely, if η is a nonzero holomorphic one-form and $g\colon \Sigma \to \mathbb{C} \cup \{\infty\}$ is a nonconstant meromorphic function on a Riemann surface Σ such that the function $f\colon \Sigma \to \mathbb{R}^3$ given by the above formula is well-defined (the holomorphic one-forms in the formula have no real periods on Σ), then f is a conformal branched minimal immersion of Σ into $\mathbb{R}^3$ whose stereographically projected Gauss map is the meromorphic function g.

We are interested in understanding the space of complete embedded minimal surfaces in $\mathbb{R}^3$. All known examples of such surfaces satisfy the stronger hypothesis given in the next definition. Recall that a surface M has *more than one end* if it contains a smooth compact subdomain such that the complement of the interior of this domain in M has more than one noncompact component.

Definition 3.1. An immersion $f\colon M \to \mathbb{R}^3$ is *proper* if for every compact ball B, $f^{-1}(B)$ is compact in M. Let $\mathcal{P}$ denote the space of all properly embedded connected minimal surfaces in $\mathbb{R}^3$ and let $\mathcal{M} \subset \mathcal{P}$ be the subspace of examples with more than one end. The topology on $\mathcal{P}$ is the topology of smooth convergence on compact subsets of $\mathbb{R}^3$.

Definition 3.2. A Riemannian manifold M with nonempty boundary is *parabolic* if every bounded harmonic function on M is determined by its boundary values.

Definition 3.3. Given a Riemannian manifold M with nonempty boundary and a point $p \in \text{Int}(M)$, one can define the *harmonic* or *hitting* measure μ_p of an interval $I \subset \partial M$ as the probability that a Brownian path, beginning at p, exits the boundary ∂M somewhere on the interval I.

Instead of defining the harmonic measure μ_p of $I \subset \partial M$ in terms of probability and Brownian motion, one can also define it as follows. Consider a compact exhaustion $I \subset \partial M_1 \subset M_1 \subset M_2 \subset \dots$ of M. Let $h_n\colon M_n \to [0,1]$ be the bounded harmonic function with boundary values 1 on $\text{Int}(I)$ and 0 on the interior of $\partial M_n - I$. Since h_n is an increasing sequence of harmonic functions on M bounded by the constant function 1, h_n has a unique limit harmonic function h. In this case $\mu_p(I) = h(p)$.

The following useful Proposition is an elementary consequence of the definition of *harmonic* or *hitting* measure.

PROPOSITION 3.2. *Suppose M is a Riemannian manifold with nonempty boundary. The following are equivalent:*

1. *M is parabolic;*
2. *Bounded harmonic functions on M are determined by their boundary values;*
3. *For some $p \in Int(M)$, the measure μ_p is full on ∂M, i.e., $\int_{\partial M} \mu_p = 1$;*
4. *Given any $p \in Int(M)$ and any bounded harmonic function $f\colon M \to \mathbb{R}$, then $f(p) = \int_{\partial M} f(x)\mu_p$;*
5. *There exists a proper positive superharmonic function on M.*

The property of a Riemannian manifold with boundary being parabolic is closely related to the following notion of recurrence for Brownian motion. See [**32**] for an excellent survey of recurrence and Brownian motion on Riemannian manifolds.

Definition 3.4. A Riemannian manifold M (without boundary) is *recurrent* or *recurrent for Brownian motion* if and only if for any given point $p \in M$, almost all continuous paths $\alpha\colon [0,\infty) \to M$ with $\alpha(0) = p$ are dense in M.

It is well known [**32**] that $\mathbb{R}^n$ is recurrent for Brownian motion if and only if $n \leq 2$.

The following Lemma makes clear the relationship between the concepts of recurrence and of parabolicity.

LEMMA 3.3. *A connected Riemannian manifold M without boundary is recurrent if and only if for any nonempty open set $O \subsetneq M$, $M - O$ is parabolic.*

PROOF. Suppose M is recurrent and $O \subsetneq M$ is a nonempty open subset. Let C be a component of $M - O$ and $p \in \text{Int}(C)$. Since almost all Brownian paths beginning at p are dense in M, almost all Brownian paths beginning at p must enter O. But, in order to enter O, such a path must cross ∂C which means that the hitting measure μ_p on ∂C is full and so Proposition 3.2 implies that C is parabolic.

We now prove the converse statement. Suppose that for any nonempty open set $O \subsetneq M$, $M - O$ is parabolic. Let $p, q \in M$ and let $\alpha \subset M$ be a Brownian path starting at p. For any open ball $B(q, \varepsilon)$ centered at q, $M - B(q, \varepsilon)$ is parabolic and so, with probability 1, the path α will enter the closed ball $\overline{B}(q, \varepsilon)$. Since ε is arbitrary, with probability 1 the closure of α in M is all of M.

□

Recent work of Meeks and Rosenberg [**74, 76**] proves:

THEOREM 3.4. *An $M \in \mathcal{P}$ of finite topology is conformally diffeomorphic to a finitely punctured compact Riemann surface. In particular, such a minimal surface is recurrent for Brownian motion.*

In order to understand generalizations of the above theorem, we now discuss the topological notion of "ends" of a surface.

Definition 3.5. Suppose M is a noncompact connected manifold. The *space of ends* of M, denoted by $\mathcal{E}(M)$, is the set of equivalence classes of proper arcs $\alpha\colon [0, \infty) \to M$ where α_1 is equivalent to α_2 if for every smooth compact subdomain $C \subset M$, α_1 and α_2 intersect the same component of $M - \text{Int}(C)$ in a noncompact set.

A basis for the topology of $\mathcal{E}(M)$ is defined as follows. For each compact set $C \subset M$, define the basis open set $B(C) \subset \mathcal{E}(M)$ to be those equivalence classes of proper arcs in M which have representatives contained in $M - C$.

It can be shown that $\mathcal{E}(M)$ is a totally disconnected compact Hausdorff space that embeds as a subspace of the unit interval (this is not difficult to prove once one knows that it is true but I do not have a reference for it). Conversely, every compact totally disconnected subset C of the unit interval corresponds to the space of ends of a noncompact surface; namely, consider C to lie on an Jordan curve in S^2, then $S^2 - C$ has C as its space of ends. It is also interesting to note that two connected genus-zero surfaces are homeomorphic if and only if their spaces of ends are homeomorphic.

Definition 3.6. An end $e \in \mathcal{E}(M)$ is a *simple* end of M if it is an isolated point in $\mathcal{E}(M)$. Note that in dimension two that e is simple if and only if there exists a representative proper arc $\alpha \in e$ and a proper subdomain $W \subset M$ containing α such that W is homeomorphic to $S^1 \times [0, \infty)$ or to $S^1 \times [0, \infty)$ connected sum with an infinite number of tori where the nth connected sum occurs at the point $(1, n) \in S^1 \times [0, \infty)$. In the first case we refer to e as an *annular* end and in the second case we refer to e as a *simple end of infinite genus.* We say that the domain W *represents* the end e.

Definition 3.7. An end $e \in \mathcal{E}(M)$ is a *limit* end of $\mathcal{E}(M)$ if it is not a simple end. In other words, e is a limit end if it is a limit point of $\mathcal{E}(M)$. As in the previous definition, a limit end has *genus-zero* if it can be represented by a proper domain $W \subset M$ with compact boundary and the genus of W is zero. If a limit end e does not have genus-zero, then we say that e has *infinite genus*; in this case every proper subdomain with compact boundary representing e has infinite genus.

Definition 3.8. *The limit tangent plane at infinity* of a properly embedded minimal surface M in $\mathbb{R}^3$ with more than one end is the plane passing through the origin whose normal vector equals the normal vector of some end of a noncompact properly embedded minimal surface $\Sigma \subset (\mathbb{R}^3 - M)$ with compact boundary and finite total curvature (see Theorem 7.1); see **[6]** for further details on the existence and uniqueness of the limit tangent plane at infinity when $M \in \mathcal{M}$. We say that the limit tangent plane at infinity is *horizontal* if it is the x_1x_2-plane.

THEOREM 3.5. **[30]** *[Ordering Theorem] Suppose $M \in \mathcal{M}$ has horizontal limit tangent plane at infinity. Then the ends of M can be linearly ordered geometrically by their relative heights over the x_1x_2-plane. Furthermore, this ordering is a topological ordering in the following sense. If M is properly isotopic to a properly embedded minimal surface M' with horizontal limit tangent plane at infinity, then the associated ordering of the ends of M' either agrees with or is opposite to the ordering coming from M.*

Definition 3.9. Consider the ordering on the ends $\mathcal{E}(M)$ given by the above theorem. The end $e_T \in \mathcal{E}(M)$ is the *top* end of M if it is the unique end in $\mathcal{E}(M)$ that is maximal in the ordering. The top end e_T exists since $\mathcal{E}(M)$ is compact. The end $e_B \in \mathcal{E}(M)$ is the *bottom* end of M if it is the unique minimal element in the ordering of the ends. An end of M that is neither the top nor the bottom end of M is called a *middle* end of M.

4. Eight classical examples of minimal surfaces.

1. **Plane: P** $= x_2x_3$-plane. Weierstrass data: $\Sigma = \mathbb{C},\ \eta = dz,\ g(z) = 1$.

   A twisted plane would be a helicoid which is ruled by straight lines which rotate around the axis of the helicoid.
2. **Helicoid: H** $= \{(t\ \cos(s), t\ \sin(s), s) \mid t, s \in \mathbb{R}\}$.
   Weierstrass data: $\Sigma = \mathbb{C},\ \eta = idz,\ g(z) = e^z$.

   By taking the conjugate surface to the helicoid H, using the conjugate harmonic coordinate functions, we obtain an image surface which is a surface of revolution called a catenoid.
3. **Catenoid: C** $= \{(x_1, x_2, x_3) \mid x_1^2 + x_2^2 = \cosh^2(x_3)\}$.
   Weierstrass data: $\Sigma = \mathbb{C} - \{0\},\ \eta = \frac{1}{z}dz,\ g(z) = z$.
   Weierstrass data on the universal cover of **C** : $\Sigma = \mathbb{C},\ \eta = dz,\ g(z) = e^z$.

   In 1835, Scherk **[98]** defined five new minimal surfaces. Two of these examples, 4 and 5 below, have had an important influence on the theory of minimal surfaces. Scherk's singly-periodic minimal surface $\mathbf{S}_\theta$ defined below is asymptotic, away from the x_3-axis, to two planes P_1, P_2 containing the x_3-axis and which make an angle of 2θ with each other. In particular, under homothetic shrinkings, $\frac{1}{t}\mathbf{S}_\theta \to P_1 \cup P_2$ as $t \to \infty$. For t small, $\frac{1}{t}\mathbf{S}_\theta$ has the appearance of an embedded minimal surface which approximates $P_1 \cup P_2$ with the self-intersection curve $(P_1 \cap P_2) \subset (P_1 \cup P_2)$ desingularized by adding small one-handles along it. The example $\mathbf{S}_{\frac{\pi}{4}}$ can be defined implicitly by the formula $\sin(z) = \sinh(x)\sinh(y)$.

Scherk's doubly-periodic minimal surface $\hat{\mathbf{S}}_\theta$ defined below, which is the conjugate surface to $\mathbf{S}_\theta$, is asymptotic, away from the x_1x_2-plane, to families of equally spaced vertical parallel halfplanes in the respective halfspaces, with the halfplanes in $\{x_3 > 0\}$ and the halfplanes in $\{x_3 < 0\}$ making an angle of 2θ with each other. These doubly-periodic minimal surfaces are invariant under translation by a rombus lattice in $\mathbb{R}^2 \times \{0\}$. The example $\hat{\mathbf{S}}_{\frac{\pi}{4}}$ can be defined implicitly by the formula $e^z \cos y = \cos x$.

4. **Scherk's Singly-Periodic Surfaces:** $\mathbf{S}_\theta$, $0 < \theta \leq \frac{\pi}{4}$.
   Weierstrass data: $\Sigma = \mathbb{C} \cup \{\infty\} - \{\pm e^{\pm i\theta}\}$, $\eta = \frac{zidz}{\Pi(z \pm e^{\pm i\theta})}$, $g(z) = z$.

5. **Scherk's Doubly-Periodic Surfaces:** $\hat{\mathbf{S}}_\theta$, $0 < \theta \leq \frac{\pi}{4}$.
   Weierstrass data: $\Sigma = \mathbb{C} \cup \{\infty\} - \{\pm e^{\pm i\theta}\}$, $\eta = \frac{zdz}{\Pi(z \pm e^{\pm i\theta})}$, $g(z) = z$.

After Scherk's discovery, the next important examples were found by Riemann [**95**] who classified all of the minimal surfaces in $\mathbb{R}^3$ which are foliated by single circles and lines in horizontal planes. He wrote down a one-parameter family $\mathcal{R}_t$ of these examples defined for $t \in (0, \infty)$. The $\mathcal{R}_t$ converge to catenoids as $t \to 0$ and to helicoids as $t \to \infty$ (when appropriately normalized). Up to scaling by a homothety, $\mathcal{R}_t$ intersects the horizontal planes at integer heights in lines parallel to the x_1-axis and intersects other horizontal planes in circles symmetric with respect to reflection in the x_2x_3-plane. Each $\mathcal{R}_t$ has two limit ends with planar horizontal middle ends.

We will use the Weierstrass data to define the surfaces $\mathcal{R}_t$, up to a possible rotation by $\frac{\pi}{2}$ around the x_3-axis. Let $\mathbb{T}_t$ be the rectangular elliptic curve $\mathbb{T}_t = \mathbb{C}/\Lambda_t, \Lambda_t = \{m + tni \mid m, n \in \mathbb{Z}\}$. Let $\mathcal{P}_t$ be the meromorphic function on $\mathbb{T}_t$ with a double zero at 0 and a double pole at $\frac{1+ti}{2}$ and with value $\mathcal{P}_t(\frac{1+ti}{4}) = i$. Let $\bar{\Sigma}_t = \mathbb{T}_t - (Z_t \cup P_t)$, where Z_t, P_t are the zeros and poles of $\mathcal{P}_t$. Consider the infinite cyclic cover $\pi\colon \mathbb{C}/ti\mathbb{Z} \to \mathbb{C}/\Lambda_t$ and let $\Sigma_t = \pi^{-1}(\bar{\Sigma}_t)$ and $\tilde{\mathcal{P}}_t = \mathcal{P}_t \circ \pi$. Then the Riemann example $\mathcal{R}_t$ has the following Weierstrass data, when considered to be a periodic minimal surface in $\frac{\mathbb{R}^3}{<V>}$, where V is some vector in the x_1x_3-plane.

6. **Riemann Minimal Examples $\mathcal{R}_t$:**
   Weierstrass data: $\Sigma = \Sigma_t$, $\eta = dz$, $g_t(z) = \tilde{\mathcal{P}}_t(z)$.

In 1982 Costa [**17**] wrote down, in terms of elliptic functions on the square torus $\mathbb{T} = \mathbb{C}/\mathbb{Z}^2$, for $\Sigma = \mathbb{T} - \{0, \frac{1}{2}, \frac{1+i}{2}\}$, a conformal harmonic immersion $f\colon \Sigma \to \mathbb{R}^3$. Costa proved that $f(\Sigma)$ was an embedded surface outside of a ball in $\mathbb{R}^3$. Later Hoffman and Meeks [**44**] proved that the Costa surface was embedded and constructed for every positive integer k a related properly embedded minimal surface Σ_k in $\mathbb{R}^3$ of genus k with three ends, where Σ_1 is Costa's surface. We call this sequence of minimal surfaces:

7. **Costa-Hoffman-Meeks Examples:** Weierstrass data: for some $\lambda > 0$, $\Sigma_k = \{(z,w) \in \mathbb{C}^2 \mid w^{k+1} = z^k(z^2-1)\} - \{(1,0),(1,0)\}, \eta = \frac{dz}{z^2-1}, g = \frac{\lambda}{w}$.

   A couple of years after the discovery of the Costa-Hoffman-Meeks examples, Callahan, Hoffman and Meeks produced by computer graphics techniques many other new examples of finite total curvature. As a limit of one family of these finite topology examples, they wrote down the Weierstrass data for a sequence of very symmetric properly embedded minimal surfaces $M(n)$ which are invariant under vertical translation by $v = (0,0,2)$, had $2n-1$ vertical planes of symmetry containing the x_3-axis and making equal angles, had planar middle ends at integer heights with $2n-1$ horizontal lines meeting the x_3-axis at such heights, horizontal planes of symmetry at heights of the form $k+\frac{1}{2}, k \in \mathbb{Z}$, and such that horizontal planes at non-integer heights intersected $M(n)$ in simple closed curves. These examples were the first properly embedded minimal surfaces with an infinite number of ends and infinite genus.

   We refer the interested reader to [**5**] for a beautiful full page colored computer graphics rendered photo of the surface $M(1)$. Also, in [**5**], one can find pictures of the surface $M(2)$ and one of the Riemann examples. Unfortunately, these examples do not have a simple Weierstrass representation. The computer graphics pictures of these surfaces are obtained as numerical solutions to associated period problems on Riemann surfaces modelled on certain infinite cyclic branched covers of rectangular elliptic curves.
8. **Callahan-Hoffman-Meeks Examples:** $M(n), n \in \mathbb{N}$.

5. Stable minimal surfaces.

By definition, a minimal surface is locally a surface of least-area where by "local" we mean small disks on the surface. If instead we use "local" to mean in a small neighborhood of the entire surface, then we say that the minimal surface is stable. More precisely we have the following definition.

Definition 5.1. A *stable* minimal surface Σ in $\mathbb{R}^3$ is a surface such that every smooth compact subdomain $\bar{\Sigma}$ is stable in the following sense: if $\bar{\Sigma}(t)$ is a smooth family of minimal surfaces with $\partial\bar{\Sigma}(t) = \partial\bar{\Sigma}$ and $\bar{\Sigma}(0) = \Sigma$, then the second derivative of the area function $A(t)$ of the family $\bar{\Sigma}(t)$ is nonnegative at $t = 0$. We will say that Σ has *finite index* if outside of a compact subset it is stable.

Given a smooth variation $\bar{\Sigma}(t)$ of a compact minimal surface $\bar{\Sigma}$ with $\bar{\Sigma}(0) = \bar{\Sigma}$ and $\partial\bar{\Sigma}(t) = \partial\bar{\Sigma}$, one can express for t small the surfaces $\bar{\Sigma}(t)$ as normal graphs over $\bar{\Sigma}$ and so one obtains a normal variational vector field V on $\bar{\Sigma}$ which is zero on $\partial\bar{\Sigma}$. Assume that $\bar{\Sigma}$ is orientable with unit normal field N. Then $V = fN$ where $f\colon \bar{\Sigma} \to \mathbb{R}$ is a smooth function with zero boundary values. Conversely, if $f\colon \bar{\Sigma} \to \mathbb{R}$ is a smooth function with zero boundary values, then for small t one can find normal graphs $\bar{\Sigma}(t)$ which are the graphs $p+tf(p)N(p)$ over $\bar{\Sigma}$ with variational vector field fN. An elementary calculation gives the following second variational formula [**89**].

THEOREM 5.1. *(Second Variation of Area Formula) Suppose Σ is a compact oriented minimal surface and $f\colon \Sigma \to \mathbb{R}$ is a smooth function with zero boundary values. Let $\Sigma(t)$ be a variation of Σ with variational vector field fN and let $A(t)$*

be the area of $\Sigma(t)$. *Then*

$$A''(0) = -\int_{\bar{\Sigma}} f(\Delta f - 2Kf)dA,$$

where K *is the Gaussian curvature function on* Σ *and* Δ *is the Laplace operator on* Σ.

Definition 5.2. If Σ is a minimal surface, then $f\colon \Sigma \to \mathbb{R}$ is a *Jacobi function* if $\Delta f - 2Kf = 0$.

Jacobi functions f on Σ arise from normal variations $\Sigma(t)$, not necessarily with the same boundary, where the $\Sigma(t)$ are minimal surfaces with $\Sigma(0) = \Sigma$ and with variational vector field fN on Σ.

Using standard elliptic theory, it is easy to prove that an open oriented minimal surface Σ is stable if and only if it has a positive Jacobi function. Since the universal covering space of an orientable stable minimal surface is stable (it has a positive Jacobi function by composing), for many theoretical questions concerning a stable minimal Σ, we may assume Σ is simply-connected.

Suppose Σ is a minimal surface and $D \subset \Sigma$ is a geodesic disk of radius R on Σ centered at p which is stable. A short calculation (see below) by way of the second variation of area formula, using the function $f(r,\theta) = \frac{(R-t)}{R}$ defined in polar geodesic coordinates (t,θ) on D, gives a proof of the following beautiful formula of Colding-Minicozzi [**14**] for estimating the area of D.

THEOREM 5.2. *If* $D \subset \Sigma$ *is a stable minimal disk of geodesic radius* r_0 *on a minimal surface* $\Sigma \subset \mathbb{R}^3$, *then*

$$\pi r_0^2 \le \text{Area}(D) \le \frac{4}{3}\pi r_0^2.$$

PROOF. We now give the proof of the above formula, following the calculation in [**14**]. This calculation is excerpted from [**60**].

Since D has nonpositive Gaussian curvature, the area of D is at least as great as the comparison Euclidean disk of radius r_0, which implies $\pi r_0^2 \le \text{Area}(D)$. Consider a test function $f(r,\theta) = \eta(r)$ on the disk $D = D(r_0)$ that is a function of the radial coordinate r and which vanishes on ∂D. By the second variation of area formula, Green's formula and the coarea formula, we obtain:

(1)

$$0 \le \int_D -f\Delta f + 2Kf^2 = \int_D |\nabla f|^2 + 2\int_D Kf^2 = \int_0^{r_0} (\eta'(s))^2 l(s) + 2\int_0^{r_0}\left(\int_{r=s} K\right)\eta^2(s),$$

where K is the Gaussian curvature function on $D(s)$ of radius s and $l(s)$ is the length of $\partial D(s)$.

Let $K(s) = \int_{D(s)} K$. Then, by the first variation of arc length and the Gauss-Bonnet formula, we obtain:

$$l'(s) = \int_{\partial D(s)} \kappa_g = 2\pi - K(s) \Rightarrow K(s) = 2\pi - l'(s). \tag{2}$$

Since $K'(s) = \int_{r=s} K$, substituting in (1) yields:

$$0 \le \int_0^{r_0} (\eta'(s))^2 l(s) + 2\int_0^{r_0} K'(s)\eta^2(s). \tag{3}$$

Integrating (3) by parts and then substituting the value of $K(s)$ given in (2) yields:
(4)

$$0 \leq \int_0^{r_0} (\eta'(s))^2 l(s) - 2\int_0^{r_0} K(s)(\eta^2(s))' = \int_0^{r_0} (\eta'(s))^2 l(s) - 2\int_0^{r_0} (2\pi - l'(s))(\eta^2(s))'.$$

Now let $\eta(s) = 1 - \frac{s}{r_0}$ and so $\eta'(s) = \frac{-1}{r_0}$ and $(\eta^2(s))' = \frac{-2}{r_0}(1 - \frac{s}{r_0})$. Substituting these functions in (4) and then rearranging gives the following inequality:

$$(5) \qquad -\frac{1}{r_0^2}\int_0^{r_0} l(s) + \frac{4}{r_0}\int_0^{r_0} l'(s)(1 - \frac{s}{r_0}) \leq \frac{8\pi}{r_0}\int_0^{r_0} (1 - \frac{s}{r_0}) = 4\pi.$$

Integration of (5) by parts followed by an application of the coarea formula yields:

$$\frac{-1}{r_0^2}\int_0^{r_0} l(s) + \frac{4}{r_0^2}\int_0^{r_0} l(s) = \frac{3}{r_0^2}\int_0^{r_0} l(s) = \frac{3}{r_0^2}\text{Area}(D) \leq 4\pi \Rightarrow \text{Area}(D) \leq \frac{4}{3}\pi r_0^2.$$

□

We now apply the above area estimate for stable minimal disks to give a short proof of the famous classical result of do Carmo and Peng [**20**] and of Fischer-Colbrie and Schoen [**26**] which states:

THEOREM 5.3. *The plane is the only complete stable orientable minimally immersed surface in* $\mathbb{R}^3$.

PROOF. If Σ is a complete orientable stable minimal surface in $\mathbb{R}^3$, then the universal covering space of Σ composed with the inclusion of Σ in $\mathbb{R}^3$ is also a complete minimally immersed stable minimal surface in $\mathbb{R}^3$. Since Σ is a plane if and only if its universal cover is a plane, we may assume that Σ is simply-connected. Since the Gaussian curvature of Σ is nonpositive, Hadamard's theorem implies that, after picking a base point $p_0 \in \Sigma$, we obtain global geodesic polar coordinates (t, θ) on Σ centered at p_0. In these coordinates let $D(R)$ denote the disk of radius R centered at p_0.

Let $A(R)$ be the area of $D(R)$ and note that $A(R)$ is a smooth function of R. The first derivative of $A(R)$ is equal to

$$A'(R) = \text{Length}(\partial D(R)).$$

Also it is easy to see by the first variation of arc length that

$$A''(R) = \int_{\partial D(R)} \kappa_g,$$

where κ_g is the geodesic curvature of $\partial D(R)$. By the Gauss-Bonnet formula, we obtain

$$A''(R) = 2\pi - \int_{D(R)} K dA,$$

and so $A''(R)$ is monotonically increasing as a function of R. Since $A''(R)$ is monotonically increasing and $A(D(R)) \leq \frac{4}{3}\pi R^2$, then $A''(R) \leq \frac{8}{3}\pi$ and so $-\int_{D(R)} K dA$ is less than $\frac{2}{3}\pi$. Thus, Σ has total absolute Gaussian curvature which is finite and at most $\frac{2}{3}\pi$. At this point one obtains a contradiction in any of several different ways. One way is to appeal to a theorem of Osserman (Theorem 7.1) that states that the total curvature of a complete orientable nonplanar minimal surface is an integer multiple of -4π. Since the absolute total curvature of Σ is at most $\frac{2}{3}\pi$, its total curvature must be zero and we conclude Σ is a plane.

□

REMARK 5.4. Let $D(2R_0)$ be a stable minimal geodesic disk of radius $2R_0$ and $D(R_0)$ be the subdisk of radius R_0. The calculations and estimates used in the proof of Theorem 5.3 easily yield an upper bound of $\frac{8}{3}\pi$ for the total absolute curvature of $D(R_0)$. Since $A''(R) \leq 2\pi + \frac{8}{3}\pi = \frac{14}{3}\pi$ in the range $0 \leq R \leq R_0$, one obtains the estimate of $\frac{14}{3}\pi R$ for the Length$(\partial D(R)) = A'(R)$ for $0 \leq R \leq R_0$. We will use these estimates in the proof of Theorem 5.6. Theorem 5.3 is also an immediate consequence of Theorem 5.6 below for which we will give a self-contained proof that does not appeal to Osserman's Theorem.

Modifications of the arguments in the previous two Theorems (using different cut-off functions, etc. [**60**]) show that a related Theorem holds for complete orientable minimal surfaces of finite index; this result is the following theorem of Fischer-Colbrie [**25**].

THEOREM 5.5. *If Σ is a complete orientable minimal surface with compact boundary and finite index, then Σ has finite topology and finite total curvature.*

An important consequence of Theorem 5.3, using a blow-up argument, is that orientable minimally immersed stable surfaces with boundary in $\mathbb{R}^3$ have curvature estimates up to their boundary of the form given in the next Theorem. These curvature estimates by Schoen play an important role in numerous applications.

THEOREM 5.6. [**99**] *There exists a constant $c > 0$ such that for any stable orientable minimally immersed surface Σ in $\mathbb{R}^3$ and p a point in Σ of intrinsic distance $\mathbf{d}(p)$ from the boundary of Σ, then the absolute Gaussian curvature of Σ at p is less than $\frac{c}{(\mathbf{d}(p))^2}$.*

The above theorem by way of the same blow-up argument implies a similar estimate for stable minimal surfaces in a Riemannian three-manifold N^3 with injectivity radius bounded from below and which is uniformly locally quasi-isometric to Euclidean space; in particular, one obtains a similar curvature estimate for any compact Riemannian three-manifold M, where the constant c depends on M. We now give a sketch of the construction of the aforementioned blow-up argument, which gives a different proof for Theorem 5.6 from the argument given by Schoen [**99**].

PROOF. Suppose the desired curvature estimate were to fail. By taking universal covering spaces, we may assume that the stable minimal surfaces we are considering are simply-connected. We may assume that there exists a sequence of points $p(n) \in \Sigma(n)$ in the interior of stable orientable simply-connected minimal surfaces $\Sigma(n)$ such that the absolute Gaussian curvature at $p(n)$ is at least $\frac{n}{(d(p(n),\partial\Sigma(n))^2}$. Let $D(p(n))$ be the geodesic disk in $\Sigma(n)$ centered at $p(n)$ of radius $d(p(n), \partial\Sigma(n))$. Let $q(n) \in D(p(n))$ be a point in $D(p(n))$ where the function $\mathbf{d}^2|K|\colon D \to [0, \infty)$ has a maximum value; here $\mathbf{d}$ is the distance function to the boundary of D and $|K|$ is the absolute value of the Gaussian curvature (for simplicity we omit the dependence of d, D and $|K|$ on n). Let $D(n) \subset D(p(n))$ be the geodesic disk of radius $\frac{\mathbf{d}(q(n))}{2}$ centered at $q(n)$. Let $\tilde{D}(n)$ be the disk obtained by first translating $D(n)$ so that $q(n)$ is moved to the origin in $\mathbb{R}^3$ and then homothetically expanding the translated disk by the scaling factor $\sqrt{|K(q(n))|}$. The normalized disks $\tilde{D}(n)$ have Gaussian curvature -1 at the origin, Gaussian curvature bounded from below by -4, and the radii $r(n)$ of the $\tilde{D}(n)$ go to infinity as $n \to \infty$. A standard compactness result (see for example [**74**] or [**93**]) shows that a subsequence of the $\tilde{D}(n)$ converges smoothly as subsets to a complete simply-connected immersed minimal surface $D(\infty)$ passing through the origin of bounded Gaussian curvature and with no boundary. It is straightforward to show that the limit of stable minimal surfaces is stable and so $D(\infty)$ is stable. By Theorem 5.3, $D(\infty)$ is a plane but by construction $D(\infty)$ has Gaussian curvature -1 at the origin since each of the $\tilde{D}(n)$ have this property. This contradiction proves the desired curvature estimate of Schoen.

For the sake of completeness we give a self-contained modification of the end of the proof of Theorem 5.6 that does not depend on the stated compactness result or on the statement of Theorem 5.3. A slight modification of these same arguments can be used to give a simple complete proof (see [**60**] or [**107**]) of Osserman's Theorem, which is Theorem 7.1 and was used in our proof of Theorem 5.3, without appealing to the theorem of Huber (see the paragraph following the statement of Theorem 7.1).

A standard compactness argument shows the following: There exists an $\varepsilon > 0$ such that for any minimal disk E in $\mathbb{R}$ of geodesic radius at least 1 and center p with $|K(p)| = 1$ and $|K|\colon E \to [0,4]$, the image $G(E) \subset S^2$ of the Gauss map contains a geodesic cap centered at $G(p)$ of radius ε. An important application and immediate consequence of this result on the size of the Gauss map, together with a slight variation of the blow-up argument in the previous paragraph, is the following curvature estimate: For all $\eta > 0$, there exists a $\delta > 0$ such that if the total absolute curvature of a minimal disk D centered at p of geodesic radius at least 1 is less than δ, then $|K(p)| < \eta$. (See [**60**] for simple proofs of these and other related results.) We will now make use of both of these elementary results in order to complete the proof of Schoen's curvature estimate.

Recall that $r(n)$ is the radius of the disk $\tilde{D}(n)$ and the $r(n) \to \infty$ as $n \to \infty$. For each $r, 0 < r < r(n)$, let $\tilde{D}(r,n)$ be the geodesic subdisk of radius r and that each $D(n)$ is contained in a stable minimal disk of radius $2r(n)$. From the Remark 5.4, $\tilde{D}(r,n)$ has absolute total curvature at most $\frac{8}{3}\pi$ and the length of $\partial\tilde{D}(r,n)$ is less than $\frac{14}{3}\pi r$. Since $r(n) \to \infty$ as $n \to \infty$ and the total absolute curvature of each $\tilde{D}(n)$ is at most $\frac{8}{3}\pi$, there exist positive integers $k(n)$ such that $2^{k(n)+2} < r(n)$ and such that the total curvatures $C(n)$ of the annuli $\tilde{A}(n)$ bounded by $\partial\tilde{D}(2^{k(n)}, n)$ and $\partial\tilde{D}(2^{k(n)+2}, n)$ satisfy $C(n) \to 0$ as $n \to \infty$.

Now consider the new geodesic disks $\hat{D}(n)$ obtained by homothetically scaling $\tilde{D}(n)$ by the factor $2^{-k(n)}$ and let $\hat{A}(n) \subset \hat{D}(n)$ be the correspondingly scaled annuli. Let $\partial\hat{D}(2,n) \subset \hat{D}(n)$ be the circle of geodesic radius 2 in $\hat{D}(n)$. Since for any point of $\partial\hat{D}(2,n)$, the geodesic disk of radius 1 in $\hat{D}(n)$ centered at such a point is contained in $\hat{A}(n)$ and so has total absolute curvature approaching zero as $n \to \infty$. Our previous curvature estimate implies that the Gaussian curvature of the $\hat{D}(n)$ uniformly approach zero along $\partial\hat{D}(2,n)$ as $n \to \infty$. Since the length of $\partial\hat{D}(2,n)$ is less than $\frac{28}{3}\pi$, it follows that as $n \to \infty$, the length of $G(\partial\hat{D}(2,n))$ in S^2 approaches zero, where G is the Gauss map of $\hat{D}(2,n)$.

By our previous discussion, there exists an $\varepsilon > 0$ such that the Gaussian image $G(\tilde{D}(1,n))$ contains a spherical cap of radius ε centered at the value of G at the center of $\tilde{D}(1,n) \subset \tilde{D}(n)$. It follows that $G(\hat{D}(2,n))$ contains the same spherical cap. Since the Gauss map of $\hat{D}(2,n)$ is an open mapping, $G(\hat{D}(2,n))$ contains a fixed size spherical cap and Length $(G(\partial\hat{D}(2,n))) \to 0$ as $n \to \infty$, for n large the image by G of $\hat{D}(2,n)$ must have area approaching the area of S^2 which is 4π. But this contradicts the fact that the total curvature of $\hat{D}(2,n)$ is at most $\frac{8}{3}\pi$. This contradiction proves the desired curvature estimate.

□

6. The Plateau problem and the Meeks-Yau barrier construction.

The classical Plateau problem is the following:

Classical Plateau problem: Suppose Γ is a smooth embedded simple closed curve in $\mathbb{R}^3$. Does there exist a smooth map $f\colon D \to \mathbb{R}^3$ where D is the unit disk such that $f|\partial D$ is a parametrization of Γ and f has least-area with respect to all such mappings?

The answer to the above question is yes. One particularly natural solution to this problem was given by Douglas [**21**] who gave a solution $f\colon D \to \mathbb{R}^3$ which minimizes the energy over all harmonic maps (harmonic coordinate functions) whose boundaries monotonically parametrize the curve Γ. Later Morrey [**85**] solved the similar Plateau problem in Riemannian manifolds which satisfy a technical condition called homogeneously regular.

Some years after Douglas solved the classical Plateau problem, geometers became interested in solving the following related least-area question where Γ is a finite collection of pairwise disjoint smooth simple closed curves in $\mathbb{R}^3$.

Area-minimizing Plateau Problem:
Does Γ bound a smooth least-area surface Σ with $\partial\Sigma = \Gamma$?

The answer to this problem can be found in [**24**] and is yes, up to a question of boundary regularity. In other words, there exists an open surface Σ which is smooth and embedded such that the homological boundary of Σ is Γ and any other rectifiable two-chain which has homological boundary Γ with $\mathbb{Z}_2$-coeficients has area at least as big as Σ. If Σ has finitely generated fundamental group, then a neighborhood of $\partial\Sigma$ is orientable. It follows from the next Theorem and boundary regularity [**36**] that such a finite topology Σ attaches to its boundary in a smooth way.

THEOREM 6.1. *(Hardt-Simon* [**35**]*)*

1. Γ *is the boundary of a smooth immersed orientable surface of least-area;*
2. *Every such least-area surface is embedded with finite topology;*
3. *There are a finite number of such solutions.*

In general, the classical Douglas solution of least-area is not embedded. However, by an important result of Osserman [**91**], it has no interior branch points. In certain cases it can be shown that this least-area disk is an immersion along its boundary as well. One basic open problem in the classical theory is to prove that the Douglas solution has no boundary branch points. When Γ is analytic, then this result is a theorem of Gulliver and Lesley [**34**]. When Γ is extremal (lies on the boundary of its convex hull), then one has the following regularity theorem for the Douglas solution.

THEOREM 6.2. *(Meeks-Yau* [**82**]*) If Γ is extremal, then every Douglas solution to the classical Plateau problem is a smooth injective immersion. In particular, every solution is a smooth embedded disk.*

Meeks and Yau also proved, using the Morrey solution, the regularity of the classical Plateau in the following setting.

THEOREM 6.3. *(*[**83**]*) Suppose Γ is a smooth simple closed curve on the boundary of a homogeneously regular Riemannian three-manifold N^3 such that the boundary of N^3 has nonnegative mean curvature. If Γ is homotopically trivial in N^3, then there exists a Morrey disk $f\colon D \to N^3$ of least-energy and every such disk is a smooth embedding.*

In the above situation the nonnegative mean curvature condition makes the boundary into a good barrier for solving Plateau problems in N^3, including the previous possibly nonorientable and the Hardt-Simon solutions. By using a minimal surface as a barrier against itself, one can often prove the existence of least-area minimal surfaces in the complement of a given minimal surface. An important case of this barrier argument is the theorem in [**78**] which states that if Σ_1, Σ_2 are two properly immersed minimal surfaces in $\mathbb{R}^3$ which are disjoint, then Σ_1 and Σ_2 are contained in closed halfspaces of $\mathbb{R}^3$. In this case one proves that there is a properly embedded least-area surface Σ which separates Σ_1 and Σ_2; Σ is a plane by Theorem 5.3. Later Hoffman and Meeks [**47**] proved that a properly immersed minimal surface contained in a closed halfspace of $\mathbb{R}^3$ is a plane. Therefore, the original Σ_1, Σ_2 we were considering must be planes. This result, called the Strong Halfspace Theorem, has many important applications. In a different direction, if Γ is an extremal simple closed curve in $\mathbb{R}^3$ which does not bound a unique compact branched minimal surface, then Meeks and Yau [**83**] prove that Γ is the boundary of two stable embedded minimal disks; here one uses the union of two minimal surfaces bounding Γ as a barrier to solving the classical Plateau problem in certain mean convex three-manifolds that lie in the convex hull of Γ. This result by Meeks and Yau, together with a disk uniqueness theorem of Nitsche [**88**], has the following corollary. In the proof of Nitsche's theorem, Nitsche assumes the boundary curve is analytic because of the consideration of boundary branch points. When Γ is extremal there are never boundary branch points as shown in [**83**].

THEOREM 6.4. [**83**] *If Γ is a smooth extremal curve with total curvature at most 4π, then Γ is the boundary of a unique compact branched minimal surface and this surface is a smooth embedded minimal disk of least-area.*

7. Minimal surfaces of finite topology.

The deepest results in classical minimal surface theory concern the geometry of properly embedded minimal surfaces in $\mathbb{R}^3$ with finite topology. An important subcollection of these surfaces are the examples which have finite total Gaussian curvature. In this regard, one has the following classical theorem of Osserman [**90**]. (See [**7**], [**60**], and [**107**], for the n-dimensional version of Osserman's theorem.) It follows from this Theorem that such minimal surfaces are defined analytically in terms of meromorphic data on a closed Riemann surface.

THEOREM 7.1. *Suppose M is a complete oriented minimal surface in $\mathbb{R}^3$ with finite total Gaussian curvature $C(M) = \int_M K dA$. Then:*

1. *$C(M)$ is a integer multiple of -4π;*
2. *M has finite conformal type, which means M is conformally diffeomorphic to a compact Riemann surface $\bar{M}$ punctured in a finite number of points;*
3. *The meromorphic Gauss map $g\colon M \to C \cup \{\infty\}$ extends to a meromorphic function on $\bar{M}$;*
4. *The holomorphic one-form $\eta = dx_3 + idx_3^*$ on M given in the Weierstrass representation extends to meromorphic one-form on $\bar{M}$.*

The proof of the above theorem is straightforward if one assumes the result of Huber [**49**] that a complete Riemannian surface M of nonpositive curvature and finite total curvature is conformally $\bar{M} - \{p_1, \cdots, p_n\}$ where $\bar{M}$ is a compact Riemann surface. In this case the meromorphic Gauss map $g\colon \bar{M} - \{p_1, \cdots, p_n\} \to \mathbb{C} \cup \{\infty\}$ has finite area $-C(M)$ counted with multiplicity. Picard's theorem implies g extends analytically across the punctures to a meromorphic function $\bar{g}\colon \bar{M} \to \mathbb{C} \cup \{\infty\}$ of integer degree k. Thus, $C(M) = -4\pi k$, since the area of the unit sphere S^2 is 4π. The theorem then follows rather easily from these observations. We refer the reader to [**60**] for a simple short proof of Osserman's Theorem, which does not assume Huber's or Picard's theorem.

Until 1982 there were only three known examples of properly embedded minimal surfaces of finite topology: they are the plane, helicoid and catenoid. This situation changed radically after this date with the discovery of many new examples of positive genus with finite total curvature. From the pioneering work by Ros [**96**] (also see [**92, 93, 94**]), we now understand reasonably well the structure of the moduli spaces of examples with some bound on the genus and number of ends. In particular, we know that these moduli spaces are real semi-analytic varieties and we understand something about the degeneration of sequences of examples which diverge to points in the boundary of the spaces.

In Section 4 we explained how the Cost-Hoffman-Meeks minimal surfaces $\Sigma(g)$ of genus $g, g \geq 1$, with three ends could be constructed by using the classical Weierstrass representation. Hoffman-Meeks (unpublished) proved that these surfaces could each be deformed analytically through embedded minimal surfaces $\Sigma(g,t)$ of finite total curvature whose middle end is a graph with logarithmic growth t. By the maximum principle these surfaces are embedded in the parameter t, beginning at $t = 0$, until the logarithmic growth of the middle end is equal to the logarithmic growth of one of the other two ends. In [**39**] Hoffman and Karcher proved that for $g \geq 2$, the logarithmic growth of the middle end of $\Sigma(g,t)$ is always less than the logarithmic growth of the other two ends and so the $\Sigma(g,t)$ are embedded for all t when $g \geq 2$.

Using computer graphics techniques, Calahan, Hoffman and Meeks constructed many other properly embedded minimal surfaces of finite total curvature with more than three ends. Within a short time, it became clear that there were probably examples with an arbitrarily large number of ends. While they could not give a rigorous proof of the existence of such surfaces, they were able to give a proof of the existence of properly embedded minimal surfaces which are limits of the expected finite topology examples with more and more ends. These are the Callahan-Hoffman-Meeks [**5**] examples $M(n)$ discussed in Section 4. The $M(n)$ are periodic with infinite genus and an infinite number of planar-type ends; in other words, their middle ends are asymptotic to planes. Since these surfaces are periodic, they have two limit ends which are the top and bottom ends of the surfaces.

For a few years computer graphics ruled the existence part of the theory as geometers constructed more and more intricate examples of properly embedded minimal surfaces of finite positive genus and at least three ends. All of these constructions supported the conjecture of Hoffman and Meeks (see Conjecture 3 in Section 14) that when an example Σ has at least two ends, then $e(\Sigma) \leq g(\Sigma) + 2$, where $e(\Sigma)$ is the number of ends of Σ and $g(\Sigma)$ is the genus of Σ.

In the past decade there has been much success in creating new theoretical methods for constructing properly embedded minimal surfaces which are not obtained by the Weierstrass representation. Shortly after Hoffman and Meeks gave a proof of the existence of the $\Sigma(g)$ examples [**41**] using the Weierstass representation, they gave an abstract minimax construction of these examples [**42**]. In [**45**] Meeks and Hoffman proved that, when properly normalized, the surfaces $\Sigma(g)$ of Costa-Hoffman-Meeks converge as $g \to \infty$ to the union of a vertical catenoid $\mathbf{C}$ and the x_1x_2-plane P and that, on the scale of the maximum curvature along the forming intersection curve $\mathbf{C} \cap P$, the surfaces converge to Scherk's one-periodic minimal surface $\mathbf{S}_{\frac{\pi}{4}}$ described in Section 4. These results motivated the general question of whether two transversely intersecting minimal surfaces could be desingularized by sewing in a "curve" of "Scherk" surfaces. This proposed desingularization became known as the procedure of "minimal" surgery. The theoretical procedure of minimal surgery was given a rigorous basis by the work of Kapouleas [**51**]. Kapouleas was able to prove that if $C_1, C_2, \ldots, C_n$ are a finite collection of catenoids with axes the x_3-axis and of varying logarithmic growth, then the union of these catenoids can be approximated by properly embedded minimal surfaces with $2n$ catenoid type ends and large genus and which approximate scaled down Scherk surfaces near the intersection curves as the genus approaches infinity. Actually one can also take C_1 to be the x_1x_2-plane, and one then obtains examples with $2n-1$ ends and large genus.

More recently Weber and Wolf [**106**] have combined the Weierstrass representation with new conformal methods to produce properly immersed minimal surfaces of every possible odd $e \geq 3$ and genus g which satisfy the Hoffman-Meeks inequality. These surfaces are almost certainly embedded but a rigorous proof of embeddedness seems difficult. In a different direction, which uses an algebraic-geometric type implicit function theorem, Traizet [**104**] has been able to verify the existence of many properly embedded minimal surfaces which satisfy the Hoffman-Meeks inequality but his methods fall short of proving the existence part of it holds in general. Traizet obtains many families of examples of varying dimensions depending on the genus and the number of ends under consideration.

During the 1980's, there were a number of partial results on what became known as the generalized Nitsche Conjecture. Nitsche's original conjecture [**86**] was that if Σ is a complete minimal surface which is the union of simple closed curves in parallel planes, then Σ is a catenoid. The generalized Nitsche conjecture states that if Σ is a properly embedded minimal surface of finite topology $\mathbb{R}^3$ with more than one end, then Σ has finite total Gaussian curvature $C(\Sigma) = \int_\Sigma K dA$. Partial results on the generalized Nitsche conjecture were obtained by Hoffman and Meeks [**43**] and by Meeks and Rosenberg [**76**]. This conjecture was finally proven by Collin in 1997.

THEOREM 7.2. [**15**] *If $\Sigma \in \mathcal{M}$, then each annular end of Σ is asymptotic to the end of a plane or catenoid. In particular, if Σ also has finite topology, then* *by the formula of Jorge and Meeks* [**50**], Σ *has total absolute curvature* $C(\Sigma) = -4\pi(g(\Sigma)+e(\Sigma)-1)$, *where* $g(\Sigma)$ *is the genus of* Σ *and* $e(\Sigma)$ *is the number of ends of* Σ.

Finally, in the case where $\Sigma \in \mathcal{P}$ has one end which is annular and Σ is not a plane, Meeks and Rosenberg [**74**] proved that Σ is asymptotic to a helicoid. Their work is based on recent important pioneering work of Colding and Minicozzi which we will discuss in the next section. Putting together the results of Collin [**15**], of Meeks and Rosenberg [**74**], we have the following theorem:

THEOREM 7.3. *Suppose* $\Sigma \in \mathcal{P}$ *has finite topology and* Σ *is not a plane. Then:*

1. Σ *is conformally a compact Riemann surface* $\bar{\Sigma}$ *punctured in a finite number of points;*
2. *The embedding of* Σ *into* $\mathbb{R}^3$ *via the Weierstrass representation can be obtained in terms of meromorphic data on* $\bar{\Sigma}$*;*
3. *The moduli space of examples in* $\mathcal{P}$ *with the same topology as* Σ *is an semi-analytic variety;*
4. *If* Σ *has more than one end, then each end of* Σ *is asymptotic to the end of a plane or catenoid;*
5. *If* Σ *has one end, then* Σ *is asymptotic to a helicoid.*

In certain cases it turns out that knowing that a $\Sigma \in \mathcal{P}$ of finite topology can be described as in Theorem 7.3, implies that Σ must have a particular topology. The first result of this type was proven by Jorge and Meeks [**50**] where they showed that the sphere S^2 punctured in 3, 4 or 5 points can not be properly minimally embedded in $\mathbb{R}^3$ with finite total curvature. Next Schoen [**100**] proved that a $\Sigma \in \mathcal{M}$ of finite total curvature and two ends is a catenoid. Then Lopez and Ros [**56**] generalized the Jorge-Meeks obstruction by proving that every $\Sigma \in \mathcal{M}$ with finite topology and genus-zero is a catenoid. Next, Meeks and Rosenberg [**74**] proved that if $\Sigma \in \mathcal{P}$ is simply-connected, then Σ is a plane or a helicoid. Recently, Meeks, Perez and Ros [**69**] have shown that there is an upper bound on the number of ends of $\Sigma \in \mathcal{M}$ with finite topology and fixed genus. Putting these results together, Theorem 7.3 implies:

THEOREM 7.4. *If* $\Sigma \in \mathcal{P}$ *has finite topology, then:*

1. *If* Σ *has genus-zero, then* Σ *is a plane, a helicoid or a catenoid;*
2. *If* Σ *has two ends, then* Σ *is a catenoid;*
3. *For every genus* g*, there exists an integer* $e(g)$ *such that if* Σ *has genus* g*, then the number of ends of* Σ *is at most* $e(g)$.

8. The local theory of properly embedded minimal surfaces.

In a recent series of papers, Colding and Minicozzi [**9, 10, 11, 12, 13**], have attempted to describe the basic structure of compact embedded minimal surfaces M of fixed genus which are contained in the unit ball B and which have their boundary on the boundary of B. The most important case of their structure theorem is when M is a disk which passes through the origin where its Gaussian curvature is large. In this case Colding and Minicozzi prove that M has the appearance, in a smaller ball $B(\varepsilon)$ centered at the origin, of a multisheeted graph with many sheets and an axis similar to the axis of a helicoid. They then use this local picture to prove the following beautiful compactness theorem:

THEOREM 8.1. *If $M(n) \subset B, n \in \mathbb{N}$, is a sequence of properly embedded minimal disks in the interior of B with $\partial M(n) \subset \partial B$ and the curvature of the family $M(n)$ is unbounded at the origin, then a subsequence of the $M(n)$ converges to a minimal lamination $\mathcal{L}$ by minimal disks of the interior of B and which is a foliation in a neighborhood of the origin. Furthermore, the convergence of this subsequence is smooth except along a connected Lipschitz curve $S(\mathcal{L})$ passing through the origin.*

Meeks and Rosenberg [**74**] then applied this local structure theorem to prove that the plane and the helicoid are the only properly embedded simply-connected minimal surfaces in $\mathbb{R}^3$. Meeks [**61**] recently applied this uniqueness of the helicoid theorem to prove that the singular curve $S(\mathcal{L})$ in the above theorem is of class $C^{1,1}$ and is orthogonal to the leaves of $\mathcal{L}$. Of class $C^{1,1}$ means that $S(\mathcal{L})$ is of class $C^1$and the unit tangent vector field to $S(\mathcal{L})$ extends to an ambient Lipschitz vector field.

These results of Colding and Minicozzi, of Meeks and of Meeks and Rosenberg involve a geometric analysis of the local geometry of embedded minimal surfaces $M(n)$ at a sequence of points of large normalized curvature, a concept that we now define.

Definition 8.1. A *sequence of points of large normalized curvature* is a sequence $p(n) \in M(n) \subset B$ such that:

1. $\lambda(n) := \sqrt{|K_{M(n)}|(p(n))}$ tends to ∞ as $n \to \infty$;
2. $B(p(n), \frac{\tau(n)}{\lambda(n)}) \subset B$ for each n for some positive $\tau(n)$, where $\tau(n) \to \infty$ as $n \to \infty$;
3. There exists $c > 0$ such that $|K_{M(n)}| \leq c\lambda(n)^2$ in $M(n) \cap B(p(n), \frac{\tau(n)}{\lambda(n)})$.

The definition of points of large normalized curvature is made so that the minimal surfaces $M(n) \cap B(p(n), \frac{\tau(n)}{\lambda(n)})$ translated by $-p(n)$ and then scaled homothetically by the factor $\lambda(n)$, are embedded minimal surfaces with Gaussian curvature -1 at the origin, properly embedded in balls of radii $\tau(n) \to \infty$ and in these balls the surfaces have uniformly bounded Gaussian curvature. It follows from [**74**] that a subsequence of these related normalized minimal surfaces converges with multiplicity one to a connected properly embedded minimal surface Σ in $\mathbb{R}^3$ with bounded nonzero Gaussian curvature.

Suppose $M(n)$ is a sequence of properly embedded minimal disks in B which converges to $\mathcal{L}$ defined above with singular curve $S(\mathcal{L})$. For any point $q \in S(\mathcal{L})$ there exist a sequence of points $q(n) \in M(n)$ of large normalized curvature that converge to q. It follows from the results in [**74**] that a subsequence of the surfaces $\tilde{M}(n)$, obtained by translating $M(n)$ by $-q(n)$ and then homothetically expanding this surface by the factor $\sqrt{|K(q(n))|}$, converges with multiplicity one to a surface $M(\infty)$, called a *normalized blow-up* of the $M(n)$, which is a properly embedded minimal surface in $\mathbb{R}^3$ of bounded nonzero absolute curvature. Suppose now the $M(n)$ are not necessarily disks. It follows from the convex hull property for minimal surfaces that the limit $M(\infty)$ has genus less than or equal to any upper bound of the genus of the $M(n)$ and also that $M(\infty)$ has at most as many generators in its fundamental group as the $M(n)$ have. Thus, when the surfaces $M(n)$ are simply-connected, $M(\infty)$ is simply-connected, and then by [**74**], $M(\infty)$ is a helicoid. Hence, in a small neighborhood of a point of $M(n)$ of very large normalized curvature, $M(n)$ has the appearance of a homothetically shrunk helicoid with a large number of sheets.

These results lead to basic curvature estimates for embedded minimal surfaces. These curvature estimates have proven useful in describing the local geometry of a sequence of properly embedded minimal surfaces $M(n)$ of fixed genus, but not necessary with a bound on the number of generators of their fundamental groups, near points of large normalized curvature. The desired description can be obtained from the special case where the genus of $M(n)$ is zero but $M(n)$ is not necessarily simply-connected. If at such a point p of large normalized curvature the components of $M(n)$ in a small ball centered at p are simply-connected, then the previous analysis implies that in some smaller neighborhood of p, $M(n)$ has the appearance of a homothetically shrunk helicoid with many sheets. It is convenient to make the following definition.

Definition 8.2. A sequence $M(n) \subset B$ is **not** *uniformly-locally-simply-connected* if at some point x in the interior of B there exists a sequence $\varepsilon(n) \to 0$ such that for some large $k(n)$, $M(n + k(n)) \cap B(x, \varepsilon(n))$ contains at least one component which is not simply-connected.

It follows that if the $M(n)$ are not uniformly-locally-simply-connected in B, then there exists a point x in the interior of B and a subsequence $n(k)$ with associated points $p(n(k)) \in M(n(k))$ which converge to x and which are points of large normalized curvature. In particular such a sequence of minimal surfaces has a normalized blow-up. Under the assumption that the $M(n)$ have fixed finite genus, there is a conjecture as to what are the possible normalized blow-ups of such a sequence. Since in this case the normalized blow-up is a properly embedded minimal surface in $\mathbb{R}^3$ of finite genus and bounded curvature, the next more general conjecture explains what the possible normalized blow-ups should be. Note that all of these surfaces actually occur as normalized blow-ups.

CONJECTURE 8.1. *(Meeks, Perez, Ros) If $M \in \mathcal{P}$ has finite genus and is not a plane, then:*

1. *M has bounded curvature;*
2. *If M has one end, then M is asymptotic to a helicoid;*
3. *If M has a finite number of ends greater than one, then M has finite total curvature;*
4. *If M has an infinite number of ends, then M has two limit ends, each of which is asymptotic as $x_3 \to \infty$ to translated limit ends of one of the classical Riemann minimal examples (see Section 4 and also* [**70**] *for a description of these beautiful singly-periodic minimal surfaces of genus-zero which are foliated by circles and lines in horizontal planes);*
5. *If M has genus-zero, then M is a helicoid, a catenoid, or a Riemann minimal example.*

In the case M has finite topology, the conjecture holds for M by Theorem 7.4. If M has an infinite number of ends, then the results in [**16**] show that M can have at most two limit ends (see Section 2 for definitions and Section 9 for results). Meeks, Perez and Ros have shown that a finite genus M cannot have one limit end. An important partial result on the above conjecture is the next Theorem.

THEOREM 8.2. [**66, 67**] *If M is a properly embedded minimal surface in $\mathbb{R}^3$ with finite genus and an infinite number of ends, then:*

1. *M has bounded curvature;*

2. *M has two limit ends.*

If $\Sigma(n)$ is a sequence of properly embedded minimal surfaces in a Riemannian three-manifold N^3 which has a lower bound on its injectivity radius and a bound on its sectional curvature, then it makes sense to talk about a sequence of points $p(n) \in \Sigma(n)$ of large normalized curvature, and one obtains by the arguments in [**74**] a normalized blow-up $\Sigma(\infty)$ as a limit of some subsequence of the $\Sigma(n)$ around $p(n), \Sigma(\infty)$ being a properly embedded minimal surface in $\mathbb{R}^3$ with bounded Gaussian curvature. As the limit $\Sigma(\infty)$ is nonflat, it can be shown that the convergence $\Sigma(n) \to \Sigma(\infty)$ has multiplicity one. Our previous discussion implies that local bounds on the genus or the numbers of generators of the fundamental group in a local neighborhood of $p(n)$ on $\Sigma(n)$ give the same bounds on genus and number of generators of the fundamental group of $\Sigma(\infty)$.

THEOREM 8.3. *([**66, 67**]) If a sequence of properly embedded minimal surfaces $\Sigma(n) \subset N^3$ has uniformally-locally-bounded-genus and has a normalized blow-up $\tilde{M} \subset \mathbb{R}^3$, then the sequence $\Sigma(n)$ has another normalized blow-up M satisfying:*

1. *M is a helicoid;*
2. *M has a finite number of ends greater than one and M has finite total curvature;*
3. *M has two limit ends and genus-zero.*

9. Minimal surfaces of infinite topology.

Before about 1980, there were only three known classical examples in $\mathcal{P}$ which had finite topology; these surfaces are the plane, the helicoid and the catenoid. The remainder of the classical examples were periodic and not simply-connected and so had infinitely generated fundamental groups. Except for the Riemann minimal examples, all of the remaining known examples in $\mathcal{P}$ had infinite genus and one end. Most of these examples were doubly or triply-periodic and so, by the following Theorem 9.1, they had infinite genus and one end.

THEOREM 9.1. [**6**] *If $\Sigma \in \mathcal{P}$ is not a plane and it is doubly-periodic or triply-periodic, then Σ has infinite genus and one end.*

Later it was shown by Frohman and Meeks [**31**] that given two surfaces in $\mathcal{P}$ with one end and the same genus, there is a diffeomorphism of $\mathbb{R}^3$ which takes one to the other. Thus, for example, one of Scherk's singly-periodic and one of Scherk's doubly-periodic differ by a diffeomorphism of $\mathbb{R}^3$, even though their geometric appearance is completely different.

In [**5**] Callahan, Hoffman and Meeks constructed many new singly-periodic examples in $\mathcal{P}$ with two limit ends and infinite genus; these examples are described in Section 4 and have middle ends which are annular and are asymptotic to horizontal planes. Thus, these new infinite ended examples all have end representatives contained in horizontal slabs and these representatives have asymptotic area growth of πR^2. It turns out that some related properties hold for any example in $\mathcal{P}$ with two limit ends, which we now explain.

Definition 9.1. A surface $M \subset \mathbb{R}^3$ has *quadratic area growth* if there exists a positive constant c such that for all large positive R, the area of M inside the ball $B(R)$ of radius R centered at the origin is less than cR^2.

By the proof of the ordering theorem [**30**], a middle end of a properly embedded minimal surface M with horizontal limit tangent plane at infinity can be represented by a proper subdomain $W \subset M$ with compact boundary such that W "lies between two catenoids." This means that W is contained in a neighborhood S of the x_1x_2-plane, S being topologically a slab, whose width grows logarithmically with the distance from the origin.

One of the fundamental results in the global theory of properly embedded minimal surfaces is that the middle ends of an $M \in \mathcal{M}$ are never limit ends. This is shown by first proving that if W is a properly immersed minimal surface with compact boundary and contained between two catenoids, then W has quadratic area growth [**16**]. By the monotonicity formula for area [**101**], every subend of a representative of a limit end has limiting area growth at least πR^2. Hence, a limit end never has a representative with quadratic area growth. Thus:

THEOREM 9.2. [**16**] *If $M \in \mathcal{M}$, then a limit end of M must be a top or bottom end of M. In particular, M can have at most two limit ends. Furthermore, each middle end has limiting area growth which is approximately like a positive integer times πR^2; the* parity *of a middle end is the parity of this integer.*

In fact, a theorem from [**16**] states that when $\Sigma \in \mathcal{P}$ has two limit ends and horizontal limit tangent plane at infinity, then there exists a proper family $\{P_n \mid n \in \mathbb{Z}\}$ of horizontal planes ordered by their relative heights, each of which intersects Σ transversely in a compact family of curves. Furthermore, the slab determined by P_n and P_{n+1} intersects Σ in a proper subdomain which represents the nth middle end of Σ.

In [**106**] Weber and Wolf consider a method which proves the existence of a sequence $M(n) \subset \mathbb{R}^3$ of properly immersed minimal surfaces of odd genus n with $n + 2$ horizontal planar ends. Computer graphics pictures of these surfaces for n relatively large indicate that they are all embedded and it is believed that, when properly normalized, sequences of these surfaces apparently converge to the properly embedded periodic minimal surfaces $M(1)$ of Callahan, Hoffman and Meeks in Section 4. Assuming that these surfaces are embedded, a slightly different normalization by a vertical translation should yield as a limit a properly embedded minimal surface with a bottom catenoid end, middle planar ends and a top limit end. Such a limit minimal surface would then have infinite genus and one limit end. One should compare the probable existence of this infinite genus minimal surface with one limit end to Theorem 8.2 in Section 14, which implies that a one limit end example must have infinite genus.

10. The topological classification theorem for minimal surfaces.

In 1992, Meeks and Yau [**84**] proved that properly embedded minimal surfaces of finite topology in $\mathbb{R}^3$ are unknotted in the sense that any two such homeomorphic surfaces are properly ambiently isotopic. Later Frohman [**28**] proved that any two triply periodic minimal surfaces are properly ambiently isotopic. Recall that a *handlebody* is a three-manifold with boundary which is homeomorphic to the closed regular neighborhood of a connected properly embedded one-dimensional CW-complex in $\mathbb{R}^3$. A surface in a three-manifold is called a *Heegaard surface* if it separates the three-manifold into two closed complements which are handlebodies. More recently Frohman and Meeks [**31**] proved that a properly embedded minimal surface in $\mathbb{R}^3$ with one end is a Heegaard surface in $\mathbb{R}^3$ and that Heegaard surfaces of

$\mathbb{R}^3$ with the same genus are unknotted; hence, properly embedded minimal surfaces in $\mathbb{R}^3$ with one end are unknotted even when the genus is infinite. These topological uniqueness theorems of Meeks and Yau, Frohman, and Frohman and Meeks are special cases of the following general classification theorem which was conjectured in [**31**] and which represents the final solution to the topological classification problem. The space of ends of a properly embedded minimal surface in $\mathbb{R}^3$ has a natural linear ordering which is determined up to reversal by Theorem 3.5 and the middle ends in this ordering have a parity (even or odd) according to Theorem 9.2.

THEOREM 10.1. [**29**] *(Topological Classification Theorem for Minimal Surfaces) Two properly embedded minimal surfaces in $\mathbb{R}^3$ are properly ambiently isotopic if and only if there exists a homeomorphism between the surfaces that preserves the ordering of their ends and preserves the parity of their middle ends.*

The constructive nature of the proof of the Topological Classification Theorem provides an explicit description of the topological embedding of any properly embedded minimal surface in terms of the ordering of the ends, the parity of the middle ends, the genus of each end - zero or infinite - and the genus of the surface. This topological description depends on several major advances in the classical theory of minimal surfaces. First, associated to any properly embedded minimal surface M with more than one end is a unique plane passing through the origin called the limit tangent plane at infinity of M (see Definition 3.8). Furthermore, the ends of M are geometrically ordered over its limit tangent plane at infinity and this ordering is a topological property of the ambient isotopy class of M by Theorem 3.5. Second, the proof of the classification theorem depends on the nonexistence of middle limit ends for properly embedded minimal surfaces given in Theorem 9.2. Third, the proof relies heavily on a topological description of the complements of M in $\mathbb{R}^3$; this topological description of the complements was carried out by Frohman and Meeks [**31**] when M has one end and by Freedman [**27**] in the general case.

Here is an outline of the proof of the classification theorem. The first step is to construct a proper family $\mathcal{F}$ of topologically parallel standardly embedded planes in $\mathbb{R}^3$ such that the closed slabs and halfspaces determined by $\mathcal{F}$ each contains exactly one end of M and each plane in $\mathcal{F}$ intersects M transversely in a simple closed curve. The next step is to reduce the global classification problem to a tractable topological-combinatorial classification problem for "Heegaard" decompositions of closed slabs or half spaces in $\mathbb{R}^3$.

Recently Meeks and Rosenberg have been able to generalize some of the above arguments to prove the following unknotted theorem:

THEOREM 10.2. [**73**] *(Unknotted Theorem) Suppose S^2 is a two sphere endowed with a Riemannian metric with no stable simple closed geodesics. Then:*

1. *If Σ is a noncompact properly embedded minimal surface in $S^2 \times \mathbb{R}$, then Σ is a Heegaard surface for $S^2 \times \mathbb{R}$;*
2. *Every Heegaard surface for $S^2 \times \mathbb{R}$ has two ends and if Σ is a connected orientable surface with two ends, then Σ embedds in $S^2 \times \mathbb{R}$ as a Heegaard surface;*
3. *Heegaard surfaces of $S^2 \times \mathbb{R}$ are unknotted in the sense that if two such surfaces are diffeomorphic, then there exists an orientation preserving diffeomorphism of $S^2 \times \mathbb{R}$ which takes one surface to the other.*

11. The conformal structure of properly embedded minimal surfaces.

By Theorem 7.3, if $\Sigma \in \mathcal{P}$ has finite topology, then it is conformally a compact Riemann surface $\bar{\Sigma}$ punctured in a finite number of points and the Weierstrass representation for Σ can be expressed in terms of meromorphic data on $\bar{\Sigma}$. In the case that Σ has more than one end and finite topology, the conformal structure of being $\bar{\Sigma}$ punctured in a finite number of points was first proven in [**76**]. The case when Σ has one end appears in [**74**]. The proof in [**74**] that a simply connected Σ is conformally $\mathbb{C}$ proceeds by first proving that there exists a plane that intersects Σ transversely in a single proper arc. The result on the conformal structure of Σ, then follows from the next theorem [**71**].

THEOREM 11.1. *If Σ is a properly immersed minimal surface of finite topology and one end which intersects some plane transversely in a finite number of immersed (possibly noncompact) curves, then Σ is conformally a compact Riemann surface punctured in one point.*

The main tool for proving the above theorem is Fatou's Lemma on the almost everywhere radial limits of a bounded harmonic function on the open disk, which is applied in conjunction with the next general theorem.

THEOREM 11.2. [**16**] *If Σ is a properly immersed minimal surface with boundary (possibly empty) in $\mathbb{R}^3$, then every component of the intersection of a closed halfspace with Σ is a parabolic Riemannian surface with boundary.*

The proof of the above basic result introduces an important new definition to the subject; we will give the proof here of the special case when Σ has no boundary.

Definition 11.1. A function $f\colon \Omega \to \mathbb{R}$ defined on a domain $\Omega \subset \mathbb{R}^3$ is called a *universal superharmonic* function if its restriction to any minimal surface Σ in Ω is superharmonic, i.e., $\Delta(f|\Sigma) \leq 0$.

Examples of universal superharmonic functions on all of $\mathbb{R}^3$ include coordinate functions such as x_1 or the function $-x_1^2$. In the proof of the quadratic area growth property of middle ends, one uses the universal superharmonic function $\ln(\sqrt{x_1^2+x_2^2}) - x_3\tan^{-1}(x_3) + \frac{1}{2}\ln(1+x_3^2)$ on a certain region of $\mathbb{R}^3$.

Recall by Proposition 3.1 that a Riemannian surface M with boundary is parabolic if and only if there exists a proper positive superharmonic function on M. We now use this defining property of parabolicity and the universal superharmonic function $\ln(\sqrt{x_1^2+x_2^2}) - \frac{1}{2}x_3^2$ on the complement of the cylinder $\{(x_1,x_2,x_3) \mid x_1^2+x_2^2 < 1\}$ to prove Theorem 11.2 for a properly immersed minimal surface M without boundary; the proof of the case when M has boundary is a small modification of the proof of the empty boundary case.

PROOF. We will show that $M(+) = \{(x_1,x_2,x_3) \in M \mid x_3 \geq 0\}$ is parabolic. Assume that $M(+)$ is connected; the general case can be obtained by proving each component of $M(+)$ is parabolic. For each positive integer n define $M(n) = \{(x_1,x_2,x_3) \in M \mid 0 \leq x_3 \leq n\}$ and let $M(n,*) = \{(x_1,x_2,x_3) \in M(n) \mid 1 \leq x_1^2 + x_2^2\}$. Let h_n be the restriction of the universal superharmonic function $\ln(\sqrt{x_1^2+x_2^2}) - \frac{1}{2}x_3^2$ to $M(n,*)$ and note that $h_n\colon M(n,*) \to \mathbb{R}$ is proper and bounded from below. (This function is superharmonic on $M(n,*)$ by the easy to calculate estimate [**16**], $\Delta\ln(x) \leq \frac{|\nabla x_3|^2}{r^2}$ where $r = \sqrt{x_1^2+x_2^2}$ and Δ is the surface

Laplacian). Hence, $M(n,*)$ is parabolic. Since $M(n)$ is the union of a compact surface with $M(n,*)$, then $M(n)$ is also parabolic.

For n large choose a $p \in M(n)$ such that $x_3(p) = 1$. Let $\partial(n)$ denote the part of $\partial M(n)$ at x_3-height n and let $\partial(0)$ denote the part of $\partial M(n)$ at height zero. Since $M(n)$ is parabolic and $x_3|M(n)$ is a bounded harmonic function on $M(n)$, for the hitting measure $\mu_p(n)$ on $\partial M(n)$, we have by Proposition 3.2,

$$1 = x_3(p) = \int\limits_{\partial M(n)} x_3(x)\mu_p(n) = n \int\limits_{\partial(n)} \mu_p(n).$$

Hence, $\int\limits_{\partial(n)} \mu_p(n) = \frac{1}{n}$ and since $M(n)$ is parabolic, $\int\limits_{\partial(0)} \mu_p(n) = 1 - \frac{1}{n}$. Hence, for all positive integers n,

$$\int\limits_{\partial M(+)} \mu_p = \int\limits_{\partial(0)} \mu_p \geq \int\limits_{\partial(0)} \mu_p(n) = 1 - \frac{1}{n}.$$

Therefore, $\int\limits_{\partial M(+)} \mu_p = 1$, which proves that $M(+)$ is parabolic.

□

COROLLARY 11.3. *If M is a properly immersed minimal surface in $\mathbb{R}^3$ and M intersects some plane in a compact set, then M is recurrent for Brownian motion. In particular, by the discussion after Theorem 9.2, if $M \in \mathcal{P}$ has two limit ends, then M is parabolic.*

Theorem 8.2 and the above Corollary imply the next Theorem.

THEOREM 11.4. *Every $\Sigma \in \mathcal{P}$ of finite genus is conformally a closed Riemann surface $\bar{\Sigma}$ punctured in a closed countable set of points W and when this set of points is infinite, then W has exactly two limit points on $\bar{\Sigma}$. In particular, Σ is recurrent for Brownian motion.*

For some related results see [**55**].

12. Periodic minimal surfaces.

In [**75**] Meeks and Rosenberg developed the classical theory of properly embedded doubly-periodic minimal surfaces in $\mathbb{R}^3$. Theoretical questions concerning the geometry of properly embedded doubly-periodic minimal surfaces are usually most easily approached by studying the quotient surface Σ in $\mathbb{T} \times \mathbb{R}$ where $\mathbb{T}$ is a flat two dimensional torus. One of the main theorems in [**75**] is that such a Σ has total Gaussian curvature $c(\Sigma) = 2\pi\chi(\Sigma)$ where $\chi(\Sigma)$ is the Euler characteristic of Σ; thus, Σ has finite total curvature if it has finite topology. This finite total curvature property for finite topology Σ leads to strong restrictions on the geometry and the topology of such surfaces and forces each annular end of Σ to be asymptotic to the end of a flat annulus in $\mathbb{T} \times \mathbb{R}$.

Later Meeks [**65**] generalized these results by proving that any properly embedded minimal surface Σ in $\mathbb{T} \times \mathbb{R}$ has a finite number of ends and that if the genus of Σ is finite, then Σ has finite topology and linear area growth. Since any complete Riemannian surface Σ with at most linear area growth and nonpositive curvature has total curvature $c(\Sigma) = 2\pi\chi(\Sigma)$, Meeks' theorem gave a new proof of the total curvature formula of Meeks and Rosenberg. More importantly, Meeks' theorems identified properly embedded minimal surfaces in $\mathbb{T} \times \mathbb{R}$ of finite total curvature with those surfaces of finite genus. In particular, if Σ has genus-zero, then Σ has finite total curvature. This result, together with some other constraints finite total curvature planar domains in $\mathbb{T} \times \mathbb{R}$ satisfy [**75**], was then applied by Lazard-Holly and Meeks [**53**] to prove the deep result:

THEOREM 12.1. *A genus-zero properly embedded minimal $\Sigma \subset \mathbb{T} \times \mathbb{R}$ is the quotient of one of the classical doubly-periodic examples defined by Scherk* [**98**] *in 1835. (See Section 4).*

Recently, Meeks and Wolf [**81**] have been able to prove the following related uniqueness theorem for the conjugate surfaces to the minimal surfaces in Theorem 12.1. The proof of these theorems are similar in approach but quite different in their details. They are hopeful that they will be able to prove the same result without the symmetry assumption.

THEOREM 12.2. *If M is a connected minimal surface with area less than $2\pi R^2$ in balls of radius R and the symmetry group of M is infinite, then M is a singly-periodic Scherk surface $\mathbf{S}_\theta$ described in Section 4, M is a catenoid or M is a plane.*

Another important uniqueness theorem for periodic minimal surfaces is the uniqueness of the Riemann minimal examples first defined by Riemann in [**95**]. See Section 4 for a description of these beautiful singly-periodic minimal surfaces of genus-zero which are foliated by circles and lines in horizontal planes. The following uniqueness of the Riemann examples was proved by Meeks, Perez and Ros in [**70**]. They are actively working on proving this theorem without the hypothesis of infinite symmetry group. See Conjecture 8.1 and Section 8 for some related discussion on the Riemann examples.

THEOREM 12.3. *The plane, catenoid, helicoid and Riemann minimal examples are the only properly embedded minimal surfaces in $\mathbb{R}^3$ of genus zero with infinite symmetry group.*

More generally, Meeks and Rosenberg [**77**] prove the following theorem for singly-periodic minimal surfaces whose quotient surfaces have finite topology.

THEOREM 12.4. *A properly embedded minimal surface Σ in a nonsimply connected complete flat three-manifold N^3 has finite total curvature if and only if it has finite topology. If $N^3 = \mathbb{R}^3/S_\theta$, where S_θ is a screw motion symmetry with angle $\theta, 0 \leq \theta \leq \pi$, with axis being the x_3-axis and $\Sigma \subset N^3$ is a properly embedded minimal surface with finite topology, then Σ has finite conformal type and can be defined analytically in terms of meromorphic data on its conformal compactification. Furthermore, each annular end of Σ is asymptotic to a horizontal plane in N^3, a vertical half-plane in N^3 or an end of a helicoid in N^3.*

As we already remarked, Meeks' theorem [**65**] states that a properly embedded minimal surface in $\mathbb{T} \times \mathbb{R}$ always has a finite number of ends. On the other hand, there exist properly embedded minimal surfaces of finite genus in $\mathbb{R}^2 \times S^1$ and $\mathbb{R}^3/S_\pi$ with an infinite number of ends (see [**58**]). For example, a singly-periodic quotient $\bar{\mathbf{S}}_{\frac{\pi}{4}} \subset \mathbb{R}^2 \times S^1$ of a doubly-periodic Scherk surface can have an infinite number of ends. The following theorem shows that these flat three-manifolds with infinite cyclic fundamental groups are special cases.

THEOREM 12.5. [**58**] *If* $\Sigma \subset \mathbb{R}^3/S_\theta, \theta \neq 0$ *or* π, *is a properly embedded minimal surface with more than one end, then* Σ *has at most quadratic area growth and has a finite number of ends. If* $\Sigma \subset \mathbb{R}^3/S_\theta, \theta \neq 0$ *or* π, *has finite genus, then* Σ *has at most quadratic area growth, finite topology and finite total curvature.*

The author refers the interested reader to [**65**] for a detailed survey of the classical theory of periodic minimal surfaces and to [**64**] for the more specialized theory of triply-periodic minimal surfaces.

13. Minimal surfaces in $M \times \mathbb{R}$.

In Section 12 we discussed briefly some of the theoretical results of Meeks [**65**] and of Meeks and Rosenberg [**75**] concerning minimal surfaces in $\mathbb{T}\times\mathbb{R}$ where $\mathbb{T}$ is a flat two-dimensional torus; these surfaces are just the quotients of doubly-periodic minimal surfaces in $\mathbb{R}^3$. When $\Sigma \subset \mathbb{T}\times\mathbb{R}$ is a properly embedded minimal surface, then, by the theorem of Meeks [**65**], Σ has a finite number of ends and if Σ also has finite genus, then it has bounded curvature, linear area growth, total curvature $2\pi\chi(\Sigma)$ and finite index with respect to the stability operator. In [**72**] and [**73**] Meeks and Rosenberg generalized many of their results for $\mathbb{T}\times\mathbb{R}$ to the case $M\times\mathbb{R}$ where M is a compact Riemannian surface. In particular, they prove that if M is endowed with a metric of nonpositive curvature, then the just described result of Meeks for a properly embedded minimal Σ in $M \times \mathbb{R}$ holds (see the Finiteness of Ends Theorem and the Bounded Curvature Theorem at the end of this Section). The four main theorems in these papers - The Linear Area Growth Theorem, the Stability Theorem, the Finiteness of Ends Theorem and the Bounded Curvature Theorem - all represent surprisingly strong theoretical results which will likely have an impact on research in other areas of classical surface theory.

In part because of the possible applications of these results to the study of constant mean curvature surfaces in $\mathbb{R}^3$ and S^3, we will briefly go over their statements and some of the ideas behind these proofs. In all of these theorems M denotes a compact Riemannian surface.

Given a properly immersed minimal surface Σ in $M \times \mathbb{R}$, we define the *flux of* Σ to be the flux of the gradient ∇h across $\Sigma \cap (M \times \{0\})$ where $h\colon \Sigma \to \mathbb{R}$ is the harmonic height function $h(p,t) = t$. Since h is a proper harmonic function, the flux of Σ is the flux of ∇h across any level set of h, not just the level set at height zero. The invariance of the flux of Σ plays a crucial role in the proofs of many of these theorems, including the following.

THEOREM 13.1. [**73**] *(Linear Area Growth Theorem) If* Σ *is a properly embedded noncompact minimal surface in* $M \times \mathbb{R}$ *of bounded curvature, then* Σ *has a finite number of ends and linear area growth, in the sense that* $c_1 t \leq \text{Area}(\Sigma \cap (M \times [-t,t])) \leq c_2 t$ *where* $c_1 > 0$ *depends only on the injectivity radius of* M *and* c_2 *depends only on the geometry of* M, *a lower bound of the flux of* Σ *and an upper bound on the absolute Gaussian curvature of* Σ.

Every sequence of properly embedded minimal surfaces in a three-dimensional Riemannian manifold, which intersect a compact domain and satisfy uniform local area and local curvature estimates, has a subsequence that converges to another properly embedded minimal surface with local area and local curvature estimates (see for example [**74**]). A simple consequence of this compactness result and Theorem 13.1 is that every noncompact properly embedded minimal surface in $M \times \mathbb{R}$

with bounded curvature is quasiperiodic in the following sense. A properly embedded surface Σ in a Riemannian three-manifold M is *quasiperiodic* if there exists a discrete infinite closed subset $S = \{T_n \mid n \in \mathbb{N}\}$ of the isometry group of M such that $T_n(\Sigma)$ converges on compact subsets of M to a properly embedded surface.

COROLLARY 13.2. *If Σ is a properly embedded noncompact minimal surface of bounded curvature in $M \times \mathbb{R}$, then Σ is quasiperiodic. In fact, any sequence of vertical translations of Σ in $M \times \mathbb{R}$ contains a convergent subsequence to another properly embedded minimal surface with the same bound on its curvature.*

By the curvature estimates of Schoen [**99**], every properly embedded stable minimal surface in $M \times \mathbb{R}$ has bounded curvature. Therefore, every properly embedded noncompact stable minimal surface in $M \times \mathbb{R}$ is quasiperiodic. This quasiperiodicity property is essential in proving the next theorem.

THEOREM 13.3. [**72**] *(Stability Theorem) Suppose that Σ is a connected properly embedded stable orientable minimal surface in $M \times \mathbb{R}$. Then Σ is one of the surfaces described in (1)-(4) below:*

1. *Σ is compact and $\Sigma = M \times \{t\}$ for some $t \in \mathbb{R}$;*
2. *Σ is of the form $\gamma \times \mathbb{R}$ where γ is a simple closed stable geodesic in M;*
3. *Σ is periodic and has a quotient $\bar{\Sigma}$ in $M \times S^1(r)$ where r is the circumference of the circle. In this case, for every $p \in M, \{p\} \times S^1(r)$ intersects $\bar{\Sigma}$ transversely in a single point and the orbit of the natural action of $S^1(r)$ on $M \times S^1(r)$ gives rise to a product minimal foliation of $M \times S^1(r)$. In particular, $\bar{\Sigma}$ is homeomorphic to M and is area minimizing in its integer homology class;*
4. *Σ is a graph over an open connected subdomain of M bounded by a finite number of stable geodesics with each end of Σ asymptotic to the end of one of the flat vertical annuli described in (2);*
5. *The moduli space of examples described in (3) in the case M is orientable is naturally parametrized by $P(H_1(M)) \times \mathbb{R}^+$ where $P(H_1(M))$ consists of the primitive (non-multiple) elements in the first homology group of M. Given an example $\bar{\Sigma} \subset M \times S^1(r)$ we obtain the corresponding element $([\bar{\Sigma}] \cap [M \times \{*\}], r) \in P(H_1(M)) \times \mathbb{R}^*$, where $\cap$ is the intersection pairing of the associated homology classes in $M \times S^1(r)$ and $*$ is a base point on $S^1(r)$.*

Theorem 13.1 states, among other things, that a properly embedded minimal surface of bounded curvature in $M \times \mathbb{R}$ must have a finite number of ends. The next theorem demonstrates that the bounded curvature hypothesis on the surface can be dropped and one still obtains the finite number of ends conclusion; Lemmas and Assertions used in the proof of Theorem 13.3 play a fundamental role in proving this more general result.

THEOREM 13.4. [**73**] *(Finiteness of Ends Theorem) If Σ is a properly embedded minimal surface in $M \times \mathbb{R}$, then Σ has a finite number of ends.*

We next focus our attention on the case when the properly embedded minimal surface Σ in $M \times \mathbb{R}$ has finite genus. By Theorem 13.4, such a surface Σ has a finite number of ends and so each end of Σ is an annulus and Σ has finite topology. Meeks and Rosenberg then use this finite topology property of Σ to prove that Σ has bounded curvature and so, by Theorem 13.1, Σ has linear area growth. The proof that Σ has bounded curvature is difficult and uses some of the recent results of Meeks, Perez and Ros [**66**] on the local structure of properly embedded minimal surfaces in three manifolds with bounded genus in a neighborhood of a point of large curvature; these results depend on recent curvature estimates of Colding and Minicozzi [**11, 12, 13**], and results of Meeks [**61**] and of Meeks and Rosenberg in [**74**]. See Section 8 for a more detailed discussion of these topics. With some further geometric analysis, Meeks and Rosenberg obtain the following theorem which significantly generalizes their earlier stated results.

THEOREM 13.5. [**73**] *(Bounded Curvature Theorem) Suppose Σ is a properly embedded minimal surface of finite genus in $M \times \mathbb{R}$. Then:*

1. *Σ has finite topology, finite conformal type, bounded curvature, and linear area growth;*
2. *If M has nonpositive curvature, then Σ has finite index with respect to the stability operator and Σ has total curvature $2\pi\chi(\Sigma)$;*
3. *If M has nonpositive curvature and M is not a torus, then each end of Σ is asymptotic to $\gamma \times \mathbb{R}$ where γ is a stable simple closed geodesic in M.*

In [**73**] Meeks and Rosenberg discuss several general methods for constructing minimal surfaces of finite topology in $M \times \mathbb{R}$, in particular the minimal graphs described below.

THEOREM 13.6. *If M is an orientable Riemannian surface of genus at least one and M is not a torus endowed with a metric which admits a foliation by closed geodesics, then there exists an infinite number of non-isotopic domains in M bounded by a finite number of stable simple closed geodesics and proper minimal graphs in $M \times \mathbb{R}$ over these domains.*

In the case M is a two sphere S^2 endowed with a metric of constant positive curvature, Meeks and Rosenberg write down a two-parameter family $\mathcal{A}$ of properly embedded minimal annuli in $S^2 \times \mathbb{R}$ that are closely related to the infinite-ended periodic Riemann examples of genus-zero in $\mathbb{R}^3$. The surfaces in $\mathcal{A}$ coincide with the properly embedded minimal annuli in $S^2 \times \mathbb{R}$ foliated by circles, one in each $S^2 \times \{t\}$. This family is defined in [**73**] in terms of meromorphic functions on rectangular elliptic curves and is closely related to a family of "tori" of constant mean curvature in $\mathbb{R}^3$ defined by Abresh [**1**].

14. Sixteen of my favorite conjectures.

In this section the author will present sixteen fundamental conjectures in the classical theory of minimal surfaces. For the most part these conjectures are motivated by the author's own research and are not widely known except to classical minimal surface specialists. Hopefully, the presentation of these problems and suggestions for a plan of attack on solving them will speed up their solution and stimulate further interest in this beautiful subject. In the statement of each conjecture the author has included a suggested expected time frame for a solution; only time will tell how accurate this time frame is. The author has listed in the statement of each conjecture the principal researchers to whom the conjecture might be attributed. These conjectures are listed approximately in order according to the author's interest in them or by his personal feeling of their general importance or deepness.

Most of these problems and many others appear in [**59**] along with further discussion; also, see the author's 1978 book [**62**] for a much longer list of conjectures in the subject, some of whose solutions we have discussed in this survey. In the following discussion we again let $\mathcal{P}$ denote the space of all properly embedded connected minimal surfaces in $\mathbb{R}^3$ and let $\mathcal{M} \subset \mathcal{P}$ denote the subspace of examples with more than one end.

CONJECTURE 1. *[Convex Curve Conjecture (Meeks) Time Frame = 30 years] Two convex Jordan curves in parallel planes cannot bound a compact minimal surface of positive genus.*

There are some partial results on the Convex Curve Conjecture under the assumption of some symmetry on the curves (see [**79, 97, 100**]). Also, the results in [**79, 80**] indicate that the Convex Curve Conjecture probably holds in the more general case where the two convex planar curves do not necessarily lie in parallel planes but rather lie on the boundary of their convex hull, in this case the planar Jordan curves are called extremal. Recent results by Ekholm, White and Wienholtz [**22**] show that every compact orientable minimal surface that arises as a counterexample to the convex curve conjecture is embedded and that for a fixed pair of extremal convex planar curves there is a bound on the genus of such a minimal surface.

The next conjecture is motivated in part by the case where Γ is extremal (see Theorem 6.4), where it is known even in the more general case where the minimal surface is allowed to be nonorientable.

CONJECTURE 2. *[4π-Conjecture (Meeks-Yau, Nitsche) Time Frame = 20 years] If Γ is a simple closed curve in $\mathbb{R}^3$ with total curvature at most 4π, then Γ bounds a unique orientable branched minimal surface and this unique minimal surface is an embedded disk.*

There exists a conjecture by Ekholm, White and Wienholtz [**22**] that generalizes Conjecture 2, removing the orientability assumption on the minimal surface spanning Γ (these authors conjecture that besides the unique minimal disk given by Nitsche's Theorem, only one or two Möbius strips can occur).

CONJECTURE 3. *[Finite Topology Conjecture (Hoffman and Meeks) Time Frame = 100 years] A noncompact orientable surface M of finite topology with genus g and k ends, $k \neq 2$, occurs in $\mathcal{P}$ if and only if $k \leq g+2$.*

See [**39, 44, 104, 106**] and the discussion in Section 7 for partial existence results which seem to indicate that the existence implication in Finite Topology Conjecture holds when $k > 2$. There is experimental computer evidence that every orientable surface with finite genus and one end properly minimally embedds in $\mathbb{R}^3$ as a minimal surface of finite type (see [**2, 3, 74**]); also see the later Conjecture 14. Theorem 7.4 shows that for each positive genus g, there exists an upper bound $e(g)$ on the number of ends of an $M \in \mathcal{M}$ with finite topology and genus g. Results of Collin [**15**] and Schoen [**100**] imply that the only examples in $\mathcal{M}$ with finite topology and two ends are catenoids. Results of Collin [**15**] and Lopez-Ros [**56**] imply that if M has finite topology, genus zero and at least two ends, then M is a catenoid.

CONJECTURE 4. *[Liouville Conjecture (Meeks, Sullivan) Time Frame = 50 years] If $M \in \mathcal{P}$ and $h\colon M \to \mathbb{R}$ is a positive harmonic function, then h is constant.*

The above conjecture is closely related to work in [**16, 71**]. For example, from the discussion in Section 11, we know that if $M \in \mathcal{P}$ has finite genus or two limit ends, then M is recurrent for Brownian motion which implies M satisfies the Liouville Conjecture.

CONJECTURE 5. *[Properness Conjecture (Calabi, Meeks) Time Frame = 100 years] If* $f\colon M \to \mathbb{R}^3$ *is a complete injective minimal immersion, then* $M \in \mathcal{P}$.

The author has an outline for a possible proof of properness in the finite topology case. The author in conjunction with Perez and Ros have conjectured [**66**] that if a complete embedded minimal $M \subset \mathbb{R}^3$ has finite genus, then M has bounded Gaussian curvature (also see, [**74**]). It follows from work in [**66, 67, 74**] that if such an M has locally bounded Gaussian curvature in $\mathbb{R}^3$ and finite genus, then M is properly embedded.

Gulliver and Lawson [**33**] proved that if Σ is a stable orientable minimal surface with compact boundary that is properly embedded in the punctured unit ball in $\mathbb{R}^3$, then its closure is an embedded surface. If Σ is not stable, then the corresponding result is not known. Recent results in [**12, 67**] indicate that a more general result might hold.

CONJECTURE 6. *[Isolated Singularities Conjecture (Gulliver-Lawson) Time Frame = 8 years] There does not exist a properly embedded minimal surface in the punctured ball* $B - \{(0,0,0)\}$ *whose closure is not a surface at* $(0,0,0)$.

CONJECTURE 7. *[Isometry Conjecture (Meeks) Time Frame = 20 years] If* $M \in \mathcal{P}$, *then intrinsic isometries of* M *extend to ambient isometries of* $\mathbb{R}^3$. *Furthermore, if* M *is not simply-connected, then it is "minimally rigid" in the sense that any isometric minimal immersion of* M *into* $\mathbb{R}^3$ *is congruent to* M.

This Isometry Conjecture is known if $M \in \mathcal{P}$ has more than one end (see [**8**]). Results of Meeks and Rosenberg [**74**] and [**77**] imply that the isometry conjecture can only fail if M has one end and infinite genus. It is also known to hold for doubly-periodic minimal surfaces [**75**]. One way to prove the conjecture would be to prove that if $M \in \mathcal{P}$ has one end and infinite genus, then there exists a plane in $\mathbb{R}^3$ that intersects M in a set that contains a simple closed curve.

CONJECTURE 8. *[Genus-zero Conjecture (Meeks-Perez-Ros) Time Frame = 2 years] If* $M \in \mathcal{P}$ *has genus-zero, then* M *is a plane, a catenoid, a helicoid or a Riemann example. In particular,* M *is foliated by lines and circles in parallel planes.*

The above conjecture is known if M has genus-zero and finite topology by results in [**15, 56, 74**]. By the main theorem in [**16**], if M has infinite topology, then it must have one or two limit ends. Theorem 8.2 states that if M has finite genus and infinite topology, then M must have two limit ends; in fact, the main goal of [**67**] is to prove this result. If M has finite genus and two limit ends, then the curvature estimates in [**66**] show that M is quasiperiodic and the results in [**70**] imply the conjecture if M is actually periodic. See Section 8 for related discussions and an explanation of the importance of Conjecture 8.

CONJECTURE 9. *[Geometric Flux Conjecture (Meeks-Rosenberg) Time Frame = 80 years] Suppose* $M \in \mathcal{P}$ *and* $h\colon M \to \mathbb{R}$ *is a nonconstant coordinate function on* M. *Consider the set* I *of integral curves of* ∇h. *Then there exists a countable set* $C \subset I$ *such that for any integral curve* $\alpha \in I - C$, $h|\alpha\colon \alpha \to \mathbb{R}$ *is a diffeomorphism. Here we consider* $\alpha\colon \mathbb{R} \to M$ *to be, after a choice of* $p \in \alpha$, *a curve* $\alpha(t)$ *with* $\alpha(0) = p$ *and* $\alpha'(t) = \nabla h(\alpha(t))$.

One could weaken the hypothesis in the above conjecture that "except for a countable number of integral curves, h restricted to an integral curve of ∇h is a diffeomorphism with $\mathbb{R}$" to the hypothesis that "for almost-all integral curves of ∇h, h restricted to an integral curve is a diffeomorphism with $\mathbb{R}$". This weaker version of the Flux Conjecture is, for technical reasons, more likely to be solved with a suggested time frame of only two years for its solution. This weaker conjecture, via Stokes theorem, has as a consequence the recent Algebraic Flux Lemma [**59**] by Meeks. The Algebraic Flux Lemma and Theorem 11.2 imply that for any coordinate function $h\colon M \to \mathbb{R}$ on an a properly immersed minimal surface in $\mathbb{R}^3$, the flux of ∇h across a level set of h is independent of the level set. The author feels that this flux result may have important theoretical consequences.

CONJECTURE 10. *[Scherk Uniqueness Conjecture (Meeks) Time Frame = 5 years] If M is a properly immersed minimal surface in $\mathbb{R}^3$ and, in balls $B(R)$ of radius R, Area$(M \cap B(R)) < 2\pi R^2$, then M is a Scherk singly-periodic minimal surface, a catenoid or a plane.*

A related conjecture on the uniqueness of Scherk's doubly-periodic minimal surfaces was recently solved by Lazard-Holly and Meeks [**53**]; they proved that if $M \in \mathcal{P}$ is doubly-periodic and the quotient surface has genus-zero, then M is one of Scherk's doubly-periodic minimal surfaces. The basic approach used in [**53**] can be adapted to prove the above conjecture under the assumption that the surface is periodic; this result is a recent theorem of Mike Wolf and the author (see Theorem 12.2). In fact their result gives a proof of the above conjecture in the case where M has an infinite symmetry group. Their approach for solving the general conjecture is first to prove the following conjecture on the uniqueness of the limit tangent cone of M, from which it follows by unpublished work of Meeks and Ros that M has two Alexandrov-type planes of symmetry. From these planes of symmetry one can describe the Weierstrass representation of M, which hopefully would be useful in completing the proof of the conjecture. Much of the interest in the previous conjecture arises from the role that Scherk surfaces play in desingularizing two intersecting minimal surfaces (see Kapouleas [**51**]).

CONJECTURE 11. *[Unique Limit Tangent Cone Conjecture (Meeks) Time Frame = 4 years] If $M \in \mathcal{P}$ is not a plane and has quadratic area growth, then $\lim_{t\to\infty} \frac{1}{t}M$ exists and is a cone. Furthermore, if M has area less than $2\pi R^2$ in balls of radius R, then the limit tangent cone is the union of two planes or one plane of multiplicity two passing through the origin.*

CONJECTURE 12. *[Connected Graph Conjecture (Meeks) Time Frame = 8 years] A minimal graph in $\mathbb{R}^{n+1}$ with zero boundary values over a proper, possibly disconnected, domain in $\mathbb{R}^n$ can have at most two nonplanar components. If the graph also has sublinear growth, then such a graph with no planar components is connected.*

The above conjecture was made by Meeks a number of years ago. The first important partial result came out of work by Meeks and Rosenberg on the uniqueness of the helicoid (see [**74**]). They proved, under the additional hypothesis of gradient estimates, that such a graph can only have a finite number of nonplanar components. Spruck [**102**] has given some related results and Li and Wang [**54**] have recently proven finiteness of the number of nonflat components without assuming gradient bounds.

CONJECTURE 13. *[Quadratic Area Growth Conjecture (Meeks) Time Frame = 9 years] $M \in \mathcal{P}$ has quadratic area growth if and only if there exist a double cone C (of the form $x_3^2 = \lambda(x_1^2 + x_2^2)$ and possibly rotated) that intersects M in a compact set.*

It follows from computations in [**16**] that $M \in \mathcal{P}$ has quadratic area growth if M intersects the union of the negative end of a vertical catenoid and a positive vertical cone in a compact set. If the conclusion of the previous unique limit tangent cone conjecture holds for an $M \in \mathcal{P}$ with quadratic area growth, then for such an M there exists a double cone that intersects M in a compact set. Hence, the validity of the unique limit tangent cone conjecture would give one of the implications in the quadratic area growth conjecture.

CONJECTURE 14. *[One-ended Conjecture (Meeks and Rosenberg) Time Frame = 20 years] For every nonnegative integer g, there exists a unique nonplanar $M \in \mathcal{P}$ with genus g and one end.*

There are some partial results on the above conjecture. Meeks and Rosenberg have shown that every example $M \in \mathcal{P}$ of finite genus and one end has a special analytic representation which makes M into a "surface of finite type" (see [**74**]). In the case of genus-zero, Meeks and Rosenberg proved that the plane and the helicoid are the only genus-zero examples. Based on an earlier computational proof of the existence of a helicoid with a handle by Hoffman, Karcher and Wei [**40**], Hoffman, Weber and Wolf [**48**] have given a rigorous mathematical proof of the existence of the genus-one helicoid. Work in progress by Martin and Weber [**57**] indicates that this genus-one helicoid is unique. Other computational results in [**2, 3, 103**] indicate the conjecture is true for genus 2, 3, 4 and 5. We believe that a recent theorem of Traizet and Weber [**105**] will eventually lead to a proof of the existence part of the above conjecture.

The Finite Topology Conjecture of Hoffman and Meeks and the One-ended Conjecture of Meeks and Rosenberg together propose the precise topological conditions under which a noncompact orientable surface of finite topology would properly minimally embed in $\mathbb{R}^3$. What about the case where the noncompact orientable surface M has infinite topology; i.e., either M has infinite genus or M has an infinite number of ends? By Theorem 9.2, such an M can have at most one or two limit ends. Theorem 8.2 states that such an M cannot have one limit end and finite genus. The following conjecture is nothing more than the claim that these restrictions are the only ones.

CONJECTURE 15. *[Infinite Topology Conjecture (Meeks) Time Frame = 50 years] A noncompact orientable surface of infinite topology occurs in $\mathcal{P}$ if and only if it has at most one or two limit ends and when it has one limit end, then it also has infinite genus.*

Meeks and Rosenberg [**74**] and Meeks, Perez and Ros [**68**] have obtained some partial results on the following conjecture (See Section 13). It is closely related to Conjecture 8.

CONJECTURE 16. *[Uniqueness of the $\mathcal{A}$-family Conjecture (Meeks, Rosenberg) Time Frame = 2 years.] Let $S^2 \subset \mathbb{R}^3$ be the unit sphere centered at the origin. A properly embedded minimal annulus in $S^2 \times \mathbb{R}$ is in the family $\mathcal{A}$ described in Section 13. In particular, every such minimal annulus is foliated by circles in level set spheres.*

William H. Meeks, III at bill@gang.umass.edu, research supported by NSF grant DMS - 0104044

References

[1] U. Abresh. Constant mean curvature tori in terms of elliptic functions. *J. Reine Angew. Math.*, 374:169–192, 1987.

[2] A. I. Bobenko. Helicoids with handles and Baker-Akhiezer spinors. *Math. Z.*, (1):9–29, 1998.

[3] A. I. Bobenko and M. Schmies. Computer graphics experiments for helicoids with handles. Personal communication.

[4] C. V. Boys. *Soap bubbles: Their colours and the forces which mold them.* Dover Publications, New York, 1959.

[5] M. Callahan, D. Hoffman, and W. H. Meeks III. Embedded minimal surfaces with an infinite number of ends. *Invent. Math.*, 96:459–505, 1989.

[6] M. Callahan, D. Hoffman, and W. H. Meeks III. The structure of singly-periodic minimal surfaces. *Invent. Math.*, 99:455–481, 1990.

[7] J. Choe. Index, vision number, and stability of complete minimal surfaces. *Arch. Rational Mech. Anal.*, 109(3):195–212, 1990.

[8] T. Choi, W. H. Meeks III, and B. White. A rigidity theorem for properly embedded minimal surfaces in $\mathbb{R}^3$. *J. of Differential Geometry*, 32:65–76, 1990.

[9] T. H. Colding and W. P. Minicozzi II. The space of embedded minimal surfaces of fixed genus in a 3-manifold I; estimates off the axis for disks. Preprint.

[10] T. H. Colding and W. P. Minicozzi II. The space of embedded minimal surfaces of fixed genus in a 3-manifold II; multi-valued graphs in disks. Preprint.

[11] T. H. Colding and W. P. Minicozzi II. The space of embedded minimal surfaces of fixed genus in a 3-manifold III; locally simply-connected. Preprint.

[12] T. H. Colding and W. P. Minicozzi II. The space of embedded minimal surfaces of fixed genus in a 3-manifold IV; planar domains. Preprint.

[13] T. H. Colding and W. P. Minicozzi II. The space of embedded minimal surfaces of fixed genus in a 3-manifold V; fixed genus. Preprint.

[14] T. H. Colding and W. P. Minicozzi II. Estimates for parametric elliptic integrands. *Int. Math. Res. Not.*, (6):291–297, 2002.

[15] P. Collin. Topologie et courbure des surfaces minimales de $\mathbb{R}^3$. *Annals of Math. 2nd Series*, 145–1:1–31, 1997.

[16] P. Collin, R. Kusner, W. H. Meeks III, and H. Rosenberg. The geometry, conformal structure and topology of minimal surfaces with infinite topology. Preprint.

[17] C. Costa. *Imersões minimas en* $\mathbb{R}^3$ *de gênero un e curvatura total finita.* PhD thesis, IMPA, Rio de Janeiro, Brasil, 1982.

[18] R. Courant. Soap film experiments with minimal surfaces. *Amer. Math. Monthly*, 47:167–174, 1940.

[19] Dr. Crypton. Shapes that eluded discovery. *Science Digest*, pages 50–55, 78, April 1986.

[20] M. do Carmo and C. K. Peng. Stable minimal surfaces in $\mathbb{R}^3$ are planes. *Bulletin of the AMS*, 1:903–906, 1979.

[21] J. Douglas. Minimal surfaces of higher topological structure. *Annals of Math.*, 40:205–298, 1939.

[22] T. Ekholm, B. White, and D. Wienholtz. Embeddedness of minimal surfaces with total curvature at most 4π. *Annals of Math*, 155(1):209–234, 2002.

[23] L. Euler. *Methodus inveniendi lineas curvas maximi minimive propietate gaudeates sive solutio problematis isoperimetrici latissimo sensu accepti.* Harvard Univ. Press, Cambridge, MA, 1969. A source book in mathematics, partially translated by D. J. Struik, see pages 399–406.

[24] H. Federer. *Geometric measure theory.* Springer Verlag, Berlin-Heidelberg, New York, 1969.

[25] D. Fischer-Colbrie. On complete minimal surfaces with finite Morse index in 3-manifolds. *Invent. Math.*, 82:121–132, 1985.

[26] D. Fischer-Colbrie and R. Schoen. The structure of complete stable minimal surfaces in 3-manifolds of non-negative scalar curvature. *Comm. on Pure and Appl. Math.*, 33:199–211, 1980.

[27] M. Freedman. An unknotting result for complete minimal surfaces in $\mathbb{R}^3$. *Invent. Math*, 109(1):41–46, 1992.

[28] C. Frohman. The topological uniqueness of triply-periodic minimal surfaces in $\mathbb{R}^3$. *J. of Differential Geometry*, 31:277–283, 1990.

[29] C. Frohman and W. H. Meeks III. The topological classification of minimal surfaces in $\mathbb{R}^3$. Preprint.

[30] C. Frohman and W. H. Meeks III. The ordering theorem for the ends of properly embedded minimal surfaces. *Topology*, 36(3):605–617, 1997.

[31] C. Frohman and W. H. Meeks III. The topological uniqueness of complete one-ended minimal surfaces and Heegaard surfaces in $\mathbb{R}^3$. *J. of the Amer. Math. Soc.*, 10(3):495–512, 1997.

[32] A. Grigoryan. Analytic and geometric background of recurrence and non-explosion of Brownian motion on Riemannian manifolds. *Bull. of A.M.S*, 36(2):135–249, 1999.

[33] R. Gulliver and H. B. Lawson. The structure of minimal hypersurfaces near a singularity. *Proc. Symp. Pure Math.*, 44:213–237, 1986.

[34] R. Gulliver and F. Lesley. On boundary branch points of minimizing surfaces. *Arch. Rational Mech. Anal.*, 52:20–25, 1973.

[35] R. Hardt and L. Simon. Boundary regularity and embedded minimal solutions for the oriented Plateau problem. *Annals of Math.*, 110:439–486, 1979.

[36] S. Hildebrandt. Boundary behavior of minimal surfaces. *Archive Rational Mech. Anal.*, 35:47–81, 1969.

[37] S. Hildebrandt. The calculus of variations today. *Mathematical Intelligencer*, 11(4):50–60, 1989.

[38] D. Hoffman. The computer-aided discovery of new embedded minimal surfaces. *Mathematical Intelligencer*, 9(3):8–21, 1987.

[39] D. Hoffman and H. Karcher. Complete embedded minimal surfaces of finite total curvature. In R. Osserman, editor, *Encyclopedia of Mathematics, Vol. 90, Geometry V*, pages 5–93. Springer Verlag, 1997.

[40] D. Hoffman, H. Karcher, and F. Wei. The genus one helicoid and the minimal surfaces that led to its discovery. In *Global Analysis and Modern Mathematics*. Publish or Perish Press, 1993. K. Uhlenbeck, editor, p. 119–170.

[41] D. Hoffman and W. H. Meeks III. A complete embedded minimal surface in $\mathbb{R}^3$ with genus one and three ends. *J. of Differential Geometry*, 21:109–127, 1985.

[42] D. Hoffman and W. H. Meeks III. A variational approach to the existence of complete embedded minimal surfaces. *Duke J. of Math.*, 57(3):877–894, 1988.

[43] D. Hoffman and W. H. Meeks III. The asymptotic behavior of properly embedded minimal surfaces of finite topology. *J. of the Amer. Math. Soc.*, 2(4):667–681, 1989.

[44] D. Hoffman and W. H. Meeks III. Embedded minimal surfaces of finite topology. *Annals of Math.*, 131:1–34, 1990.

[45] D. Hoffman and W. H. Meeks III. Limits of minimal surfaces and Scherk's Fifth Surface. *Arch. Rat. Mech. Anal.*, 111(2):181–195, 1990.

[46] D. Hoffman and W. H. Meeks III. Minimal surfaces based on the catenoid. *Amer. Math. Monthly, Special Geometry Issue*, 97(8):702–730, 1990.

[47] D. Hoffman and W. H. Meeks III. The strong halfspace theorem for minimal surfaces. *Invent. Math.*, 101:373–377, 1990.

[48] D. Hoffman, M. Weber, and M. Wolf. The existence of the genus-one helicoid. Preprint.

[49] A. Huber. On subharmonic functions and differential geometry in the large. *Comment. Math. Helvetici*, 32:181–206, 1957.

[50] L. Jorge and W. H. Meeks III. The topology of complete minimal surfaces of finite total Gaussian curvature. *Topology*, 22(2):203–221, 1983.

[51] N. Kapouleas. Complete embedded minimal surfaces of finite total curvature. *J. Differential Geom.*, 47(1):95–169, 1997.

[52] J. L. Lagrange. Essai d'une nouvelle methode pour determiner les maxima et les minima des formules integrales indefinies. *Miscellanea Taurinensia 2*, 325(1):173–199, 1760.

[53] H. Lazard-Holly and W. H. Meeks III. The classification of embedded doubly-periodic minimal surfaces of genus zero. *Invent. Math.*, 143:1–27, 2001.

[54] P. Li and J. Wang. Finiteness of disjoint minimal graphs. Preprint.

[55] F. J. Lopez and J. Perez. Parabolicity and Gauss map of minimal surfaces. *Indiana J. of Math.*, 2003.

[56] F. J. Lopez and A. Ros. On embedded complete minimal surfaces of genus zero. *J. of Differential Geometry*, 33(1):293–300, 1991.
[57] F. Martin and M. Weber. The uniqueness of the genus-one helicoid. Work in progress.
[58] W. H. Meeks III. The geometry and topology of singly-periodic minimal surfaces. Preprint.
[59] W. H. Meeks III. Global problems in classical minimal surface theory. To appear in the Clay Minimal Surface Conference Proceedings.
[60] W. H. Meeks III. Proofs of some classical results in minimal surface theory. Preprint.
[61] W. H. Meeks III. The regularity of the singular set in the Colding and Minicozzi lamination theorem. To appear in Duke Journal of Math.
[62] W. H. Meeks III. *Lectures on Plateau's Problem.* Instituto de Matematica Pura e Aplicada (IMPA), Rio de Janeiro, Brazil, 1978.
[63] W. H. Meeks III. A survey of the geometric results in the classical theory of minimal surfaces. *Bol. Soc. Brasil Mat.*, 12:29–86, 1981.
[64] W. H. Meeks III. The theory of triply-periodic minimal surfaces. *Indiana Univ. Math. J.*, 39(3):877–936, 1990.
[65] W. H. Meeks III. The geometry, topology, and existence of periodic minimal surfaces. *Proceedings of Symposia in Pure Math.*, 54:333–374, 1993. Part I.
[66] W. H. Meeks III, J. Pérez, and A. Ros. The geometry of minimal surfaces of finite genus I; curvature estimates and quasiperiodicity. Preprint.
[67] W. H. Meeks III, J. Pérez, and A. Ros. The geometry of minimal surfaces of finite genus II; nonexistence of one limit end examples. Preprint.
[68] W. H. Meeks III, J. Pérez, and A. Ros. The geometry of minimal surfaces of finite genus IV; Jacobi fields and uniqueness for small flux. Work in progress.
[69] W. H. Meeks III, J. Pérez, and A. Ros. The geometry of minimal surfaces of finite genusIII; bounds on the topology and index of classical minimal surfaces. Preprint.
[70] W. H. Meeks III, J. Pérez, and A. Ros. Uniqueness of the Riemann minimal examples. *Invent. Math.*, 131:107–132, 1998.
[71] W. H. Meeks III and H. Rosenberg. Maximum principles at infinity with applications to minimal and constant mean curvature surfaces. Preprint.
[72] W. H. Meeks III and H. Rosenberg. Stable minimal surfaces in $M \times \mathbb{R}$. Preprint.
[73] W. H. Meeks III and H. Rosenberg. The theory of minimal surfaces in $M \times \mathbb{R}$. Preprint.
[74] W. H. Meeks III and H. Rosenberg. The uniqueness of the helicoid and the asymptotic geometry of properly embedded minimal surfaces with finite topology. Preprint.
[75] W. H. Meeks III and H. Rosenberg. The global theory of doubly periodic minimal surfaces. *Invent. Math.*, 97:351–379, 1989.
[76] W. H. Meeks III and H. Rosenberg. The geometry and conformal structure of properly embedded minimal surfaces of finite topology in $\mathbb{R}^3$. *Invent. Math.*, 114:625–639, 1993.
[77] W. H. Meeks III and H. Rosenberg. The geometry of periodic minimal surfaces. *Comment. Math. Helvetici*, 68:538–578, 1993.
[78] W. H. Meeks III, L. Simon, and S. T. Yau. The existence of embedded minimal surfaces, exotic spheres and positive Ricci curvature. *Annals of Math.*, 116:221–259, 1982.
[79] W. H. Meeks III and B. White. Minimal surfaces bounded by convex curves in parallel planes. *Comment. Math. Helvetici*, 66:263–278, 1991.
[80] W. H. Meeks III and B. White. The space of minimal annuli bounded by an extremal pair of planar curves. *Communications in Analysis and Geometry*, 1(3):415–437, 1993.
[81] W. H. Meeks III and M. Wolf. The solution of the Scherk uniqueness conjecture for minimal surfaces of infinite symmetry. Work in progress.
[82] W. H. Meeks III and S. T. Yau. The classical Plateau problem and the topology of three-dimensional manifolds. *Topology*, 21(4):409–442, 1982.
[83] W. H. Meeks III and S. T. Yau. The existence of embedded minimal surfaces and the problem of uniqueness. *Math. Z.*, 179:151–168, 1982.
[84] W. H. Meeks III and S. T. Yau. The topological uniqueness of complete minimal surfaces of finite topological type. *Topology*, 31(2):305–316, 1992.
[85] C. B. Morrey. The problem of Plateau in a Riemannian manifold. *Annals of Math.*, 49:807–851, 1948.
[86] J. C. C. Nitsche. A characterization of the catenoid. *J. of Math. Mech.*, 11:293–302, 1962.
[87] J. C. C. Nitsche. On new results in the theory of minimal surfaces. *Bull. Amer. Math. Soc.*, 71:195–270, 1965.

[88] J. C. C. Nitsche. A new uniqueness theorem for minimal surfaces. *Arch. Rat. Mech. Anal.*, 52:319–329, 1973.

[89] J. C. C. Nitsche. *Lectures on Minimal Surfaces*, volume 1. Cambridge University Press, 1989.

[90] R. Osserman. Global properties of minimal surfaces in E^3 and E^n. *Annals of Math.*, 80(2):340–364, 1964.

[91] R. Osserman. A proof of the regularity everywhere to Plateau's problem. *Annals of Math*, 91(2):550–569, 1970.

[92] J. Pérez and A. Ros. The space of properly embedded minimal surfaces with finite total curvature. *Indiana Univ. Math. J.*, 45(1):177–204, 1996.

[93] J. Pérez and A. Ros. Properly embedded minimal surfaces. In *The Global Theory of Minimal Surfaces in Flat Spaces-LN M-1775*, pages 15–66. Springer Verlag, 1999. G. P. Pirola, editor.

[94] J. Pérez and A. Ros. The space of complete minimal surfaces with finite total curvature as Lagrangian submanifold. *Trans. Amer. Math. Soc.*, 351(10):3935–3952, 1999.

[95] B. Riemann. *Ouevres Mathématiques de Riemann.* Gauthiers-Villars, Paris, 1898.

[96] A. Ros. Compactness of spaces properly embedded minimal surfaces with finite total curvature. *Indiana Univ. Math. J.*, 44(1):139–152, 1995.

[97] A. Ros. Embedded minimal surfaces: forces, topology and symmetries. *Calc. Var.*, 4:469–496, 1996.

[98] H. F. Scherk. Bemerkungen über die kleinste Fläche innerhalb gegebener Grenzen. *J. R. Angew. Math.*, 13:185–208, 1835.

[99] R. Schoen. *Estimates for Stable Minimal Surfaces in Three Dimensional Manifolds*, volume 103 of *Annals of Math. Studies.* Princeton University Press, 1983.

[100] R. Schoen. Uniqueness, symmetry, and embeddedness of minimal surfaces. *J. of Differential Geometry*, 18:791–809, 1983.

[101] L. Simon. Lectures on geometric measure theory. In *Proceedings of the Center for Mathematical Analysis*, volume 3, Canberra, Australia, 1983. Australian National University.

[102] J. Spruck. Two-dimensional minimal graphs over unbounded domains. Preprint.

[103] M. Traizet. The genus 2 helicoid. Personal Communication.

[104] M. Traizet. An embedded minimal surface with no symmetries. *J. Differential Geometry*, 60(1):103–153, 2002.

[105] M. Traizet and M. Weber. Hermite polynomials and helicoidal minimal surfaces. Preprint.

[106] M. Weber and M. Wolf. Teichmuller theory and handle addition for minimal surfaces. *Annals of Math*, 156(3):713–795, 2002.

[107] B. White. Complete surfaces of finite total curvature. *J. Diff. Geom.*, 26:315–326, 1987.

On global existence of wave maps with critical regularity

Andrea Nahmod

ABSTRACT. We survey recent work on Wave maps from Minkowski space $\mathbb{R}^{1+n}$ into (compact) Riemannian manifolds. We focus on the results obtained and some of the methods from harmonic analysis and gauge theory used.

0. Introduction

This article arose from a lecture given by the author at the JDG Conference held at Harvard University, Cambridge Massachusetts in May 2002 and celebrating Karen Uhlenbeck's sixtieth birthday. The article is expository in nature and surveys recent joint work with A. Stefanov and K.Uhlenbeck on wave maps from Minkowski space $\mathbb{R}^{1+n}, n \geq 4$ into compact Riemannian manifolds. We mainly focus on the results obtained and some of the methods from harmonic analysis and gauge theory used. The results and techniques presented actually work on any constant curvature complete Riemannian manifold (e.g. Lie groups and their symmetric Riemannian spaces). It is also probable that they can be further extended to bounded geometry complete Riemannian manifolds, for example but we do not pursue the latter extension here. The paper is organized as follows. We first describe the wave map problem and some of the literature. We then present some of the relevant tools from harmonic analysis and their use in the study of a related nonlinear wave equation. Finally we briefly comment on the difficulties that arise when passing from 4 to 3 dimensions. The ideas and results on wave maps presented here by the author are in collaboration with A. Stefanov and K. Uhlenbeck.

Let $\mathbb{R}\times\mathbb{R}^n = \mathbb{R}^{1+n}$ be Minkowski space with flat metric $\eta = (\eta_{\alpha\beta}) = \mathrm{diag}(-1, 1, \ldots, 1)$ Then $\eta^{\alpha\beta} = (\eta_{\alpha\beta})^{-1} = \eta_{\alpha\beta}$ in our case . By $\partial_\alpha, \partial^\alpha \quad \alpha = 0, 1, \ldots, n$ we denote the usual derivatives with respect to the Minkowski metric. Then $\partial^\alpha = \eta^{\alpha\beta}\partial_\beta$, $\Box := \partial^\alpha\partial_\alpha = \partial_t^2 - \Delta$ is the D'Lambert operator; and

$$\frac{1}{2}\langle\partial^\alpha u, \partial_\alpha u\rangle = \frac{1}{2}(|u_t|^2 - |\nabla u|^2)$$

is the 'Lagrangian density' of u.

Wave maps are the natural Minkowskian analogue of harmonic maps.

1991 *Mathematics Subject Classification.* Primary 35J10, Secondary 45B15, 42B35 .

The author was partially supported by NSF grant DMS 0202139.

For a map $\Phi : \mathbb{R}\times\mathbb{R}^n \to M$, where $(M, \mathfrak{g})$ is -for example- a compact Riemannian manifold, the critical points of the Euler-Lagrangian functional

$$\Phi \to \int_{\mathbb{R}^{n+1}} \langle \partial_\alpha \Phi, \partial^\alpha \Phi \rangle_{\mathfrak{g}}\, d\sigma = \int\int |\frac{\partial \Phi}{\partial t}|^2_{\mathfrak{g}} - |\nabla_x \Phi|^2_{\mathfrak{g}} dx dt$$

are the solutions to the *Wave Map equation.* This equation can be regarded as given through covariant derivatives and in coordinate free notation is given by:

$$\Phi^*\nabla_0 \frac{\partial \Phi}{\partial t} - \sum_{j=1}^{n} \Phi^*\nabla_j \frac{\partial \Phi}{\partial x^j} = 0. \tag{1}$$

where $\Phi^*\nabla$ are the covariant derivatives (corresponding to the pullback of the Levi-Civita connection on M via the map Φ). This is a nonlinear wave equation with a non-polynomial nonlinearity including derivatives.

For example, when $M = \mathbb{S}^{m-1} \hookrightarrow \mathbb{R}^m$, equation (1) looks like

$$\Box\Phi = -\frac{\partial^2 \Phi}{\partial t^2} + \Delta\Phi = -\Phi\left(|\nabla_x \Phi|^2 - |\frac{\partial \Phi}{\partial t}|^2\right).$$

And in general a (classical) wave map is a (smooth) map $\Phi : \mathbb{R}\times\mathbb{R}^n \to M$ satisfying,

$$\Box\Phi = -\Phi(\partial_\alpha\Phi)^T \partial^\alpha \Phi \tag{WM}$$

(with, say, Φ constant outside the finite union of light cones). One key feature of wave maps is that of *energy conservation*:

$$E(\Phi(t)) := \frac{1}{2}||D\Phi(t)||^2_{L^2(\mathbb{R}^n)} = \frac{1}{2}||D\Phi(0)||^2_{L^2(\mathbb{R}^n)}.$$

The study of well-posedness of the Cauchy problem with initial data in the Sobolev spaces $H^s \times H^{s-1}(\mathbb{R}^n; TM)$, $s \geq 1$ seeks answers to the following questions.

Local in time Existence and Uniqueness: for what values of s does the initial value problem admits a unique local solution ?

Local well posedness: in addition to the above, does the solution depend continuously on the initial data ?

Global well-posedness: for what values of s does this solution extend for all time ?

Global regularity: does the solution corresponding to smooth initial data stays smooth for all times ?

Wave maps have a natural scale invariance of the form

$$\Phi(x,t) \to \Phi(\frac{x}{\lambda}, \frac{t}{\lambda}).$$

As a consequence, the Sobolev norms $||\Phi||_{\dot{H}^s}$ become dimensionless when $s = n/2$.

This number is referred to as the *critical -regularity- exponent* or alternatively $||\cdot||_{\dot{H}^{n/2}}$ as the *critical* or *scale invariant* norm. One expects well posedness for the Cauchy problem with data in a Sobolev space with exponent above the critical one (i.e. subcritical case).

In short, wave maps have a natural scaling from where the *critical index* for the Cauchy problem with data in the Sobolev spaces is $s_c = \frac{n}{2}$.

Classical energy estimates for equations of the form $\Box u = F(u, \partial u)$ - hence for wave maps- imply local well-posedness of the Cauchy problem in H^s, $s > \frac{n}{2}+1$. The special structure associated to the nonlinearity in the wave map system however, allows for improvement.

0.1. Some background.

• When $n = 1$ global existence and regularity of wave maps with smooth data into complete Riemannian manifolds was established by Gu Chao-Hao [**12**] and Ginibre-Velo [**10**]. The idea was to use *characteristic coordinates* in $\mathbb{R}^{1+1}$,

$$\eta = t + r, \qquad \xi = t - r$$

to rewrite the wave map system in the form

$$-u_{\eta\,\xi} = A(u)(\partial_\eta u, \partial_\xi u).$$

The latter form allowed them to use L^∞ estimates to obtain g.w.p. of finite energy solutions.

• Keel and Tao [**16**] studied the one dimensional Cauchy problem with data in $H^s \times H^{s-1}$ further and in particular established local well-posedness. for $s > 1/2$ and global well-posedness when the target is a sphere and $s > 3/4$.

• In higher dimensions there are several special existence type results; such as the global well-posedness for smooth Cauchy data close to a geodesic and more. (c.f. [**4**] [**5**] [**31**])

• When $n \geq 3$ Shatah [**27**]; Cazenave, Shatah, Tahvildar-Zadeh [**3**] showed that solutions to the Cauchy problem for wave maps may blow up in finite time. Singularities can form from large data even when data is smooth and rotationally symmetric. Targets could be quite general as well (e.g. in $n \geq 4$ could be convex manifolds).

• When $n = 2$ (energy critical case) the first results were for equivariant maps. Equivariant maps give rise to semilinear wave equations in $\mathbb{R}^{n+3}$ spatial dimensions with critical growth. The structure of the nonlinearity is determined by the geometry of the target manifold.

Christodoulou and Shatah [**6**] obtained regularity of equivariant maps to hyperbolic space. Christodoulou and Tahvildar-Zadeh [**7**] studied the regularity of spherically symmetric wave maps assuming a convexity condition in the target manifold. Shatah and Tahvildar-Zadeh [**30**] studied the regularity of equivariant wave maps into two-dimensional rotationally symmetric and geodesically convex Riemannian manifolds. More work in this area has been done by M. Grillakis [**11**] and more recently by M. Struwe [**34**] who relaxed the convexity assumption.

• When $n \geq 2$ Klainerman and Machedon [**17**] [**18**] and Klainerman and Selberg [**22**] obtained the 'almost optimal' local well posedness results for the Cauchy problem for wave maps. That is, for data $(\Phi, \partial_t \Phi) \restriction_{t=0} = (f, g) \in H^s \times H^{s-1}$ with $s > n/2$, they showed that the Wave Map system is locally well posed. The so called '*null form* structure of the nonlinearity played an important role in their work [**19**][**20**].

0.2. Two pivotal breakthroughs at critical level. The first one came from Tataru when $n \geq 5$ [**38**] and for $n = 2, 3\ldots$ afterward [**39**]. His results established that for $n \geq 2$ and Cauchy data $(\Phi, \partial_t \Phi)\restriction_{t=0} = (f, g) \in \dot{B}^{n/2}_{2,1} \times \dot{B}^{n/2-1}_{2,1}$ *sufficiently small* there is global well posedness, regularity and scattering for wave maps.

An important reason for studying well posedness in Besov spaces instead was that $\dot{B}^{n/2}_{2,1} \hookrightarrow L^\infty$ while $\dot{H}^{n/2}$ does not embed into L^∞. The latter is a big problem; for starters one cannot localize to small coordinate patches, and loses worse as well.

The second one came shortly afterward by T. Tao , once again first for $n \geq 5$ [**35**] and later for $n = 2, 3, \ldots$ [**36**]. Tao studied the *global regularity for small data* problem, asking whether solutions to (WM) corresponding to smooth initial data, *small* in $\| \cdot \|_{s_c}$ stay smooth for all time. He proved the following.

For $G = \mathbb{S}^m$, $n \geq 2$ and Cauchy data $(\Phi, \partial_t \Phi)\restriction_{t=0} = (f, g) \in H^s \times H^{s-1}$ with $s > n/2$ and critical norm $\dot{H}^{n/2} \times \dot{H}^{n/2-1}$ *sufficiently small*, WM have global regularity. Furthermore, for s close to $n/2$ have global bounds

$$\| (\Phi, \partial_t \Phi) \|_{L_t^\infty (H_x^s \times H_x^{s-1})} \lesssim \|(f, g)\|_{H_x^s \times H_x^{s-1}}.$$

0.3. Recent developments. For spatial dimensions $n \geq 5$, similar results to those of Tao were obtained by Klainerman and Rodnianski [**21**] for target manifolds admitting a parallelizable structure (e.g. general Lie groups) . Roughly at the same time and independently, Shatah and Struwe [**28**] on the one hand and on the other Nahmod, Stefanov and Uhlenbeck [**25**] established the following result.

Let M be a compact Lie group or Riemannian symmetric space (e.g. $\mathbb{S}^m$)
Main Theorem *Let $n \geq 4$, $\Phi : \mathbb{R} \times \mathbb{R}^n \to M$. Suppose the Cauchy data $(\Phi, \partial_t \Phi)\restriction_{t=0} = (f, g)$ has sufficiently small norm in $\dot{H}^{n/2} \times \dot{H}^{n/2-1}$. Then there exists a unique global solution to the WM problem such*

$$\|((\Phi, \partial_t \Phi)\|_{L_t^\infty (\dot{H}_x^{n/2} \times \dot{H}_x^{n/2-1})} \lesssim \|(f, g)\|_{\dot{H}_x^{n/2} \times \dot{H}_x^{n/2-1}}.$$

Moreover, there is global regularity; i.e. if in addition $(f, g) \in H^s \times H^{s-1}$ with $s > n/2$ then the solution $(\Phi, \partial_t \Phi)$ belongs to $H^s \times H^{s-1}$ for all time and satisfies global bounds

$$\| (\Phi, \partial_t \Phi) \|_{L_t^\infty (H_x^s \times H_x^{s-1})} \lesssim \|(f, g)\|_{H_x^s \times H_x^{s-1}}.$$

Remarks

• Shatah and Struwe's result is more general. In their case the target is any complete Riemannian manifold with bounded geometry.

The restriction in [**25**] to compact manifolds stemmed from their use of a Nash embedding into an Euclidean space with bounded geometry.

The authors of [**25**] learned later from Shatah and Struwe's work that Matthias Günther ([**13**] and references therein) extended Nash's embedding theorem to hold for any complete Riemannian manifold with bounded geometry (bounded second fundamental form) -not necessarily compact-; showing that it too is isometrically embedded into an Euclidean space with bounded geometry. Using this embedding, the results in [**25**] would extend to any complete Riemannian manifold with constant curvature.

• On the other hand Fabrice Planchon has pointed out that certain multiplications theorems for Besov spaces and L^∞ are sufficient to include variable curvature in our proof in [**25**]. Thus *a posteriori* our results seem might also hold in the case of variable bounded curvature $R(\Phi)(x)$ as well.

• The method in [**25**] combines both delicate techniques from harmonic analysis with fairly standard global gauge theoretic geometric methods. Both [**25**] and [**28**] works use the same gauge change : the Hodge or Coulomb gauge. The analytic approach is significantly different as Shatah-Struwe base their results on Lorentz spaces and we use Besov spaces. Besov spaces are contained in Lorentz spaces -for appropriate indexes-. Lorentz spaces seem better behaved under coordinate transformations.

• One indeed has well posedness for the *gauged map* but there are no estimates available on *differences* for the *original wave map* itself at the critical level. One cannot obtain any continuous dependence of the map on the data in the coordinate setting. The problem stems in that *well posedness* is not a gauge invariant notion; it is not even necessarily true that uniqueness in one coordinate system implies uniqueness in another directly. Hence, in none of the works above is possible to obtain (strong) well posedness at the critical level for the wave map itself.

We now proceed to explain the ideas behind [**21**].

1. Wave Maps for $n \geq 4$

1.1. Beginning of the Proof. Regard the wave map equation as an equation given through covariant derivatives.

$$\Phi : \mathbb{R} \times \mathbb{R}^n \to M \qquad d\Phi : \mathcal{T}(\mathbb{R} \times \mathbb{R}^n) \to \mathcal{T}M$$

where M is an arbitrary Riemannian manifold and $\mathcal{T}(\mathbb{R}\times\mathbb{R}^n) = (\mathbb{R}\times\mathbb{R}^n)\times(\mathbb{R}\oplus\mathbb{R}^n)$. If we let $\Phi^*\nabla$ be the pullback of the Levi-Civita connection on M to $\Phi^*\mathcal{T}M$ via the map Φ then in coordinate free notation, where we have set $t = x^0$.

$$\Phi^*\nabla_0 \frac{\partial \Phi}{\partial t} - \sum_{j=1}^{n} \Phi^*\nabla_j \frac{\partial \Phi}{\partial x^j} = 0 \tag{1}$$

• The Levi-Civita connection on M is torsion free; i.e. if we set $t = x^0$,

$$\Phi^*\nabla_j \frac{\partial \Phi}{\partial x^k} = \Phi^*\nabla_k \frac{\partial \Phi}{\partial x^j} \qquad \text{for } j = 0, 1, \ldots, n, \; k = 1, \ldots, n. \tag{2}$$

We also have control on the curvature of $\Phi^*\nabla$ via the equation

$$[\Phi^*\nabla_j, \Phi^*\nabla_k] = R(\Phi)(\frac{\partial \Phi}{\partial x^j}, \frac{\partial \Phi}{\partial x^k}) \tag{3}$$

The *wave map system* (1)-(3) is overdetermined. We assume the map $\Phi : \mathbb{R}^n \cup \{\infty\} \to M$ is topologically trivial. Hence, $\Phi^*\mathcal{T}M$ is the trivial bundle $(\mathbb{R}^1 \times \mathbb{R}^n) \times \mathbb{R}^m$. By our choice of target M , we have $R(\Phi) \equiv R$ constant.

Next, under smallness assumptions on $\Phi \in L^\infty_t \dot{W}^{1,n/2}_x$, we obtain a unique choice of coordinates for $\Phi^*\mathcal{T}M$. This follows from the existence and uniqueness of a *gauge change* g under suitable hypothesis on the space-time curvature $F_A = dA + [A, A]$ and connection A. The following result is proved following the methods used by K. Uhlenbeck in [**40**].

Theorem 1 (Existence of a good gauge) *Let $d + A$ be a smooth connection with compact structure groups G over $\mathbb{R}\times\mathbb{R}^n$ or $I\times\mathbb{R}^n$. Assume $A \sim 0$ at spatial infinity and let $F_A = dA + [A, A]$ be the space-time curvature. Then there exists a positive constant $\epsilon = \epsilon(n, G)$ such that if the mixed space-time Lebesgue norm*

$$\|F_A\|_{L^\infty_t L^{n/2}_x} < \epsilon,$$

then, there exits a unique smooth gauge change g, $g \sim I$ at spatial infinity, such that if $\tilde{A} = gAg^{-1} - dg\, g^{-1}$ we have,

1. $\|\tilde{A}\|_{L_t^\infty \dot{W}_x^{1,n/2}} \leq c(n,G)\|F_A\|_{L_t^\infty L_x{}^{n/2}}$

2. $\sum_{j=1}^n \frac{\partial}{\partial x^j} \tilde{A}_j = 0$

Corollary *The above remains true if $A \in L_t^\infty \dot{W}_x^{1,n/2}$ and $F_A \in L_t^\infty L_x^{n/2}$.*

We now apply this gauge change in the wave map system (1)-(3). If we denote by $b = d\Phi$ schematically we get :

$$b_j \to g b_j g^{-1} := \tilde{b}_j$$

$$g s^* \nabla_j g^{-1} \to \frac{\partial}{\partial x_j} + \frac{1}{2} g b_j g^{-1} - dg g^{-1} := \frac{\partial}{\partial x_j} + a_j := D_j$$

In the new gauge, we have that $\operatorname{div} a = 0$ and the *same* equations (1)-(3) albeit with D_j replacing $\Phi^* \nabla_j$ and $\tilde{b}_j$, b_j.

More precisely, let $b = d\Phi$. Then the equations themselves are written

$$D_0 \tilde{b}_0 - \sum_{j=1}^n D_j \tilde{b}_j = 0.$$

$$D_k \tilde{b}_j = D_j \tilde{b}_k\,, \quad k = 0, 1, \ldots, n\,,\ j = 1, 2, \ldots, n.$$

This is a non-linear first order hyperbolic system. Moreover we also have

$$da + [a,a] + R[\tilde{b}, \tilde{b}] \;=\; 0.$$

All in all we have

$$\begin{array}{ll} \text{(a)} & \frac{\partial}{\partial t}\tilde{b}_0 - \sum \frac{\partial}{\partial x_j}\tilde{b}_j + (a \cdot \tilde{b})_{\text{space-time}} = 0 \\ \text{(b)} & d\tilde{b} + a \wedge \tilde{b} = 0 \\ \text{(c)} & da + [a,a] = R[\tilde{b}, \tilde{b}] \\ \text{(d)} & \sum_{j=1}^n \frac{\partial}{\partial x^j} a_j = 0 \end{array}$$

We refer to this as the *Gauged Wave Map* system.

This system however, is still not as nice to work with because of the presence of the $\frac{\partial a}{\partial t}$ terms. So we go one step further and convert it to a single equation using Hodge theory. From now on we abuse notation and call $\tilde{b}$ just b and let $b = d\phi + d^*\psi$. Then we can show **[25]** that the gauged wave map system can be rewritten as

$$\begin{array}{ll} \text{(a)} & \Box\phi + (a,b) = 0 \\ \text{(b)} & \Box\psi + a \wedge b = 0 \\ \text{(c)} & b = d\phi + d^*\psi \\ \text{(d)} & da + [a,a] = R[b,b] \\ \text{(e)} & \sum_{j=1}^n \frac{\partial}{\partial x^j} a_j = 0. \end{array}$$

The initial data on ϕ and ψ can be taken to be

$$\begin{array}{ll} \phi(0,x) = 0\,, & \psi(0,x) = 0 \\ \frac{\partial\phi}{\partial t}(0,x) = b_0(0,x) & \frac{\partial \psi}{\partial t}_{0,j}(0,x) = b_j(0,x) \\ \frac{\partial}{\partial t}\psi_{j,k}(0,x) = 0 & j,\, k \neq 0 \end{array}$$

Since the gauged wave map system is still overdetermined we consider a subset of it. Incorporating (e) into (d) we obtain the following.

Theorem 2 *Under our assumptions on the target manifold M, a* subset *of the* gauged wave map equations *(a)- -(e) has a structure of a non-linear wave system of integral differential operators. Namely,*

$$(a) \qquad \Box\phi + (a, b) = 0$$

$$(b) \qquad \Box\psi + a \wedge b = 0$$

$$(c) \qquad b = d\phi + d^*\psi$$

$$(d) \qquad \Delta a_j + \sum_{j=1}^{n} \frac{\partial}{\partial x^k}[a_k, a_j] + \frac{\partial}{\partial x^k}[b_k, b_j] = 0,\ j = 0, 1, \ldots n$$

We refer to this system as the Modified Wave Map system - or simply MWM .

The existence and uniqueness of *wave maps* will follow from the next Theorems 3-5. In section 1.2 we explain how the global well posedness and higher regularity of the MWM, together with a stronger uniqueness result come together to give the *Main Theorem* above.

Theorem 3 (Well posedness of the MWM) *There exists $\varepsilon > 0$ such that whenever the initial data*

$$\|(f, g)\|_{\dot{H}^{n/2} \times \dot{H}^{n/2-1}} < \varepsilon,$$

the system above has a unique global solution $v = (\phi, \psi)$ which belongs both to

- $L^\infty(\mathbb{R}; \dot{H}_x^{n/2}) \cap L^2(\mathbb{R}; \dot{B}^1_{2n,2})$ *and*
- $W^{1,\infty}(\mathbb{R}; \dot{H}_x^{n/2-1}) \cap W^{1,2}(\mathbb{R}; \dot{B}^0_{2n,2})$.

Moreover, there is stability; i.e.

$$esssup_t \|v_1 - v_2\|_{\dot{H}^{n/2}} \lesssim \|(f_1, g_1) - (f_2, g_2)\|_{\dot{H}^{n/2} \times \dot{H}^{n/2-1}}$$

provided the r.h.s. is small enough.

Remark The theorem above gives the *existence* part of our Main Theorem on *wave maps.* The uniqueness of solutions to the MWM however is *solely* in the Besov spaces which is **not enough** to claim the solution to the MWM system came from a wave map. Thus an additional argument is needed. The following is a stronger uniqueness result which will indeed suffice to return to the wave map (ie. will give uniqueness of the *original* wave map).

Theorem 4 (Uniqueness) *Suppose (v_1, a_1) and (v_2, a_2) are two solutions to*

$$\Box v + B(a, dv) = 0$$

$$\Delta a + divB(a, a) + divB(dv, dv) = 0.$$

such that $dv_j = b_j$, for $j = 1, 2$ are small in $L_t^\infty L_x^n$. Suppose that $dv_j = b_j \in L_t^2 L_x^{2n}$ for $j = 1, 2$. Assume in addition that $a_1 = a_1(v_1) \in L_t^1 L_x^\infty$. Then $v_1 = v_2$.

The smallness of dv_j in $L_t^\infty L_x^n$ is the necessary condition to solve the 'gauged' equation. The proof follows a scheme devised by Shatah-Struwe to establish uniqueness. Essentially follows via *energy estimates* under minimal assumptions on the solution. If $w = v_1 - v_2$, then $\exists\, h \in L_t^1(\mathbb{R})$ such that

$$\frac{1}{2}\partial_t \|Dw(t)\|_{L_x^2}^2 = \int_{\mathbb{R}^n} \langle \Box w, w_t \rangle dx \lesssim h(t) \|Dw(t)\|_{L_x^2}^2.$$

Gronwall's Lemma then implies $w \equiv 0$.

Finally, by differentiating the MWM system and observing that the resulting nonlinearity has the same bilinear structure -for which the necessary 'multiplication estimates' hold- the following regularity result follows.

Theorem 5 (Higher Regularity)*Suppose the initial data (f, g) to (MWM) is in $H^{n/2+1} \times H^{n/2}$ and has sufficiently small $\dot{H}^{n/2} \times \dot{H}^{n/2-1}$ norm. Then the solution v to the Cauchy problem (MWM) with initial data (f, g) can be continued in $H^{n/2+1} \times H^{n/2}$ globally in time. Furthermore, we have the global bounds*

$$\|v\|_{L_t^\infty(\mathbb{R};\dot{H}_x^{n/2+1})} \lesssim \|(f,g)\|_{\dot{H}_x^{n/2+1} \times \dot{H}_x^{n/2}}.$$

1.2. The Return to the Map. The well-posedness results on the modified wave map apply to a larger class of formal solutions (a, b) to the equation than those which come from wave maps. Our method of using the results on the modified wave map equation to show existence of wave maps is similar to the idea we used for non-linear Schrödinger and not very different from the technique used by Shatah-Struwe Roughly: regularize the data to the WM system. Then it has a *local smooth* solution. If *in addition* the $\dot{H}^{n/2} \times \dot{H}^{n/2-1}$ norm of the data is *sufficiently small* the local smooth solution is *global and smooth* and satisfies the a priori global estimates satisfied by the solution to the MWM system. This a priori estimate is now used to pass to the limit. The translation depends on the compactness of M (or certain bounds on the isometric Nash embedding of a non-compact M in an Euclidean space).

2. Basic Littlewood-Paley theory

We now introduce and develop the techniques and tools from harmonic analysis that enter in the proof of the well posedness result of the MWM system in $n \geq 4$.

2.1. Background of the classical theory.

Let $f(x)$ be a function on $\mathbb{R}^n$ and $\hat{f}(\xi)$ its Fourier transform. Consider $m(\xi)$ to be a non-negative radial bump function supported on the ball $|\xi| \leq 2$ and equal to 1 on the ball $|\xi| \leq 1$. Then for each integer k let $P_k(f)$ be the *Littlewood-Paley* projection operator onto *frequencies* $|\xi| \lesssim 2^k$. This is defined by

$$\widehat{P_k(f)}(\xi) := m(2^{-k}\xi)\hat{f}(\xi).$$

• $P_k \to 0$ as $k \to -\infty$ and $P_k \to I$ as $k \to \infty$ in any reasonable sense (e.g. L^2). The function $P_k(f)(x)$ is a (smoothed) average of f localized to *physical scales* $\lesssim 2^{-k}$. By the uncertainty principle one expects $P_k(f)$ to be essentially constant at scales much smaller than 2^{-k}. The operator Q_k is the projection onto the *frequency annulus* $|\xi| \sim 2^k$ given by the formula,

$$Q_k := P_k - P_{k-1}.$$

Hence $\psi(\xi) := m(\xi) - m(2\xi)$ is supported on the annulus $1/2 \leq |\xi| \leq 2$, for all $\xi \neq 0$, $\sum_{k\in\mathbb{Z}} \psi(2^{-k}\xi) \equiv 1$, and

$$\widehat{Q_k(f)}(\xi) = \psi(2^{-k}\xi)\hat{f}(\xi).$$

The Littlewood-Paley projections are bounded operators in all the Lebesgue spaces. In fact, Q_k is given by a convolution kernel whose L^p-norm equals $2^{(kn)(1-1/p)}$,

$1 \leq p \leq \infty$. In particular its L^1- norm is identically 1 for all $k \in \mathbb{Z}$. It is essentially constant on physical scales $<< 2^{-k}$ and it has mean zero at scales $\lesssim 2^{-(k-10)}$. In fact, on a ball in physical space of radius $O(2^{-k})$ the function $Q_k(f)$ is *smooth* at physical scales $<< 2^{-k}$ and all moments of $Q_k(f)$ vanish: $\partial_\xi^\alpha(\widehat{Q_k(f)})(0) = 0$.

By telescoping the series we have the *Littlewood-Paley decomposition*

$$f := \sum_{k\in\mathbb{Z}} Q_k(f) \quad \text{in the sense of } L^2$$

or for any locally integrable function with decay at infinity. We have thus written f as a superposition of functions $Q_k(f)$, each of which has frequency of magnitude $\sim 2^k$. *Lower values* of k represent *low frequency* components of f. *Higher values* represent *high frequency* components.

The Haar system on $\mathbb{R}$ given by $h_I(x) = 2^{k/2}h(2^k - m)$ as $I = [2^{-k}m, 2^{-k}(m+1))$, $k, m \in \mathbb{Z}$ and where $h(x) = 1$ for $0 \leq x < 1/2$; $h(x) = -1$ for $1/2 \leq x < 1$ and $h(x) = 0$ otherwise is a good 'model' to bear in mind. This is the Walsh analogue of the Littlewood-Paley decomposition.

To relate the Littlewood-Paley pieces $Q_k(f)$ back to the function f itself, suppose $f \in L^2$. By construction and Plancherel,

$$||f||_2 \sim \Big(\sum_k ||Q_k(f)||_2^2\Big)^{1/2} \sim ||\Big(\sum_k |Q_k(f)(\cdot)|^2\Big)^{1/2}||_2$$

The *function* $S(f)(x) := \left(\sum_k |Q_k(f)(x)|^2\right)^{1/2}$ is known as the Littlewood-Paley *Square Function.* In general, for any $1 < p < \infty$, on has the *Littlewood-Paley Inequality:*

$$||Sf||_p \sim ||f||_p$$

The proof relies on standard harmonic analysis and follows in a straightforward fashion from Calderón-Zygmund theory.

We thus have a nice characterization of the Lebesgue spaces in terms of very friendly building blocks. From the PDE viewpoint one of the advantages of using Littlewood-Paley theory lies in the simple equivalence,

$$||\nabla Q_k(f)||_p \sim 2^k||Q_k(f)||_p\,, \qquad 1 \leq p \leq \infty$$

Roughly, ∇ is multiplication by $2\pi i\xi$ and $|\xi| \sim 2^k$ on the support of $Q_k(f)$. So -morally- one can decompose a derivative as a linear combination of its LP pieces, $2^k\,Q_k(f)$. We see the the effect of a derivative on a function f is to *accentuate* the *high frequencies* and *diminish* the *low frequencies.* A similar principle applies, of course, to other differentiation or pseudo-differential operators such as $(-\Delta)^{s/2}$.

Thus Littlewood-Paley is nicely adapted to dealing with spaces which combine L^p-type norms with derivatives: *Sobolev spaces, Besov spaces, Hölder spaces, etc.* For example, the Sobolev spaces $W^{s,p}$, consisting of those functions f (distributions) such that f and the first s derivatives of f are in L^p can be very simply characterized using Littlewood-Paley decompositions. Indeed,

$$||f||_{W^{s,p}} \sim ||f||_p + |||\nabla|^s f||_p \quad \text{where} \quad \mathcal{F}(|\nabla|^s f)(\xi) =: (2\pi|\xi|)^s \hat{f}(\xi).$$

Then,

$$||f||_{W^{s,p}} \sim ||f||_p + ||\Big(\sum_k |2^{ks}Q_k(f)|^2\Big)^{1/2}||_p$$

Once again we note that $|\nabla|^s$ accentuates the kth frequency piece of f by a factor of 2^{ks}; ie. $|\nabla|^s Q_k(f) \sim 2^{ks}Q_k(f)$.

2.2. Products and Product Estimates.

Let f and g be two nice functions. By splitting them using Littlewood-Paley decompositions we analyze *bilinear* expressions such as the pointwise product $B(f,g)(x) = f(x)g(x)$. Since $\widehat{(f\,g)}(\xi) = \int \widehat{f}(\xi-\eta)\widehat{g}(\eta)\,d\eta$ we have that

$$\text{supp}\ \widehat{(f\,g)} \subseteq \text{ supp}\ \widehat{f} + \text{ supp}\ \widehat{g}$$

We write,

$$\begin{aligned} fg &= \textstyle\sum_{k,j} Q_k(f)Q_j(g) = \sum_{k\geq j} Q_k(f)Q_j(g) + \sum_{k<j} Q_k(f)Q_j(g) \\ &= \textstyle\sum_{k\in\mathbb{Z},m\geq 0} Q_k(f)Q_{k-m}(g) + \sum_{j\in\mathbb{Z},m>0} Q_{j-m}(f)Q_j(g) \end{aligned}$$

Further splitting gives,

$$\begin{aligned} fg &= \textstyle\sum_l Q_l(fg) \\ &= \textstyle\sum_{l,k}\sum_{m\geq 0} Q_l(Q_k(f)Q_{k-m}(g)) + \sum_{l,j}\sum_{m>0} Q_l(Q_{j-m}(f)Q_j(g)). \end{aligned}$$

By inspecting the Fourier supports we find that :

$$(1) \qquad \text{supp}\,(\widehat{Q_k(f)Q_{k-m}(g)}) \subseteq \{|\xi| \lesssim 2^k\}.$$

Hence

$$Q_l(Q_k(f)Q_{k-m}(g)) = 0 \text{ unless } k \geq l$$

$$(2) \quad \text{supp}\,(\widehat{Q_k(f)Q_{k-m}(g)}) \cap \{|\xi| << 2^{k-m}\} = \emptyset \text{ if } m > 5.$$

Hence,

$$\begin{aligned} Q_l(Q_k(f)Q_{k-m}(g)) \quad &= 0 \quad \text{unless :} \\ &l = k \text{ and } m > 5 \qquad \text{or} \\ &l < k \text{ and } 0 \leq m \leq 5 \end{aligned}$$

All in all we only have three types of sums :

$$\begin{aligned} fg &= \textstyle\sum_l \sum_{m\geq 5} Q_l(Q_l(f)Q_{l-m}(g)) + \sum_{m=0}^5 \sum_l \sum_{k>l} Q_l(Q_k(f)Q_{k-m}(g)) \\ &+ \textstyle\sum_l \sum_{m\geq 5} Q_l(Q_{l-m}(f)Q_l(g)) + \sum_{m=0}^5 \sum_l \sum_{j>l} Q_l(Q_{j-m}(f)Q_j(g)) \end{aligned}$$

This could be re-written as well as :

$$\begin{gathered} fg = \textstyle\sum_\ell P_{\ell+1}(f)P_{\ell+1}(g) - P_\ell(f)P_\ell(g) \\ = \textstyle\sum_\ell Q_\ell(f)P_\ell(g) + \sum_\ell P_\ell(f)Q_\ell(g) + \sum_\ell Q_\ell(f)Q_\ell(g) \\ \text{Paraproduct} \ + \ \text{Paraproduct} \ + \ \text{Diagonal} \end{gathered}$$

In various applications the *high-low* interactions are 'easily dealt with' ($\sim$ paraproducts). It is the *high-high* interactions what usually may account for energy cascade effects; they are subtler to analyze.

We finish this brief introduction to Littlewood-Paley theory with two well known results that can be derived as applications of the above.

The Div-Curl Lemma

Convergence of approximating solutions for partial differential equations is not clear when weak continuity is not available. The notion of compensated compactness was developed to overcome these difficulties for non-linear equations in elasticity and fluid flow by exploiting cancellation properties of certain nonlinear quantities, usually bilinear, which arise naturally in studying the existence of global solutions. In this context, the decomposition above gives a very simple proof of the div-curl lemma used in compensated compactness.

If f, g are L^2 functions fg is only in L^1 which is not enough for weak continuity arguments since passing to weak limits is a discontinuous operation. If one considers $\varphi \in C_0^\infty$, then $\varphi e^{in\cdot} = f_n \rightharpoonup 0$ and $\varphi e^{-in\cdot} = g_n \rightharpoonup 0$. But $f_n g_n = \varphi^2$ does not converge weakly to 0. The *renormalization of the product* however, given by $fg - \sum_\ell Q_\ell(f)Q_\ell(g))$ is smoother than fg , and has *more cancellation.* Thus it is in a much 'regular' space, namely in the Hardy space $\mathcal{H}^1 \subset L^1$ which allows for weak continuity arguments to ensure convergence of approximating solutions. The point is that when $\operatorname{div} f = 0$ and $\operatorname{curl} g = 0$ one can essentially replace fg by its renormalization; thus the div-curl lemma of Murat and Tartar follows ([**8**]).

The Leibniz Rule for Fractional Derivatives

Let $s \in \mathbb{R}$, $s > 0$. Then for any $1 < p, q, r < \infty$ such that $\frac{1}{r} = \frac{1}{p} + \frac{1}{q}$,

$$\||\nabla|^s(fg)\|_r \lesssim \||\nabla|^s f\|_p \|g\|_q + \|f\|_p \||\nabla|^s g\|_q$$

This estimate has played a fundamental role in many nonlinear estimates such as those arising in well-posedeness problems below the energy norm. It essentially follows from the work of R. Coifman and Y. Meyer [**9**] and it is closely related to the Kato-Ponce commutator estimate [**14**].

3. Function Spaces and Multiplication Estimates

3.1. Set up and Function Spaces. For the linear wave equation,

$$\Box u = 0 \qquad u(0,x) = f(x), \quad \partial_t u(0,x) = g(x)$$

in higher dimensions it is simple to find an explicit formula by solving the equation using the Fourier transform:

$$\phi(t,x) \sim \int e^{ix\cdot\xi}\big(\cos(t|\xi|)\hat{f}(\xi) + \frac{\sin(t|\xi|)}{|\xi|}\hat{g}(\xi)\big)d\xi$$

These formulas can be rewritten by setting $\phi = \phi^+ + \phi^-$ where

$$\phi(t,x)^\pm \sim \int e^{ix\cdot\xi} e^{\pm it|\xi|}\hat{f_\pm}(\xi)d\xi$$

where

$$\hat{f_\pm}(\xi) = \hat{f}(\xi) \pm i\frac{\hat{g}(\xi)}{|\xi|}.$$

In other words,

$$\phi^\pm = e^{\pm it|\nabla|}\frac{1}{2}(f \mp i|\nabla|^{-1}g)$$

where we have denoted by $|\nabla|$ the pseudodifferential operator whose symbol is $m(\xi) = |\xi|$

In other words this shows that the *space- time* Fourier transform of ϕ defined by

$$\tilde{\phi}(\tau,\xi) := \int_{\mathbb{R}}\int_{\mathbb{R}^n} e^{-i(t\tau + x\cdot\xi)}\phi(t,x)dxdt$$

is a *distribution* supported on the *light cone* $|\tau| = |\xi|$ in $\mathbb{R}\times\mathbb{R}^n$. More precisely,

$$\tilde{\phi}^{\pm}(\tau,\xi) \sim \delta(\tau \mp |\xi|)\hat{f}_{\pm}(\xi)$$

are supported on the positive cone $\tau = |\xi|$ and negative cone $\tau = -|\tau|$ respectively. These formulas say that ϕ may be viewed as the (adjoint to the) *restriction* of the Fourier transform to the light cone. One may then try to perform a similar analysis than the one carried out by E. Stein and P. Thomas to studied the restriction to the sphere of the Fourier transform. The cone however is a non-compact manifold with one degenerate direction along which the curvature vanishes. Still, there are $n-1$ non-zero principal curvatures at each point; and this is actually enough to obtain good decay estimates for $\|\phi(t,\cdot)\|_{L^\infty_x}$ (i.e. ' the dispersive inequality') by stationary phase methods. ([**33**] and references therein).

For the inhomogeneous problem for the wave equation with zero initial data,

$$\Box\psi = F, \quad \psi(0,x) = 0, \quad \partial_t\psi(0,x) = 0.$$

We have by the *Duhamel's principle* that

$$\psi(t,x) \sim \int_0^t \int_{\mathbb{R}^n} e^{ix\cdot\xi}\frac{\sin((t-s)|\xi|)}{|\xi|}\widehat{F}(s,\xi)d\xi ds$$

Definition *Let $1 \le q, r \le \infty$. The space-time mixed Lebesgue spaces $L^q_tL^r_x$ is the set of functions ϕ on $\mathbb{R}\times\mathbb{R}^n$ with*

$$\Big(\int_{\mathbb{R}}\Big(\int_{\mathbb{R}^n}|\phi(t,x)|^r dx\Big)^{q/r} dt\Big)^{1/q} := \|\phi\|_{L^q_tL^r_x} < \infty$$

Strichartz Estimates for the Wave Equation *Let $n \ge 2$ and let u be a solution to the problem*

$$\Box u(t,x) = F(t,x), \quad t > 0, \quad x \in \mathbb{R}^n$$

with initial data

$$u(0,x) = f(x), \qquad \partial_t u(0,x) = g(x)$$

Then we have the estimate

$$\|u\|_{L^q_tL^r_x} \lesssim \|f\|_{\dot{H}^s} + \|g\|_{\dot{H}^{s-1}} + \|F\|_{L^{\tilde{q}'}_tL^{\tilde{r}'}_x}$$

provided

(1) the norms are dimensionally balanced (ie. they scale properly)

$$\frac{1}{q} + \frac{n}{r} = \frac{n}{2} - s = \frac{1}{\tilde{q}'} + \frac{n}{\tilde{r}'} - 2$$

(2) $\frac{1}{\tilde{q}} + \frac{1}{\tilde{q}'} = 1$ *and* $\frac{1}{\tilde{r}} + \frac{1}{\tilde{r}'}$

(3) The pairs (q,r) and $(\tilde{q},\tilde{r})$ are wave <u>*admissible*</u> *pairs. In other words they both belong to the set*

$$\mathcal{A} := \{(q,r)\,:\, 2 \le q,\, r \le \infty,\, \frac{1}{q} + \frac{n-1}{2r} \le \frac{n-1}{4}\}\setminus\{(2,\infty) \text{ when } n = 3\}$$

For future reference, we say that $(q,r) \in \mathcal{A}$ is <u>*sharp admissible*</u> if $\frac{1}{q} + \frac{n-1}{2r} = \frac{n-1}{4}$.

Let $\phi(t,x)$ be now a function on $\mathbb{R} \times \mathbb{R}^n$ and $\hat{\phi}(t,\xi)$ be its *spatial* Fourier transform.

Just as before, for each integer k let $P_k(\phi)$ be the usual Littlewood-Paley projection operator onto spatial frequencies $|\xi| \lesssim 2^k$ and let the operator Q_k be the projection onto the *spatial* frequency annulus $|\xi| \sim 2^k$. We denote by $\phi_k(t,x) := Q_k(\phi)(t,x)$.

Introduce $\mathcal{S}_k$ the *localized Strichartz space* at frequency 2^k, as the set of functions ϕ_k whose space-time norm is given by:

$$||\phi_k||_{\mathcal{S}_k} := \sup_{(q,r)\in\mathcal{A}} 2^{k(\frac{1}{q}+\frac{n}{r})}(||\phi_k||_{L^q_t L^r_x} + 2^{-k}||\partial_t \phi_k||_{L^q_t L^r_x}).$$

where $\mathcal{A}$ is, as before, the set of wave admissible Strichartz exponents.

The space $\mathcal{S}$ is now defined as an l^2 based Besov space, i.e.

$$\mathcal{S} = \oplus_{l^2}\mathcal{S}_k = \{f : ||f||_{\mathcal{S}} = \left(\sum_k ||f||^2_{\mathcal{S}_k}\right)^{1/2}\} .$$

We also introduce the space $\mathcal{S}^{(-1)}$, which is defined so that

$$\phi \in \mathcal{S} \quad \text{if and only if} \quad \partial_x \phi \in \mathcal{S}^{(-1)}.$$

If we denote by $b = \partial_x \phi$ then,

$$||b_k||_{\mathcal{S}^{(-1)}_k} := \sup_{(q,r)\in\mathcal{A}} 2^{k(\frac{1}{q}+\frac{n}{r}-1)}(||b_k||_{L^q_t L^r_x} + 2^{-k}||\partial_t b_k||_{L^q_t L^r_x}).$$

- The spaces $\mathcal{S}$ are $\dot{H}^{n/2}$- normalized while the $\mathcal{S}^{(-1)}$ are $\dot{H}^{n/2-1}$ -normalized:

$$\begin{aligned}||\partial_x^{n/2} Q_k(f)||_{L^\infty_t L^2_x} &\sim 2^{n/2k}||Q_k(f)||_{L^\infty_t L^2_x} \\ &\leq ||Q_k(f)||_{\mathcal{S}_k}\end{aligned}$$

Hence by taking ℓ^2 norms both sides

$$\Big(\sum_{k\in\mathbb{Z}} 2^{2k(n/2)}||Q_k(f)||^2_{L^\infty_t L^2_x}\Big)^{1/2} \lesssim ||f||_{\mathcal{S}}.$$

In other words,

$$\mathcal{S} \hookrightarrow L^\infty_t \dot{H}^{n/2}_x.$$

Moreover for $n \geq 4$ one has for example the estimates:

$$\begin{aligned}||Q_k(\phi)||_{L^\infty_t L^2_x} &\leq 2^{-\frac{n}{2}k}\,||Q_k(\phi)||_{\mathcal{S}_k} \\ ||Q_k(\phi)||_{L^2_t L_x^{\frac{2(n-1)}{(n-3)}}} &\leq 2^{k(\frac{n}{(n-1)}-\frac{(n+1)}{2})}\,||Q_k(\phi)||_{\mathcal{S}_k} \\ ||Q_k(\phi)||_{L^2_t L^\infty_x} &\leq 2^{-\frac{k}{2}}\,||Q_k(\phi)||_{\mathcal{S}_k};\end{aligned}$$

from where we notice that to control high frequencies one should use $L_t^q L_x^r$ with small r; while large r is instrumental to control the low frequencies.

The Strichartz estimates in this context now read,

Strichartz Estimates Revisted (T. Tao; Keel-Tao) *Let $k \in \mathbb{Z}$ and let $Q_k(\phi)(t,x)$ be any function on $\mathbb{R} \times \mathbb{R}^n$ with spatial Fourier support on the annulus $|\xi| \sim 2^k$.*

(SE)

$$\|Q_k(\phi)\|_{S_k} \lesssim \|Q_k(\phi)(0,\cdot)\|_{\dot{H}_x^{n/2}} + \|\partial_t Q_k(\phi)(0,\cdot)\|_{\dot{H}_x^{n/2-1}} + 2^{k(\frac{n}{2}-1)}\|\Box Q_k(\phi)\|_{L_t^1 L_x^2}.$$

3.2. Multiplication Estimates. Denote by $\mathcal{C}$, $\mathcal{D}$, $\mathcal{E}$, $\mathcal{G}$ the following sets of pairs.

$$\mathcal{C} =: \{(q,p) \in \mathcal{A} : \frac{1}{q} + \frac{n}{p} \leq 1^-\}$$

$$\mathcal{D} =: \{(q,p) : \frac{1}{2q} + \frac{n-1}{4p} \leq (\frac{n-1}{4})^-\}$$

$$\mathcal{E} =: \{(q,p) : \frac{1}{q} = \frac{1}{q_1} + \frac{1}{q_2};\ \frac{1}{p} = \frac{1}{p_1} + \frac{1}{p_2} \quad \text{with } (q_1,p_1) \in \mathcal{A} \text{ and } (q_2,p_2) \in \mathcal{C}\}$$

where $\mathcal{A}$ is the set of all wave admissible pairs. Finally, let

$$\mathcal{A} \subset \mathcal{G} := \mathcal{D} \cap \mathcal{E}.$$

Definition *Let $\mathcal{S}_+^{(-1)}$ be the space of functions on $\mathbb{R} \times \mathbb{R}^n$ whose norm is given by*

$$\|\phi\|_{\mathcal{S}_+^{(-1)}} := \sum_{k \in \mathbb{Z}} \|Q_k(\phi)\|_{S_k{}_+^{(-1)}}$$

where

$$\|\Phi\|_{S_k{}_+^{(-1)}} := \sup_{(q,p)\in\mathcal{G}} 2^{k(\frac{1}{q}+\frac{n}{p}-1)}(\|\Phi\|_{L_t^q L_x^p} + 2^{-k}\|\partial_t \Phi\|_{L_t^q L_x^p})$$

Let us denote by $|\nabla|^{-1} := \nabla\Delta^{-1}$ the pseudo-differential operator of order -1; i.e.

$$\widehat{|\nabla|^{-1} f}(t,\xi) = \frac{1}{|\xi|}\hat{f}(t,\xi).$$

The principal technical result in our main theorem is the 'multiplication estimate' :

$$\||\nabla|^{-1}(f \cdot g)\|_{\mathcal{S}_+^{(-1)}} \lesssim \|f\|_{\mathcal{S}^{(-1)}} \|g\|_{\mathcal{S}^{(-1)}},$$

The space $\mathcal{S}_+^{(-1)}$ is a similar to $\mathcal{S}^{(-1)}$ but with a larger norm that controls a *larger collection* of space-time Lebesgue norms than those in $\mathcal{A}$. For example when $n \geq 4$ one has the embeddings,

$$\mathcal{S}_+^{(-1)} \hookrightarrow \mathcal{S}^{(-1)} \cap L_t^q \dot{B}_{p,2}^s \quad \text{where} \quad q,p \geq 2;\ s = \tfrac{1}{q} + \tfrac{n}{p} - 1$$

$$\mathcal{S}_+^{-1} \hookrightarrow \{f : \textstyle\sum_k \|Q_k(f)\|_{L_t^1 L_x^\infty} < \infty\},$$

which are *crucial in closing the estimates.* In fact in **[25]** we show the following.

Main Multiplication Estimate

$$|\nabla|^{-1} : \mathcal{S}^{(-1)} \times \mathcal{S}^{(-1)} \longrightarrow \mathcal{S}_{+}^{(-1)}$$

In other words, for $f,\, g \in \mathcal{S}^{(-1)}$ we have

$$\sum_{l\in\mathbb{Z}} \sup_{(\tilde{q},\tilde{p})\in\mathcal{G}} 2^{l(1/\tilde{q}+n/\tilde{p}-1)} \quad \|Q_l(|\nabla|^{-1}(f\cdot g))\|_{L_t^{\tilde{q}}L_x^{\tilde{p}}}$$
$$\lesssim \|f\|_{\mathcal{S}^{(-1)}}\|g\|_{\mathcal{S}^{(-1)}}$$

Perform Littlewood-Paley decomposition on f, g, and the product $f\cdot g$ as we discuss earlier. Denote $f_k := Q_k(f)$. By studying the supports of the L-P projections and symmetry considerations the result follows from the following two estimates.

• High-High into Low:

$$\sum_{l\in\mathbb{Z}} \sup_{(\tilde{q},\tilde{p})\in\mathcal{G}} 2^{-l}\, 2^{l(1/\tilde{q}+n/\tilde{p}-1)} \sum_{k>l} \quad \|Q_l((f_k\cdot g_k))\|_{L_t^{\tilde{q}}L_x^{\tilde{p}}}$$
$$\lesssim \|f\|_{\mathcal{S}^{(-1)}}\|g\|_{\mathcal{S}^{(-1)}}$$

• High-Low into High:

$$\sum_{l\in\mathbb{Z}} \sup_{(\tilde{q},\tilde{p})\in\mathcal{G}} 2^{-l}\, 2^{l(1/\tilde{q}+n/\tilde{p}-1)}\|Q_l\big(\sum_{m>10} f_l\cdot g_{l-m}\big)\|_{L_t^{\tilde{q}}L_x^{\tilde{p}}}$$
$$\lesssim \|f\|_{\mathcal{S}^{(-1)}}\|g\|_{\mathcal{S}^{(-1)}}.$$

The dimension restriction enters only to control the *high-high* term.

In order to shed some light into the above we will present an alternative route which though longer might be more enlightening as to the action of $|\nabla|^{-1}$ on products of 'solutions'. It also establishes a somewhat stronger multiplication result. We start with an auxiliary Lemma.

Lemma 6 *Let $n \geq 4$ and let f be a function on $\mathcal{S}^{(-1)}$. For any $q \geq 2$, and p' defined by $\frac{1}{q}+\frac{n}{p'}=1$ we have that*

$$(i) \qquad \big(\sum_{k\in\mathbb{Z}} \|Q_k(f)\|^2_{L_t^qL_x^{p'}}\big)^{1/2} \lesssim \|f\|_{\mathcal{S}^{(-1)}}$$

In addition, we also have

$$(ii) \qquad \|f\|_{L_t^\infty \dot{W}^{n/2-1,2}} \lesssim \big(\sum_{k\in\mathbb{Z}} 2^{2k(n/2-1)}\|Q_k(f)\|^2_{L_t^\infty L_x^2}\big)^{1/2} \lesssim \|f\|_{\mathcal{S}^{(-1)}}$$

The same conclusions hold for $2^{-k}\|\partial_t Q_k(f)\|$ replacing $\|Q_k(f)\|$.

Proof. Let $f_k = Q_k(f)$. Clearly,

$$\|\partial_x^{n/2-1} f_k\|_{L_t^\infty L_x^2} \sim 2^{k(n/2-1)}\|f_k\|_{L_t^\infty L_x^2} \leq \|f_k\|_{\mathcal{S}_k^{(-1)}}.$$

Then (ii) follows by taking ℓ^2 norms both sides.

To prove (i) we proceed as follows. Given $q \geq 2$, let p' be defined by $\frac{1}{q}+\frac{n}{p'}=1$. In particular we have that $4 \leq n \leq p' \leq 2n$. Since $n \geq 4$ we can now choose $2 \leq r < p'$ such that (q, r) is sharp admissible and

$$\|f_k\|_{L^q_t L^{p'}_x} \lesssim 2^{k\gamma}\|f_k\|_{L^q_t L^r_x}$$

by the Sobolev embedding where γ is given by and $\frac{1}{p'} = \frac{1}{r} - \frac{\gamma}{n}$.

In particular then $0 < \gamma = \frac{n}{r} - \frac{n}{p'} = \frac{n}{r} + \frac{1}{q} - 1$. From where we can conclude that

$$\|f_k\|_{L^q_t L^{p'}_x} \lesssim \|f_k\|_{S^{(-1)}_k}$$

since by Sobolev embedding , (c.f. Lemma 2.7 in [**25**]) we have that

$$\sup_{(q,r)-\text{admissible}} 2^{k(\frac{1}{q}+\frac{n}{r}-1)}\|Q_k(f)\|_{L^q_t L^r_x} = \sup_{(q,r)-\text{sharp admissible}} 2^{k(\frac{1}{q}+\frac{n}{r}-1)}\|Q_k(f)\|_{L^q_t L^r_x}.$$

The desired conclusion follows by taking ℓ^2 norms both sides □.

First Multiplication estimate *For $1 \leq q < \infty$, let $\mathcal{L}^1_t \dot{B}^0_{\infty,q}$ be the Banach space of functions on $\mathbb{R} \times \mathbb{R}^n$ whose norm is given by*

$$\|f\|_{\mathcal{L}^1_t \dot{B}^0_{\infty,q}} := \big(\sum_{k\in\mathbb{Z}} \|Q_k(f)\|^q_{L^1_t L^\infty_x}\big)^{1/q}.$$

Then

$$|\nabla|^{-1} : \mathcal{S}^{(-1)} \times \mathcal{S}^{(-1)} \longrightarrow \mathcal{L}^1_t \dot{B}^0_{\infty,1}.$$

Proof. Let f and g be in $\mathcal{S}^{(-1)}$ and let $f_k = Q_k(f)$ and $g_j = Q_j(g)$ be their corresponding Littlewood-Paley projections. We write

$$\begin{aligned} |\nabla|^{-1}(f \cdot g) &= |\nabla|^{-1}\big(\sum_{k,j\in\mathbb{Z}} f_k \cdot g_j\big) \\ &= |\nabla|^{-1}\big(\sum_{k,j\in\mathbb{Z}:k\geq j} f_k \cdot g_j\big) + |\nabla|^{-1}\big(\sum_{k,j\in\mathbb{Z}:k<j} f_k \cdot g_j\big). \end{aligned}$$

By symmetry of the sums, it is enough to consider only one of them. The proof for the other is identical after exchanging k and j. Hence we need to estimate

$$\begin{aligned} &\sum_{l\in\mathbb{Z}} \|Q_l(|\nabla|^{-1}(\sum_{k,j\in\mathbb{Z}:k\geq j} f_k \cdot g_j))\|_{L^1_t L^\infty_x} \\ &= \sum_{l\in\mathbb{Z}} \|Q_l(|\nabla|^{-1}(\sum_{k\in\mathbb{Z}} \sum_{m\geq 0} f_k \cdot g_{k-m}))\|_{L^1_t L^\infty_x} \end{aligned}$$

Since supp $(\widehat{f_k \cdot g_{k-m}}) \subseteq \{\xi : |\xi| \leq 2^k\}$ we have that $Q_l(f_k \cdot g_{k-m}) \equiv 0$ unless $k \geq l$.

Therefore we can make the last sum less than or equal to

$$\sum_{l\in\mathbb{Z}} 2^{-l} \sum_{k\geq l} \sum_{m\geq 0} \|Q_l(f_k \cdot g_{k-m})\|_{L^1_t L^\infty_x}.$$

On the other hand, we have that supp $(\widehat{f_k \cdot g_{k-m}}) \cap \{\xi : |\xi| << 2^{k-m}\} = \emptyset$ if $m > 5$ Hence, $Q_l(f_k \cdot g_{k-m}) \equiv 0$ unless $l = k$ and $m > 5$ or $m \leq 5$ and $l < k$.

We must then have that the above sum is

$$\lesssim \sum_{0\leq m\leq 5}\sum_{l\in\mathbb{Z}} 2^{-l}\sum_{k>l}\|Q_l(f_k\cdot g_{k-m})\|_{L^1_tL^\infty_x}+\sum_{m>5}\sum_{l\in\mathbb{Z}}2^{-l}\|Q_l(f_l\cdot g_{l-m})\|_{L^1_tL^\infty_x}.$$

We consider the first sum first.

$$\sum_{0\leq m\leq 5}\sum_{l\in\mathbb{Z}} 2^{-l}\sum_{k>l}\|Q_l(f_k\cdot g_{k-m})\|_{L^1_tL^\infty_x}$$
$$\lesssim \sum_{0\leq m\leq 5}\sum_{l\in\mathbb{Z}} 2^{-l}\sum_{k>l}(2^{nl})^{\frac{n-3}{n-1}}\|f_k\|_{L^2_tL^{2p}_x}\|g_{k-m}\|_{L^2_tL^{2p}_x},$$

by Young's inequality with $p=\frac{n-1}{n-3}$, $\frac{1}{p}+\frac{1}{p'}=1+\frac{1}{\infty}$ and Cauchy-Schwartz inequality.

The endpoint Strichartz estimates (2.1) now yield the bound

$$\textstyle\sum_{0\leq m\leq 5}\sum_{l\in\mathbb{Z}}2^{-l}\sum_{k>l}(2^{nl})^{\frac{n-3}{n-1}}2^{2k(1+\frac{n}{(n-1)}-\frac{(n+1)}{2})}\|f_k\|_{\mathcal{S}^{(-1)}_k}\|g_{k-m}\|_{\mathcal{S}^{(-1)}_k}$$
$$\textstyle\sim\sum_{0\leq m\leq 5}\sum_{k\in\mathbb{Z}}\sum_{l\leq k}2^{l\frac{w}{n-1}}2^{-k\frac{w}{n-1}}\|f_k\|_{\mathcal{S}^{(-1)}_k}\|g_{k-m}\|_{\mathcal{S}^{(-1)}_k}$$

where $w=(n-2)^2-3$ which is positive provided $n\geq 4$. Hence by summing first in l and then applying Cauchy-Schwartz to the sum in k we get that the above is $\lesssim\|f\|_{\mathcal{S}^{(-1)}}\|g\|_{\mathcal{S}^{(-1)}}$ as desired.

We proceed next with the second sum.

$$\sum_{m>5}\sum_{l\in\mathbb{Z}}2^{-l}\|Q_l(f_l\cdot g_{l-m})\|_{L^1_tL^\infty_x}$$
$$\textstyle\lesssim\sum_{m>5}\sum_{l\in\mathbb{Z}}2^{-l}2^{nl/r}\|(f_l\cdot g_{l-m})\|_{L^1_tL^r_x}$$

by Young's inequality with $r=2n$. Now, by Hölder's inequality we can bound the last sum by

$$\sum_{m>5}\sum_{l\in\mathbb{Z}}2^{-l}2^{nl/r}\|f_l\|_{L^2_tL^{2r}_x}\|g_{l-m}\|_{L^2_tL^{2r}_x}.$$

Since the pair $(2,2r)$ is admissible we have by the Strichartz estimates that the above sum is up to a constant less than or equal to

$$\textstyle\sum_{m>5}\sum_{l\in\mathbb{Z}}2^{-l}2^{nl/r}2^{2l(1/2-n/(2r))}2^{-m(1/2-n/(2r))}\|f_l\|_{\mathcal{S}^{(-1)}_l}\|g_{l-m}\|_{\mathcal{S}^{(-1)}_l}$$
$$\textstyle\lesssim\sum_{m>5}2^{-m(1/2-n/(2r))}\sum_{l\in\mathbb{Z}}\|f_l\|_{\mathcal{S}^{(-1)}_l}\|g_{l-m}\|_{\mathcal{S}^{(-1)}_l}$$
$$\textstyle\lesssim\sum_{m>5}2^{-m(1/2-n/(2r))}\big(\sum_{l\in\mathbb{Z}}\|f_l\|^2_{\mathcal{S}^{(-1)}_l}\big)^{1/2}\big(\sum_{l\in\mathbb{Z}}\|g_{l-m}\|^2_{\mathcal{S}^{(-1)}_l}\big)^{1/2}$$
$$\lesssim\|f\|_{\mathcal{S}^{(-1)}}\|g\|_{\mathcal{S}^{(-1)}}$$

since by our choice of r, $1/2-n/(2r)=1/4>0$ □.

Second Multiplication estimate. Let $q\geq 2$ and $p<2$ such that $\dfrac{1}{q}+\dfrac{n}{p}=\dfrac{n}{2}+1$. Then

$$|\nabla|^{-1}:\mathcal{S}^{(-1)}\times\mathcal{S}^{(-1)}\longrightarrow L^q_t\dot{B}^s_{\tilde{p},2}$$

for any $\tilde{p}\geq p$ and $s=\frac{1}{q}+\frac{n}{\tilde{p}}-1$.

In particular, we have that $|\nabla|^{-1} : \mathcal{S}^{(-1)} \times \mathcal{S}^{(-1)} \longrightarrow \mathcal{L}_t^2 \dot{B}^s_{\tilde{p},2}$; that is

$$\Big(\sum_k 2^{2ks}\|Q_k(|\nabla|^{-1}(f \cdot g))\|^2_{L^2_t L^{\tilde{p}}_x}\Big)^{1/2} \lesssim \|f\|_{\mathcal{S}^{(-1)}}\|g\|_{\mathcal{S}^{(-1)}}$$

for any $2 \le \tilde{p} < 2n$ and $s = \frac{n}{\tilde{p}} - \frac{1}{2}$.

Proof Let f and g be two function on $\mathcal{S}^{(-1)}$.

Let $q \ge 2$ and let p' and p be defined by

$$\frac{1}{q} + \frac{n}{p'} = 1 \quad \text{and} \quad \frac{1}{p} = \frac{1}{2} + \frac{1}{p'}.$$

Claim:

$$\||\nabla|^{-1}(f \cdot g)\|_{L^q_t \dot{W}^{n/2,p}} \lesssim \|f\|_{\mathcal{S}^{(-1)}}\|g\|_{\mathcal{S}^{(-1)}}.$$

Assuming the claim we note that since $p < 2$ and $\frac{1}{q} + \frac{n}{p} = 1 + \frac{n}{2}$

$$L^q_t \dot{W}^{n/2,p} \hookrightarrow L^q_t \dot{B}^{n/2}_{p,2} \hookrightarrow L^q_t \dot{B}^s_{\tilde{p},2}$$

provided $s = \frac{1}{q} + \frac{n}{\tilde{p}} - 1$ and $\tilde{p} \ge p$ as desired.

To prove the claim we first note that by Lemma 6 we have, in particular, the following two 'endpoint estimates'.

(*i*) $$\|f\|_{L^\infty_t \dot{W}^{n/2-1,2}} \lesssim \|f\|_{\mathcal{S}^{(-1)}}.$$

(*ii*) $$\|f\|_{L^q_t L^{p'}_x} \lesssim \|f\|_{L^q_t \dot{B}^0_{p',2}} \lesssim \|f\|_{\mathcal{S}^{(-1)}}.$$

since $p' \ge 4 > 2$.

If $n = 4$, the above two estimates suffice. For then,

$$\|f \cdot g\|_{L^q_t \dot{W}^{n/2-1,p}} \lesssim \|f\|_{\mathcal{S}^{(-1)}}\|g\|_{\mathcal{S}^{(-1)}} \text{ where } \frac{1}{p} = \frac{1}{2} + \frac{1}{p'}$$

since $p' \ge n$. In turn, this implies that

$$|\nabla|^{-1} : \mathcal{S}^{(-1)} \times \mathcal{S}^{(-1)} \longrightarrow L^q_t \dot{W}^{n/2,p}$$

as desired.

In the case $n \ge 5$ however, we need to prove additional estimates. We consider separately the cases when n is *even* first and then indicate the necessary modifications when n is *odd*.

More precisely, let $n \ge 5$ be even. Given q, p' as above and $1 \le j \le \frac{n}{2} - 2$ let $0 < \theta_j < 1$ be defined by $\theta_j = \frac{j}{(n/2-1)}$. Next, let q_j, $q_{n/2-1-j}$ and p'_j, $p'_{n/2-1-j}$ be any solutions to the following equations:

$$\frac{1}{q_j} + \frac{n}{p'_j} = 1 \quad \text{and} \quad \frac{1}{q_{n/2-1-j}} + \frac{n}{p'_{n/2-1-j}} = 1$$

$$\frac{1-\theta_j}{q_j} + \frac{\theta_j}{q_{n/2-1-j}} = \frac{1}{q}, \qquad \frac{1-\theta_j}{p'_j} + \frac{\theta_j}{p'_{n/2-1-j}} = \frac{1}{p'}$$

Next, let $\frac{1}{\tilde{q}_j} = \frac{1-\theta_j}{q_j}$ and $\frac{1}{\tilde{p}'_j} = \frac{1-\theta_j}{p'_j} + \frac{\theta_j}{2}$.

We claim that

$$(iii) \qquad \|\partial_x^j f\|_{L_t^{\tilde q_j} L^{\tilde p'_j}} \lesssim \|\partial_x^j f\|_{L_t^{\tilde q_j} \dot B^0_{\tilde p'_j,2}} \lesssim \|f\|_{S^{(-1)}}.$$

Indeed, the first inequality follows from the embeddings between Sobolev and Besov spaces since $\tilde p'_j \geq 2$. For the second one we have that

$$\begin{aligned} \|\partial_x^j f_k\|_{L_t^{\tilde q_j} L^{\tilde p'_j}} &\sim 2^{kj} \|f_k\|_{L_t^{\tilde q_j} L^{\tilde p'_j}} \\ &= 2^{k(\theta_j(n/2-1))} \|f_k\|_{L_t^{\tilde q_j} L^{\tilde p'_j}} \\ &= 2^{k(\frac{1}{\tilde q_j}+\frac{n}{\tilde p'_j}-1)} \|f_k\|_{L_t^{\tilde q_j} L^{\tilde p'_j}} \lesssim \|f_k\|_{S_k^{(-1)}} \end{aligned}$$

whence the second inequality follows by taking ℓ^2-norms.

Finally we put together (i), (ii) and (iii) to obtain that

$$\begin{aligned} \|f\cdot g\|_{L_t^q \dot W^{n/2-1,p}} &= \|\partial_x^{n/2-1}(f\cdot g)\|_{L_t^q L_x^p} \\ &\lesssim \sum_{j=0}^{n/2-1} \|\partial_x^j f\|_{L_t^{\tilde q_j} L_x^{\tilde p'_j}} \|\partial_x^{n/2-1-j} g\|_{L_t^{\tilde q_{n/2-1-j}} L_x^{\tilde p'_{n/2-1-j}}} \\ &\lesssim \|f\|_{S^{(-1)}} \|g\|_{S^{(-1)}}, \end{aligned}$$

as desired.

We indicate now the technicalities needed when $n \geq 5$ is odd. Given q, p' as above and $0 \leq j \leq [\frac{n}{2}] - 1$ let $0 < \theta_j < 1$ to be defined in a moment. As before, let q_j, $q_{n/2-1-j}$ and p'_j, $p'_{n/2-1-j}$ be any solutions of the equations as above for θ_j and as before let $\frac{1}{\tilde q_j} = \frac{1-\theta_j}{q_j}$ and $\frac{1}{\tilde p'_j} = \frac{1-\theta_j}{p'_j} + \frac{\theta_j}{2}$. Now,

$$\begin{aligned} \|f\cdot g\|_{L_t^q \dot W^{n/2-1,p}} &= \|\partial_x^{n/2-1}(f\cdot g)\|_{L_t^q L_x^p} \\ &\lesssim \sum_{j=0}^{[n/2]-1} \Big(\|\partial_x^{j+1/2} f\|_{L_t^{\tilde q_j} L_x^{\tilde p'_j}} \|\partial_x^{n/2-j-3/2} g\|_{L_t^{\tilde q_{n/2-1-j}} L_x^{\tilde p'_{n/2-1-j}}} + \\ &\quad + \|\partial_x^j f\|_{L_t^{\tilde q_j} L_x^{\tilde p'_j}} \|\partial_x^{n/2-j-1} g\|_{L_t^{\tilde q_{n/2-1-j}} L_x^{\tilde p'_{n/2-1-j}}} \Big) \end{aligned}$$

For the first term inside the big sum we take $\theta_j = \dfrac{j+1/2}{(n/2-1)}$; note that $\theta_{n/2-j-1} = 1 - \theta_j = \dfrac{n/2-j-3/2}{n/2-1}$. Then for $0 \leq j \leq [n/2]-1$ we have that

$$\|\partial_x^{j+1/2} f\|_{L_t^{\tilde q_j} L_x^{\tilde p'_j}} \lesssim \|\partial_x^{j+1/2} f\|_{L_t^{\tilde q_j} \dot B^0_{\tilde p'_j,2}} \lesssim \|f\|_{S^{(-1)}};$$

while

$$\|\partial_x^{n/2-j-3/2} g\|_{L_t^{\tilde q_{n/2-1-j}} L_x^{\tilde p'_{n/2-1-j}}} \lesssim \|\partial_x^{n/2-j-3/2} g\|_{L_t^{\tilde q_{n/2-j-1}} \dot B^0_{\tilde p'_{n/2-j-1},2}} \lesssim \|g\|_{S^{(-1)}}$$

For the second we take $\theta_j = \dfrac{j}{(n/2-1)}$ as in (iii), and the needed estimates follow just as in the even case.

All in all we have that

$$\|f \cdot g\|_{L^q_t \dot{W}^{n/2-1,p}} \lesssim \|f\|_{\mathcal{S}^{(-1)}} \|g\|_{\mathcal{S}^{(-1)}},$$

as desired $\square$.

4. The Modified Wave Map System

The general scheme to find a solution to a nonlinear wave equation

$$\begin{aligned} \Box u &= \mathcal{N}(u) \\ u(\cdot, 0) &= f \\ \partial_t u(\cdot, 0) &= g, \end{aligned}$$

relies on Picard iteration. We denote by $u_{-1} \equiv 0$ and let u_0 be the solution to the homogeneous problem

$$\Box u_0 = 0 \qquad u_0(\cdot, 0) = f, \quad \partial_t u_0(\cdot, 0) = g.$$

Subsequent iterates u_m, $m \geq 1$ are obtained by solving

$$\Box u_m = \mathcal{N}(u_{m-1}) \qquad u_m(\cdot, 0) = f \quad \partial_t u_m(\cdot, 0) = g.$$

In other words, formally

$$u_m = u_0 + \Box^{-1} \mathcal{N}(u_{m-1}), \quad m \geq 1$$

where by $\Box^{-1}$ we really mean the Duhamel operator giving the solution to the inhomogeneous linear problem with zero data as described in section 3.1.

To find a solution then we need to identify two Banach spaces $\mathcal{X}$, the '*solution space*' and $\mathcal{Y}$ the '*nonlinearity space*' such that:

- Free solutions $\subset \mathcal{X}$ with norm controlled by that of the initial data
- $u_{m-1} \in \mathcal{X}$ implies that $u_m \in \mathcal{X}$ (for $u \in \mathcal{X}$, $\mathcal{N}(u) \in \mathcal{Y}$ and $\Box^{-1}\mathcal{Y} \subset \mathcal{X}$)
- $\{u_m\}$ is bounded in $\mathcal{X}$
- $\{u_m\}$ is Cauchy in $\mathcal{X}$ (estimates on differences).

We will follow the above scheme to study the well-posedness of the MWM system. Since $n \geq 4$ the actual form of the bilinear nonlinearity on the right hand side of the MWM plays no role: estimates on products are sufficient. Thus for simplicity we denote by $B(a, b)$ any finite linear combination of functions $a \in \mathcal{S}_+^{(-1)}$ and $b \in \mathcal{S}^{(-1)}$ of the form $\sum_{\kappa,\ell} c_{\kappa\ell} a_\kappa b_\ell$ where $a_\kappa \in \mathcal{S}_+^{(-1)}$, $b_\ell \in \mathcal{S}^{(-1)}$ and $c_{\kappa\ell} \in \mathbb{C}$.

Thus consider the MWM system of coupled wave equations in R^{n+1}, $n \geq 4$.

$$\begin{aligned} \Box v &= B(a, b) \sim a \cdot b \\ v(x, 0) &= f(x) \\ v_t(x, 0) &= g(x) \end{aligned}$$

where $v = (\varphi, \psi)$, $b = d\varphi + div_{(sp.t)}\, \psi$ and

$$da = [a, a] + [b, b], \quad i.e. \quad a = |\nabla|^{-1}[a, a] + |\nabla|^{-1}[b, b]$$

In our case the space $\mathcal{X}$ will be $\mathcal{S}$. The nonlinearity space $\mathcal{Y}$ is determined by the following result:

Theorem 7 (Main Nonlinear Estimate) *Let $a \in \mathcal{S}_+^{(-1)}$ and $b \in \mathcal{S}^{(-1)}$ then*

$$\Big(\sum_{k\in\mathbb{Z}} 2^{2k(n/2-1)}\|Q_k(a\cdot b)\|^2_{L^1_tL^2_x}\Big)^{1/2} \lesssim \|a\|_{\mathcal{S}_+^{(-1)}}\|b\|_{\mathcal{S}^{(-1)}}$$

The proof starts by performing a Littlewood-Paley decomposition of a and b to obtain

$$\begin{aligned}\textstyle\sum_{k\in\mathbb{Z}} 2^{2k(n/2-1)}\|Q_k(a\cdot b)\|^2_{L^1_tL^2_x} \lesssim\\ \textstyle\sum_{k\in\mathbb{Z}} 2^{2k(n/2-1)}\|\sum_{m>5}\big(Q_k(Q_k(a)\cdot Q_{k-m}(b))\big)\|^2_{L^1_tL^2_x} +\\ \textstyle\sum_{k\in\mathbb{Z}} 2^{2k(n/2-1)}\|\sum_{m>5}\big(Q_k(Q_{k-m}(a)\cdot Q_k(b))\big)\|^2_{L^1_tL^2_x} +\\ \textstyle\sum_{k\in\mathbb{Z}} 2^{2k(n/2-1)}\|\sum_{k<l}\big(Q_k(Q_l(a)\cdot Q_l(b))\big)\|^2_{L^1_tL^2_x}.\end{aligned}$$

Now since a and b belong to different spaces we lose the 'symmetry' and need to consider all three cases separately. We show how the most delicate case of *low frequencies* in the curvature term a versus *high frequencies in b* proceeds:

$$\begin{aligned}&\textstyle\sum_{k\in\mathbb{Z}} 2^{2k(n/2-1)}\|\sum_{m>5}\big(Q_k(Q_k(b)\cdot Q_{k-m}(a))\big)\|^2_{L^1_tL^2_x}\\ &\lesssim \textstyle\sum_{k\in\mathbb{Z}} 2^{2k(n/2-1)}\|Q_k(b)\|^2_{L^\infty_tL^2_x}\big(\sum_{m>5}\|Q_{k-m}(a)\|_{L^1_tL^\infty_x}\big)^2\\ &\lesssim \|a\|^2_{\mathcal{S}_+^{(-1)}}\textstyle\sum_{k\in\mathbb{Z}}\|Q_k(b)\|^2_{\mathcal{S}_k^{(-1)}}\\ &\lesssim \|a\|^2_{\mathcal{S}_+^{(-1)}}\|b\|^2_{\mathcal{S}^{(-1)}}\end{aligned}$$

invoking the fact that

$$a \in \mathcal{S}_+^{(-1)} \hookrightarrow \{f : \sum_k \|Q_k(f)\|_{L^1_tL^\infty_x} < \infty\}$$

Lemma 8 (A priori estimate) *Let $a \in \mathcal{S}_+^{(-1)}$ and $b \in \mathcal{S}^{(-1)}$. Then the solution to the MWM system with initial data $(f,g) \in \dot{H}^{n/2}\times\dot{H}^{n/2-1}$ satisfies*

$$\|v\|_{\mathcal{S}} \lesssim \|f\|_{\dot{H}^{n/2}} + \|g\|_{\dot{H}^{n/2-1}} + \|a\|_{\mathcal{S}_+^{(-1)}}\|b\|_{\mathcal{S}^{(-1)}}$$

Proof Let us denote by $v_k = Q_k(v)$. By the Strichartz's estimates (SE) at the end of section 3.1 , we have that

$$\|v_k\|_{\mathcal{S}_k} \lesssim \|f_k\|_{\dot{H}^{n/2}} + \|g_k\|_{\dot{H}^{n/2-1}} + 2^{k(n/2-1)}\|B(a,b)\|_{L^1_tL^2_x}.$$

The nonlinear estimate above then gives that

$$\begin{aligned}\|v\|_{\mathcal{S}} &= \big(\textstyle\sum_{k\in\mathbb{Z}}\|v_k\|^2_{\mathcal{S}_k}\big)^{1/2}\\ &\lesssim \|f\|_{\dot{H}^{n/2}} + \|g\|_{\dot{H}^{n/2-1}} + \big(\textstyle\sum_{k\in\mathbb{Z}} 2^{2k(n/2-1)}\|B(a,b)\|^2_{L^1_tL^2_x}\big)^{1/2}\\ &\lesssim \|f\|_{\dot{H}^{n/2}} + \|g\|_{\dot{H}^{n/2-1}} + \|a\|_{\mathcal{S}_+^{(-1)}}\|b\|_{\mathcal{S}^{(-1)}}\end{aligned}$$

as desired. □

We turn now to the proof of global well posedness of the MWM with small data. The *a priori* estimate we just proved together with the multiplication estimates allow us to control $\|a\|_{\mathcal{S}_+^{(-1)}}$ by $\|b\|_{\mathcal{S}^{(-1)}}$ and close the estimates as follows. We proceed proceeds by Picard's iteration relying on the smallness of the data.

Suppose $||(f,g)||_{\dot{H}^{n/2}\times\dot{H}^{n/2-1}} = \delta$ and let v_0 be the solution to

$$\Box v_0 = 0; \qquad v_0(0,\cdot) = f \quad \partial_t v_0(0,\cdot) = g.$$

By the Strichartz's estimates

$$||v_0||_{\mathcal{S}} \le c_1 ||(f,g)||_{\dot{H}^{n/2}\times\dot{H}^{n/2-1}} = c_1\delta.$$

Now, $v_0 = (\varphi_0, \psi_0)$ produces $b_0 = d\,\varphi_0 + div_{(sp.t)}\,\psi_0$ with $||b_0||_{\mathcal{S}^{(-1)}} \le c_2||v_0||_{\mathcal{S}} \le c_3\delta$.

Next, the multiplication estimates allow one to perform a fixed point argument to produce a_0 from b_0 by solving

$$a_0 = |\nabla|^{-1}[a_0, a_0] + |\nabla|^{-1}[b_0, b_0].$$

Moreover, $||a_0||_{\mathcal{S}_+^{(-1)}} \le c_4 ||b_0||^2_{\mathcal{S}^{(-1)}} \le c_5\delta^2$

Let v_1 be the solution of

$$\Box v_1 = B(a_0, b_0) \qquad v_1(0,\cdot) = f \quad \partial_t v_1(0,\cdot) = g.$$

By the *a priori* estimate,

$$||v_1||_{\mathcal{S}} \le c_0\big(\delta + ||a_0||_{\mathcal{S}_+^{(-1)}} ||b_0||_{\mathcal{S}^{(-1)}}\big) \le 2c_0\delta$$

provided δ is small enough.

We proceed next by induction to show :

• For any $j \ge 0$, $\quad ||b_j||_{\mathcal{S}} \le 2c_2c_o\delta$ and $||a_j||_{\mathcal{S}} \le c_5\delta^2$.

Hence if

$$\Box v_{j+1} = B(a_j, b_j) \qquad v_{j+1}(0,\cdot) = f \quad \partial_t v_{j+1}(0,\cdot) = g,$$

by the *a priori* estimates we have that

$$||v_{j+1}||_{\mathcal{S}} \le 2c_0\delta,$$

provided $\delta > 0$ is small enough (indep. of j). Moreover we have estimates for the differences,

$$||b_{j+1} - b_j||_{\mathcal{S}^{(-1)}} \le c_2 ||v_{j+1} - v_j||_{\mathcal{S}}$$

and

$$||a_{j+1} - a_j||_{\mathcal{S}_+^{(-1)}} \le c\delta ||b_{j+1} - b_j||_{\mathcal{S}^{(-1)}} \le c\delta ||v_{j+1} - v_j||_{\mathcal{S}}.$$

Whence all in all, by choosing δ small enough we have that

$$||v_{j+2} - v_{j+1}||_{\mathcal{S}} \le \frac{1}{2}||v_{j+1} - v_j||_{\mathcal{S}}.$$

Hence v_j is Cauchy in $\mathcal{S}$, thus establishing existence and uniqueness. For the stability result one proceeds in the same fashion as in the proof of being Cauchy; thus concluding the proof of the theorem. □.

5. Remarks on wave maps in $\mathbb{R}^{1+3}$ with critical regularity

We now turn to the question of extending the results in the previous sections to spatial dimension $n = 3$. In **[36]** T. Tao established global regularity in $n = 2, 3$ when the target is any sphere, $\mathbb{S}^m$. His result has recently been extended by J. Krieger for $n = 3$ and when the target is the hyperbolic space $\mathbb{H}^2$ instead **[24]**.

In this section we describe some ideas and estimates developed jointly with A. Stefanov and K. Uhlenbeck and contained in [**26**] to study the problem of establishing existence and uniqueness of solutions to the wave map equations from Minkowski space $\mathbb{R}^{3+1}$ into -say- a constant curvature complete Riemannian manifold.

The first point is that when $n = 3$ and unlike the case in higher dimensions the precise dependence of the curvature term a on b, the derivative of the solution, is needed. If one uses the Coulomb gauge as we did in higher dimensions the curvature term looks essentially like $|\nabla|^{-1}(b{\cdot}b)$ and the lack of any structure whatsoever makes it impossible to prove good a priori estimates to control the nonlinearity; *even* if the 'missing' Strichartz estimate, $L^2_t L^\infty_x$ in dimension 3 were true. If one considers the very simple 'model problem' $\Box u = |\nabla|^{-1} u \cdot du$ it is clear that the nonlinearity is dangerous at low-high interactions. More precisely for $m > 0$,

$$(|\nabla|^{-1}u)_{k-m} \cdot (du)_k \sim 2^m u_{k-m} u_k$$

which in low dimensions this may cause blow up. A similar situation occurs in low dimensions for the 'model equation' $\Box v = |\nabla|^{-1}(dv \cdot dv) \cdot dv$. Dangerous low-high interactions in the nonlinearity occur for example when $m > 0$ and $k-m << l << k$ in

$$(|\nabla|^{-1}((dv)_l \cdot (dv)_l))_{k-m} \cdot (dv)_k \sim 2^m 2^{2l} v_l v_l v_k.$$

However, in general it is possible to obtain 'extra structure' in which case there is some hope to rule out blow up. For example if the curvature term in nonlinearity of the latter model problem has a 'null form' $Q_{ij}(u, v) := \partial_i u \partial_j v - \partial_j u \partial_i v$ instead of just a product, then the curvature term at low frequencies gains a factor of $2^{(k-m-l)}$; that is now

$$(|\nabla|^{-1}(Q_{ij}(u_l, v_l)))_{k-m} \cdot (dw)_k \sim 2^k 2^l v_l v_l v_k.$$

The principle behind this is very simple. The Fourier transform of a solution to the free wave equation is a distribution supported on the light cone Λ. As we have seen in section 2 the support of the Fourier transform of the product of u_1 and u_2, two such solutions, lies in the algebraic sum of the support of $\hat{u}_1$ and $\hat{u}_2$. On the other hand the sum of two 'vectors' on Λ is close to the light cone if and only if they are collinear; e.g. if u_1 and u_2 are two waves (wave packets) traveling in the same direction. The 'null form' has a symbol that vanishes precisely when the two arguments are collinear and on Λ; i.e.

$$\widehat{Q_{ij}(u, v)}(\xi) = \int \langle \xi, \eta^\perp \rangle \hat{u}_1(\xi - \eta) \hat{u}_2(\eta)\, d\eta$$

In other words the 'null form' precisely helps with parallel interactions. This idea follows from the works of J. Bourgain, Klainerman-Machedon, Klainerman-Tataru, T. Wolff and T. Tao ([**2**], [**17**][**23**], [**41**][**37**]) who realized that a null form structure or some angular separation between the Fourier transform of the two factors in a product helps rule out parallel interactions; thus allowing for the range of possible $L^q_t L^r_x$ bilinear estimates to be enlarged considerably.

5.1. Gauge fixing and the equations for curvature. Following K. Uhlenbeck's ideas we now describe a gauge fixing for $n = 3$ that yields a better structure for the curvature term; as described above [**26**]. The following holds on an appropriate band in time-space. However, we ignore this issue in what follows and leave the localization in time out of the present discussion [**26**].

Given the (derivative) solution space is in $L^4_t L^4_x$, we assume that $b := du$ is sufficiently small in $L^4_t L^4_x$. Then the curvature $F = [b, b]$ term is small in $L^2_t L^2_x$. Thus the "good gauge theorem" above ([**25**] [**40**]) says that for any target we can fix a gauge ψ so that the *elliptic space-time* divergence of $a = (a_0, a_1, a_2, a_3)$ vanishes. In other words so that,

$$d^* a = \frac{\partial a_0}{\partial t} + \sum_{j=1}^{3} \frac{\partial a_j}{\partial x_j} = 0 = \sum_{j=0}^{3} \frac{\partial a_j}{\partial x_j}$$

by letting $t = x_0$. Moreover we have the bound

$$\|a\|_{H^1(\mathbb{R}\times\mathbb{R}^3)} \lesssim \|b\|^2_{L^4_{x,t}}$$

Now, depending on whether the target is abelian (e.g. $\mathbb{S}^2$, $\mathbb{H}^2$) or non-abelian, we have respectively that

$$da = [b, b] \quad \text{or} \quad da = [a, a] + [b, b]$$

where d is in $\mathbb{R}^4$; e.g. d acting on 0-forms is the $\nabla = (\partial_t, \nabla_x)$. To get 'a null form' structure in the equation for a, we 'approximate' b by q as follows. We write $b = (b_0, b_{sp})$ and do a Hodge decomposition of the spatial part, $b_{sp} = d_{sp}q + d^*_{sp}r$ where now (d_{sp}, d^*_{sp}) are the exterior differentiation and its dual over $\mathbb{R}^3$. Then if let $r_0 := b_0 - \partial_t q$ and $R = (r_0, d^* r)$ we get that

$$b := (b_0, b_{sp}) \sim (\partial_t q + r_0, d_{sp}q + d^*_{sp}r) \sim dq + R$$

Revisiting then the equations for a,

$$\begin{aligned} d^* a \quad &= 0 \\ da \quad &= [b, b] \quad \text{or} \quad da = [a, a] + [b, b]] \end{aligned}$$

we find that the second equation now becomes

$$da = dq \wedge dq \; + \; 2[dq, R] \; + \; [R, R] \quad (\text{ in the abelian case}) \text{ or}$$

$$da = dq \wedge dq \; + \; 2[dq, R] \; + \; [R, R] + [a, a] \quad (\text{ in the non-abelian case})$$

where now the first term on the r.h.s is elliptic in $3+1$ and is the only term that has a 'null form' structure. The last terms despite not having any special structure, do possess better regularity properties and hence are overall somewhat better behaved. For example, the term $[R, R]$ behaves like the nonlinearity in the $4 + 1$-dimensions wave map equation.

If we denote the bilinear forms above by $B(\alpha, \gamma) := \sum c_{ij}\alpha_i\gamma_j$,with $c_{ij} \in \mathbb{C}$ and

$$Q_{ij}(u, v) := \partial_i u \partial_j v - \partial_j u \partial_i v,$$

Note that $Q_{ii} = 0$ and that $i, j = 0$ signify as usual, the time derivatives.
Then we can schematically write the second equations for the curvature term as

$$da = Q_{i,j}(q, q) + B(R, dq) + B(R, R) + B(a, a) \tag{3}$$

Heuristically, at least in the estimates, $R \sim |\nabla|^{-1}(a, b)$ and hence R will satisfy the same *a priori* estimates $|\nabla|^{-1}(b, b)$ satisfies.

Thus from now on we abuse notation and simply write the equations for a as

$$
\begin{aligned}
d^* a &= 0 \\
d a &= Q_{ij}(u,u) + B(du,a) + B(a,a)
\end{aligned} \tag{4}
$$

where u is the solution and $du = b$.

5.2. Null forms and a priori curvature estimates. First, a technical lemma establishing the boundedness of products of Riesz transforms in the mixed Lebesgue spaces $L_t^q L_x^r$. This is of course well known in the case of ordinary Lebesgue spaces (i.e. $q = r$), but seems to be missing from the literature in this general context.

Lemma 9[**26**]*Let* $1 \le q, r \le \infty$ *and let*

$$\|\phi\|_{\mathcal{L}_t^q L_x^r} := \left(\sum_k \|\phi_k\|^2_{L_t^q L_x^r} \right)^{1/2}.$$

Then each of the products of Riesz transforms in $T_i = \partial_i \nabla_{x,t} |\Delta_{xt}|^{-1}$, $i = 0,1,2,3$ *is a bounded mapping from* $\mathcal{L}_t^q L_x^r$ *to itself.*

A proof of this lemma can be found in [**26**]. Recently, Stefanov and Torres studied Calderón-Zygmund operators in mixed Lebesgue space-time norms [**32**] thus extending this boundedness result to a large class of operators.

Remark: In order to include the endpoints $q = \infty$ or $r = \infty$, we consider the action of the operators only on spatially frequency localized pieces. We remark that even in the diagonal case $q = r$, one does not have in general $R_j = \partial_j |\nabla|^{-1} : L^\infty \to L^\infty$.

The so called null form structure is given by the bilinear form

$$Q_{ij}(u,v) = \partial_i u \partial_j v - \partial_j u \partial_i v,$$

for $i, j = 0, \ldots 3$. The operator can be written as a Fourier multiplier operator as follows

$$Q_{ij}(u,v)(x) = \int (\xi_i \eta_j - \xi_j \eta_i) \hat{u}(\xi) \hat{v}(\eta) e^{i(\xi+\eta)x} d\xi d\eta.$$

Note that ξ_0 and η_0 signify the time components of the corresponding vectors.

The special structure of the symbol is exploited in the following manner.

Lemma 10 *For all* $1 \le q, r, q_1, q_2, r_1, r_2 \le \infty$, *with*

$$
\begin{aligned}
& 1/q_1 + 1/q_2 = 1/q \ , \quad 1/r_1 + 1/r_2 = 1/r \\
& \|\nabla_{xt} |\Delta_{xt}|^{-1} Q_{ij}(u,v)\|_{L_t^q L_x^r} \\
\lesssim \ & \min(\|\partial u\|_{L_t^{q_1} L_x^{r_1}} \|v\|_{L_t^{q_2} L_x^{r_2}}, \|u\|_{L_t^{q_1} L_x^{r_1}} \|\partial v\|_{L_t^{q_2} L_x^{r_2}}).\|\nabla_{xt} |\Delta_{xt}|^{-1} Q_{ij}(u,v)\|_{L_t^q L_x^r} \\
\lesssim \ & \min(\|\partial u\|_{L_t^{q_1} L_x^{r_1}} \|v\|_{L_t^{q_2} L_x^{r_2}}, \|u\|_{L_t^{q_1} L_x^{r_1}} \|\partial v\|_{L_t^{q_2} L_x^{r_2}}).
\end{aligned}
$$

Proof Write $\xi_i \eta_j - \xi_j \eta_i = (\xi_i + \eta_i)\eta_j - (\xi_j + \eta_j)\eta_i$. Thus

$$\frac{\nabla_{xt}}{\Delta_{xt}} Q_{ij}(u,v) = c \frac{\nabla_{x,t} \partial_i}{\Delta_{xt}} (u \partial_j v) - c \frac{\nabla_{xt} \partial_j}{\Delta_{xt}} (u \partial_i v).$$

The result follows from Lemma 9 and the Hölder inequality.

Remark: Note that if the 'missing' Strichartz estimate $L_t^2 L_x^\infty$ in dimension 3 *were* true then the Lemma above would yield the needed $\mathcal{L}_t^1 \dot{B}^0_{\infty,1}$ estimate for the delicate portion of the curvature term. Just as in the higher dimensional case such an estimate would then be enough to prove existence and uniqueness of wave maps into constant curvature complete Riemannian manifolds in $\mathbb{R}^{1+3}$.

Because of the failure of such Strichartz estimate; the largest range of *a priori* estimates in the mixed-Lebesgue norm that one can establish for the curvature term are contained in the following two Lemmata. First we need to modify the definition of the set $\mathcal{E}$ according to the exclusion of the 'missing' Strichartz pair in dimension 3. Let $\kappa > 0$ be small and fixed. Define the set of exponents $\mathcal{E}$ to be the convex hull of the points,

$$(0,0)\,,\ (0, 5/6 - \kappa) \quad \text{and} \quad \{(q,r) : 1 - \kappa > 1/q > 1/4 + \kappa \ \ 1/q + 1/r \leq 1\}\,.$$

In particular, $\mathcal{A} \subset \mathcal{E}$.

Lemma 11 [**26**] *For every* $(q,r) \in \mathcal{E}$ *we have*

$$\sum_k 2^{k(1/q+3/r-1)} \|(d^*_{xt} \Delta^{-1}_{xt} Q_{ij}(u,v))_k\|_{L^q_t L^r_x} \leq C_\kappa \|u\|_{\mathcal{S}} \|v\|_{\mathcal{S}}.$$

Lemma 12 [**26**] *Suppose that* $u \in \mathcal{S}$ *with* $\|u\|_{\mathcal{S}}$ *sufficiently small. Then the solution* a *of the problem*

$$\begin{aligned} d^* a &= 0 \\ da + B(a,a) + B(a,du) &= Q_{ij}(u,u), \end{aligned}$$

satisfies the a priori estimate

$$\sum_k 2^{k(1/q+3/r-1)} \|a_k\|_{L^q_t L^r_x} \lesssim \|u\|^2_{\mathcal{S}}$$

for every $(q,r) \in \mathcal{E}$. *In particular, if we denote by*

$$\|a\|_{\mathcal{S}'} := \sup_{(q,r)\in\mathcal{E}} \sum_k 2^{k(1/q+3/r-1)} \ \|a_k\|_{L^q_t L^r_x}$$

we have that

$$\begin{aligned} \sum_k \sum_{l>k} 2^{k(1/q+3/r-1)} \| |\nabla|^{-1} (a_l du_l)_k \|_{L^q_t L^r_x} &\lesssim \|a\|_{\mathcal{S}'} \|u\|_{\mathcal{S}}, \\ \sum_{m>0} \sum_k 2^{k(1/q+3/r-1)} \| |\nabla|^{-1} (a_{k-m} du_k) \|_{L^q_t L^r_x} &\lesssim \|a\|_{\mathcal{S}'} \|u\|_{\mathcal{S}}, \\ \sum_k \sum_{l>k} 2^{k(1/q+3/r-1)} \| |\nabla|^{-1} (a_l b_l)_k \|_{L^q_t L^r_x} &\lesssim \|a\|_{\mathcal{S}'} \|b\|_{\mathcal{S}'}, \\ \sum_k 2^{k(1/q+3/r-1)} \| |\nabla|^{-1} (a_{k-m} b_k) \|_{L^q_t L^r_x} &\lesssim \|a\|_{\mathcal{S}'} \|b\|_{\mathcal{S}'}. \end{aligned}$$

The proof of the previous two lemmata can be found in [**26**].

The estimates in Lemma 12 however are not sufficient to find good a priori estimates for the nonlinearity $\mathcal{N}(u) := a \cdot du$, where $d\,a = Q_{i,j}(q,q) + B(R,dq) + B(R,R) + B(a,a)$ as above. In other words it is not possible any longer to place the solution of $\Box u = a \cdot du$ in the $\mathcal{S}$ space and close the estimates in $L^1_t \dot{H}^{1/2}_x$ for example. The majority of nonlinear terms are indeed controllable mostly in the mixed-Lebesgue norms. But not all of them. The main obstacle is the contribution of the low-high interactions term $a_{<k} du_k$ in the non-linearity. As it is clear from its form, the derivative lands on the high frequency part u_k, which makes the control of that term problematic in the mixed Lebesgue spaces (mainly due to the "missing" Strichartz estimate at $L^2_t L^\infty_x$).

If one ignores the quadratic correction term $B(a,a)$ in the second equation for the curvature term a in Lemma 10, and applies the space-time divergence operator d^*_{xt}, one gets,

$$\Delta_{xt} a = d^* da = d^*_{xt} Q_{ij}(u,u).$$

Thus, for the purposes of estimates in this case, the non-linearity (essentially) looks like

$$(|\nabla_{tx}|^{-1} Q_{ij}(u,u))_{<k} du_k.$$

At this point roughly what one wishes is to find (local in time) Banach spaces $\mathcal{X}$ and $\mathcal{Y}$ replacing $\mathcal{S}$ and $L^1_t \dot{H}^{1/2}_x$ respectively such that :

1. $\mathcal{X} \subset L^\infty_t \dot{H}^{3/2}_x \qquad L^\infty_t \dot{H}^{5/2}_x \subset \mathcal{X} \subset \mathcal{S}$
2. Free solutions with data in $\dot{H}^{3/2} \times \dot{H}^{1/2}$ are in $\mathcal{X}$.
3. For u a solution of $\Box u = F$, $u(x,0) = f$, $u_t(x,0) = g$.

$$\|u\|_{\mathcal{X}} \lesssim \|(f,g)\|_{\dot{H}^{3/2} \times \dot{H}^{1/2}} + \|F\|_{\mathcal{Y}}$$

4. $\|\mathcal{N}(u)\|_{\mathcal{Y}} \lesssim \|u\|^3_{\mathcal{X}}$. In particular, the following estimates hold
   - (high-high interactions in the null form)

$$\|(|\nabla_{tx}|^{-1} Q_{ij}(u_l, v_l))_{<k} dw_k\|_{\mathcal{Y}} \lesssim \|u_l\|_{\mathcal{X}} \|v_l\|_{\mathcal{X}} \|w_k\|_{\mathcal{X}}$$

   - (high-low interactions in the null form) For $m > 0$ and $l < k - m$,

$$\||\nabla_{tx}|^{-1} Q_{ij}(u_l, v_{k-m}) dw_k\|_{\mathcal{Y}} \lesssim 2^{-(k-m-l)/2} \|u_l\|_{\mathcal{X}} \|v_{k-m}\|_{\mathcal{X}} \|w_k\|_{\mathcal{X}}.$$

Remarks. (i) It is important to note that the a priori estimates needed to control the problematic term in the nonlinearity are really trilinear –and not bilinear–. That is they come from considering truly quatrilinear forms and where also the precise dependence of the curvature term a on b, the derivative of the solution, is needed. T. Tao [**36**] and J. Krieger [**24**] had to deal with similar type of trilinear estimates .

(ii) The scheme outline in this section provides a path to establish global existence, uniqueness and regularity of wave maps with small data in $\dot{H}^{3/2}(\mathbb{R}^3) \times \dot{H}^{1/2}(\mathbb{R}^3)$. Once a solution space $\mathcal{X}$ and a nonlinearity space $\mathcal{Y}$ are found so that (1)–(4) hold, the theorem giving the existence of a local in time gauge alluded above and in [**26**] produces a local in time solution. A global regularity result then implies this solution turns out to be a global solution, which is the weak limit in $L_t^\infty(\dot{H}_x^{3/2} \times \dot{H}_x^{1/2})$ of smooth solutions. Uniqueness holds in the sense that if $\tilde{u} \in L_t^\infty \dot{H}_x^{3/2} \cap \dot{W}^{1,4}$ is a 'small' solution to the wave map problem, then there exists a gauge transformation such that $\tilde{u} = u$ where $u \in \mathcal{X}$ is the local solution constructed by the special choice of gauge [**26**].

(iii) Finally we note that a possible approach to this problem is to use for $\mathcal{X}$ and $\mathcal{Y}$ the corresponding solution and nonlinear spaces T. Tao introduced in [**36**]. This approach would of course mean one is assuming most of Tao's difficult and delicate paper at this point. It is possible however that in dimensions 3 one could find alternative spaces $\mathcal{X}$ and $\mathcal{Y}$ that would allow one to establish (1)–(4) above more directly. This is part of subject under investigation in [**26**]. [1]

References

[1] J. Bergh and J. Löfström, *Interpolation Spaces,* (Springer Verlag. (1976.)

[2] J. Bourgain, *Estimates for cone multipliers,* Operator Theory: Advances and Applications, **77** (1995) 41–60

[3] T. Cazenave, J. Shatah and A. Tahvildar-Zadeh, *Harmonic maps of the hyperbolic space and develpment of singularities in wave maps and Yang Mills fields.* , Preprint (1996.)

[4] Y. Choquet-Bruhat, *Global existence theorems for hyperbolic harmonic maps,* Ann. Inst. H. Poincaré Phys. Théor., **46** (1987) 97–111

[5] Y. Choquet-Bruhat and C.H. Gu, *Existence globale d'applications harmoniques sur l'espace-temps de Minkowski M_3,* C. R. Acad. Sci. Paris, Ser. I, **308** (1)989 167–170

[6] D. Christodoulou and J. Shatah, Personal communication; unpublished (1988.)

[7] D. Christodoulou and A. Tahvildar-Zadehi, *On the regularity of spherically symmetric wave maps,,* Comm. Pure Appl. Math., **46** (1993) 1041–1091

[8] R. Coifman, P.L.Lions, Y.Meyer and S. Semmes, *Compensated compactness and Hardy spaces,* J. Math. Pures Appl., **72, No. 3** (1993) 247-286

[9] R. Coifman and Y. Meyer, *Au-delà des opérateurs pseudo-differentiels.,* (Astérisque, 57 Société Math. de France. (1978.)

[10] J. Ginibre and G. Velo, *The Cauchy problem for the $O(N), \mathbb{C}P(N-1)$ and $G\mathbb{C}(N,P)$ models.,* Ann. Physics, **142** (1982) 393–415

[11] M. Grillakis, *Classical solutions for the equivariant wave map in $1+2$-dimensions,* Indiana Univ. Math. J., to appear

[12] Gu, Chao-Hao, *On the Cauchy problem for harmonic maps defined on two-dimensional Minkowski space,* Comm. Pure Appl. Math, **33** (1980) 727–737

[13] Günther, Matthias, *Isometric Embeddings of Riemannian Manifolds,* Proceedings of the ICM in Kyoto, Japan 1990 (Math. Soc. Japan), **I** (1991) 1137–1143

[14] T. Kato and G. Ponce, *Commutator estimates and the Euler and Navier-Stokes equations.,* Comm. Pure Appl. Math., **41, No. 7** (1988) 891–907

[15] M. Keel and T. Tao, *Endpoint Strichartz estimates,* Amer. Math. J., **120** (1)998 955–980

[16] M. Keel and T. Tao, *Local and global well-posedness of wave maps in R^{1+1} for rough data,* Internat. Math. Res. Notices, **21** (1998) 1117–1156

[17] S. Klainerman and M. Machedon, *Smoothing estimates for null forms and applications,* Duke Math J., **81** (1995) 99–133

[1] At the time this paper went into print, D. Tataru posted a proof of the global existence and uniqueness of $2-d$ wave maps into complete Riemannian manifolds with bounded geometry and small initial data in $H^1 \times L^2$.

[18] S. Klainerman and M. Machedon, *On the optimal local regularity for gauge fields theories,* Diff. and Integral Eqs., **10** (1997) 1019–1030

[19] S. Klainerman and M. Machedon, *Space-time estimates for null forms and the local existence theorem,* Comm. Pure Appl. Math., **46** (1993) 1221–1268

[20] S. Klainerman and M. Machedon, *Estimates for null forms and the spaces $H_{s,\delta}$,* Internat. Math. Res. Notices, **17** (1996) 853–865

[21] S. Klainerman and I. Rodnianski, *On the global regularity of wave maps in the critical Sobolev norm,* Internat. Math. Res. Notices, **13** (2001)

[22] S. Klainerman and S. Selberg, *Remarks on the optimal regularity for equations of wave maps type,* Comm. PDE, **22** (1997) 901–918

[23] S. Klainerman and D. Tataru, *On the optimal local regularity for Yang-Mills equations in* $\mathbb{R}^{4+1}$, J. Amer. Math.Soc., **12** (1999) 93–116

[24] J. Krieger, *Global regularity of wave maps from* $\mathbb{R}^{3+1}$ *to* $\mathbb{H}^2$, Preprint (2002)

[25] A. Nahmod, A. Stefanov and K. Uhlenbeck, *On the well posedness of the wave map problem in high dimensions,* Comm. Anal. Geom., **11, No. 1** (2003) 49–83

[26] A. Nahmod, A. Stefanov and K. Uhlenbeck, *On global existence of wave maps in three dimensions with critical regularity,* in preparation

[27] J. Shatah, *Weak solutions and development of singularities in* $SU(2)$ *σ- model,* Comm. Pure Appl. Math, **41** (1988) 459–469

[28] J. Shatah and M. Struwe, *The Cauchy problem for wave maps,* IMRN, **11** (2002) 555–571

[29] J. Shatah and M. Struwe, *Geometric wave equations,* (Courant Lecture Notes in Mathematics 2. (1998)

[30] J. Shatah and A. Tahvildar-Zadeh, *On the stability of stationary wave map,* Comm. Math. Phys., **185, No. 1** (1997) 231–256

[31] T. Sideris, *Global existence of harmonic maps in Minkowski space,* Comm. Pure Appl. Math., **42** (1989) 1–13

[32] A. Stefanov and R. Torres, *Calderón-Zygmund operators on mixed Lebesgue spaces and applications to null forms,* preprint (2003)

[33] E. Stein, *Harmonic Analysis: Real variable methods, orthogonality and oscillatory integrals,* (Princeton University Press. (1993)

[34] M. Struwe, *Radially symmetric wave maps from* $(1+2)$*-dimensional Minkowski space to a sphere,* Math. Z., **242** (2002) 407–414

[35] T. Tao, *Global regularity of wave maps I. Small critical Sobolev norm in high dimension,* IMRN, **7** (2001) 299–328

[36] T. Tao, *Global regularity of wave maps II. Small energy in two dimensions,* Comm. Math. Phys., **224** (2001) 443-544

[37] T. Tao, *Endpoint bilinear restriction theorems for the cone and some sharp null form estimates,* Math. Z., **238, No. 2** (2001) 215–268

[38] D. Tataru, *Local and global results for wave maps I,* Comm. in PDE, **23** (1998) 1781-1793

[39] D. Tataru, *On global existence and scattering for the wave maps equation,* Amer. J. Math, **123, No. 1** (2001) 37–77

[40] K. Uhlenbeck, *Connections with L^p bounds on curvature* , Comm. Math. Phys., **83** (1982) 31–42

[41] T. Wolff, *A sharp bilinear restriction estimate,* Ann. of Math (2), **153, No. 3** (2001) 661–698

A. Nahmod, Department of Mathematics and Statistics, Lederle GRT, University of Massachusetts, Amherst, MA 01003-4515

E-mail address: `nahmod@math.umass.edu`

Discreteness of minimal models of Kodaira dimension zero and subvarieties of moduli stacks

Eckart Viehweg and Kang Zuo

Contents

1. Introduction

Let $f : V \to U$ be a smooth projective morphism with connected fibres over a complex quasi-projective manifold U.

Definition 1.1.

i. $\mathrm{Var}(f)$ is the smallest integer η for which there exists a finitely generated subfield K of $\overline{\mathbb{C}(U)}$ of transcendence degree η over $\mathbb{C}$, a variety F' defined over K, and a birational equivalence

$$V \times_U \mathrm{Spec}(\overline{\mathbb{C}(U)}) \sim F' \times_{\mathrm{Spec}(K)} \mathrm{Spec}(\overline{\mathbb{C}(U)}).$$

ii. $f : V \to Y$ is birationally isotrivial if $\mathrm{Var}(f) = 0$, hence if there exists some generically finite covering $U' \to U$, a projective manifold F', and a birational map

$$V \times_U U' \sim U' \times F'.$$

This work has been supported by the "DFG-Schwerpunktprogramm Globale Methoden in der Komplexen Geometrie". The second named author is supported by a grant from the Research Grants Council of the Hong Kong Special Administrative Region, China (Project No. CUHK 4239/01P).

iii. $f : V \to U$ is (biregulary) isotrivial if there exists a generically finite covering $U' \to U$, a projective manifold F and an isomorphism

$$V \times_U U' \simeq U' \times F.$$

So the variation of a morphisms counts the number of parameters controlling the birational structure of the fibres of F.

Maehara has shown in [**6**] that under the assumption that ω_F is semi-ample and big for a general fibre F of f, a family is birationally isotrivial, if and only if it is biregulary isotrivial. In different terms, for families of minimal models of complex manifolds of general type, $\mathrm{Var}(f)$ measures the number of directions where the structure of F varies.

We will slightly extend the methods used to prove [**8**], Theorem 6.24, to show that for families with $\omega_F^\delta = \mathcal{O}_F$, for some $\delta > 0$, the same holds true. We will show that for a given projective manifold F' the set of minimal models is discrete, hence that there are no non-trivial families of minimal models.

Let us fix some polarization $\mathcal{L}$ of $f : V \to U$, with Hilbert polynomial h. If $\omega_{V/U}$ is f-ample we will choose $\mathcal{L} = \omega_{V/U}^\rho$, for some $\rho > 0$. By [**8**] there exists a quasi-projective moduli schemes M_h, parameterizing polarized manifolds $(F, \mathcal{L})$ with ω_F semiample and with $h(\nu) = \chi(\mathcal{L}^\nu)$. The family $f : V \to U$ together with $\mathcal{L}$ induces a map

$$\varphi : U \to M_h$$

Since we require $\varphi : U \to M_h$ to be induced by a family it factors through the moduli stack $\mathcal{M}_h$.

THEOREM 1.2. *Let $f : V \to U$ be a family of polarized manifolds. Assume that $\omega_F^\delta = \mathcal{O}_F$, for some $\delta > 0$ (or that all fibres F of f are canonically polarized). Let $\varphi : U \to M_h$ be the induced morphism to the moduli scheme. Then* $\mathrm{Var}(f) = \dim(\varphi(U))$.

If in Theorem 1.2 the morphism $f : V \to U$ is birationally isotrivial, $\varphi(U)$ must be zero dimensional, hence f is biregulary isotrivial.

It seems reasonable to conjecture that Theorem 1.2 remains true for families of polarized manifolds with $\omega_{X/Y}$ f-semiample. Here however one should replace the moduli scheme M_h by the moduli scheme P_h of polarized manifolds, up to numerical equivalence (see [**8**]).

PROBLEM 1.3. Let $f : V \to U$ be a family of polarized manifolds with $\omega_{V/U}$ f-semiample, and let $\psi : U \to P_h$ be the morphism to the moduli scheme of polarized manifolds up to numerical equivalence. Does this imply $\mathrm{Var}(f) = \dim(\psi(U))$?

Theorem 1.2 allows to extends some of the results obtained in [**10**] for canonically polarized manifolds to families of manifolds F with $\omega_F^\delta = \mathcal{O}_F$ (see Theorem 6.3, i) and ii) and Section 7). This is done in the second half of this article, a continuation of [**10**]. What methods are concerned, the reader familiar with [**10**] will find nothing new. In fact we just sketch the changes needed to extend some of the results to this case.

In the final Section 8 we will state a criterion for the rigidity of non-isotrivial families over curves, and its translation to curves in the moduli stack of minimal polarized manifolds of Kodaira dimension zero, or of canonically polarized manifolds. This criterion is implicitly used in [**10**], Proof of 6.4 and 6.5, but it was not explicitly stated there.

A slightly weaker statement (8.2) extends to all families with ω_F semiample. A similar criterion has been shown by S. Kovács and, for families of Calabi-Yau manifolds by K. Liu, A. Todorov, S.-T. Yau and the second named author in [**5**]. As a corollary one obtains (see 8.4):

COROLLARY 1.4. *Let M_h be either the moduli scheme of canonically polarized manifolds or the moduli scheme of polarized manifolds F with $\omega_F^\delta = \mathcal{O}$ for some $\delta > 0$. There are only finitely many morphisms $\varphi : U \to M_h$ which are induced by a smooth family $f : V \to U$ with:*

For a general fibre F of f the n-th wedge product

$$0 \neq \wedge^n \xi \in H^n(F, \omega_F^{-1}),$$

where $\xi \in H^1(F, T_F)$ denotes the Kodaira Spencer class corresponding to the deformation $f : V \to U$ of F.

The way 1.2 is formulated, it is a trivial statement for canonically polarized manifolds. Maehara showed in [**6**] that problem 1.3 has an affirmative answer if the fibres of f are of maximal Kodaira dimension. This implies, that geometric properties of submanifolds of moduli stacks M_h of canonically polarized manifolds carry over to those of the moduli scheme P_h of polarized minimal manifolds of general type. To make notations not more complicated than they are already, we leave the necessary changes to the reader.

The first half of this article presents a proof of Theorem 1.2, hopefully of interest independently of the applications to subvarieties of moduli stacks. In the first section, we will show, that the proof of Theorem 1.2 can be reduced to families over a curve. Next we recall and strengthen a Positivity Theorem from [**8**]. It allows to reprove Maehara's Result and thereby the first part of 1.2 in section 4. The case of minimal models of Kodaira dimension zero is handled in Section 5.

2. Reduction to families over curves

In Definition 1.1, i), we may choose a finitely generated subfield L of $\overline{\mathbb{C}(U)}$ which contains $\mathbb{C}(U)$ and K. Let U' be the normalization of U in L, and let T be a smooth quasi-projective variety with function field $\mathbb{C}(T) = K$. Replacing T by some open subscheme, we may assume that there exists a smooth projective morphism $g : Z \to T$ with general fibre F', and replacing U' by some open subscheme one finds morphisms $\tau : U' \to U$ and $\pi : U' \to T$ fitting into a diagram

$$\begin{array}{ccccccc} V & \longleftarrow & V' & \xrightarrow{\sim} & Z' & \longrightarrow & Z \\ {\scriptstyle f}\downarrow & & {\scriptstyle f'}\downarrow & & {\scriptstyle g'}\downarrow & & {\scriptstyle g}\downarrow \\ U & \xleftarrow{\tau} & U' & \xrightarrow{=} & U' & \xrightarrow{\pi} & T, \end{array}$$

where $V' \to Z'$ is a birational equivalence, and where the right and left hand squares are fibre products. For a point $t \in T$ in general position, one has

$$\dim(\pi^{-1}(t)) = \dim(U') - \dim(T) = \dim(U) - \mathrm{Var}(f).$$

If under the assumptions made in Theorem 1.2 $\mathrm{Var}(f) < \dim(\varphi(U))$, for a point $\eta \in \varphi(U)$ in general position,

$$\dim(\tau^{-1}\varphi^{-1}(\eta)) = \dim(U') - \dim(\varphi(U)) < \dim(\pi^{-1}(t)),$$

hence there exists a curve C in $\pi^{-1}(t)$ with $\tau \circ \varphi|_C$ finite. In order to prove Theorem 1.2 one just has to show that such a curve can not exist. Theorem 1.2 follows from

PROPOSITION 2.1. *Let U be a non-singular irreducible curve and let F' be a projective manifold. Let $f : V \to U$ be a family of polarized manifolds. Assume that there exists a birational equivalence $V \sim U \times F'$ over U. If either the fibres F of f are canonically polarized, or if $\omega_F^\delta = \mathcal{O}_F$, for some $\delta > 0$, then the induced morphism $\varphi : U \to M_h$ is constant.*

In particular, $f : V \to U$ is biregulary isotrivial.

In order to prove Proposition 2.1 we may replace U by a finite covering. Doing so one can assume that $f : V \to U$ extends to a semistable morphism $f : X \to Y$ of projective manifolds, hence that $\Delta = f^{-1}(S)$ is a reduced normal crossing divisor, for $S = Y \setminus U$.

Moreover, we may replace the given polarization by some power. In fact, the corresponding map of the moduli schemes M_h is a finite map. This allows to assume that for all fibres F of $V \to U$ the polarization $\mathcal{L}$ is very ample, and without higher cohomology.

3. Positivity of direct image sheaves

Recall that a locally free sheaf $\mathcal{E}$ on a projective non-singular curve Y is numerically effective (nef), if for all finite morphisms $\tau : Z \to Y$ and for all invertible quotients $\mathcal{L}$ of $\tau^*(\mathcal{E})$ the degree $\deg(\mathcal{L}) \geq 0$.

Fujita's positivity theorem (today an easy corollary of Kollár's vanishing theorem) says that $f_*\omega_{X/Y}$ is nef. By [**9**], 2.3, one obtains as a direct consequence.

LEMMA 3.1. *Let $f : X \to Y$ be a morphism from a normal projective variety X to a curve Y, with connected fibres. Assume that X has at most rational double points as singularities. Let $\mathcal{N}$ be an invertible sheaf on X and Γ an effective divisor. Assume that for some $N > 0$ there exists a nef locally free sheaf $\mathcal{E}$ on Y and a surjection*

$$f^*\mathcal{E} \longrightarrow \mathcal{N}^N(-\Gamma).$$

Then

$$f_*\left(\mathcal{N} \otimes \omega_{X/Y}\left\{-\frac{\Gamma}{N}\right\}\right)$$

is nef.

Here $\omega_{X/Y}\left\{-\frac{\Gamma}{N}\right\}$ denotes the (algebraic) multiplier sheaf (see for example [**3**], 7.4, or [**8**], section 5.3). If $\tau : X' \to X$ is any blowing up with $\Gamma' = \tau^*\Gamma$ a normal crossing divisor, then

$$\omega_{X/Y}\left\{-\frac{\Gamma}{N}\right\} = \tau_*\left(\omega_{X'/Y}\left(-\left[\frac{\Gamma'}{N}\right]\right)\right).$$

As in [**3**], § 7 and [**8**], section 5.3, we are mainly interested in the case where the multiplier sheaf on a general fibre F is isomorphic to ω_F. The corresponding threshold is defined for any effective divisor Π or any invertible sheaf $\mathcal{L}$ on F with $H^0(F, \mathcal{L}) \neq 0$.

$$e(\Pi) = \mathrm{Min}\left\{N \in \mathbb{N} - \{0\};\ \omega_F\left\{-\frac{\Pi}{N}\right\} = \omega_F\right\} \qquad \text{and}$$

$$e(\mathcal{L}) = \mathrm{Max}\left\{e(\Pi);\ \Pi \text{ the zero set of } \sigma \in H^0(F, \mathcal{L}) - \{0\}\right\}.$$

For smooth morphisms $f : X \to Y$ and for an f-ample sheaf $\mathcal{L}$ on X we obtained in [**8**], 6.24 and 7.20, strong positivity theorems. Their proof, in case Y is a curve, can easily be extended to semistable morphisms $f : X \to Y$.

THEOREM 3.2. *Let Y be a curve, let $f : X \to Y$ be a semistable morphism between projective manifolds with connected fibres, and let $\mathcal{M}$ be an invertible sheaf on X. Let $U \subset Y$ be an open dense subscheme with $V = f^{-1}(U) \to U$ smooth. Assume that for all fibres F of $V \to U$ the canonical sheaf ω_F is semiample, that $\mathcal{M}|_F$ is very ample and without higher cohomology. Then for*

$$e \geq c_1(\mathcal{M}|_F)^{\dim(F)} + 2, \qquad r = \operatorname{rank}(f_*\mathcal{M})$$
$$\textit{and} \quad r(\nu) = \operatorname{rank}(f_*(\mathcal{M}^\nu \otimes \omega_{X/Y}^{e\cdot\nu})) :$$

a. *For all $\nu > 0$*

$$\Big(\bigotimes^r f_*(\mathcal{M}^\nu \otimes \omega_{X/Y}^{e\cdot\nu})\Big) \otimes \det(f_*\mathcal{M})^{-\nu}$$

is nef.

b. *If the invertible sheaf $\det(f_*(\mathcal{M}^\nu \otimes \omega_{X/Y}^{e\cdot\nu})) \otimes \det(f_*\mathcal{M})^{-\nu r(\nu)}$ is ample for some $\nu > 0$,*

$$\Big(\bigotimes^r f_*(\mathcal{M} \otimes \omega_{X/Y}^{e})\Big) \otimes \det(f_*\mathcal{M})^{-1}$$

is ample.

c. *If for all $\nu > 0$ the degree of*

$$\det(f_*(\mathcal{M}^\nu \otimes \omega_{X/Y}^{e\cdot\nu})) \otimes \det(f_*\mathcal{M})^{-\nu r(\nu)}$$

is zero, then $f : V \to U$ is biregulary isotrivial as a family of polarized manifolds, i.e. there exists some finite covering $U' \to U$, a projective manifold F', invertible sheaves $\mathcal{L}'$ on F' and $\mathcal{B}$ on U', and an isomorphism

$$\pi : V' = X \times_Y U' \to F' \times U'$$

with

$$pr_1^*\mathcal{M} = \pi^*(pr_1^*\mathcal{L} \otimes pr_2^*\mathcal{B}).$$

PROOF. As indicated already, the proof of parts a) and b) will follow the arguments used in [**8**], 194–196, to prove 6.20. We just have to take care, that for a semistable family over a curve the sheaves are at most getting larger. So we repeat the arguments.

For a) let us fix some $\nu > 0$. For b) we assume that

$$\det(f_*(\mathcal{M}^\nu \otimes \omega_{X/Y}^{e\cdot\nu})) \otimes \det(f_*\mathcal{M})^{-\nu\cdot r(\nu)}$$

is ample.

The semicontinuity of the threshold, shown in [**8**], 5.17 for example, allows to find some $\gamma \geq e \cdot \nu$ with

$$e(\mathcal{M}|_F^{\nu\cdot e} \otimes \omega_F^{e\cdot\nu\cdot(e-1)}) \leq \gamma \tag{3.2.1}$$

for all fibres F of $V \to U$.

For b) we will show that

$$S^\gamma\Big(\Big(\bigotimes^{r\cdot r(\nu)} f_*(\mathcal{M} \otimes \omega_{X/Y}^{e})\Big) \otimes \det(f_*\mathcal{M})^{-r(\nu)}\Big) \otimes$$
$$\det(f_*(\mathcal{M}^\nu \otimes \omega_{X/Y}^{e\cdot\nu}))^{-r\cdot(e-1)} \otimes \det(f_*\mathcal{M})^{\nu\cdot r\cdot(e-1)\cdot r(\nu)}$$

is nef. Hence for both, a) or b), it is sufficient to prove the corresponding statements for the pullback of the sheaves to any finite covering Y' of Y. Since we assumed $X \to Y$ to be semistable, the fibre product $X' = X \times_Y Y'$ is a normal variety with at most rational double points. Flat base change allows to replace Y by such a covering and $(f : X \to Y, \mathcal{M})$ by a desingularization of the pullback family.

Doing so, we may assume that $\det(f_*\mathcal{M})$ is the r-th power of an invertible sheaf, and since all the sheaves occurring in a), b) or c) are compatible with changing the polarization by the pullback of an invertible sheaf on Y, we can as well assume that $\det(f_*\mathcal{M}) = \mathcal{O}_Y$. Under this additional assumption we have to verify in a) that

$$f_*(\mathcal{M}^\nu \otimes \omega_{X/Y}^{e\cdot\nu})$$

is nef. For part b) we may assume in addition that

$$\det(f_*(\mathcal{M}^\nu \otimes \omega_{X/Y}^{e\cdot\nu}))^{r\cdot(e-1)} = \mathcal{O}_Y(\gamma \cdot H)$$

for some effective divisor H supported in U. We have to prove that

$$\Big(\bigotimes^{r\cdot r(\nu)} f_*(\mathcal{M} \otimes \omega_{X/Y}^{e})\Big) \otimes \mathcal{O}_Y(-H)$$

is nef.

Let $f^s : X^s \to Y$ be the s-fold fibre product. X^s is normal with at most rational double points (see [**7**], page 291, for example). Consider

$$\mathcal{P} = \bigotimes_{i=1}^{s} pr_i^*\mathcal{M}.$$

By flat base change one obtains

$$f_*^s(\mathcal{P}^\alpha \otimes \omega_{X^s/Y}^{\beta}) = \bigotimes^{s} f_*(\mathcal{M}^\alpha \otimes \omega_{X/Y}^{\beta})$$

for all α, β. The restriction of $\mathcal{P}^\nu \otimes \omega_{X^s/Y}^{e\cdot\nu-\iota}$ to $f^{s-1}(U) = V^s$ is f^s-ample for all $\iota \leq e \cdot \nu$. Let us write $\epsilon = e \cdot \nu$ or $\epsilon = e \cdot \nu - 1$, where ν may be any positive integer.

If Γ is the zero divisor of a section of $\mathcal{P}$, which does not contain any fibre F^s of $V^s \to U$ the compatibility of the threshold with products and its semicontinuity imply (see [**8**], 5.14 and 5.21)

(3.2.2) $$e(\Gamma|_{F^s}) \leq e(\mathcal{P}|_{F^s}) = e(\mathcal{M}|_F) < e \quad \text{and} \quad e(\Gamma|_{V^s}) < e.$$

In fact, as shown in [**8**], 5.11), one has

$$e(\mathcal{M}|_F) \leq c_1(\mathcal{M}|_F)^{\dim(F)} + 1.$$

Moreover, by the choice of ϵ

(3.2.3) $$e(\nu \cdot \Gamma|_{F^s}) \leq \nu \cdot e(\Gamma|_{F^s}) \leq \nu \cdot (c_1(\mathcal{M}|_F)^{\dim(F)} + 1) \leq \epsilon.$$

Let $\mathcal{H}$ be an ample invertible sheaf on Y.

CLAIM 3.3. *Assume that for some $\rho \geq 0$, $N > 0$, $M_0 > 0$ and for all multiples M of M_0, the sheaf*

$$f_*((\mathcal{M}^\nu \otimes \omega_{X/Y}^{\epsilon})^{M\cdot N}) \otimes \mathcal{H}^{\rho\cdot\epsilon\cdot N\cdot M}$$

is nef. Then

$$f_*((\mathcal{M}^\nu \otimes \omega_{X/Y}^{\epsilon})^{N}) \otimes \mathcal{H}^{\rho\cdot(\epsilon\cdot N-1)}$$

is nef.

PROOF. Let us choose $s = r$. The determinant gives an inclusion

$$\det(f_*\mathcal{M}) = \mathcal{O}_Y \longrightarrow f^r_*\mathcal{P} = \bigotimes^r f_*\mathcal{M},$$

which splits locally. Hence the zero divisor Γ of the induced section of $\mathcal{P}$ does not contain any fibre of $V^r \to U$. For

$$\mathcal{N} = \mathcal{P}^{\nu\cdot N} \otimes \omega_{X^r/Y}^{\epsilon\cdot N-1} \otimes f^{r*}\mathcal{H}^{\rho\cdot(\epsilon\cdot N-1)\cdot r}$$

one obtains that the restriction of

$$\mathcal{N}^{\epsilon}(-\nu\cdot\Gamma) = (\mathcal{P}^{\nu} \otimes \omega_{X^r/Y}^{\epsilon} \otimes f^{r*}\mathcal{H}^{\rho\cdot\epsilon\cdot r})^{(\epsilon\cdot N-1)}$$

to V^s is f^r-ample. If M' is a positive integer, divisible by $M_0\cdot N$ the sheaf

$$f^r_*(\mathcal{N}^{\epsilon}(-\nu\cdot\Gamma)^{M'}) = \bigotimes^r (f_*(\mathcal{M}^{\nu}\otimes\omega_{X/Y}^{\epsilon})^{(\epsilon\cdot N-1)\cdot M'} \otimes \mathcal{H}^{\rho\cdot\epsilon\cdot r(\epsilon\cdot N-1)\cdot M'})$$

is nef. Choose M' such that

$$f^*f_*((\mathcal{M}^{\nu}\otimes\omega_{X/Y}^{\epsilon})^{(\epsilon\cdot N-1)\cdot M'}) \longrightarrow (\mathcal{M}^{\nu}\otimes\omega_{X/Y}^{\epsilon})^{(\epsilon\cdot N-1)\cdot M'}$$

is surjective over U.

3.1 implies that the subsheaf $f^r_*(\mathcal{N}\otimes\omega_{X^r/Y}\{-\frac{\nu\cdot\Gamma}{\epsilon}\})$ of

$$f^r_*(\mathcal{N}\otimes\omega_{X^r/Y}) = \bigotimes^r (f_*(\mathcal{M}^{\nu\cdot N}\otimes\omega_{X/Y}^{\epsilon\cdot N})\otimes\mathcal{H}^{\rho\cdot(\epsilon\cdot N-1)})$$

is nef. On the other hand, (3.2.2) and (3.2.3) imply that both sheaves coincide on U. □

Choose some $N_0 > 0$ such that for all multiples N of N_0 and for all $M > 0$ the multiplication maps

$$m : S^M(f_*(\mathcal{M}^{\nu\cdot N}\otimes\omega_{X/Y}^{\epsilon\cdot N})) \longrightarrow f_*(\mathcal{M}^{\nu\cdot N\cdot M}\otimes\omega_{X/Y}^{\epsilon\cdot N\cdot M})$$

are surjective over U. Define

$$\rho = \mathrm{Min}\{\mu > 0;\ f_*(\mathcal{M}^{\nu\cdot N}\otimes\omega_{X/Y}^{\epsilon\cdot N})\otimes\mathcal{H}^{\mu\cdot\epsilon\cdot N} \text{ is nef}\}.$$

The surjectivity of m implies that

$$f_*(\mathcal{M}^{\nu\cdot N\cdot M}\otimes\omega_{X/Y}^{\epsilon\cdot N\cdot M})\otimes\mathcal{H}^{\rho\cdot\epsilon\cdot N\cdot M}$$

is nef for all $M > 0$. By 3.3

$$f_*(\mathcal{M}^{\nu\cdot N}\otimes\omega_{X/Y}^{\epsilon\cdot N})\otimes\mathcal{H}^{\rho\cdot(\epsilon\cdot N-1)}.$$

is nef, hence by the choice of ρ

$$(\rho-1)\cdot\epsilon\cdot N < \rho\cdot(\epsilon\cdot N-1)$$

or equivalently $\rho < \epsilon\cdot N$. Then

$$f_*(\mathcal{M}^{N}\otimes\omega_{X/Y}^{\epsilon\cdot N})\otimes\mathcal{H}^{\epsilon^2\cdot N^2}$$

is nef. This remains true if one replaces Y by any finite covering, and by [**9**], 2.2, one obtains that $f_*(\mathcal{M}^N\otimes\omega_{X/Y}^{\epsilon\cdot N})$ is nef. Applying 3.3 a second time, for the numbers (N', N_0) instead of (N, M_0) and for $\rho = 0$, one finds

CLAIM 3.4. For $\nu > 0$ and $\epsilon = e\cdot\nu$ or $\epsilon = e\cdot\nu - 1$ and for all $N' > 0$ the sheaf

$$f_*((\mathcal{M}^{\nu}\otimes\omega_{X/Y}^{\epsilon})^{N'})$$

is nef.

In particular, choosing $N' = 1$ and $\epsilon = \nu e$ one obtains a).

For b) we consider the s-fold product $f^s : X^s \to Y$ for $s = r \cdot r(\nu)$. One has natural inclusions, splitting locally,

$$\mathcal{O}_Y = \det(f_*\mathcal{M})^{r(\nu)} \longrightarrow f^s_*\mathcal{P} = \bigotimes^s f_*\mathcal{M}$$

and

$$\det(f_*(\mathcal{M}^\nu \otimes \omega_{X/Y}^{e\cdot\nu}))^r \longrightarrow f^s_*(\mathcal{P}^\nu \otimes \omega_{X^s/Y}^{e\cdot\nu}) = \bigotimes^s f_*(\mathcal{M}^\nu \otimes \omega_{X/Y}^{e\cdot\nu}).$$

If Δ_1 and Δ_2 denote the corresponding zero-divisors on X^s then $\Delta_1 + \Delta_2$ does not contain any fibre of $V^s \to U$. Then

$$\mathcal{P}^{e\cdot\nu} \otimes \omega_{X^s/Y}^{e\cdot\nu\cdot(e-1)} = f^{s*}\det(f_*(\mathcal{M}^\nu \otimes \omega_{X/Y}^{e\cdot\nu}))^{r\cdot(e-1)} \otimes \mathcal{O}_{X^s}((e-1)\cdot\Delta_2 + \nu\cdot\Delta_1),$$

and

$$\mathcal{P}^\gamma \otimes \omega_{X^s/Y}^{\gamma\cdot(e-1)} = (\mathcal{P} \otimes \omega_{X^s/Y}^{e-1})^{\gamma-\nu e} \otimes \mathcal{O}_X(\gamma\cdot f^{s*}H + (e-1)\cdot\Delta_2 + \nu\cdot\Delta_1)).$$

By 3.4 the sheaf

$$f^s{}_*((\mathcal{P} \otimes \omega_{X^s/Y}^{e-1})^{\gamma-\nu\cdot e})^M = \bigotimes^s f_*((\mathcal{P} \otimes \omega_{X/Y}^{e-1})^{\gamma-\nu\cdot e})^M$$

is nef for all $M > 0$. 3.1 implies that

$$\mathcal{P} \otimes \omega_{X^s/Y}^{e-1} \otimes \omega_{X^s/Y}\left\{-\frac{\gamma\cdot f^{s*}H + (e-1)\cdot\Delta_2 + \nu\cdot\Delta_1}{\gamma}\right\}$$

is nef. By (3.2.2) and (3.2.1)

$$e(((e-1)\cdot\Delta_2 + \nu\cdot\Delta_1)|_{F^s}) \leq e(\mathcal{P}|_{F^s}^{\nu\cdot e} \otimes \omega_{F^s}^{e\cdot\nu\cdot(e-1)}) =$$
$$e(\mathcal{M}|_F^{\nu\cdot e} \otimes \omega_F^{e\cdot\nu\cdot(e-1)}) \leq \gamma$$

for all fibres F of $V \to U$. Hence the cokernel of

$$\omega_{X^s/Y}\left\{\frac{-\gamma\cdot f^{s*}H + (e-1)\cdot\Delta_2 + \nu\cdot\Delta_1}{\gamma}\right\} \to \omega_{X^s/Y}(-f^{s*}H)$$

lies in $X^s \setminus V^s$, and thereby

$$f^s{}_*(\mathcal{P} \otimes \omega_{X^s/Y}^e) \otimes \mathcal{O}_Y(-H) = \Big(\bigotimes^{r\cdot r(\nu)} f_*(\mathcal{M} \otimes \omega_{X/Y}^e)\Big) \otimes \mathcal{O}_Y(-H)$$

is nef.

Part c) follows from part a) and Kollar's ampleness criterion (see [**8**], 4.34). Again we may assume that $\det(f_*\mathcal{M}) = \mathcal{O}_Y$. By part a) for all $\eta > 0$ the sheaf $\mathcal{E} = f_*(\mathcal{M}^\eta \otimes \omega_{X/Y}^{e\cdot\eta})$ is nef. Choose $\nu > 0$ such that the multiplication map

$$\mu : S^\nu(\mathcal{E}^\eta) \to f_*(\mathcal{M}^{\nu\cdot\eta} \otimes \omega_{X/Y}^{e\cdot\nu\cdot\eta})$$

is surjective over U. By [**8**], 4.34, $\det(\mathrm{Im}(\mu))$ is ample, if the kernel $\mathcal{K}$ of the multiplication map is of maximal variation. Let us recall the definition. For a point $y \in U$ choose a local trivialization of $\mathcal{E}$. Then $\mathcal{K}_y = \mathcal{K} \otimes \mathbb{C}(y)$ as a subvectorspace of $S^\nu(\mathbb{C}^{r(\eta)})$, defines a point $[\mathcal{K}_y]$ in the Grassmann variety

$$\mathbb{G}r = \mathrm{Grass}(r(\nu\cdot\eta), S^\nu(\mathbb{C}^{r(\eta)})).$$

The group $G = \mathrm{Sl}(r(\eta), \mathbb{C})$ acts on $S^\nu(\mathbb{C}^{r(\eta)})$, hence on $\mathbb{G}r$. Let G_y denote the orbit of $[\mathcal{K}_y]$. The kernel has maximal variation, if

$$\{z \in Y;\ G_z = G_y\}$$

is finite, as well as the stabilizer of $[\mathcal{K}_y]$.

The second condition holds true for η and ν sufficiently large. In fact, $\mathcal{K}_y$ determines the fibre $f^{-1}(y)$ as a subvariety of

$$\mathbb{P}(H^0(f^{-1}(y), (\mathcal{M}^\eta \otimes \omega_{X/Y}^{e\cdot\eta})|_{f^{-1}(y)})),$$

and $[\mathcal{K}_y]$ is nothing but the point of the Hilbert scheme Hilb parameterizing subvarieties of this projective space. By [**8**], 7.2, the stabilizer of such a point is finite.

The assumption in c) implies that $\mathcal{K}$ is not of maximal variation. Hence for all points z in a neighborhood U_y of y, the orbits G_z coincide. In different terms the images of $z \in U_y$ in Hilb all belong to the same G-orbit. Since M_h is a quotient of a subscheme of Hilb by the G-action, the morphism $\varphi : U \to M_h$ is constant, as claimed in c). □

4. Families of canonically polarized manifolds

For families with ω_F big and semi-ample the equivalence of birational and biregular isotriviality has been shown by Maehara in [**6**]. For families of canonically polarized manifolds, one just has to use, that the fibres are their own canonical model.

Or, to formulate he proof parallel to the one given below in the Kodaira dimension zero case, one could argue in the following way. Assume that U is a curve, and choose a semistable compactification $f : X \to Y$ of $V \to U$. By assumption X is birational to the trivial family $F' \times Y$ over Y, hence $f_*\omega_{X/Y}^\nu$ is a direct sum of copies of $\mathcal{O}_Y$, for all ν. Obviously, if $\mathcal{M}$ is some power of $\omega_{X/Y}$ this implies that

$$\det(f_*(\mathcal{M}^\nu \otimes \omega_{X/Y}^{e\cdot\nu})) \otimes \det(f_*\mathcal{M})^{-\nu r(\nu)} = \mathcal{O}_Y,$$

and by 3.2, c), one finds $V \to U$ to be biregulary isotrivial.

5. Families of manifolds of Kodaira dimension zero

Let U be a curve and $f : V \to U$ be a family of polarized manifolds F with $\omega_F^\delta = \mathcal{O}_F$. In order to prove 2.1 we may replace U by some finite cover, and we may choose a compactification $f : X \to Y$ satisfying the assumptions made in Theorem 3.2.

By assumption, $\lambda = f_*\omega_{X/Y}^\delta$ is an invertible sheaf, and the natural map $f^*\lambda \to \omega_{X/Y}^\delta$ is an isomorphism over V. Let E be the zero divisor of this map, i.e.

$$\omega_{X/Y}^\delta = f^*\lambda \otimes \mathcal{O}_X(E).$$

E is supported in $\Delta = X \setminus V$ and, since the fibres of f are reduced divisors, E can not contain a whole fibre. Hence for all $\mu > 0$ $f_*\mathcal{O}_X(\mu E) = \mathcal{O}_Y$ and $f_*\omega_{X/Y}^{\mu\cdot\delta} = \lambda^\mu$.

Let $\mathcal{M}'$ be any polarization which is very ample and without higher cohomology on the fibres of $V \to U$. The sheaf

$$f_*(\mathcal{M}' \otimes \mathcal{O}_X(*E)) = f_*(\mathcal{M}' \otimes (\lim_{\mu>0} \mathcal{O}(\mu E)))$$

is coherent, hence locally free.

In fact, locally étale or locally analytic we can choose over a neighborhood $\mathcal{U}$ of $s \in S = Y \setminus U$ a section $\sigma : \mathcal{U} \to X$ with image C, not meeting the support of E. If I denotes the ideal sheaf of C, for some $\rho \gg 0$

$$f_*(\mathcal{M}'|_{f^{-1}(\mathcal{U})} \otimes I^\rho) = 0.$$

Since direct images are torsion free, and since E is supported in fibres,

$$f_*((\mathcal{M}' \otimes \mathcal{O}_X(*E))|_{f^{-1}(\mathcal{U})} \otimes I^\rho) = 0.$$

Then $f_*(\mathcal{M}' \otimes \mathcal{O}_X(*E))|_{\mathcal{U}}$ is a torsion free subsheaf of

$$f_*((\mathcal{M}' \otimes \mathcal{O}_X(*E))|_{f^{-1}(\mathcal{U})} \otimes \mathcal{O}_{f^{-1}(\mathcal{U})}/I^\rho) = f_*(\mathcal{M}'|_{f^{-1}(\mathcal{U})} \otimes \mathcal{O}_{f^{-1}(\mathcal{U})}/I^\rho),$$

hence coherent.

Let $\mathcal{M}$ be the reflexive hull of the image of

$$f^*f_*(\mathcal{M}' \otimes \mathcal{O}_X(*E)) \longrightarrow \mathcal{M}' \otimes \mathcal{O}_X(*E).$$

$\mathcal{M}$ is again coherent, and it must be contained in $\mathcal{M}' \otimes \mathcal{O}_X(\alpha \cdot E)$ for some α. Since it is reflexive, it is an invertible sheaf. By construction $\mathcal{M}|_V \simeq \mathcal{M}'|_V$ and

$$f_*(\mathcal{M}' \otimes \mathcal{O}_X(*E)) = f_*f^*f_*(\mathcal{M}' \otimes \mathcal{O}_X(*E)) \subset \\ f_*\mathcal{M} \subset f_*(\mathcal{M} \otimes \mathcal{O}_X(*E)) \subset f_*(\mathcal{M}' \otimes \mathcal{O}_X(*E)),$$

hence all those sheaves coincide. We found an invertible sheaf $\mathcal{M}$ satisfying the assumptions made in 3.2 with the additional condition

$$f_*(\mathcal{M} \otimes \omega^e_{X/Y}) = f_*(\mathcal{M} \otimes \mathcal{O}_X(\frac{e}{\delta} \cdot E) \otimes f^*\lambda^{\frac{e}{\delta}}) = f_*\mathcal{M} \otimes \lambda^{\frac{e}{\delta}},$$

for all multiples e of δ. For those e

$$(5.0.1)\quad (\bigotimes^r f_*(\mathcal{M} \otimes \omega^e_{X/Y})) \otimes \det(f_*\mathcal{M})^{-1} = \\ (\bigotimes^r (f_*\mathcal{M}) \otimes \lambda^{\frac{e}{\delta}}) \otimes \det(f_*\mathcal{M})^{-1}.$$

PROOF OF 2.1 FOR $\omega^\delta_F = \mathcal{O}_F$.
By assumption $f : X \to Y$ is birational over Y to the trivial family $pr_2 : F' \times Y \to Y$, hence

$$\lambda = f_*\omega^\delta_{X/Y} = \mathcal{O}_Y.$$

So the sheaf in (5.0.1) is

$$(\bigotimes^r (f_*\mathcal{M})) \otimes \det(f_*\mathcal{M})^{-1}.$$

Since its determinant is of degree zero, it can not be ample. (5.0.1) and 3.2, b), imply that for no $\nu > 0$ the sheaf

$$\det(f_*(\mathcal{M}^\nu \otimes \omega^{e\cdot\nu}_{X/Y})) \otimes \det(f_*\mathcal{M})^{-\nu\cdot r(\nu)}$$

is ample. By 3.2, a), it is of non negative degree, and 3.2, c), implies that the family $V \to U$ is biregulary isotrivial. □

REMARK 5.1. Assume that $\delta = 1$, hence that $\omega_{V/U} = f^*\lambda|_U$. The argument used in the proof of 2.1 shows in this particular case that "f non-isotrivial" implies that on the compactification Y of U

$$\deg(f_*\omega_{X/Y}) > 0.$$

Since the same holds true for all finite coverings of U, one obtains that the fibres of the period map from M_h to the period domain classifying the corresponding variations of Hodge structures can not contain a quasi-projective curve. Of course this is a well known consequence of the local Torelli Theorem for manifolds with a trivial canonical bundle.

6. Kodaira-Spencer maps

Recall first the following definition, replacing of nef and ample, on projective manifolds Y of higher dimension.

DEFINITION 6.1. Let $\mathcal{F}$ be a torsion free coherent sheaf on a quasi-projective normal variety Y and let $\mathcal{H}$ be an ample invertible sheaf.

a) $\mathcal{F}$ is generically generated if the natural morphism
$$H^0(Y,\mathcal{F}) \otimes \mathcal{O}_Y \longrightarrow \mathcal{F}$$
is surjective over some open dense subset U_0 of Y. If one wants to specify U_0 one says that $\mathcal{F}$ is globally generated over U_0.

b) $\mathcal{F}$ is weakly positive if there exists some dense open subset U_0 of Y with $\mathcal{F}|_{U_0}$ locally free, and if for all $\alpha > 0$ there exists some $\beta > 0$ such that
$$S^{\alpha\cdot\beta}(\mathcal{F}) \otimes \mathcal{H}^{\beta}$$
is globally generated over U_0. We will also say that $\mathcal{F}$ is weakly positive over U_0, in this case.

c) $\mathcal{F}$ is big if there exists some open dense subset U_0 in Y and some $\mu > 0$ such that
$$S^{\mu}(\mathcal{F}) \otimes \mathcal{H}^{-1}$$
is weakly positive over U_0. Underlining the role of U_0 we will also call $\mathcal{F}$ ample with respect to U_0.

Here, as in [**8**] and [**10**], we use the following convention: If $\mathcal{F}$ is a coherent torsion free sheaf on a quasi-projective normal variety Y, we consider the largest open subscheme $i : Y_1 \to Y$ with $i^*\mathcal{F}$ locally free. For
$$\Phi = S^{\mu}, \quad \Phi = \bigotimes^{\mu} \quad \text{or} \quad \Phi = \det$$
we define
$$\Phi(\mathcal{F}) = i_*\Phi(i^*\mathcal{F}).$$

Again, $f : V \to U$ denotes a smooth family of manifolds over a quasi-projective manifold U, which is allowed to be of dimension larger than one. We choose non-singular projective compactifications Y of U and X of V, such that both $S = Y \setminus U$ and $\Delta = X \setminus V$ are normal crossing divisors and such that f extends to $f : X \to Y$. As usual η will denote a closed point in sufficient general position on U and X_η the fibre of f over η. We will write $T^i_{X_\eta}$ (or $T^i_{X/Y}(-\log\Delta)$ ...) for the i-th wedge product of T_{X_η} (or of $T_{X/Y}(-\log\Delta) = \Omega^1_{X/Y}(\log\Delta)^\vee$ ...).

Let T_η denote the restriction $T_U \otimes \mathbb{C}$ of the tangent sheaf of U to η. The Kodaira-Spencer map
$$T_\eta \longrightarrow H^1(X_\eta, T_{X_\eta})$$
gives rise to
$$\bigotimes^{\nu} T_\eta \longrightarrow \bigotimes^{\nu} H^1(X_\eta, T_{X_\eta}) \longrightarrow H^\nu(X_\eta, T^\nu_{X_\eta}).$$

The composite map factors through

$$S^\nu(T_\eta) \longrightarrow H^\nu(X_\eta, T^\nu_{X_\eta}).$$

One defines

$$\mu(f) = \mathrm{Max}\{\nu \in \mathbb{N} - \{0\};\ S^\nu(T_\eta) \longrightarrow H^\nu(X_\eta, T^\nu_{X_\eta}) \text{ is non zero}\}.$$

Of course, $\mu(f) \leq n = \dim(X_\eta)$. We do not know any criterion, implying that for $f : V \to U$ one has $\mu(f) = \dim(V) - \dim(U)$.

For example, if U is a curve and $f : V \to U$ a family of polarized manifolds, restricting the tautological sequence to $X_\eta = f^{-1}(\eta)$ one obtains an extension

$$0 \longrightarrow T_{X_\eta} \longrightarrow T_X|_{X_\eta} \longrightarrow \mathcal{O}_{X_\eta} \longrightarrow 0$$

and the induced class $\xi_\eta \in H^1(X_\eta, T_{X_\eta})$. Then $\mu(f) = \mu$ if and only if for η in general position, the wedge product $\wedge^\mu \xi_\eta \in H^\mu(X_\eta, T^\mu_{X_\eta})$ is non-zero, whereas $\wedge^{\mu+1}\xi_\eta \in H^{\mu+1}(X_\eta, T^{\mu+1}_{X_\eta})$ is zero.

PROBLEM 6.2. Are there properties of X_η which imply that for all families $V \to U$ over a curve U, with general fibre $X_\eta = f^{-1}(\eta)$ the class

$$\wedge^n \xi_\eta \in H^n(X_\eta, \omega^{-1}_{X_\eta})$$

is non-zero?

Being optimistic, one could try in 6.2 the condition "$\Omega^1_{X_\eta}$ ample".

A slight extension of the main result of [**10**] says:

THEOREM 6.3. *Assume that for a general fibre X_η of $f : X \to Y$ either ω_{X_η} is ample, or $\omega^\delta_{X_\eta} = \mathcal{O}_{X_\eta}$, for some δ.*

i. *Then for some $m > 0$ the sheaf $S^m(\Omega^1_Y(\log S))$ contains an invertible subsheaf $\mathcal{M}$ of Kodaira dimension* $\mathrm{Var}(f)$.
ii. *If* $\mathrm{Var}(f) = \dim(U)$ *the sheaf $S^{\mu(f)}(\Omega^1_Y(\log S))$ contains a big coherent subsheaf $\mathcal{P}$.*
iii. *Let Z be a submanifold of Y such that $S_Z = S \cap Z$ remains a normal crossing divisor, and such that $W = X \times_Y Z$ is non-singular. For the induced family $h : W \to Z$ assume that $\mu(f) = \mu(h)$. Then, if* $\mathrm{Var}(f) = \dim(U)$*, the restriction of the sheaf $\mathcal{P}$ from part ii) to $S^{\mu(f)}(\Omega^1_Z(\log S_Z))$ is non trivial.*
iv. *Assume in iii) that $h : W \to Z$ is a desingularization of the pullback of a family $h' : W' \to Z'$ under $\pi : Z \to Z'$, with Z' non-singular and with h' smooth over $Z' \setminus S_{Z'}$ for a normal crossing divisor $S_{Z'}$. Then then the restriction of the sheaf $\mathcal{P}$ to $S^{\mu(f)}(\Omega^1_Z(\log S_Z))$ lies in $S^{\mu(f)}(\pi^*(\Omega^1_{Z'}(\log S_{Z'})))$.*

PROOF. Parts i) and ii) have been shown in [**10**], 1.4, for canonically polarized manifolds with $\mu(f)$ replaced by the fibre dimension n. We will just sketch the changes which allow to extend the arguments used in [**10**] to cover 6.3, ii), iii) and iv), for canonically polarized manifolds. Next we will try to convince the reader, that the same proof goes through for minimal models of Kodaira dimension zero.

As in [**10**] we drop the assumption that Y is projective. Leaving out a codimension two subscheme, we may assume that f is flat and that Δ is a relative normal crossing divisor. Then we have the tautological exact sequence

$$0 \longrightarrow f^*\Omega^1_Y(\log S) \longrightarrow \Omega^1_X(\log \Delta) \longrightarrow \Omega^1_{X/Y}(\log \Delta) \longrightarrow 0$$

and the wedge product sequences

$$(6.3.1)\quad 0 \longrightarrow f^*\Omega^1_Y(\log S)\otimes \Omega^{p-1}_{X/Y}(\log \Delta) \longrightarrow \mathfrak{gr}(\Omega^p_X(\log \Delta)) \longrightarrow \Omega^p_{X/Y}(\log \Delta) \longrightarrow 0,$$

where

$$\mathfrak{gr}(\Omega^p_X(\log \Delta)) = \Omega^p_X(\log \Delta)/f^*\Omega^2_Y(\log S)\otimes \Omega^{p-2}_{X/Y}(\log \Delta).$$

For the invertible sheaf $\mathcal{L} = \Omega^n_{X/Y}(\log \Delta)$ we consider the sheaves

$$F^{p,q} := R^q f_*(\Omega^p_{X/Y}(\log \Delta)\otimes \mathcal{L}^{-1})$$

together with the edge morphisms

$$\tau_{p,q} : F^{p,q} \longrightarrow F^{p-1,q+1}\otimes \Omega^1_Y(\log S),$$

induced by the exact sequence (6.3.1), tensored with $\mathcal{L}^{-1}$. As explained in [**10**], Proof of 4.4 iii), over U the edge morphisms $\tau_{p,q}$ can also be obtained in the following way. Consider the exact sequence

$$0 \longrightarrow T_{V/U} \longrightarrow T_V \longrightarrow f^*T_U \longrightarrow 0,$$

and the induced wedge product sequences

$$0 \longrightarrow T^{n-p+1}_{V/U} \longrightarrow \tilde{T}^{n-p+1}_V \longrightarrow T^{n-p}_{V/U}\otimes f^*T_U \longrightarrow 0,$$

where $\tilde{T}^{n-p+1}_V$ is a subsheaf of T^{n-p+1}_V. One finds edge morphisms

$$\tau^\vee_{p,q} : (R^q f_* T^{n-p}_{V/U})\otimes T_U \longrightarrow R^{q+1} f_* T^{n-p+1}_{V/U}.$$

Restricted to η those are just the wedge product with the Kodaira-Spencer class. Moreover, tensoring with Ω^1_U one gets back $\tau_{p,q}|_U$. Hence $\mu(f)$ is the smallest number m for which the composite

$$\tau^m : F^{n,0} = \mathcal{O}_Y \xrightarrow{\tau_{n,0}} F^{n-1,1}\otimes \Omega^1_U \xrightarrow{\tau_{n-1,1}} F^{n-2,2}\otimes S^2(\Omega^1_U) \longrightarrow \cdots \xrightarrow{\tau_{n-m+1,m-1}} F^{n-m,m}\otimes S^m(\Omega^1_U)$$

is non-zero. Next we used that (replacing Y by some covering) there is an ample invertible sheaf $\mathcal{A}$ on Y such that the kernel $\mathcal{K}$ of

$$\mathrm{id}_{\mathcal{A}}\otimes \tau_{n-m,m} : \mathcal{A}\otimes F^{n-m,m} \longrightarrow \mathcal{A}\otimes F^{n-m-1,m+1}\otimes \Omega^1_Y(\log S)$$

is negative, or to be more precise, that its dual is weakly positive. This gives a non-trivial map

$$\upsilon : \mathcal{A}\otimes \mathcal{K}^\vee \longrightarrow S^m(\Omega^1_Y(\log S))$$

and we take for $\mathcal{P}$ its image.

The number m was used in the proof of [**10**], 1.4 ii), hence there is no harm to replace the upper bound n, used there, by the more precise number $\mu(f)$ in 6.3, ii).

The sheaves $F^{p,q}$ are compatible with restriction to the subvariety Z. The assumption $\mu(f) = \mu(h)$ implies that the restriction

$$\upsilon|_Z : \mathcal{A}|_Z\otimes \mathcal{K}^\vee|_Z \longrightarrow S^m(\Omega^1_Z(\log S_Z))$$

is non-trivial. In fact, the kernel $\mathcal{K}'$ of

$$\mathrm{id}_{\mathcal{A}_Z}\otimes \tau^Z_{n-m,m} : \mathcal{A}|_Z\otimes F^{n-m,m}|_Z \longrightarrow \mathcal{A}|_Z\otimes F^{n-m-1,m+1}|_Z\otimes \Omega^1_Z(\log S_Z)$$

contains $\mathcal{K}|_Z$, and the diagram

$$\begin{array}{ccc} \mathcal{A}|_Z \otimes \mathcal{K}^\vee|_Z & \longrightarrow & S^m(\Omega^1_Y(\log S))|_Z \\ \uparrow & & \downarrow \\ \mathcal{A}|_Z \otimes \mathcal{K}'^\vee & \longrightarrow & S^m(\Omega^1_Z(\log S_Z)) \end{array}$$

is commutative. One obtains iii).

Since the sheaves $F^{p,q}$ and the maps $\tau_{p,q}$ are compatible with pullbacks, under the additional assumptions made in iv), the image of

$$\mathcal{A}|_Z \otimes \mathcal{K}'^\vee \longrightarrow S^m(\Omega^1_Z(\log S_Z))$$

lies in $S^m(\pi^*(\Omega^1_{Z'}(\log S_{Z'})))$ and the same holds true for the restriction of $\mathcal{P}$.

If one considers the proof of [**10**], 1.4, i) and ii), the assumption that the fibres are canonically polarized is used twice. First of all, since we apply in the proof of 4.4, iv), the Akizuki-Kodaira-Nakano vanishing theorem to the restriction of ω_F to a smooth multicanonical divisor B. If some power of ω_F is trivial the divisor B is empty, and there is nothing to show.

The second time is in the proof of [**10**], 4.8. We use the diagram (2.8.1) and the fact that the morphism $Z^\# \to Y^\#$ considered there is of maximal variation. The construction of (2.8.1) just uses the existence of the moduli scheme M_h, and it provides a morphism $Z^\# \to Y^\#$ induced by a generically finite morphism $Y^\# \to M_h$.

This construction works in particular for the moduli scheme of polarized manifolds F with $\omega_F^\delta = \mathcal{O}_F$, and 1.2 implies that the variation of the morphism $Z^\# \to Y^\#$ is again maximal.

The rest of the arguments, as given on page 311–313 of [**10**] remain unchanged, and one obtains 6.3, i), ii) and iii). □

For families $f : X \to Y$ with ω_{X_η} semiample, and with $\mu(f) = n$ one can add to [**10**], 1.4, a statement similar to 6.3, iii) and iv). Since the later will not be used, we omit it.

THEOREM 6.4. *Assume ω_{X_η} is semiample, and* $\mathrm{Var}(f) = \dim(f)$.

i. *There exists a non-singular finite covering $\psi : Y' \to Y$ and a big coherent subsheaf $\mathcal{P}'$ of $\psi^* S^m(\Omega^1_Y(\log S))$, for some $m \leq \mu(f)$.*
ii. *If $\mu(f) = n$, then one finds $m = n$ in i).*
iii. *Let Z be a submanifold of Y such that $S_Z = S \cap Z$ remains a normal crossing divisor, and such that $W = X \times_Y Z$ is non-singular. For the induced family $h : W \to Z$ assume that $\mu(f) = \mu(h) = n$. Then one can choose the covering ψ such that $\psi^{-1}(Z)$ is non-singular, $\psi^{-1}(S_Z)$ a normal crossing divisor and such that the image of the sheaf $\mathcal{P}'$ from part i) in $\psi^* S^n(\Omega^1_Z(\log S_Z))$ is non trivial.*

PROOF. We keep the notations from the sketch of the proof of 6.3. For part i) we replaced in [**10**], page 309 and 310, the sheaves $F^{p,q}$ (in fact a twist of those by some invertible sheaf on Y) by some quotient sheaves. But then $\mu(f)$ remains an upper bound for the number m, used there, and one obtains 6.3, i), as stated.

However, one has no control on the behavior of m under restriction to subvarieties. So for part ii) and iii) we have to recall the construction in more detail. To get the weak positivity of the kernels $\mathcal{K}$ one has to replace (over some covering Y' of Y whose ramification divisor is in general position) $\mathcal{A} \otimes F^{n-m,m}$ by its image $\mathcal{A} \otimes \tilde{F}^{n-m,m}$ in some larger sheaf $E^{n-m,m}$. Here

$$(\bigoplus_{p+q=n} E^{p,q}, \bigoplus_{p+q=n} \theta_{p,q})$$

is again a Higgs bundle, and $\theta_{p,q}$ is compatible with $\tau_{p,q}$.

As stated in the proof of [**10**], 4.4, iv) the kernel and cokernel of the map

$$\mathcal{A} \otimes F^{n-m,m} \to E^{n-m,m}$$

are direct images of the $n-m-1$-forms of a multicanonical divisor. If $n = m$ there are no such forms, and $\mathcal{A} \otimes F^{0,n}$ is a subsheaf of $E^{0,n}$. Hence $\tau^n \neq 0$ implies that the corresponding map for $\mathcal{A} \otimes \tilde{F}^{n-m,m}$ is non-zero. The compatibility with restrictions follows by the argument used in the proof of 6.3, iii). □

7. Subvarieties of the moduli stack of polarized manifolds of Kodaira dimension zero

Theorem 6.3 has a number of geometric implication for manifolds U mapping to moduli stacks of polarized manifolds, i.e. for morphisms $\varphi : U \to M_h$ induced by a family $f : V \to U$. Those had been shown in [**10**] for the moduli stack of canonically polarized manifolds. The proves are all based on vanishing theorems for logarithmic differential forms, and they do not refer to the type of fibres of f, once 6.3, i) and ii), is established.

Using 1.2 we extended 6.3, i) and ii), to a larger class of families of polarized manifolds. Hence the geometric implications carry over to this larger class, i.e. to the moduli stack of polarized manifolds with $\omega_F^\delta = \mathcal{O}_F$, for some $\delta > 0$. For the readers convenience we recall the statements below.

For polarized manifolds with ω_F semiample, the lack of an affirmative answer to Problem 1.3 still forces us to assume that $\mathrm{Var}(f) = \dim(U)$ to obtain similar results.

THEOREM 7.1 (see [**10**], 5.2, 5.3, 7.2, 6.4, and 6.7). *Let M_h be the moduli scheme of canonically polarized manifolds, or of polarized manifolds F with ω_F^δ trivial for some $\delta > 0$.*

I. *Assume that U satisfies one of the following conditions*

a) *U has a non-singular projective compactification Y with $S = Y \setminus U$ a normal crossing divisor and with boundary $T_Y(-\log S)$ weakly positive.*

b) *Let $H_1 + \cdots + H_\ell$ be a reduced normal crossing divisor in $\mathbb{P}^N$, and $\ell < \frac{N}{2}$. For $0 \leq r \leq l$ define*

$$H = \bigcap_{j=r+1}^{\ell} H_j, \quad S_i = H_i|_H, \quad S = \sum_{i=1}^{r} S_i,$$

and assume $U = H \setminus S$.

c) *$U = \mathbb{P}^N \setminus S$ for a reduced normal crossing divisor*

$$S = S_1 + \cdots + S_\ell$$

in $\mathbb{P}^N$, *with* $\ell < N$.

Then a morphism $U \to M_h$, *induced by a family, must be trivial.*

II. *For* $Y = \mathbb{P}^{\nu_1} \times \cdots \times \mathbb{P}^{\nu_k}$ *let*

$$D^{(\nu_i)} = D_0^{(\nu_i)} + \cdots + D_{\nu_i}^{(\nu_i)}$$

be coordinate axes in $\mathbb{P}^{\nu_i}$ *and*

$$D = \sum_{i=1}^{k} D^{(\nu_i)}.$$

Assume that $S = S_1 + \cdots S_\ell$ *is a divisor, such that* $D + S$ *is a reduced normal crossing divisor, and* $\ell < \dim(Y)$. *Then there exists no morphism* $\varphi : U = Y \setminus (D + S) \to M_h$ *with*

$$\dim(\varphi(U)) > \mathrm{Max}\{\dim(Y) - \nu_i;\ i = 1, \ldots, k\}.$$

III. *Let* U *be a quasi-projective variety and let* $\varphi : U \to M_h$ *be a quasi-finite morphism, induced by a family. Then* U *can not be isomorphic to the product of more than* $\mu(f)$ *varieties of positive dimension.*

8. Rigidity

Again, $f : V \to U$ denotes a smooth family of manifolds with $\omega_{V/U}$ f-semiample and with $\mathrm{Var}(f) = \dim(U) > 0$. We say that f is rigid, if there exists no non-trivial deformation over a non-singular quasiprojective curve T.

Here a deformation of f over T, with $0 \in T$ a base point, is a smooth projective morphism

$$g : \mathcal{V} \to U \times T$$

for which there exists a commutative diagram

$$\begin{array}{ccccc} V & \xrightarrow{\simeq} & g^{-1}(U \times \{0\}) & \xrightarrow{\subset} & \mathcal{V} \\ {\scriptstyle f}\downarrow & & \downarrow & & \downarrow{\scriptstyle g} \\ U & \xrightarrow{\simeq} & U \times \{0\} & \xrightarrow{\subset} & U \times T \end{array} .$$

If the fibres F of f are canonically polarized, or if some power of ω_F is trivial, this says that morphisms from U to the moduli stack do not deform.

PROPOSITION 8.1. *Assume either that* ω_{X_η} *is ample, or that* $\omega_{X_\eta}^{\delta} = \mathcal{O}_{X_\eta}$, *for some* δ. *Assume that* $\mathrm{Var}(f) = \dim(U) > 0$. *Let* T *be a non-singular quasi-projective curve. Let* $g : \mathcal{V} \to U \times T$ *be a deformation of* f. *If* $\mu(f) = \mu(g)$, *then* $\mathrm{Var}(g) = \dim(U)$.

PROOF. Suppose that $\mathrm{Var}(g) > \dim(U)$. Then

$$\dim(U) + 1 = \dim U \times T \geq \mathrm{Var}(g) > \dim(U),$$

hence, $\dim(U \times T) = \mathrm{Var}(g)$. Let $\bar{T}$ be a non-singular compactification of T, $S_{\bar{T}} = \bar{T} \setminus T$. Correspondingly we write $S_{Y \times \bar{T}}$ for the complement of $U \times T$ in $Y \times \bar{T}$.

By Theorem 6.3, ii, one finds a big coherent subsheaf $\mathcal{P}$ of

$$S^{\mu(g)}(\Omega^1_{Y \times \bar{T}}(\log S_{Y \times \bar{T}})),$$

and by 6.3, iii) the image of $\mathcal{P}$ in

$$S^{\mu(g)}(\Omega^1_{Y \times \{0\}}(\log S_{Y \times \{0\}})) = pr_1^*(S^{\mu(g)}(\Omega^1_Y(\log S)))|_{Y \times \{0\}}$$

is non-zero. Then, the image of $\mathcal{P}$ in

$$pr_1^*(S^{\mu(g)}(\Omega^1_Y(\log S)))$$

is non-zero, and for a point $y \in Y$ in general position, the image of $\mathcal{P}$ under

$$S^{\mu(g)}(\Omega^1_{Y\times\bar{T}}(\log S_{Y\times\bar{T}}))|_{\{y\}\times T} \longrightarrow pr_1^*(S^{\mu(g)}(\Omega^1_Y(\log S)))|_{\{y\}\times T}$$

is not zero. Note that any non-zero quotient of the coherent sheaf $\mathcal{P}|_{\{y\}\times T}$ for y in general position must be big. In fact, if $\mathcal{P}$ is ample over some open dense subset W_0 of $Y \times \bar{T}$, one just has to make sure that $\{y\} \times T$ meets W_0. Since $pr_1^*(S^{\mu(g)}(\Omega^1_Y(\log S)))|_{\{y\}\times T}$ is a direct sum of copies of $\mathcal{O}_{\{y\}\times T}$ this is not possible. □

Using 6.4 instead of 6.3 one obtains a similar result for families with $\omega_{V/U}$ f semi-ample, whenever $\mu(f) = n$.

PROPOSITION 8.2. *Assume that* ω_{X_η} *is semiample, that*

$$\mathrm{Var}(f) = \dim(U) > 0$$

and that

$$\mu(f) = \dim(X_\eta) = n.$$

Let T be a non-singular quasi-projective curve, and let $g : \mathcal{V} \to U \times T$ be a deformation of f. Then $\mathrm{Var}(g) = \dim(U)$.

PROOF. If $\mathrm{Var}(g) > \dim(U)$, again one finds $\dim(U\times T) = \mathrm{Var}(g)$. Let us keep the notations from the proof of 8.1. By Theorem 6.4, ii, one finds a finite covering $\psi : Y' \to Y$ and a big coherent subsheaf $\mathcal{P}'$ of

$$\psi^*(S^{\mu(g)}(\Omega^1_{Y\times\bar{T}}(\log S_{Y\times\bar{T}}))),$$

and by 6.4, iii) the image of $\mathcal{P}'$ in

$$\psi^*(S^{\mu(g)}(\Omega^1_{Y\times\{0\}}(\log S_{Y\times\{0\}}))) = \psi^* pr_1^*(S^{\mu(g)}(\Omega^1_Y(\log S))|_{Y\times\{0\}})$$

is non-zero. Then, for a point $y \in Y$, in general position, the image of $\mathcal{P}'$ under

$$\psi^*(S^{\mu(g)}(\Omega^1_{Y\times\bar{T}}(\log S_{Y\times\bar{T}}))|_{\{y\}\times T}) \longrightarrow \psi^*(pr_1^*(S^{\mu(g)}(\Omega^1_Y(\log S)))|_{\{y\}\times T})$$

is not zero. Again, since the sheaf on the right hand side is trivial, one obtains a contradiction. □

Let $h : \mathcal{X} \to Z$ be a polarized family of manifolds F with ω_F semiample and of maximal variation, over a non-singular quasi-projective manifold Z. Assume that $\bar{Z}$ is a projective compactification of Z, such that $\bar{Z} \setminus Z$ is a normal crossing divisor. Assume in addition, that there is an open dense subscheme Z_0 such that for all subvarieties $\mathcal{U}$ of Z meeting Z_0

$$\mathrm{Var}(\mathcal{X} \times_Z \mathcal{U} \to \mathcal{U}) = \dim(\mathcal{U}).$$

Let Y be a non-singular projective curve and let $U \subset Y$ be open and dense. Let us write

$$\mathbf{H} = \mathbf{Hom}((Y,U),(\bar{Z},Z))$$

for the scheme parameterizing non-trivial morphisms $\psi : Y \to \bar{Z}$ with $\psi(U) \subset Z$ and

$$\mathbf{H}_{Z_0} = \mathbf{Hom}((Y,U),(\bar{Z},Z);Z_0) \subset \mathbf{H}$$

for those with $\psi(U) \cap Z_0 \neq \emptyset$. Based on the bounds obtained in [**9**] we have shown in [**10**] that $\mathbf{H}_{Z_0}$ is of finite type.

COROLLARY 8.3.

I. a. *Let $\psi : U \to Z$ be a morphism and $f : V \to U$ the pull back family. Assume that $\psi(U) \cap Z_0 \neq \emptyset$ and that*

$$\mu(f : V \to U) = \dim(F) = n$$

Then the point $[\psi : Y \to \bar{Z}]$ is isolated in $\mathbf{H}_{Z_0}$.

b. *Assume for all fibres F of $h^{-1}(Z_0) \to Z_0$ and for all $\xi \in H^1(F, T_F)$*

$$0 \neq \wedge^n \xi \in H^n(F, \omega_F^{-1}).$$

Then $\mathbf{H}_{Z_0}$ is a finite set of points.

II. *Assume that the fibres F of $h : \mathcal{X} \to Z$ are either canonically polarized, or of Kodaira dimension zero.*

a. *Let $\psi : U \to Z$ be a morphism and $f : V \to U$ the pull back family. Assume that $\psi(U) \cap Z_0 \neq \emptyset$ and that*

$$\mu(f : V \to U) = \mu(h : \mathcal{X} \to Z).$$

Then the point $[\psi : Y \to \bar{Z}]$ is isolated in $\mathbf{H}_{Z_0}$.

b. *Assume there exists a constant μ such that for all fibres F of $h^{-1}(Z_0) \to Z_0$ and for all $\xi \in H^1(F, T_F)$*

$$0 \neq \wedge^\mu \xi \in H^\mu(F, T_F^\mu)$$

but

$$0 = \wedge^{\mu+1} \xi \in H^{\mu+1}(F, T_F^{\mu+1}).$$

Then $\mathbf{H}_{Z_0}$ is a finite set of points.

PROOF. In both cases b) follows from a). For the latter assume that $[\psi]$ lies in a component of $\mathbf{H}$ of dimension larger than zero. Let T be a curve in $\mathbf{H}$, containing the point $[\psi]$. Then one has a non-trivial deformation $\Psi : U \times T \to Z$ of ψ, hence a non-trivial deformation $g : \mathcal{V} \to U \times T$ of $f : V \to U$. By 8.1 in case II) or by 8.2 in case I)

$$\mathrm{Var}(g) = \mathrm{Var}(f) < \dim(U \times T) = \dim(U) + 1,$$

contradicting the assumption made on Z_0 and $\mathcal{X} \to Z$. □

Corollary 8.3, II), should imply certain finiteness results for curves in the moduli scheme M_h of canonically polarized manifolds, or the moduli scheme of minimal models of Kodaira dimension zero meeting an open subscheme W where the assumption corresponding to the one in 8.3, II), b), holds true. However, one would have to show, that morphisms φ which factor through the moduli stack, are parameterized by some coarse moduli scheme. Hopefully this can be done extending the methods used in [**2**] for moduli of curves to moduli of higher dimensional manifolds.

Here we will show a slightly weaker statement, which coincides with 1.4 for $\mu = n$.

COROLLARY 8.4. *Let M_h be either the moduli scheme of canonically polarized manifolds or the moduli scheme of polarized manifolds F with $\omega_F^\delta = \mathcal{O}$ for some $\delta > 0$. Let $0 < \mu \leq n = \dim(F)$ be a constant such that for all $(F, \mathcal{L})$, and for all $\xi \in H^1(F, T_F)$*

$$0 = \wedge^{\mu+1} \xi \in H^{\mu+1}(F, T_F^{\mu+1}).$$

Then for a quasi-projective non-singular curve U there are only finitely many morphisms $\varphi : U \to M_h$ which are induced by a smooth family $f : V \to U$ with $\mu(f) = \mu$.

PROOF. Let us choose any projective compactification $\bar{M}_h$ of M_h, and an invertible sheaf $\mathcal{H}$ on $\bar{M}_h$ which is ample with respect to M_h. As usual, Y will be a non-singular projective curve containing U. We write s for the number of points in $S = Y \setminus U$ and $g(Y)$ for the genus of Y.

To show that there are only finitely many components of the scheme

$$\mathbf{Hom}((Y, U), (M_h, \bar{M}_h))$$

which contain a morphism $\varphi : U \to M_h$ factoring through the moduli stack, one has to find an upper bound for $\varphi^*\mathcal{H}$. To this aim one may assume that $\bar{M}_h$ is reduced. The proof for the boundedness follows the line of the proof of [**10**], 6.2.

Kollar and Seshadri constructed (see [**8**], 9.25) a finite covering of M_h which factors through the moduli stack.

Consider any finite morphism $\pi : Z \to M_h$ with this property. We choose a projective compactifications $\bar{Z}$ of Z such that π extends to $\pi : \bar{Z} \to \bar{M}_h$. So $\pi^*\mathcal{H}$ is again ample with respect to $\pi^{-1}(M_h)$. Let M_0 be a non-singular subvariety of $\pi(Z) \cap M_h$ with $Z_0 = \pi^{-1}(M_0)$ non singular.

Recall that for a family of projective varieties we constructed in [**10**], 2.7, a good open subset of the base space. Applying this construction to the restriction of the universal family to Z_0, we may assume furthermore, that Z_0 coincides with this subset.

By induction on the dimension of Z, we may assume that we have found an upper bound for $\varphi^*(\mathcal{H})$ whenever $\varphi(Y) \subset \pi(Z) \setminus M_0$. Hence it is sufficient to find such a bound under the assumptions that $\varphi(Y) \subset \pi(Z)$ and $\varphi(Y) \cap M_0 \neq \emptyset$. There exists a finite covering Y' of Y of degree $d \leq \deg(Z/\pi(Z))$, such that

$$Y' \xrightarrow{\sigma} Y \xrightarrow{\varphi} \pi(Z)$$

factors through $\varphi' : Y' \to Z$, and it is sufficient to bound the degree of $\sigma^*\varphi^*\mathcal{H}$. For simplicity, we assume that $\varphi : Y \to \pi(Z)$ factors through $\varphi' : Y \to Z$.

By [**10**], 2.6 and 2.7, blowing up Z with centers in $Z \setminus Z_0$ we may assume that Z is non singular, that there exists a certain invertible sheaf λ_ν on Z, and a constant $N_\nu > 0$, such that

$$\deg(\varphi'^*\lambda_\nu) \leq N_\nu \cdot \deg(\det(f_*\omega^\nu_{X/Y})),$$

where, as usual, $f : X \to Y$ is an extension of $V \to U$ to a projective manifold X. By the explicit description of λ_ν in [**10**], 2.6, d), and by [**10**], 3.4, the sheaf λ_ν is ample with respect to Z_0 for some $\nu > 1$. Hence it is sufficient to give an upper bound for $\deg(\varphi'^*\lambda_\nu)$, or for $\deg(\det(f_*\omega^\nu_{X/Y}))$.

By [**9**] (see also [**1**] and [**4**]) there exists a constant e, depending only on the Hilbert polynomial h, with

$$\deg(\det(f_*\omega^\nu_{X/Y})) \leq (n \cdot (2g(Y) - 2 + s) + s) \cdot \nu \cdot \mathrm{rank}(f_*\omega^\nu_{X/Y}) \cdot e,$$

and we found the bound we were looking for.

It remains to show, that the points

$$[\varphi : Y \to \bar{M}_h] \in \mathbf{Hom}((Y, U), (M_h, \bar{M}_h))$$

which are induced by a family $f : X \to Y$ with $\mu(f) = \mu$, are discrete. If not, one finds a positive dimensional manifold T and a generically finite morphism to

$\mathbf{Hom}((Y, U), (M_h, \bar{M}_h))$ whose image contains a dense set of points where the corresponding morphism is induced by a family. Let us choose a smooth projective compactification $\bar{T}$ with $S_{\bar{T}} = \bar{T} \setminus T$ a normal crossing divisor.

The induced morphism $Y \times T \to \bar{M}_h$ is not necessarily factoring through the moduli stack, but using again the Kollár Seshadri construction again, we find a generically finite morphism $\pi : Z \to Y \times \bar{T}$ which over $\pi^{-1}(U \times T)$ is induced by a smooth family. Assume that Z is non-singular and that $S_Z = Z \setminus U \times T$ is a normal crossing divisor. Write $p : Z \to \bar{T}$ for the induced morphism.

Applying 6.3, ii), one obtains a big coherent subsheaf

$$\mathcal{P} \subset S^{\mu}(\Omega^1_Z(\log S_Z)).$$

By part iii), is image $\mathcal{P}'$ in $S^{\mu}(\Omega_{Z/T}(\log S_Z))$ is non zero, and iv) implies that for a dense set of points $t \in T$ the restriction $\mathcal{P}'|_{p^{-1}(t)}$ lies in

$$\pi^* S^{\mu}(\Omega^1_{Y \times \{t\}}(\log(S \times \{t\}))).$$

This is only possible, if $\mathcal{P}'$ is a big subsheaf of

$$\pi^* S^{\mu}(\Omega^1_{Y \times \bar{T}}(\log(S \times \bar{T}))).$$

Restricting to $\pi^{-1}(\{y\} \times \bar{T}$, for general $y \in Y$ one obtains as in the proof of 8.2 a big subsheaf of a trivial sheaf, a contradiction. □

Needless to say, Corollary 8.4 is sort of empty, as long as we do not know any answer to Problem 6.2.

References

[1] Bedulev, E., Viehweg, E.: On the Shafarevich conjecture for surfaces of general type over function fields. Invent. Math. **139** (2000) 603–615

[2] Caporaso, L.: On certain uniformity properties of curves over function fields. Compos. Math. **130** (2002) 1–19

[3] Esnault, H., Viehweg, E.: Lectures on Vanishing Theorems. DMV-Seminar **20** (1992), Birkhäuser, Basel-Boston-Berlin

[4] Kovács, S.: Logarithmic vanishing theorems and Arakelov-Parshin boundedness for singular varieties. preprint (AG/0003019), to appear in Comp. Math.

[5] Liu, K., Todorov A., Yau, S.-T., Zuo, K.: The Analogue of Shafarevich's Conjecture for Some CY Manifolds. preprint 2002

[6] Maehara, K.: A finiteness property of varieties of general type. Math. Ann. **262** (1983) 101–123.

[7] Mori, S.: Classification of higher dimensional varieties. In: Algebraic Geometry. Bowdoin 1985, Proc. Symp. Pure Math. **46** (1987) 269 - 331

[8] Viehweg, E.: Quasi-projective Moduli for Polarized Manifolds. Ergebnisse der Mathematik, 3. Folge **30** (1995), Springer Verlag, Berlin-Heidelberg-New York

[9] Viehweg, E., Zuo K.: On the isotriviality of families of projective manifolds over curves. J. Alg. Geom. **10** (2001) 781–799

[10] Viehweg, E., Zuo K.: Base spaces of non-isotrivial families of smooth minimal models. In: Complex Geometry (Collection of Papers dedicated to Hans Grauert), Springer Verlag (2002) 279–328

[11] Zuo, K.: On the negativity of kernels of Kodaira-Spencer maps on Hodge bundles and applications. Asian J. of Math. **4** (2000) 279–302

Universität Essen, FB6 Mathematik, 45117 Essen, Germany

E-mail address: `viehweg@uni-essen.de`

The Chinese University of Hong Kong, Department of Mathematics, Shatin, Hong Kong

E-mail address: `kzuo@math.cuhk.edu.hk`

Geometry of the Weil-Petersson completion of Teichmüller space

Scott A. Wolpert

1. Introduction

Let R be a Riemann surface of genus g with n punctures, $3g-3+n>0$, and $\mathcal{T}$ the Teichmüller space of R. The Weil-Petersson (WP) metric for $\mathcal{T}$ is a Kähler metric with negative sectional curvature [**4, 35, 36, 41**]. With the WP metric $\mathcal{T}$ is a unique geodesic space [**42**]: for each pair of points there is a unique distance-realizing joining curve. The *augmented Teichmüller space* $\overline{\mathcal{T}}$, a stratified non locally compact space, is the space of marked noded Riemann surfaces and is a bordification of $\mathcal{T}$ in the style of Baily-Borel, [**2, 5**]. For $(g,n)=(1,1)$, $\overline{\mathcal{T}}$ is the bordification $\mathbb{H}\cup\mathbb{Q}$ of the upper half-plane with the horoball-neighborhood topology. The augmented Teichmüller space is in fact the WP metric completion of the Teichmüller space [**30**]. The strata of $\overline{\mathcal{T}}$ are lower-dimensional Teichmüller spaces; each stratum with its natural WP metric isometrically embeds into the completion $\overline{\mathcal{T}}$.

Our purpose is to present a view of the current understanding of the geometry of the WP geodesics on $\overline{\mathcal{T}}$. The behavior of geodesics in-the-large has significant consequences for the action of the mapping class group; see [**7, 13, 31, 42, 47**] and Section 7 below. The behavior of geodesics is also an important consideration for the harmonic map problem, as well as the study of rigidity of homomorphisms of lattices in Lie groups to the mapping class group [**11, 12, 13, 18, 26, 47**]. Furthermore, the behavior of geodesics is a consideration for the *rank* of $\mathcal{T}$ [**9**]. We begin by mentioning a collection of recent results [**8, 7, 9, 13, 31, 32, 47**]. Recall that for a hyperbolic surface, the length of the unique geodesic in a prescribed free homotopy class provides a function, the *geodesic-length*, on $\overline{\mathcal{T}}$ valued in $[0,\infty)$. A general fact is that geodesic-length functions are strictly convex along WP geodesics [**42**].

The work of C. McMullen provides a prelude [**32**]. Recall that a Bers embedding $\beta_S:\mathcal{T}\rightarrow\mathbf{T}^*_S\mathcal{T}$ is a biholomorphic map of the Teichmüller space to a domain in a cotangent space; from the Nehari estimate the image is bounded independent of S in terms of the Teichmüller and the WP co-metrics. Observe for S_0 fixed, $-\beta_S(S_0)$ is a section of the cotangent bundle $\mathbf{T}^*\mathcal{T}$, a differential 1-form $\theta_{WP}(S)$ on $\mathcal{T}$. McMullen showed [**32**, Thrm. 1.5] that $d(i\theta_{WP})=\omega_{WP}$ is the WP symplectic form. An application is a positive lower bound for the WP Rayleigh-Ritz quotient. He then introduced a smooth modification of the WP metric by including the complex Hessians of the small-valued geodesic-length functions. He combined the above and estimates for geodesic-length derivatives to show that the modification is a *Kähler hyperbolic metric* for the moduli space of Riemann surfaces that is comparable to the Teichmüller metric [**32**, Thrm. 1.1]. As applications McMullen found: a positive lower bound for the Teichmüller Rayleigh-Ritz quotient, a complex submanifold isoperimetric inequality, and the alternating sign of the orbifold Euler characteristic for the moduli spaces [**32**].

J. Brock has established important results on the large-scale behavior of WP distance,[**8**]. Brock considered the metric space, the *pants graph* $C_{\mathbf{P}}(F)$, having vertices the distinct pants decompositions of F and joining edges of unit-length for pants decompositions differing by a single elementary move,[**8**]. He showed that the 0-skeleton of $C_{\mathbf{P}}(F)$ is quasi-isometric to $\mathcal{T}$ with the WP metric. In particular by an observation of L. Bers there is a constant L such that each hyperbolic surface has a pants decomposition by geodesics of length at most L. For a pants decomposition $\mathcal{P}$, denote by $V(\mathcal{P}) \subset \mathcal{T}$ the subset of surfaces with the designated decomposition. The union $\cup_{\mathcal{P}} V(\mathcal{P})$ provides an open cover for $\mathcal{T}$. Brock found that WP distance records the configuration of the open sets $V(\mathcal{P})$ with the 0-skeleton of $C_{\mathbf{P}}(F)$ as the metric model. An important consequence of Brock's result is the correspondence between *quasi-geodesics* (quasi length-minimizing paths) on $\mathcal{T}$ and *quasi-geodesics* on $C_{\mathbf{P}}(F)$. He further showed for $p, q \in \mathcal{T}$ that the corresponding quasifuchsian hyperbolic three-manifold has convex-core volume comparable to $d_{WP}(p, q)$. At large-scale WP distance and convex-core volume are approximately combinatorially determined. He also showed that the first eigenvalue of the hyperbolic manifold and corresponding Hausdorff dimension of the limit set are estimated in terms of WP distance.

J. Brock and B. Farb used the correspondence to study the *rank* of $\mathcal{T}$ in the sense of M. Gromov [**9**]. A notion for the rank of a metric space is the maximal dimension of a quasi-flat, a quasi-isometric embedding of a Euclidean space. Brock and Farb found that $C_{\mathbf{P}}(F)$ contains quasi-flats of dimension $g - 1 + \lfloor \frac{g+n}{2} \rfloor$. It follows from application of Brock's quasi-isometry that the WP rank is likewise bounded. Gromov-hyperbolic metric spaces have rank one and thus the bound provides for $\dim \mathcal{T} > 2$, that $\mathcal{T}$ is not Gromov-hyperbolic [**9**, Thrm. 1.1]. The authors further found for $\dim \mathcal{T} \leq 2$ that $C_{\mathbf{P}}(F)$ and thus $\mathcal{T}$ are Gromov-hyperbolic [**9**, Thrm. 5.1]. Yamada and M. Bestvina had also considered the maximal dimension of a flat, [**46**]. Z. Huang has discovered further new asymptotic flatness [**19**]. Variation of independent plumbing parameters t prescribes planes with WP curvature $O((-\log|t|)^{-1})$.

W. Ballman and P. Eberlein posed a group-theoretic notion of the rank [**21**]. For discrete cofinite isometry groups of complete simply connected Riemannian manifolds with non positive curvature bounded from below the Ballman-Eberlein notion coincides with the geometrically defined rank. N. Ivanov has shown that mapping class groups have rank one [**21**]. N. Ivanov and independently B. Farb, A. Lubotzky and Y. Minsky further proved that any infinite-order element in the mapping class group has linear growth in the word metric; at least $O(n)$ generators of the group are required to write the nth iterate of an element of infinite order [**14, 23**]. Rank-1 lattices in simple Lie groups have the $O(n)$ writing-property, while higher-rank lattices do not have the property.

An important discovery of Sumio Yamada was the non refraction of WP geodesics: a geodesic on $\overline{\mathcal{T}}$ at most changes strata at its endpoints; see [**47**, Thrm. 2], [**13**, Lemma 3.6] and Propositions 11 and 12 below. A second important observation was that the strata of $\overline{\mathcal{T}}$ are geodesically convex. Yamada refined the original WP expansion of Masur [**30**] to present a third-order remainder expansion of the metric in the C^1-category. A key ingredient was the use of an improved estimate for degenerating families of hyperbolic metrics. The considerations were based on the relatively technical work of Wolf [**37**] and Wolf-Wolpert [**38**]. Yamada used the expansion to study the behavior of geodesics in neighborhoods of the bordification. He considered the WP Levi-Civita connection and one-dimensional harmonic maps to investigate the non refraction. Yamada then used the convexity of geodesic-length functions and the negative WP curvature to find that $\overline{\mathcal{T}}$ is a $CAT(0)$ space; see [**13**], the attribution to B. Farb in [**31**] and Theorem 14 below. He further noted that geodesic convexity of strata is an immediate consequence of the convexity of geodesic-length functions [**47**, Thrm. 1]. He applied the statements to give consideration of fixed-points and realizing translation lengths for mapping classes. Yamada also presented that irreducible elements of the mapping class group have positive translation length and a unique axis. The work has served as an inspiration for the work of Daskalopoulos and Wentworth [**13**], as well as the author.

The geometry of $CAT(0)$ spaces is developed in Bridson-Haefliger [**6**]. A *geodesic triangle* is prescribed by a triple of points and a triple of joining length-minimizing curves. A characterization of curvature for metric spaces is provided in terms of distance-comparisons for geodesic triangles. In a $CAT(0)$ space the distance and angle measurements for a triangle are bounded by the corresponding measurements for a Euclidean triangle with the corresponding edge-lengths [**6**, Chap. II.1, Prop. 1.7].

G. Daskalopoulos and R. Wentworth gave an independent treatment of the WP expansion, the non refraction, the $CAT(0)$ result and a more extensive consideration of actions of mapping classes [**13**]. The authors obtained a C^0-category expansion by applying the cut-and-paste based estimates for degenerating families of hyperbolic metrics from [**44**]. Scaling considerations were used for the energy of a parameterized curve to establish non refraction. The authors proved that irreducible mapping classes have positive translation length and a unique axis [**13**, Thrm. 1.1]. Previously G. Daskalopoulos, L. Katzarkov and R. Wentworth studied the finite energy equivariant harmonic map problem for the target $\mathcal{T}$, [**12**]. In general a condition on an isometric action is required for the existence of an energy minimizing equivariant map. In the case of a symmetric space target the action should be *reductive*. For $\mathcal{T}$, the authors [**12, 13**] propose *sufficiently large* as the counterpart of the reductive hypothesis. A subgroup of the mapping class group is sufficiently large provided it contains two irreducible mapping classes acting with distinct fixed points on the space of projective measured foliations. Daskalopoulos and Wentworth established [**13**, Thrm. 6.2] divergence of the axes for two as above independent irreducible mapping classes. The authors applied their considerations and studied equivariant maps from universal covers of finite volume complete Riemannian manifolds with finitely generated fundamental groups. They showed that if there is a finite energy map with sufficiently large image of the fundamental group, then there is a finite energy equivariant harmonic map [**13**, Cor. 1.3].

B. Farb and H. Masur established general higher rank superrigidity for the mapping class group as image. For an irreducible lattice in a semisimple Lie group of $\mathbb{R}$-rank at least two, a homomorphism to the mapping class group has finite image [**15**, Thrm. 1.1]. The authors also considered homomorphisms from $SL_n(\mathbb{Z})$ to the group of homeomorphism of a surface. They showed that all homomorphisms are trivial for n greater than an explicit bound in the genus.

H. Masur and M. Wolf established the WP-analogue of H. Royden's celebrated result: for $3g-3+n>1$ and $(g,n)\neq(1,2)$, every WP isometry of $\mathcal{T}$ is induced by an element of the extended mapping class group. They considered the asymptotic WP geometry to reduce the matter to considering the restriction of an isometry to $\overline{\mathcal{T}}-\mathcal{T}$. In particular an isometry of $\mathcal{T}$ extends to the completion $\overline{\mathcal{T}}$; an isometry of $\overline{\mathcal{T}}$ preserves the strata structure and following an approach of N. Ivanov agrees with a mapping class on the maximally noded surfaces. They then established that the set of maximally noded surfaces forms a uniqueness set for WP isometries [**31**].

Brock has also studied the family of WP geodesic rays based at a point, the WP visual sphere, [**7**]. Rays are considered with the topology of convergence of initial segments. He established that the action of the mapping class group does not extend continuously to an action on the WP visual spheres, and that the rays to noded surfaces are dense in the visual spheres. An additional discovery was that convergence of initial segments in general does not provide for convergence of entire rays; see [**7**] and Section 7 below.

The purpose of this paper is to continue the study in detail of the geometry of WP geodesics on $\overline{\mathcal{T}}$. We provide an independent treatment of the WP expansion based on the less technical approach of [**44**]. We then use the opportunity to give a range of new applications including: a thorough treatment of the strata structure, a classification of locally Euclidean subspaces of $\overline{\mathcal{T}}$, for the Masur-Wolf theorem a new proof based on a convex hull property, and a classification of limits of WP geodesics.

We find that $\overline{\mathcal{T}}$ is a *stratified* unique geodesic space with the strata intrinsically characterized by the metric geometry (see Theorem 13), [**47**]. For a reference surface F and $C(F)$, the partially ordered set *the complex of curves*, consider Λ the natural labeling function from $\overline{\mathcal{T}}$ to $C(F)\cup\{\emptyset\}$. For a marked noded Riemann surface (R,f) with $f:F\rightarrow R$, the labeling $\Lambda((R,f))$ is the simplex of free homotopy classes on F mapped to the nodes on R. The level sets of Λ are the strata of $\overline{\mathcal{T}}$. The unique WP geodesic $\widehat{pq}$ connecting $p,q\in\overline{\mathcal{T}}$ is contained in the closure of the stratum with label $\Lambda(p)\cap\Lambda(q)$ (see Theorem 13). The open segment $\widehat{pq}-\{p,q\}$ is a solution of the WP geodesic differential equation on the stratum with label $\Lambda(p)\cap\Lambda(q)$. For a point p, the stratum with label $\Lambda(p)$ is the union of the open geodesic segments containing the point (see Theorem 13).

The central consideration is the expansion of the WP metric in a neighborhood of a point of a positive codimension m stratum $\mathcal{S}$. For s a general multi-index local coordinate for $\mathcal{S}$ and t a plumbing construction multi-index parameter for the transverse to $\mathcal{S}$, we show for the multi-index parameter $r=(-\log|t|)^{-1/2}$ the following expansion for the *metric symmetric-tensor* (see Corollary 4)

$$dg^2_{WP}(s,t) \;=\; \big(dg^2_{WP}(s,0)\;+\;\pi^3\sum_{k=1}^{m}(4dr_k^2+r_k^6d\arg^2 t_k)\big)\;(1+O(\|r\|^3)).$$

In particular along $\mathcal{S}$ the WP metric to third-order remainder is a *product-metric* of the WP metric of $\mathcal{S}$ and metrics $(4dr^2+r^6d\arg^2 t)$, one for each t parameter. The product-structure with higher-order remainder suggests the isometric embedding of $\mathcal{S}$ into $\overline{\mathcal{T}}$. In the transverse direction to $\mathcal{S}$ the metric is *modeled* by the surface of revolution about the x-axis of $y=(x/2)^3$. The third-order remainder suggests *higher-order flatness* for the normal along $\mathcal{S}$. We combine the above expansion, the rescaling argument for metric spaces and an elementary quadratic inequality to establish the non refraction of geodesics (see Propositions 11 and 12).

Beyond $CAT(0)$, there are important applications for the above expansion. We are able to combine the *flat triangle lemma* of A. D. Alexandrov [**6**] and Theorem 13 to study the locally Euclidean isometric subspaces (flats) of $\overline{\mathcal{T}}$. A classification is established, and it is found that the maximal dimensional flats are submanifolds of: a product of Teichmüller spaces of g once-punctured tori and $\lfloor \frac{g+n}{2} \rfloor - 1$ four-punctured spheres (see Proposition 16). The result is consistent with the conjecture of Brock-Farb regarding the rank (the maximal dimension of a quasi-isometric embedding of a Euclidean space) of the WP metric, [**9**]. Following a suggestion of Brock, the considerations also provide that for $\dim \mathcal{T} > 2$ the WP metric is not *Gromov-hyperbolic.* Flat geodesic triangles in $\overline{\mathcal{T}} - \mathcal{T}$ are uniformly approximated by geodesic triangles in $\mathcal{T}$.

We also investigate applications of the Brock result [**7**] that the geodesic rays from a point of $\mathcal{T}$ to the noded Riemann surfaces have initial tangents dense in the initial tangent space. We generalize the result and show that the geodesics connecting maximally noded Riemann surfaces have tangents dense in the tangent bundle of $\mathcal{T}$ (see Corollary 18). An immediate consequence is that $\overline{\mathcal{T}}$ is the closed WP convex hull of the subset of maximally noded Riemann surfaces (see Corollary 19). The maximally noded Riemann surfaces play a basic role for the WP $CAT(0)$ geometry. In Theorem 20 we combine the convex hull property, the intrinsic nature of the strata structure and the classification of simplicial automorphisms of $C_{\mathbf{P}}(F)$ to study WP isometries. A new proof of the Masur-Wolf result is provided: for $3g-3+n > 1$ and $(g,n) \neq (1,2)$, every WP isometry of $\mathcal{T}$ is induced by an element of the extended mapping class group.

The WP metric is mapping class group invariant. H. Masur found that the Deligne-Mumford moduli space of stable curves $\overline{\mathcal{M}}$ is the WP quotient-metric completion of the moduli space of Riemann surfaces [**30**]. We note that the WP metric for $\overline{\mathcal{M}}$ is not locally uniquely geodesic near the compactification divisor of noded Riemann surfaces (see Proposition 15). A complete, convex subset of a $CAT(0)$ space is the base for an *orthogonal projection*, [**6**, Chap. II.2]. The closure of a stratum is complete and convex. We show that the distance to a stratum $\mathcal{S}$ has an expansion in terms of the defining *geodesic-length functions.* For a positive codimension m stratum $\mathcal{S}$, defined by the vanishing of the geodesic-length sum $\ell = \ell_1 + \cdots + \ell_m$, the distance to the stratum has the simple expansion $d(\cdot, \mathcal{S}) = (2\pi\ell)^{1/2} + O(\ell^2)$ (see Corollary 21). Furthermore the vector fields $\{\operatorname{grad}(2\pi\ell_j)^{1/2}\}$ are close to orthonormal near $\mathcal{S}$.

Our final application concerns limits of sequences of geodesics. We consider the classification problem (see Proposition 23). We might expect the compactness of $\overline{\mathcal{M}}$ to be manifested in the sequential compactness of the space of geodesics. But Brock already found that convergence of initial segments in general does not provide for convergence of entire rays. In fact for each sequence of bounded length geodesics there is a subsequence of mapping class group translates that converges geometrically (sequences of products of Dehn twists are applied to subsegments of the geodesics) to a polygonal path, a curve piecewise consisting of geodesics connecting different strata (see Proposition 23). Polygonal paths were first considered by Brock in his investigation of the WP visual sphere and the action of the mapping class group, [**7**, esp. Secs. 4, 5]. We find that the limit polygonal path is unique length-minimizing amongst paths joining prescribed strata. A simple example of a polygonal path is presented in the opening of Section 7. We apply the considerations and show that a mapping class acting on $\overline{\mathcal{T}}$ either: has a fixed-point, or positive translation length realized on a closed convex set, possibly contained in $\overline{\mathcal{T}} - \mathcal{T}$ (see Theorem 25). For *irreducible* mapping classes, the positive translation length is realized on a unique geodesic within $\mathcal{T}$, [**13, 47**].

We begin our detailed considerations in the next section with a summary of the notions associated with lengths of curves in metric spaces, [**6**]. We also review the local deformation theory of noded Riemann surfaces, as well as the specification of Fenchel-Nielsen coordinates and the construction of the augmented Teichmüller space. In the third section we provide the WP expansion. We begin considerations with the exact expansion of the hyperbolic metrics for the model case $zw = t$. Then we consider in detail families of noded Riemann surfaces and their hyperbolic metrics. Beginning with Masur's description of families of holomorphic 2-differentials, we give a simple and self-contained development of the tangent-cotangent coordinate frame pairing for the local deformation space and the desired WP expansion. In the fourth section we develop the length-minimizing properties of the solutions of the WP geodesic differential equation on $\mathcal{T}$. The considerations extend the earlier treatment [**42**]. In the fifth section we develop the length-minimizing properties of curves on $\overline{\mathcal{T}}$, including the non refraction results and the main theorems. The labeling function Λ serves an important role. WP length-minimizing curves can be analyzed in terms of their strata-behavior and geodesics within strata. WP convexity of the geodesic-length functions also serves an important role. WP geodesics are confined by the level sets and sublevel sets of geodesic-length functions. Non refraction is established by a local rescaling of the metric, and an application of the strict inequality $((a+b)^2 + c^2)^{1/2} < (a^2 + c^2)^{1/2} + b$ for positive values. In the sixth section first we examine the circumstance for the WP distance between corresponding points of a pair of geodesics not strictly convex. Then we consider the locally Euclidean isometric subspaces of $\overline{\mathcal{T}}$. We also consider the distance to a stratum. In the final section we consider sequences of geodesics and establish the sequential compactness, as well as a general classification for geodesic limits. The results are applied to study the existence of axes for mapping classes.

I would like to thank Jeffrey Brock for conversations.

2. Preliminaries

We begin with a summary of the notions associated with lengths of curves in a metric space. We closely follow the exposition of Bridson-Haefliger [**6**] and commend their treatment to the reader. For a metric space (M, d) the *length* of a curve $\gamma : [a, b] \to M$ is

$$L(\gamma) = \sup_{a=t_0 \le t_1 \le \cdots \le t_n = b} \sum_{j=0}^{n-1} d(\gamma(t_j), \gamma(t_{j+1}))$$

where the supremum is over all possible partitions with no bound on n. A curve is *rectifiable* provided its length is finite. The basic properties of length are provided in [**6**, Prop. 1.20]. Length is lower semi continuous for a sequence of rectifiable curves converging uniformly to a rectifiable curve. A curve $\gamma : [a, b] \to M$ is *parameterized proportional to arc-length* provided the length of γ restricted to subintervals

$[a,t] \subset [a,b]$ is a linear function of t, [**6**, Defn. 1.21]. A space (M,d) is a *length space* provided the distance between each pair of points is equal to the infimum of the length of rectifiable curves joining the points. It is an observation that the completion of a length space is again a length space [**6**, Exer. 3.6 (3)]. A curve $\gamma : [a,b] \to M$ is *length-minimizing* provided for all $a \le t \le t' \le b$ that $L(\gamma|_{[t,t']}) = d(\gamma(t),\gamma(t'))$; we initially reserve the word *geodesic* for curves which are solutions of the geodesic differential equation on a Riemannian manifold. A space with every pair of points having a (unique) length-minimizing joining curve is a *(unique) geodesic space.* In a metric space a *geodesic triangle* is prescribed by a triple of points and a triple of joining length-minimizing curves. A geodesic triangle can be compared to a triangle in a constant-curvature space with the corresponding sides having equal lengths [**6**, Chap. II.1]. A characterization of curvature for metric spaces is provided in terms of distance-comparisons for comparison triangles [**6**, Chap. II.1].

Consider R a Riemann surface with complete hyperbolic metric having finite area. The homeomorphism type of R is given by its genus and number of punctures. Relative to a reference topological surface F, the surface R is *marked* by an orientation-preserving homeomorphism $f : F \to R$. Marked surfaces (R,f) and (R',f') are *equivalent* provided for $h : R \to R'$, h a conformal homeomorphism, $h \circ f$ is homotopic rel boundary to f'. The set of equivalence classes of the F-marked Riemann surfaces is the Teichmüller space $\mathcal{T}$, [**20**]. A neighborhood of the marked surface (R,f) is given by first specifying smooth Beltrami differentials $\nu_1,\dots,\nu_m$ spanning the Dolbeault group $H^{0,1}_{\bar\partial}(\bar R, \mathcal{E}((\kappa p_1 \cdots p_n)^{-1}))$ for κ the canonical bundle of the compactification $\bar R$ and $p_1,\dots,p_n$ the point line bundles for the punctures, [**25**]. For $s \in \mathbb{C}^m$ set $\nu(s) = \sum_j s_j\nu_j$; for s small there is a Riemann surface $R^{\nu(s)}$ and a diffeomorphism $\zeta^{\nu(s)} : R \to R^{\nu(s)}$ satisfying $\bar\partial\zeta^{\nu(s)} = \nu(s)\partial\zeta^{\nu(s)}$. The parameterization of marked surfaces $s \to (R^{\nu(s)}, \zeta^{\nu(s)} \circ f)$ is a holomorphic local coordinate for the Teichmüller space $\mathcal{T}$.

The *mapping class group* $Mod = Homeo^+(F)/Homeo_0(F)$ is the quotient of the group of orientation-preserving homeomorphisms of F fixing the punctures by the subgroup of homeomorphisms isotopic to the identity. The *extended mapping class group* is the quotient $Mod^* = Homeo(F)/Homeo_0(F)$. A mapping class $[h]$ acts on equivalence classes of marked surfaces by taking $\{(R,f)\}$ to $\{(R, f\circ h^{-1})\}$. The action of Mod on $\mathcal{T}$ is by biholomorphic maps; the quotient $\mathcal{M}$ is the *moduli space* of Riemann surfaces. The holomorphic cotangent space of $\mathcal{T}$ at the marked surface (R,f) is $Q(R) \backsimeq \check H^0(\bar R, \mathcal{O}(\kappa^2 p_1 \cdots p_n))$, the space of integrable holomorphic quadratic differentials. A co-metric for the cotangent spaces of Teichmüller space is prescribed by the Petersson Hermitian pairing $\int_R \varphi\bar\psi(dh^2)^{-1}$ for $\varphi,\psi \in Q(R)$ and dh^2 the R-hyperbolic metric, [**4**]. The dual metric is the Weil-Petersson (WP) metric. The (extended) mapping classes act on $\mathcal{T}$ as WP isometries; the WP metric projects to $\mathcal{M}$. The WP metric is Kähler with negative sectional curvature and holomorphic sectional curvature bounded away from zero, [**35, 36, 41**]. Masur estimated the metric near the compactification divisor $\mathcal{D}$ of the moduli space, [**30**]. His preliminary expansion can be used for after-the-fact insights: the metric is not complete, [**39**]; there is an almost-product structure at infinity, [**47**]; and there are submanifolds of $\mathcal{T}$ that approximate Euclidean space (see the present Section 6). The expansion provides that the WP diameter and volume of $\mathcal{M}$ are finite. In [**44**] an improved analysis was presented for the extension of the WP Kähler form considered *in the sense of currents*. The $\frac{1}{2\pi^2}$ multiple of the WP Kähler form is the pushdown of the square of the curvature of the hyperbolic metric considered on the vertical line bundle for the fibration of the universal curve $\overline{\mathcal{C}}$ over $\overline{\mathcal{M}}$. The multiple of the Kähler form is a nonsmooth characteristic class representative of the Mumford class κ_1, [**44**].

The *complex of curves* $C(F)$ is defined as follows. The vertices of $C(F)$ are (free) homotopy classes of homotopically nontrivial, nonperipheral, simple closed curves on F. An edge of the complex consists of a pair of homotopy classes of disjoint simple closed curves. A k-simplex consists of $k+1$ homotopy classes of mutually disjoint simple closed curves. A maximal set of mutually disjoint simple closed curves, a *pants decomposition*, has $3g-3+n$ elements. Brock has described the large-scale WP geometry of Teichmüller space in terms of the *pants graph* $C_{\mathbf{P}}(F)$, a complex whose vertices are the distinct pants decompositions, [**8**]. The mapping class group Mod acts on curve complexes and in particular on $C(F)$.

A free homotopy class α of a closed curve on F determines a geodesic-length function ℓ_α on $\mathcal{T}$. For a marked surface (R, f), ℓ_α is the length of the R-hyperbolic metric geodesic homotopic to $f(\alpha)$. Geodesic-length functions provide parameters for the Teichmüller space. Suitable collections provide local coordinates, [**20**]. A collection of free homotopy classes $\{\alpha_1, \dots, \alpha_q\}$ is *filling* provided for a set of representatives with minimal number of self and mutual intersections that $F - \cup_j \alpha_j$ is a union of topological discs and punctured discs. A filling geodesic-length sum $\mathcal{L} = \sum_j \ell_{\alpha_j}$ is a proper function on the Teichmüller space. The differential and the WP gradient of an ℓ_α are given by the classical Petersson theta-series for the geodesic. In [**42**] we established that the WP Hessian of ℓ_α is positive-definite: geodesic-length functions are strictly convex along WP geodesics. The convexity provides a effective way to bound the WP geometry.

The Fenchel-Nielsen coordinates include geodesic-length functions, as well as lengths of auxiliary segments, [**3, 20, 29, 40**]. A pants decomposition $\mathcal{P} = \{\alpha_1, \dots, \alpha_{3g-3+n}\}$ decomposes the topological surface F into $2g-2+n$ components (pants), each homeomorphic to a sphere with a combination of three discs or points removed. A marked Riemann surface (R, f) is likewise decomposed into pants by the geodesics representing $\mathcal{P}$. Each component pants, relative to its hyperbolic metric, has a combination of three geodesic boundaries and cusps. For each component pants the shortest geodesic segments connecting boundaries determine *designated points* on each boundary. For each geodesic in the pants decomposition of R a parameter τ is defined as the displacement along the geodesic between designated points, one for each side of the geodesic. For Riemann surfaces close to an initial reference Riemann surface, the displacement τ is simply the distance between the designated points; in general the displacement is the analytic continuation (the lifting) of the distance measurement. For α in $\mathcal{P}$ define the *Fenchel-Nielsen angle* by $\theta_\alpha = 2\pi\tau_\alpha/\ell_\alpha$. The Fenchel-Nielsen coordinates for Teichmüller space for the decomposition $\mathcal{P}$ are $(\ell_{\alpha_1}, \theta_{\alpha_1}, \dots, \ell_{\alpha_{3g-3+n}}, \theta_{\alpha_{3g-3+n}})$. The coordinates provide a real analytic equivalence of $\mathcal{T}$ to $(\mathbb{R}_+ \times \mathbb{R})^{3g-3+n}$,[**3, 20, 40**].

A bordification of Teichmüller space is introduced by extending the range of the Fenchel-Nielsen parameters. The interpretation of *length vanishing* is the key ingredient. For ℓ_α equal to zero, the angle θ_α is not defined and in place of the geodesic for α there appears a pair of cusps; f is now a homeomorphism of $F - \alpha$ to the (marked) hyperbolic surface R (curves parallel to α map to loops encircling the cusps; see the discussion of nodes in the following Section). The parameter space for the pair $(\ell_\alpha,\theta_\alpha)$ is the identification space $\mathbb{R}_{\geq 0}\times\mathbb{R}/\{(0,y)\sim(0,y')\}$. For the pants decomposition $\mathcal{P}$ a frontier set $\mathcal{F}_{\mathcal{P}}$ is added to the Teichmüller space by extending the Fenchel-Nielsen parameter ranges: for each $\alpha\in\mathcal{P}$, extend the range of ℓ_α to include the value 0, with θ_α not defined for $\ell_\alpha=0$. The points of $\mathcal{F}_{\mathcal{P}}$ parameterize (*degenerate*) Riemann surfaces with each $\ell_\alpha=0,\alpha\in\mathcal{P}$, specifying a pair of cusps. In particular for a simplex $\sigma\subset\mathcal{P}$, the σ-null stratum is $\mathcal{S}(\sigma)=\{R\mid\ell_\alpha(R)=0\text{ iff }\alpha\in\sigma\}$. The frontier set $\mathcal{F}_{\mathcal{P}}$ is the union of the σ-null strata for the subsimplices of $\mathcal{P}$. Neighborhood bases for points of $\mathcal{F}_{\mathcal{P}}\subset\mathcal{T}\cup\mathcal{F}_{\mathcal{P}}$ are specified by the condition that for each simplex $\sigma\subset\mathcal{P}$ the projection $((\ell_\beta,\theta_\beta),\ell_\alpha):\mathcal{T}\cup\mathcal{S}(\sigma)\rightarrow\prod_{\beta\notin\sigma}(\mathbb{R}_+\times\mathbb{R})\times\prod_{\alpha\in\sigma}(\mathbb{R}_{\geq 0})$ is continuous. For a simplex σ contained in pants decompositions $\mathcal{P}$ and $\mathcal{P}'$ the specified neighborhood systems for $\mathcal{T}\cup\mathcal{S}(\sigma)$ are equivalent. The *augmented Teichmüller space* $\overline{\mathcal{T}}=\mathcal{T}\cup_{\sigma\in C(F)}\mathcal{S}(\sigma)$ is the resulting stratified topological space, [**2, 5**]. $\overline{\mathcal{T}}$ is not locally compact since no point of the frontier has a relatively compact neighborhood; the neighborhood bases are unrestricted in the θ_α parameters for α a σ-null. The action of Mod on $\mathcal{T}$ extends to an action by homeomorphisms on $\overline{\mathcal{T}}$ (the action on $\overline{\mathcal{T}}$ is not properly discontinuous) and the quotient $\overline{\mathcal{T}}/Mod$ is (topologically) the compactified moduli space of stable curves (see the consideration of $\overline{\mathcal{M}}$ in the next Section), [**2**, see Math. Rev. 56 #679]. Masur noted that the WP metric extends to $\overline{\mathcal{T}}$ and is complete on $\overline{\mathcal{M}}$, [**30**, Thrm. 2, Cor. 2]. $\overline{\mathcal{T}}$ is WP complete since the quotient $\overline{\mathcal{M}}$ is compact and each point of $\overline{\mathcal{T}}$ has a neighborhood with complete closure.

3. Expansion of the WP metric about the compactification divisor

Our purpose is to provide a description of *local coordinates* for the local deformation space of a Riemann surface with nodes. We will present a modification of the standard coordinates [**5, 30**] and use the formulation to present an improved form of Masur's expansion of the WP metric. The expansion reveals that for the moduli space of stable curves $\overline{\mathcal{M}}$, along the compactification divisor $\mathcal{D}$, the WP metric behaves to third-order in distance as a product formed with the WP metric of $\mathcal{D}$.

The description begins with the *plumbing variety* $\mathcal{V}=\{(z,w,t)\mid zw=t,\ |z|,|w|,|t|<1\}$. The defining function $zw-t$ has differential $z\,dw+w\,dz-dt$. Consequences are that $\mathcal{V}$ is a smooth variety, (z,w) are global coordinates, while (z,t) and (w,t) are not. Consider the projection $\Pi:\mathcal{V}\rightarrow D$ onto the t-unit disc. Π is a submersion, except at $(z,w)=(0,0)$; we can consider $\Pi:\mathcal{V}\rightarrow D$ as a (degenerate) family of open Riemann surfaces. The t-fibre, $t\neq 0$, is the hyperbola germ $zw=t$ or equivalently the annulus $\{|t|<|z|<1,\ w=t/z\}=\{|t|<|w|<1,\ z=t/w\}$. The 0-fibre is the intersection of the unit ball with the union of the coordinate axes in $\mathbb{C}^2$; on removing the origin the union becomes $\{0<|z|<1\}\cup\{0<|w|<1\}$. Each fibre of $\mathcal{V}_0=\mathcal{V}-\{0\}\rightarrow D$ has a complete hyperbolic metric:

(1)
for $t\neq 0$, on $\{|t|<|z|<1\}$ then
$$dh_t^2=\Big(\frac{\pi}{\log|t|}\csc\frac{\pi\log|z|}{\log|t|}\Big|\frac{dz}{z}\Big|\Big)^2;$$
for $t=0$, on $\{0<|z|<1\}\cup\{0<|w|<1\}$ then
$$dh_0^2=\Big(\frac{|d\zeta|}{|\zeta|\log|\zeta|}\Big)^2\text{ for }\zeta=z,\,w.$$

The family of hyperbolic metrics (dh_t^2) is a continuous metric, degenerate only at the origin, for the vertical line bundle of $\mathcal{V}$. In particular we have the elementary expansion

$$(2)\quad \begin{aligned} dh_t^2 &= \Big(\frac{|d\zeta|}{|\zeta|\log|\zeta|}\Big)^2\Big(\Theta\csc\Theta\Big)^2 \quad \text{for} \quad \Theta = \frac{\pi\log|z|}{\log|t|} \\ &= dh_0^2\,(1+\frac{1}{3}\Theta^2+\frac{1}{15}\Theta^4+\dots). \end{aligned}$$

The parameter t is a boundary point of the annulus $\{|t|<|z|<1\}$. The boundary points t, 1 will be included in the data for gluings. To describe the variation of annuli with boundary points, we now specify a quasiconformal map ζ from the pointed t-annulus to the pointed t'-annulus $\zeta(z) = zr^{\beta(r,t')}$, $z = re^{i\theta}$, with $\frac{\partial\beta}{\partial r}$ compactly supported in the annulus. The boundary conditions are $\zeta(1) = 1$, and by specification $t|t|^{\beta(|t|,t')} = t'$. On differentiating in t' and evaluating at $(|t|,t)$ we find the boundary condition $t\log|t|\,\dot\beta(|t|,t) = 1$. More generally the infinitesimal variation of the map is the vector field $\dot\zeta(z) = z\log r\,\dot\beta(r,t)$ for $\dot\zeta,\dot\beta$ the first t-derivatives. The map ζ varies from the identity and has Beltrami differential

$$(3)\qquad \bar\partial\dot\zeta = \frac{z}{2\bar z}\frac{\partial}{\partial\log r}(\dot\beta(r,0)\log r)\,\frac{\overline{dz}}{dz}.$$

For sake of later application we evaluate the pairing with a quadratic differential $z^\alpha\big(\frac{dz}{z}\big)^2$,

$$(4)\quad \begin{aligned} \int_{\{|t|<|z|<1\}} \bar\partial\dot\zeta\, z^\alpha\big(\frac{1}{z}\big)^2 dE &= \int_{\{|t|<|z|<1\}} \frac{z^\alpha}{2z\bar z}\frac{\partial}{\partial\log r}(\dot\beta\log r)\,dE \\ &\text{where for } \alpha = 0, \text{ then} \\ &= \pi\dot\beta\log r\Big|_{|t|}^{1} = \frac{-\pi}{t}, \\ &\text{and otherwise, then} \\ &= 0, \end{aligned}$$

for dE the Euclidean area element and where we have applied the boundary condition for $\dot\beta$; the evaluation involves fixing a normalization for the Serre duality pairing and agrees with [**30**, Prop. 7.1].

We review the description of *Riemann surfaces with nodes*, [**5, 30, 44**]. A Riemann surface with nodes R is a connected complex space, such that every point has a neighborhood isomorphic to either the unit disc in $\mathbb{C}$, or the germ at the origin in $\mathbb{C}^2$ of the union of the coordinate axes. R is *stable* provided each component of $R - \{nodes\}$ has negative Euler characteristic, i.e. has a hyperbolic metric. A regular q-differential on R is the assignment of a meromorphic q-differential Θ_j for each component R_j of $R - \{nodes\}$ such that: i) each Θ_* has poles only at the punctures of R_* with orders at most q, and ii) if punctures p, p' are paired to form a node then $Res_p\Theta_* = (-1)^q\,Res_{p'}\Theta_*$, [**5**].

We review the deformation theory of Riemann surfaces with punctures and then with nodes. For a Riemann surface R with hyperbolic metric and punctures there is a natural *cusp coordinate* (with unique germ modulo rotation) at each puncture: at the puncture p, the coordinate z with $z(p) = 0$ and the hyperbolic metric of R given as $\left(\frac{|dz|}{|z|\log|z|}\right)^2$, the germ of the hyperbolic metric for the unit-disc. If the surface is uniformized by the upper half-plane with p represented by a width-one cusp at infinity then $z = e^{2\pi i\zeta}$ for ζ the uniformization variable. Now a deformation neighborhood of the marked surface R is given by specifying smooth Beltrami differentials $\nu_1, \dots, \nu_m$ spanning the Dolbeault group $H^{0,1}_{\bar\partial}(\bar R, \mathcal{E}((\kappa p_1 \cdots p_n)^{-1}))$ for κ the canonical bundle of $\bar R$ and $p_1, \dots, p_n$ the point line bundles for the punctures, [**25**]. For $s \in \mathbb{C}^m$ set $\nu(s) = \sum_k s_k\nu_k$; for s small there is a Riemann surface $R^{\nu(s)}$ and a diffeomorphism $\zeta : R \to R^{\nu(s)}$ satisfying $\bar\partial\zeta = \nu(s)\partial\zeta$. The family of surfaces $\{R^{\nu(s)}\}$ represents a neighborhood of the marked Riemann surface in its Teichmüller space. We showed in [**45**, Lemma 1.1] that the Beltrami differentials can be modified a small amount *so that* in terms of each *cusp coordinate* the diffeomorphisms $\zeta^{\hat\nu(s)}$ are simply rotations; $\zeta^{\hat\nu(s)}$is a hyperbolic isometry in a neighborhood of the cusps; $\zeta^{\hat\nu(s)}$ cannot be complex analytic in s, but is real analytic. We further note that for s small the s-derivatives of $\nu(s)$ and $\hat\nu(s)$ are close. We say that $\zeta^{\hat\nu(s)}$ *preserves cusp coordinates.* The parameterization provides a key ingredient for obtaining simplified estimates of the degeneration of hyperbolic metrics and an improved expansion for the WP metric.

We review the plumbing construction for R a Riemann surface with a pair of punctures p, p'. The data is (U, V, F, G, t) where: U and V are disjoint disc coordinate neighborhoods of p and p'; $F : U \to \mathbb{C}$, $F(p) = 0$ and $G : V \to \mathbb{C}$, $G(p') = 0$, are coordinate mappings and t is a sufficiently small complex number. Pick a constant $0 < c < 1$ such that $F(U)$ and $G(V)$ contain the disc $\{|\zeta| < c\}$. For $c' < c$ let $R^*_{c'}$ be the open surface obtained by removing from R the discs $\{|F| \le c'\} \subset U$ and $\{|G| \le c'\} \subset V$. Now we prescribe the plumbing family $\{R_t\}$ over the t-disc. Let $D_c = \{|t| < c^4\}$, $M = R^*_{c^2} \times D_c$ and $\mathcal{V}_c = \{(z, w, t) \mid zw = t, |z|, |w| < c \text{ and } |t| < c^4\}$. M and $\mathcal{V}_c$ are complex manifolds with holomorphic projections to D_c. Consider the holomorphic maps from M to $\mathcal{V}_c$: $\hat F : (q, t) \to (F(q), t/F(q), t)$ and $\hat G : (q', t) \to (t/G(q'), G(q'), t)$. The maps are consistent with the projections to D_c. The identification space $M \cup \mathcal{V}_c/\{\hat F, \hat G \text{ equivalence}\}$ is a degenerating family $\{R_t\}$ with a projection to the disc D_c. By construction the 0-fibre has a node with local model $\mathcal{V}_c$.

We are ready to describe a *local manifold cover* of the compactified moduli space $\overline{\mathcal{M}}$. For R having nodes, $R_0 = R - \{nodes\}$ is a union of Riemann surfaces with punctures. The quasiconformal deformation space of R_0, $Def(R_0)$, is the product of the Teichmüller spaces of the components of R_0. As already noted from [**45**, Lemma 1.1] for $m = \dim Def(R_0)$ there is a real analytic family of Beltrami differentials $\hat\nu(s)$, s in a neighborhood of the origin in $\mathbb{C}^m$, such that $s \to R_s = R^{\hat\nu(s)}$ is a coordinate parameterization of a neighborhood of R_0 in $Def(R)$ and the prescribed mappings $\zeta^{\hat\nu(s)} : R_0 \to R^{\hat\nu(s)}$ preserve the cusp coordinates at each puncture. Further for R with n nodes we now prescribe the plumbing data $(U_k, V_k, z_k, w_k, t_k)$, $k = 1, \dots, n$, for $R^{\hat\nu(s)}$, where z_k on U_k and w_k on V_k are cusp coordinates relative to the $R^{\hat\nu(s)}$-hyperbolic metric (the plumbing data varies with s). The parameter t_k parameterizes *opening* the k th node. For all t_k suitably small, perform the n prescribed plumbings to obtain the family $R_{s,t} = R^{\hat\nu(s)}_{t_1,\dots,t_n}$. The tuple $(s, t) = (s_1, \dots, s_m, t_1, \dots, t_n)$ provides real analytic local coordinates, the ***hyperbolic metric plumbing coordinates***, for the local manifold cover of $\overline{\mathcal{M}}$ at R, [**30, 43**] and [**44**, Secs. 2.3, 2.4]. The coordinates have a special property: for s fixed the parameterization is holomorphic in t. The property is a basic feature of the plumbing construction. The family $R_{s,t}$ parameterizes the small deformations of the marked noded surface R.

The roles of the Fenchel-Nielsen coordinates and the hyperbolic metric plumbing coordinates can be interchanged. In particular for the nodes of R given by the σ-null stratum $\{\alpha_1,\dots,\alpha_n\}$ the above local manifold cover has topological coordinates $((\ell_\beta,\theta_\beta)_{\beta\notin\sigma},(\ell_\alpha e^{i\theta_\alpha})_{\alpha\in\sigma})$. The observation can be established by expressing the Fenchel-Nielsen coordinates solely in terms of geodesic-lengths, and then applying techniques for theta-series to analyze the differentials of geodesic-lengths. Upon interchanging the roles of the coordinates, we obtain a local description of the bordification in terms of the (s,t) tuple, [**2, 5, 44**]. At the point R of the σ-null stratum in $\overline{\mathcal{T}}$ the local parameters are $(s,|t|,\arg t)$ with the *arg* valued in $\mathbb{R}$.

We review the geometry of the local manifold covers. For a complex manifold M the complexification $\boldsymbol{T}^{\mathbb{C}}M$ of the $\mathbb{R}$-tangent bundle is decomposed into the subspaces of *holomorphic* and *antiholomorphic* tangent vectors. A Hermitian metric g is prescribed on the holomorphic subspace. For a general complex parameterization $s=u+iv$ the coordinate $\mathbb{R}$-tangents are expressed as $\frac{\partial}{\partial u}=\frac{\partial}{\partial s}+\frac{\partial}{\partial \bar s}$ and $\frac{\partial}{\partial v}=i\frac{\partial}{\partial s}-i\frac{\partial}{\partial \bar s}$. For the $R_{s,t}$ parameterization the s-parameters are not holomorphic while for s-parameters fixed the t-parameters are holomorphic; $\{\frac{\partial}{\partial s_j}+\frac{\partial}{\partial \bar s_j}, i\frac{\partial}{\partial s_j}-i\frac{\partial}{\partial \bar s_j}, \frac{\partial}{\partial t_k}, i\frac{\partial}{\partial t_k}\}$ is a basis over $\mathbb{R}$ for the tangent space of the local manifold cover. For a smooth Riemann surface the dual of the space of holomorphic tangents is the space of quadratic differentials. The following is now a modification of Masur's result [**30**, Prop. 7.1].

PROPOSITION 1. *The hyperbolic metric plumbing coordinates (s,t) are real analytic and for s fixed the parameterization is holomorphic in t. Provided the modification $\hat\nu$ is small, for a neighborhood of the origin there are families in (s,t) of regular 2-differentials φ_j, ψ_j, j=1, ... ,m and η_k, k=1, ... ,n such that:*

1. *For $R_{s,t}$ with $t_k\neq 0$, all k, $\{\varphi_j,\psi_j,\eta_k, i\eta_k\}$ forms the dual basis to $\{\frac{\partial\hat\nu(s)}{\partial s_j}+\frac{\partial\hat\nu(s)}{\partial \bar s_j}, i\frac{\partial\hat\nu(s)}{\partial s_j}-i\frac{\partial\hat\nu(s)}{\partial \bar s_j}, \frac{\partial}{\partial t_k}, i\frac{\partial}{\partial t_k}\}$ over $\mathbb{R}$.*
2. *For $R_{s,t}$ with $t_k=0$, all k, the η_k, k=1, ... ,n, are trivial and the $\{\varphi_j,\psi_j\}$ span the dual of the holomorphic subspace $\boldsymbol{T}Def(R_0)$.*

Proof. The situation compares to that considered by Masur. The new element: the variation of the plumbing data is prescribed by a Schiffer variation for a gluing-function real analytically depending on the parameter s, [**43**, pg. 410]. As already noted for s fixed, plumbing produces a holomorphic family. Following Masur the families of regular 2-differentials $\{\varphi_j,\psi_j,\eta_k\}$ are obtained by starting with a local frame $\mathcal{F}$ of regular 2-differentials and prescribing the pairings with $\{\frac{\partial\hat\nu}{\partial s_j},\frac{\partial\hat\nu}{\partial \bar s_j},\frac{\partial}{\partial t_k}\}$, [**30**, Sec. 5 and Prop. 7.1]. At an initial point the basis is simply given by a linear transformation of the frame $\mathcal{F}$. The prescribed basis will then exist in a neighborhood provided the pairings are continuous. We first consider the pairings with $\frac{\partial}{\partial t_k}$. From (3) we have the Beltrami differential for the pairing with $\frac{\partial}{\partial t_k}$, $k=1,\dots,n$. In particular for a plumbing collar of $R_{s,t}$ let z (or w) be the coordinate of the plumbing. A quadratic differential φ on $R_{s,t}$ can be factored on the collar into a product of $\left(\frac{dz}{z}\right)^2$ and a function holomorphic in z. We write $\mathcal{C}_k(\varphi)$ for the constant coefficient of the Laurent expansion of the function factor. From (4) the pairing with $\frac{\partial}{\partial t_k}$ is the linear functional $-\frac{\pi}{t_k}\mathcal{C}_k$. From Masur's considerations [**30**, Sec. 5, esp. 5.4, 5.5] the pairing of $\frac{\partial}{\partial t_k}$ with the local frame $\mathcal{F}$ is continuous, and there are regular

2-differentials $\{\varphi_j, \psi_j, \eta_k^*\}$ with: $\mathcal{C}_\ell(\varphi_j) = \mathcal{C}_\ell(\psi_j) = 0$, $j = 1, \ldots, m$; $\mathcal{C}_\ell(\eta_k^*) = \delta_{k\ell}$, $k, \ell = 1, \ldots, n$. The 2-differentials $\eta_k = \frac{-t_k}{\pi}\eta_k^*$, $k = 1, \ldots, n$ have the desired pairings with $\frac{\partial}{\partial t_k}$. The final matter is to note that the pairings of $\{\varphi_j, \psi_j, \eta_k^*\}$ with $\{\frac{\partial \hat{\nu}}{\partial s_\ell}, \frac{\partial \hat{\nu}}{\partial \bar{s}_\ell}\}$ are indeed continuous in (s,t). By construction the differential $\hat{\nu}(s)$ is supported in the complement of the plumbing collars, [**45**, Lemma 1.1]. On the support of $\hat{\nu}(s)$ the 2-differentials are real analytic in (s,t). The pairings are continuous and even real analytic. The proof is complete.

We now note two general matters: the role of the coefficient functional $\mathcal{C}$, and the approximation of the hyperbolic metric. As above, for z a plumbing collar coordinate for $R_{s,t}$, a quadratic differential ψ can be factored on the collar as the product of $\left(\frac{dz}{z}\right)^2$ and a holomorphic function. $\mathcal{C}(\psi)$ denotes the constant coefficient of the Laurent expansion of the function. The surface $R_{s,t}$ is constructed by plumbing $\left(R_s\right)^*_{c^2}$ with the R_s-hyperbolic cusp coordinates. $R_{s,t}$ is the disjoint union of $\left(R_s\right)^*_c$, R_s with the cuspidal discs $|z_*|, |w_*| < c$ removed, and the annulus $\{|t|/c < |z| < c\}$. An approximate hyperbolic metric $d\omega^2$ is given by choosing the R_s-hyperbolic metric on $\left(R_s\right)^*_c$ and dh_t^2 on the annulus (see (1)). The metric $d\omega^2$ is the *model grafting* treated in detail in [**44**, Sec. 3.4.MG]; as noted in [**44**, pgs. 445, 446] for $dh^2_{s,t}$ the $R_{s,t}$-hyperbolic metric we have that $\left|d\omega^2/dh^2_{s,t} - 1\right|$ is $O\left(\sum_k (\log|t_k|)^{-2}\right)$. The approximation $d\omega^2$ will now be substituted for the construction of [**30**, Sec. 6] to obtain an improved form of the original expansion. The improved approximation of the hyperbolic metric is the new contribution. Yamada [**47**] presented a third-order expansion based on the technical work of Wolf [**37**] and Wolf-Wolpert [**38**].

THEOREM 2. *For a noded Riemann surface R the hyperbolic metric plumbing coordinates for $R_{s,t}$ provide real analytic coordinates for a local manifold cover neighborhood for $\overline{\mathcal{M}}$. The parameterization is holomorphic in t for s fixed. On the local manifold cover the WP metric is formally Hermitian satisfying:*

1. *For $t_k = 0$, $k = 1, \ldots, n$, the restriction of the metric is a smooth Kähler metric, isometric to the WP product metric for a product of Teichmüller spaces.*
2. *For the tangents $\{\frac{\partial}{\partial s_j}, \frac{\partial}{\partial \bar{s}_j}, \frac{\partial}{\partial t_k}\}$ and the quantity $\rho = \sum_{k=1}^{n} (\log|t_k|)^{-2}$ then:*

$$g_{WP}\left(\frac{\partial}{\partial t_k}, \frac{\partial}{\partial t_k}\right)(s,t) = \frac{\pi^3}{|t_k|^2(-\log^3|t_k|)}\left(1 + O(\rho)\right);$$

$g_{WP}\left(\frac{\partial}{\partial t_k}, \frac{\partial}{\partial t_\ell}\right)$ *is* $O\left((|t_k t_\ell| \log^3|t_k| \log^3|t_\ell|)^{-1}\right)$ *for* $k \neq \ell$;

and for $\mathfrak{u} = \frac{\partial}{\partial s_j}, \frac{\partial}{\partial \bar{s}_j}$; $\mathfrak{v} = \frac{\partial}{\partial s_\ell}, \frac{\partial}{\partial \bar{s}_\ell}$:

$g_{WP}\left(\frac{\partial}{\partial t_k}, \mathfrak{u}\right)$ *is* $O\left((|t_k|(-\log^3|t_k|))^{-1}\right)$ *and*

$$g_{WP}(\mathfrak{u},\mathfrak{v})(s,t) = g_{WP}(\mathfrak{u},\mathfrak{v})(s,0)\left(1 + O(\rho)\right).$$

Proof. We begin with the expansion of the dual metric for the basis provided in Proposition 1. The behavior of the φ_j, ψ_j, η_k and their contribution to the Petersson pairing $\int \alpha\overline{\beta}(d\omega^2)^{-1}$ is straightforward. On $\left(R_s\right)^*_{c^2}$ the quadratic differentials and the approximating metric are real analytic in (s,t). The contributions to the pairing are real analytic and each differential η_k, $k = 1, \ldots, n$, contributes a factor of t_k. On the plumbing collars $\{|t|/c < |z| < c\} = \{|t|/c < |w| < c\}$ each quadratic differentials is given as the product of $\left(\frac{dz}{z}\right)^2 = \left(\frac{dw}{w}\right)^2$ and a function factor. We begin with an elementary calculation

$$\int_{\{|t|/c<|z|<c\}} \left|z^{\alpha}\left(\frac{dz}{z}\right)^2\right|^2 \left(dh_t^2\right)^{-1} =$$

$$\frac{2}{\pi}\int_{|t|/c}^{c} \left(\log|t| \sin\frac{\pi \log r}{\log|t|}\right)^2 r^{2\alpha}\, d\log r$$

where for $\alpha = 0$, $\mu = \dfrac{\log r}{\log|t|}$ and $\epsilon = \dfrac{\log c}{\log|t|}$, then

$$= \frac{2}{\pi}(-\log^3|t|)\int_{\epsilon}^{1-\epsilon} \sin^2 \pi\mu\, d\mu \;=\; \frac{1}{\pi}(-\log^3|t|) + O(1),$$

and for $\alpha = 1$, since $|\sin\mu| \le |\mu|$, then

$$= O(1).$$

We are ready to consider the contribution to the Petersson pairing from the collars. Consider the contribution for the ℓth collar. By construction η_ℓ is the unique quadratic differential from the dual basis with a nonzero $\mathcal{C}_\ell$ evaluation. In particular $\mathcal{C}_\ell(\eta_\ell^*) = 1$ and the contribution to the self pairing for η_ℓ^* is $\frac{1}{\pi}(-\log^3|t_\ell|) + O(1)$. In general we note that a quadratic differential on a plumbing collar can be factored as $\left(\frac{dz}{z}\right)^2(f_z + \boldsymbol{c} + f_w)$ for f_z holomorphic in $|z| < c$, $f_z(0) = 0$; $\boldsymbol{c}$ the $\mathcal{C}$-evaluation value and f_w holomorphic in $|w| < c$, $f_w(0) = 0$. Furthermore f_z, resp. f_w, is given as the Cauchy integral of f over $|z| = c$, resp. $|w| = c$. Further from the Schwarz Lemma $|f_z| \le c'|z| \max_{|z|=c}|f|$ with a corresponding bound for $|f_w|$. The bounds are combined with the majorant bound $|\sin\mu| \le |\mu|$ to show that: for φ_j, ψ_j, η_k^* on $|z| = c + \epsilon_0$ and $|w| = c + \epsilon_0$ depending analytically on (s,t) their contribution to the Petersson pairing over the collar is also analytic in (s,t).

Combining our considerations and noting the approximation of $d\omega^2$ to the hyperbolic metric for $R_{s,t}$ we find that

$$\langle \eta_k^*, \eta_k^* \rangle_{WP} = \frac{1}{\pi}(-\log^3|t_k|)\,(1 + O(\sum_{\ell=1}^{n}(\log|t_\ell|)^{-2})),$$

$$\langle \eta_k^*, \eta_\ell^* \rangle_{WP} = O(1) \text{ for } k \neq \ell,$$

and for $\mathfrak{a} = \varphi_j, \psi_j$; $\mathfrak{b} = \varphi_\ell, \psi_\ell$:

$$\langle \mathfrak{a}, \eta_k^* \rangle_{WP} = O(1) \text{ and}$$

$$\langle \mathfrak{a}, \mathfrak{b} \rangle_{WP}(s,t) = \langle \mathfrak{a}, \mathfrak{b} \rangle_{WP}(s,0)\,(1 + O(\sum_{k=1}^{n}(\log|t_k|)^{-2})).$$

The desired expansion now follows from the following Proposition and the relations $\eta_k = -\frac{t_k}{\pi}\eta_k^*$. The proof is complete.

For $\mathcal{A}$ a symmetric $m+n \times m+n$ matrix

$$\begin{pmatrix} \lambda_1 & \dots & a_{\ell j} & \dots \\ \vdots & \lambda_k & \vdots & \vdots \\ & & \ddots & \\ a_{j\ell} & \dots & & \dots \\ \dots & \dots & \dots & B \end{pmatrix}$$

with λ_k, $1 \le k \le n$;
$a_{j\ell}$, $1 \le j \le m+n$, $j \ne \ell$, $1 \le \ell \le n$ and
$B = (b_{j\ell})$ a symmetric $m \times m$ matrix,

we consider the situation that $\lambda_1, \dots, \lambda_n$ are large compared to the $a_{j\ell}$ and $b_{j\ell}$.

PROPOSITION 3. *For* $\det B \ne 0$, *and* $\rho = \sum_{k=1}^{n} \lambda_k^{-1}$ *then:*

$$\det \mathcal{A} = \det B \prod_{k=1}^{n} \lambda_k (1 + O(\rho))$$

and $\mathcal{A}^{-1} = (\alpha_{j\ell})$ *where: for* $1 \le k \le n$, $\alpha_{kk} = \lambda_k^{-1}(1 + O(\rho))$; *for* $1 \le j < \ell \le n$, $\alpha_{j\ell}$ *is* $O((\lambda_j \lambda_\ell)^{-1})$; *for* $1 \le j \le n < \ell \le m+n$, $\alpha_{j\ell}$ *is* $O(\lambda_j^{-1})$, *and for* $1 \le j,\ \ell \le m$, $\alpha_{j+n\,\ell+n} = b^{j\ell}(1+O(\rho))$. *The constants for the O-terms are bounded in terms of* $m+n$, $\det B^{-1}$ *and* $\max\{|a_{j\ell}|, |b_{j\ell}|\}$.

Proof. We consider the general formula for the determinant as a sum over the permutation group and by the cofactor expansion. First observe that there is a dichotomy for $m+n$-fold products in the calculation of $\det \mathcal{A}$; a product either also occurs in the expansion of $\det B \prod_k \lambda_k$, or has at most $n-1$ factors $\lambda_k, 1 \le k \le n$. Products with less than n factors are bounded in terms of the cited product and $O(\rho)$. The determinant expansion is a consequence. We continue and apply the analog of the dichotomy when examining the cofactors of $\mathcal{A}$. For the cofactor for λ_ℓ we find the expansion $\det B \prod_{k \ne \ell} \lambda_k (1 + O(\rho))$. Similarly for the cofactor of $a_{j\ell}$ we find the λ-contribution to be $\prod_{k \ne j,\ell} \lambda_k (1 + O(\rho))$ for $j \ne \ell \le n$ and to be $\prod_{k \ne j} \lambda_k (1 + O(\rho))$ for $j \le n < \ell$. Finally for the cofactor of $b_{j\ell}$ we find the expansion $b^{j\ell} \det B \prod_k \lambda_k (1 + O(\rho))$ in terms of the inverse $B^{-1} = (b^{j\ell})$. The proof is complete.

By way of application we present a normal form for the quadratic form dg^2_{WP}; the result is an immediate consequence of the above Theorem.

COROLLARY 4. *For the prescribed hyperbolic metric plumbing coordinates:*

$$dg^2_{WP}(s,t) = \left(dg^2_{WP}(s,0) + \pi^3 \sum_{k=1}^{n} (4dr_k^2 + r_k^6 d\theta_k^2)\right)(1 + O(\|r\|^3))$$

for $r_k = (-\log|t_k|)^{-1/2}$, $\theta_k = \arg t_k$ *and* $r = (r_1, \dots, r_n)$.

The result provides a local expansion of the WP metric about the compactification divisor $\mathcal{D} = \{t_k = 0\}$. To the third order of approximation the WP metric is formally a product. As we will note below, a second-order approximation is already special. As already noted, the bordification $\overline{\mathcal{T}}$ has a local description in terms of the parameters $(s, |t|, \arg t)$ or equivalently in terms of $(s, (-\log|t|)^{-1/2}, \arg t)$. The above result provides the associated WP expansion.

An almost-product Riemannian metric with remainder bounded by the displacement from a submanifold is very special. We note the situation as motivation for the results of Section 5; the following considerations do not apply since

$4dr^2 + r^6 d\theta^2$ is not a Riemannian metric. Consider a product $\mathbb{R}^m \times \mathbb{R}^n$ with Euclidean coordinates x for $\mathbb{R}^m$ and y for $\mathbb{R}^n$. Consider that in a neighborhood of the origin a metric has the expansion

$$dg^2 = dg_x^2 + dg_y^2 + O_{C^1}(\|y\|^2)$$

with dg_x^2, resp. dg_y^2, a C^1-metric for $\mathbb{R}^m$, resp. for $\mathbb{R}^n$, and the remainder a C^1-symmetric tensor as indicated. The expansion provides that the *second fundamental form* of the x-axes, $\mathbb{R}^m \times \{0\}$, vanishes identically [**33**, pgs. 62, 100]. In this case the x-axes is a *totally geodesic* submanifold: a geodesic initially tangent to the x-axes is contained in the x-axes [**33**, pg. 104]. The expansion also provides that for the x-axes the *normal connection* and the *normal curvature* vanish identically [**33**, pgs. 114, 115].

4. Length-minimizing curves on Teichmüller space

We begin by developing basic facts about the behavior of WP geodesics on Teichmüller space. Although Teichmüller space is topologically a cell, the behavior of geodesics is not a consequence of general results [**24**], since the WP metric is not complete. For instance the Hopf-Rinow theorem cannot be directly applied to obtain length-minimizing curves [**10, 24, 33**], and it is necessary to show that distance is measured along geodesics. We proceed though by applying our paradigm: *a filling geodesic-length sum behaves qualitatively as the distance from a point for a complete metric.* In the following we combine the paradigm and modifications of the standard arguments to find the basic behavior of geodesics.

THEOREM 5. *The WP exponential map from a base point is a diffeomorphism from its open domain onto the Teichmüller space.*

COROLLARY 6. *Teichmüller space is a unique geodesic space. Each WP geodesic segment is the unique length-minimizing rectifiable curve connecting its endpoints.*

Proof of Corollary. Let γ be the WP geodesic connecting a pair of points p and q in the Teichmüller space. For a filling geodesic-length function $\mathcal{L}$, choose $c > 0$, such that $\gamma \subset S_c = \{\mathcal{L} < c\}$, [**42**]. Consider G the set of all rectifiable curves connecting p and q, contained in $\overline{S_c}$, and each with length at most $d(p,q) + 1$. Provided G is nonempty and the elements of G are parameterized proportional to arc-length on the interval $[0,1]$, then G constitutes an equicontinuous family of maps. In particular for $\beta \in G$ and $t, t' \in [0,1]$ by the proportional parameterization it follows that

$$|t - t'| = \frac{L(\beta[t,t'])}{L(\beta)} \geq \frac{d(\beta(t), \beta(t'))}{d(p,q) + 1}.$$

¿From the Arzelà-Ascoli Lemma [**6**, pg. 36] there exists a rectifiable length-minimizing (amongst elements of G) curve β_0 connecting p and q contained in $\overline{S_c}$.

We consider the behavior of a rectifiable length-minimizing (amongst elements of G) curve β_0 passing through an arbitrary point $r \in \overline{S_c}$ (r could lie on $\partial\overline{S_c}$). Since $\overline{S_c}$ is compact, there is a positive ϵ such that WP geodesics are uniquely length-minimizing in an ϵ-neighborhood of each point of $\overline{S_c}$. Since $\overline{S_c}$ is WP convex it follows for r', r'' on the chosen curve, close to r, with r' *before* r and r'' *after* r, that the segments $\widehat{r'r}$ and $\widehat{rr''}$ are necessarily WP geodesics. It further follows that $\widehat{r'r''}$

is a WP geodesic, since the segment is locally length-minimizing at r. By convexity of the geodesic-length function $\mathcal{L}$, its value at r' or r'' is greater than its value at r. Since $r', r'' \in \overline{S_c}$ it follows that $r \in S_c$. It now follows that a rectifiable length-minimizing (amongst elements of G) curve β_0 is a WP geodesic entirely contained in S_c.

In general given ϵ, $0 < \epsilon < 1$, there exists a curve β' connecting p and q in the Teichmüller space such that $L(\beta') < d(p,q) + \epsilon$. For c' large, $\beta' \subset \overline{S_{c'}}$ and thus the corresponding family of maps G is non empty. The length of β' bounds the length of a $S_{c'}$-length-minimizing curve β_0 connecting p and q: in particular $L(\beta') \geq L(\beta_0)$. ¿From the above paragraph and the Theorem, the unique geodesic connecting p and q is $\beta_0 = \gamma$. The inequalities now provide that $L(\gamma) < d(p,q) + \epsilon$. The proof is complete.

Proof of Theorem. First we note that the domain of the exponential map is an open set. Given a geodesic γ connecting a pair of points, select $c > 0$ such that $\gamma \subset S_c$. Since S_c is open the points in neighborhoods of the γ-endpoints are also connected by WP geodesics. In particular the domain of the exponential map is open.

We next note that the exponential map is a local diffeomorphism [**24**]. Further note that a germ of the inverse is determined by its value at a single point. We now consider the continuation of a given germ ι, with the exponential map based at p and the germ given at $q \in \mathcal{T}$. We consider the continuation of ι along α, a curve with initial point q. We argue that the continuation set is closed.

Choose a filling geodesic-length function $\mathcal{L}$ and value c such that $p, \alpha \subset \overline{S_c}$. First we observe that each WP geodesic connecting p and a point of α is contained in $\overline{S_c}$. This follows since the values of $\mathcal{L}$ at the endpoints are bounded by c and $\mathcal{L}$ is WP convex. Since $\overline{S_c}$ is compact there is an overall length bound for the WP geodesics contained in $\overline{S_c}$. As noted in Bridson-Haefliger a length-bounded family of geodesics is given by an equicontinuous family of maps, [**6**, pg. 36]. By the Arzelà-Ascoli Lemma it follows that a sequence of WP geodesics contained in $\overline{S_c}$ has a subsequence converging to a geodesic contained in $\overline{S_c}$.

Consider now that the germ ι can be continued to a sequence of points $\{q_n\}$ along α. In particular WP geodesics $\widehat{pq_n}$ are determined. A subsequence (same notation) $\widehat{pq_n}$ converges to $\widehat{pq'}$. The WP geodesic $\widehat{pq'}$ determines a germ of the inverse of the exponential map; the germ gives *exponential inverses* for the WP geodesics $\widehat{pq_n}$. The germ is the continuation of ι; the continuation set is closed. The continuation set is necessarily open; ι can be continued along every curve. On considering homotopies it is established that the continuation to the endpoint of α is path independent. Finally since the Teichmüller space is simply connected the continuations determine a global inverse for the exponential map. The proof is complete.

We are also interested in understanding the WP *join* of two sets, and in particular the distance between points on a pair of geodesics. For the WP inner product consider the Levi-Civita connection ∇ satisfying for vector fields X, Y and W the relations,[**10, 24, 33**],

$$X\langle Y,W\rangle = \langle\nabla_X Y,W\rangle + \langle Y,\nabla_X W\rangle$$
$$\nabla_X Y - \nabla_Y X = [X,Y].$$

Further consider the curvature tensor

$$R(X,Y)Z = \nabla_X\nabla_Y Z - \nabla_Y\nabla_X Z - \nabla_{[X,Y]}Z.$$

A *variation of geodesics* is a smooth map $\beta(t,s)$ from $[t_0,t_1]\times(-\epsilon,\epsilon)$ to $\mathcal{T}$ such that for each s', $\beta(t,s')$ is a WP geodesic. For the vector fields $T = d\beta(\frac{\partial}{\partial t})$ and $V = d\beta(\frac{\partial}{\partial s})$, the first variation of the geodesic V satisfies the Jacobi equation

$$\nabla_T\nabla_T V = R(T,V)T;$$

solutions are Jacobi fields, [**10**]. The Jacobi equation is a linear second-order system of ordinary differential equations for vector fields along $\beta(t,s')$. The space of solutions has dimension $2\dim\mathcal{T}$; a solution is uniquely prescribed by its initial value and its initial derivative. Furthermore since $\mathcal{T}$ has negative curvature there are no conjugate points along a geodesic and the linear map $(V|_{t=t_0},\nabla_T V|_{t=t_0})$ to $(V|_{t=t_0},V|_{t=t_1})$ is an isomorphism [**10**, pg. 19]. Solutions are uniquely prescribed by their values at the endpoints. This property is needed for the understanding of the join of sets. In particular Jacobi fields provide a mechanism for analyzing the exponential map.

The WP exponential map $(p,v)\overset{e}{\to} exp_p v$ has open domain $\mathcal{D} = \{(p,exp_p^{-1}(\mathcal{T}))\}\subset \mathbf{T}\mathcal{T}$ in the tangent bundle. We are ready to consider the behavior of geodesics.

PROPOSITION 7. *The WP exponential map $(p,v)\overset{e}{\to}(p,exp_p v)$ is a diffeomorphism from $\mathcal{D}\subset\mathbf{T}\mathcal{T}$ to $\mathcal{T}\times\mathcal{T}$. For a pair of disjoint WP geodesics parameterized proportional to arc-length, the WP distance between corresponding points is a strictly convex function.*

Proof. The map $\boldsymbol{e}$ is smooth with differential $de = (id, d\,exp)$. As already noted since the WP metric has negative curvature $d\,exp$ has maximal rank and thus $\boldsymbol{e}$ is a local diffeomorphism of $\mathcal{D}$ to $\mathcal{T}\times\mathcal{T}$. A consequence of Corollary 6 is that $\boldsymbol{e}$ is a global diffeomorphism.

We are ready to consider the distance between corresponding points of a pair of disjoint WP geodesics. From the above result a one-parameter variation of geodesics is determined $\beta(t,s)$, $(t,s)\in[t_0,t_1]\times[s_0,s_1]$. For a value $s'\in[s_0,s_1]$ we write T for the tangent field of $\beta(t,s')$ and V for its variation field; we assume $\|T\|=1$. The second variation in s at s' of the length of $\beta(t,s)$ is given by the classical formula [**10**, (1.14)]

$$(5)\quad \langle\nabla_V V,T\rangle\Big|_{t_0}^{t_1} + \int_{t_0}^{t_1}\langle\nabla_T V,\nabla_T V\rangle - \langle\nabla_T V,T\rangle\langle\nabla_T V,T\rangle - \langle R(V,T)T,V\rangle dt.$$

Observations are in order. First by hypothesis the curves $\beta(t_0,s)$ and $\beta(t_1,s)$ are geodesics with constant speed parameterization; the acceleration $\nabla_V V$ vanishes at t_0 and t_1. Second the first two terms of the integrand combine to give the length-squared of the projection $\nabla_T V$ onto the normal space of T. And the third term of the integrand is strictly positive given strictly negative curvature [**10**]. In summary the distance is a strictly convex function. The proof is complete.

We are ready to show that $\overline{\mathcal{T}}$ is a geodesic space. For points p and q of the completion let $\{p_n\}$ and $\{q_n\}$ be sequences from $\mathcal{T}$ converging to p, resp. to q. Note for the distance we have $d(p,q) = \lim_n d(p_n, q_n)$. Consider the sequence of curves $\gamma_n = \widehat{p_n q_n}$ of $\mathcal{T}$ parameterized proportional to arc-length by the unit-interval. Since the sequences $\{p_n\}$ and $\{q_n\}$ are Cauchy it follows from Proposition 7 that for each $t \in [0,1]$ the sequence $\{\gamma_n(t)\}$ is also Cauchy (*without* passing to a subsequence). The sequence $\{\gamma_n\}$ prescribes a function γ with domain the unit-interval and values in $\overline{\mathcal{T}}$. Furthermore since the γ_n are distance proportional-parameterized, for $t, t' \in [0,1]$ then

$$\frac{d(\gamma_n(t), \gamma_n(t'))}{d(p_n, q_n)} = |t - t'|.$$

It follows that $d(\gamma(t), \gamma(t')) = |t-t'|d(p,q)$; γ is a continuous function, in particular a geodesic. We summarize the considerations with the following.

PROPOSITION 8. *The completion $\overline{\mathcal{T}}$ is a geodesic space.*

5. Length-minimizing curves on the completion

For the complex of curves $C(F)$ a k-simplex is a set of $k+1$ free homotopy classes of nontrivial, nonperipheral, mutually disjoint simple closed curves for the reference surface F. A simplex σ *precedes* a simplex σ' provided $\sigma \subseteq \sigma'$; preceding is a partial ordering. With the convention that the -1-*simplex* is the null set, there is a natural function Λ from the completion $\overline{\mathcal{T}}$ to the complex $C(F) \cup \{\emptyset\}$ determined by the classes of the nodes. For a marked noded Riemann surface (R, f) with $f : F \to R$, the labeling $\Lambda((R, f))$ is the simplex of free homotopy classes on F mapped to the nodes on R. The level sets of Λ are the strata of $\overline{\mathcal{T}}$. We write $\mathcal{S}(\sigma)$ for the stratum determined by the simplex σ. The stratum for a k-simplex has complex dimension $3g(F) - 3 - k$.

We now consider first properties of length-minimizing curves on $\overline{\mathcal{T}}$. We are able to make the analysis without first establishing that a length-minimizing curve is a limit of WP geodesics. In this section we build on the following result and present an alternative approach to the basic observation of S. Yamada [**47**] that except possibly for its endpoints, a length-minimizing curve is contained in a single stratum of $\overline{\mathcal{T}}$.

PROPOSITION 9. *For a length-minimizing curve γ on $\overline{\mathcal{T}}$ the composition $\Lambda \circ \gamma$ has a left and right limit at each point. The composition is continuous at a point where the left and right limits agree.*

Proof. First observe that only a finite number of simplices precede a given simplex. There is a continuous analog for strata: in a suitable neighborhood of a point of $\overline{\mathcal{T}}$ there are only a finite number of strata, and each precedes or coincides with the stratum of the point. If a left or right limit fails to exist for $\Lambda \circ \gamma$ at t_0, then there is a monotonic convergent sequence of parameter values $\{t_n\}$, $t_n \to t_0$ with $\Lambda \circ \gamma$ having value σ on $\{t_{2n}\}$ and a different value τ on $\{t_{2n+1}\}$. We may choose that σ precedes τ and further that σ, resp. τ, is a maximal, resp. minimal, such value. Maximal connected segments of γ contained in the stratum of σ are determined by the positivity of the geodesic-length functions of the classes in $\tau - \sigma$. In particular each maximal segment is parameterized by an open parameter interval; by Corollary 6 each (closed subsegment of each) maximal segment is length-minimizing.

Consider two points $\gamma(t_{2n})$ and $\gamma(t_{2n+2})$ on different maximal segments. By Corollary 6 there exists a WP geodesic β contained in the stratum of σ connecting $\gamma(t_{2n})$ and $\gamma(t_{2n+2})$. We now compare the segments $\gamma|_{[t_{2n},t_{2n+2}]}$ and β. Assume each segment is parameterized by arc-length; the segments necessarily have the same length. On the stratum σ the curves β and the maximal segment of γ at t_{2n} are solutions of an ordinary differential equation. If the initial tangents of β and $\gamma|_{[t_{2n},t_{2n+2}]}$ coincide, then by the uniqueness of solutions and the maximality, the segments must coincide for the length of β. The coinciding contradicts $\Lambda \circ \gamma$ having different values at t_{2n} and t_{2n+1}. The alternative is that the initial (unit) tangents of β and $\gamma|_{[t_{2n},t_{2n+2}]}$ differ. In this case γ can be modified by first substituting the segment β for the parameter interval $[t_{2n}, t_{2n+2}]$ and then *smoothing the corner* (inside the stratum) at $\gamma(t_{2n})$, to obtain a new curve $\tilde{\gamma}$ of strictly smaller length, again a contradiction. A sequence $\{t_n\}$ as described cannot exist. In summary the composition $\Lambda \circ \gamma$ is locally constant to the left and right of each point of its domain.

Finally if the left and right limits have a common value at t_0 then either $\Lambda\circ\gamma(t_0)$ also has the common value, or the common value precedes $\Lambda \circ \gamma(t_0)$. In the second instance we can again construct a modification $\tilde{\gamma}$ of strictly smaller length. The proof is complete.

We are interested in a class of singular *metrics* that model the WP metric in a neighborhood of a point on the compactification divisor $\mathcal{D} \subset \overline{\mathcal{M}}$. Consider now the product $(\mathbb{R}^2)^{m+n}$ with Euclidean coordinates (x, y) for x the $2m$-tuple with Euclidean metric dx^2 and y the $2n$-tuple with Euclidean metric dy^2. We refer to $\mathbb{R}^{2m} \times \{0\}$, resp. to $\{0\} \times \mathbb{R}^{2n}$, as the x-axes, resp. the y-axes. Here the x-axes represent coordinates on a stratum of dimension $2m$ and codimension $2n$, while the y-axes represent the parameters which open nodes. We write (r_j, θ_j) for the polar coordinates for the 2-plane (y_{2j-1}, y_{2j}) and (r, θ) for the product polar coordinates for the $2n$-tuple of y-coordinates. We consider the singular metric

$$\sum_{j=1}^{n} 4\, dr_j^2 + r_j^6\, d\theta_j^2$$

for the y-axes which we simply abbreviate as $dr^2 + r^6 d\theta^2$.

DEFINITION 10. *A continuous symmetric 2-tensor ds^2 is a product cuspidal metric for a neighborhood of the origin in $(\mathbb{R}^2)^{m+n}$ provided:*

1. *ds^2 is a smooth Riemannian metric on $\bigcap_{j=1}^{n}\{r_j > 0\}$;*
2. *the restriction of ds^2 to the x-axes is a smooth Riemannian metric ds_x^2;*
3. *$ds^2 = (d\mu^2 + dr^2 + r^6 d\theta^2)(1 + O(\|r\|^2))$ for $d\mu^2$ the pullback of ds_x^2 to $(\mathbb{R}^2)^{m+n}$ by the projection onto the x-axes ,and $\|r\|$ denoting the Euclidean norm of the radius vector for the y-axes.*

We are ready to continue our consideration of a length-minimizing curve γ and a point of discontinuity t_* interior to the domain of the label composition $\Lambda \circ \gamma$. The first circumstance to consider is that $\Lambda \circ \gamma$ is continuous from one side, say the right. In particular for $\Lambda \circ \gamma$ discontinuous from the left the simplex $\sigma = \Lambda \circ \gamma(t_*^-)$ strictly precedes the simplex $\sigma' = \Lambda \circ \gamma(t_*)$. Since γ is length-minimizing, the curve is a WP geodesic in the stratum $\mathcal{S}(\sigma')$ for an initial interval to the right of t_*.

Observe by analogy that in the circumstance that ds^2, $d\mu^2$ and $dr^2 + r^2 d\theta^2$ were smooth, then the x-axes would be totally geodesic (the *second fundamental form* would be trivial; see the discussion after Corollary 4) and the suggested refracting behavior of γ would not be possible. We will now show by a scaling argument that the behavior is also not possible for a product cuspidal metric.

We observe that the individual strata of $\overline{\mathcal{T}}$ branch cover strata in $\overline{\mathcal{M}}$ and that certain curves in $\overline{\mathcal{M}}$ have unique lifts determined by their initial point. $\overline{\mathcal{M}}$ is a V-manifold; consider first the local-manifold cover $\widehat{\mathcal{U}}$ for a neighborhood of the point given by $\gamma(t_*)$ (a neighborhood of $\gamma(t_*)$ in $\overline{\mathcal{M}}$ is described as $\widehat{\mathcal{U}}/Aut(\gamma(t_*))$). From Section 2 the preimage of $\widehat{\mathcal{U}}$ in $\overline{\mathcal{T}}$ is a disjoint union of sets, including a neighborhood $\mathcal{U}$ of $\gamma(t_*)$. The local stratum $\sigma \cap \mathcal{U} \subset \overline{\mathcal{T}}$ is a covering of its image $\widehat{\sigma} \subset \widehat{\mathcal{U}}$, with covering group the lattice of Dehn twists for the set of loops $\sigma' - \sigma$. The local stratum $\sigma' \cap \mathcal{U} \subset \overline{\mathcal{T}}$ coincides with its local projection $\widehat{\sigma'}$ to $\widehat{\mathcal{U}}$. In the following paragraphs we will use the simple observation that: a rectifiable curve in $\widehat{\mathcal{U}}$ with first segment in $\widehat{\sigma}$ and second segment in $\widehat{\sigma'}$ has WP isometric lifts to $\mathcal{U}$, each uniquely determined by prescribing an initial point. We can choose coordinates so $\widehat{\mathcal{U}} \cap \widehat{\sigma}$ is given by a neighborhood of the origin in $(\mathbb{R}^2)^{m+n}$ and $\widehat{\mathcal{U}} \cap \widehat{\sigma'}$ is given by a neighborhood of the origin in the x-axes $\mathbb{R}^{2m} \times \{0\}$.

We study the WP length of an oriented curve γ having first segment off the x-axes and second segment a geodesic in the x-axes. Let $\mathbf{o}$ be the first contact point of the curve with the x-axes. Consider the Euclidean ball of radius δ about $\mathbf{o}$. From Corollary 4 we can choose δ small for the metric to have the coordinate description of a product cuspidal metric in the ball. Along γ let a be the first intersection point of γ with the δ-sphere at $\mathbf{o}$ to the left of $\mathbf{o}$. Along γ let b be the first intersection point of γ with the δ-sphere at $\mathbf{o}$ to the right of $\mathbf{o}$. We investigate the lengths of curves from a to $\mathbf{o}$ to b as δ varies. We use the coordinates (x, y) for the following constructions. Let a_x, resp. a_y, be the Euclidean projection of a to the x, resp. the y, axes. Let β be the unit-speed $d\mu^2$ geodesic in the x-axes from a_x to b. For the same arc-length parameter let $\widehat{\beta}$ be the curve from a to b whose Euclidean projection to the y-axes is a constant speed radial line. On $\widehat{\beta}$ the tensor $dr^2 + r^6 d\theta^2$ restricts to dr^2 and in particular

$$\int_{\widehat{\beta}} (ds^2)^{1/2} = \int_{\widehat{\beta}} (1 + O(\|y\|^2))(d\mu^2 + dr^2)^{1/2}.$$

Since the length of $\widehat{\beta}$ is bounded in terms of δ and the Euclidean height of $\widehat{\beta}$ is bounded by $\|a_y\|$ it follows that the length of $\widehat{\beta}$ is given as

$$\int_{\widehat{\beta}} (d\mu^2 + dr^2)^{1/2} + O(\delta \|a_y\|^2)$$

for the Euclidean norm of a_y. The integral immediately evaluates to $(|||\beta|||^2 + \|a_y\|^2)^{1/2}$ for $|||\ |||$ denoting the $d\mu^2$ length. In summary the length of $\widehat{\beta}$ is bounded above by $(|||\beta|||^2 + \|a_y\|^2)^{1/2} + O(\delta \|a_y\|^2)$.

We next consider a lower bound for the length of the segment of γ from a to b. The first expansion is provided similar to the above consideration. Select a subsequence of values δ' tending to zero such that for each δ' the maximum of the y-height $\|y\|$ on the γ segment from a to $\mathbf{o}$ actually occurs at the initial point a. For the subsequence the length of γ is

$$\int_\gamma (d\mu^2 + dr^2 + r^6 d\theta^2)^{1/2} + O(\delta' ||a_y||^2).$$

The metric $d\xi^2 = (d\mu^2 + dr^2 + r^6 d\theta^2)$ is a product and the local behavior of its geodesics is understood. Let $\widehat{\gamma}$ be a comparison curve (a $d\xi^2$ piecewise geodesic) with segments from a to $\mathbf{o}$ and from $\mathbf{o}$ to b. The projection of the first segment of $\widehat{\gamma}$ to the x-axes is the $d\mu^2$ geodesic from a_x to $\mathbf{o}$. The projection of the first segment of $\widehat{\gamma}$ to the y-axes is the constant speed radial line from a_y to $\mathbf{o}$. The second segment of $\widehat{\gamma}$ is the geodesic from $\mathbf{o}$ to b. The integral of γ is minorized by the integral of $\widehat{\gamma}$ and consequently the length of γ is minorized by

$$(|||a_x|||^2 + ||a_y||^2)^{1/2} + |||b|||^2 + O(\delta' ||a_y||^2)$$

where we have written $|||x_0|||$ for the $d\mu^2$-distance from $\mathbf{o}$ to the point x_0 of the x-axes.

We are prepared to analyze the length of γ in a small neighborhood of the point $\mathbf{o}$.

Proposition 11. *A curve having first segment off the x-axes and second segment a geodesic in the x-axes is not ds^2 length-minimizing between its endpoints. There is a shorter curve of the same description.*

Proof. We first consider the rescaling limit of a neighborhood of $\mathbf{o}$ with the substitution $\delta u = x, \delta v = r$ and $d\eta^2 = \delta^{-2} ds^2$. The curves $\widehat{\beta}$ and $\widehat{\gamma}$ considered above have radial lines as their projections to the y-axes; it suffices for length considerations to consider the projection of the y-axes to its radial component r. The rescaling limit of $(d\mu^2 + dr^2)(1 + O(||y||^2))$ is the Euclidean metric and for a subsequence the points a, b limit to points of the unit sphere (same notation). The curve $\widehat{\beta}$ limits to the chordal line connecting a to b; the curve $\widehat{\gamma}$ limits to the segmented curve of line-segments connecting a to $\mathbf{o}$ and $\mathbf{o}$ to b. If a is not antipodal to b then (on the subsequence) $\widehat{\beta}$ is strictly shorter than $\widehat{\gamma}$. On a neighborhood of $\mathbf{o}$, γ is now modified by substituting a segment of $\widehat{\beta}$ to obtain a strictly shorter curve, a desired conclusion.

It remains to consider that the rescaling limit as δ tends to zero of a is the antipode to b. In this circumstance we have that $|||a_x|||$ is comparable to δ, $|||\beta|||$ is comparable to 2δ and $||a_y||$ by hypothesis is $o(\delta)$. Pick $\epsilon < 1$ such that $\epsilon^2 |||\beta||| > |||a_x|||$ for all small δ. Now from the preliminary considerations for small δ the length of $\widehat{\beta}$ is bounded above by

$$(|||\beta|||^2 + ||a_y||^2)^{1/2} + O(\delta ||a_y||^2) \leq |||\beta||| + \frac{1}{2\epsilon |||\beta|||} ||a_y||^2 + O(\delta ||a_y||^2)$$

and for a suitable subsequence the length of γ is bounded below by

$$(|||a_x|||^2 + ||a_y||^2)^{1/2} + |||b||| + O(\delta ||a_y||^2) \geq$$
$$|||a_x||| + |||b||| + \frac{\epsilon}{2|||a_x|||} ||a_y||^2 + O(\delta' ||a_y||^2)$$

for $||a_y||(|||a_x|||)^{-1}$ sufficiently small, which is ensured for δ sufficiently small. As specified above, δ' are the special values for which the maximum of the y-height $||y||$ on the γ sement from a to $\mathbf{o}$ occurs at the initial point a. Since β is a geodesic we have that $|||\beta||| \leq |||a_x||| + |||b|||$. Observe that the coefficient of the $||a_y||^2$-term for $\widehat{\beta}$ is strictly less than that for γ. Since $||a_y||^2$ is positive for δ positive, it now follows that $\widehat{\beta}$ is strictly shorter than γ and in particular that γ is not length-minimizing in a neighborhood of $\mathbf{o}$. The proof is complete.

The second circumstance to consider is that for a length-minimizing curve γ there is an interior domain discontinuity point t_* with the label composition $\Lambda \circ \gamma$ not continuous from either side. The curve γ connects points in different strata by passing through a higher codimension stratum. In particular from Proposition 11 it follows that the values of $\Lambda \circ \gamma$ for t_*^-, t_* and t_*^+ are all distinct; furthermore the values for t_*^- and t_*^+ strictly precede the value at t_*. We have the coordinate description of the local strata $\widehat{\mathcal{U}} \cap (\mathcal{S} \circ \Lambda \circ \gamma(t_*^-) \cup \mathcal{S} \circ \Lambda \circ \gamma(t_*) \cup \mathcal{S} \circ \Lambda \circ \gamma(t_*^+))$ given in a neighborhood of the origin in $(\mathbb{R}^2)^{m+n}$. For suitable $n_- + n_+ = n$ the neighborhood is given as a neighborhood of the origin in $(\mathbb{R}^2)^{m+n_-+n_+}$ with coordinates (x, y_-, y_+). In a neighborhood of the origin the three strata are given by germs of the coordinate axes: $\mathcal{S} \circ \Lambda \circ \gamma(t_*)$ by the x-axes; $\mathcal{S} \circ \Lambda \circ \gamma(t_*^-)$ by the y_--axes; and $\mathcal{S} \circ \Lambda \circ \gamma(t_*^+)$ by the y_+-axes. Again a rectifiable curve with the prescribed behavior for $\Lambda \circ \gamma$ has WP isometric lifts to $\mathcal{U}$, each uniquely determined now by prescribing an initial and terminal point.

PROPOSITION 12. *A curve having endpoints distinct from the origin and in distinct coordinate proper subspaces of the y-axes and further having the origin as an intermediate point is not ds^2 length-minimizing between its endpoints. There is a shorter curve avoiding the origin.*

Proof. The considerations are simplified since in effect the subspaces corresponding to $\mathbb{R}^{2n_-}$ and $\mathbb{R}^{2n_+}$ are orthogonal. Choose $\epsilon > 0$ such that $(1+\epsilon) < (1-\epsilon)\sqrt{2}$; from Definition 10 the restriction of ds^2 to the y-axes is estimated above, resp. below, by the $(1+\epsilon)$, resp. $(1-\epsilon)$, multiple of $(dr^2 + r^6 d\theta^2)$. For a^- in $\mathbb{R}^{2n_-} - \{0\}$ and a^+ in $\mathbb{R}^{2n_+} - \{0\}$ the $(dr^2 + r^6 d\theta^2)$ geodesic in $\mathbb{R}^{2n}$ connecting a^- to a^+ and the piecewise geodesic connecting a^- to the origin and then to a^+ are Euclidean line segments. The line connecting a^- to a^+ has ds^2 length at most $(1+\epsilon)(||a^-||^2 + ||a^+||^2)^{1/2}$. The line segments connecting a^- to the origin to a^+ have length at least $(1-\epsilon)(||a^-|| + ||a^+||)$.

For an oriented curve γ with the prescribed strata behavior consider a Euclidean radius δ sphere at the origin and let a^-, resp. a^+, be the first intersection point along γ to the left, resp. right, of the origin. Since the radial component of the metric is comparable to the Euclidean metric, the maximum value of $||r||$ along the segments of γ is comparable to δ. Apply the above estimate for δ small to obtain the desired conclusion. The proof is complete.

We are ready to present our counterpart of S. Yamada's Theorem 2, [**47**].

THEOREM 13. *$\overline{\mathcal{T}}$ is a unique geodesic space. The length-minimizing curve connecting points $p, q \in \overline{\mathcal{T}}$ is contained in the closure of the stratum with label $\Lambda(p) \cap \Lambda(q)$. The open segment $\gamma - \{p, q\}$ is a solution of the WP geodesic equation on the stratum with label $\Lambda(p) \cap \Lambda(q)$. For a point p the stratum with label $\Lambda(p)$ is the union of the length-minimizing open segments containing p. The closure of each stratum is a convex set, complete in the induced metric.*

Proof. $\overline{\mathcal{T}}$ is a geodesic space from Proposition 8. For a length-minimizing curve γ we consider the label behavior of $\Lambda \circ \gamma$. From Proposition 9 $\Lambda \circ \gamma$ only has a finite number of discontinuities. From Propositions 11 and 12, as well as the lifting property of the indicated curves on the local-manifold covers of $\overline{\mathcal{M}}$, it follows that $\Lambda \circ \gamma$ is at most discontinuous at an endpoint. Since each stratum is a relatively open subset of its closure in $\overline{\mathcal{T}}$, it further follows that the value of Λ on the open segment of γ precedes its value at each endpoint (the open segment value is a lower bound for the partial ordering). It also follows that $\Lambda(p) \subset \Lambda(q)$ is a necessary condition for p to be an interior point of a length-minimizing curve with endpoint q.

A free homotopy class α of a simple closed curve is represented by a vertex in $\Lambda(p) \cap \Lambda(q)$. For the geodesic-length function ℓ_α, the composition $\ell_\alpha \circ \gamma$ is a continuous function. The composition vanishes at its domain endpoints and is convex on its domain interior. It follows that the composition is identically zero and consequently that the open segment of γ is contained in the stratum with label $\Lambda(p) \cap \Lambda(q)$. A stratum is a product of Teichmüller spaces. The maximal open segment of γ is a solution of the product WP geodesic equation on the stratum. Consider WP geodesics γ, γ' parameterized proportional to arc-length by the unit-interval with common endpoints. The distance between corresponding points is a continuous function, vanishing at 0 and 1, and convex on $(0,1)$ from Proposition 7. The distance is identically zero and the geodesics coincide. $\overline{\mathcal{T}}$ is a unique geodesic space.

As note above, $\Lambda(p) \subset \Lambda(q)$ is a necessary condition for p to be an interior point of a length-minimizing curve with endpoint q. Since a stratum is a product of Teichmüller spaces for which length-minimizing curves are solutions of the geodesic equation and since solutions can be extended, it follows that the condition is sufficient for extension. The final conclusion follows since $\overline{\mathcal{T}}$ is a geodesic space. The closure of a stratum is convex from the above description of geodesics. The closure of a stratum is complete in the induced metric from the completeness of $\overline{\mathcal{T}}$. The proof is complete.

We are ready to present the basic result. $CAT(0)$ is a generalized condition for a non-positively curved, uniquely geodesic space, [**6**, Chap. II.1]. With the above result there is little further need to distinguish between length-minimizing curves and solutions of the geodesic differential equation. We now also refer to length-minimizing curves parameterized proportional to arc-length as *geodesics*.

THEOREM 14. *$\overline{\mathcal{T}}$ is a CAT(0) space.*

Proof. A length-minimizing curve on $\overline{\mathcal{T}}$ is approximated by WP geodesics on $\mathcal{T}$ by choosing sequences of points converging to the endpoints, and considering the joins parameterized on the original interval. From Proposition 7 the joins converge to the designated geodesics, and for a pair of geodesics the relative distance functions converge. In particular a limit of geodesic triangles satisfying the $CAT(0)$ inequality will also satisfy the inequality, [**6**, Chap. II.1]. Since the WP metric on $\mathcal{T}$ has negative curvature, geodesic triangles satisfy the $CAT(0)$ inequality [**6**, Chap. II.1, Remark 1A.8]. The proof is complete.

The local geometry of geodesics on $\overline{\mathcal{M}}$ differs from that of $\overline{\mathcal{T}}$. A product cuspidal metric is not uniquely geodesic.

PROPOSITION 15. *$\overline{\mathcal{M}}$ is not locally uniquely geodesic at the compactification divisor and in particular is not locally a CAT(0) space.*

Proof. We show that the local manifold cover for a neighborhood of a Riemann surface having a single node is not uniquely geodesic [**6**, Chap. II.1, Prop. 1.4].

Introduce hyperbolic metric plumbing coordinates. For a reference base point $(s',t') = (s'_1,\dots,s'_m,t')$ off the s-axes, we consider the curves based at (s',t'), disjoint from the s-axes, and linking the s-axes. The base point (s',t') lifts to a point of the Teichmüller space $\mathcal{T}$. From Corollary 6 for each possible value of the linking number there is a corresponding length-minimizing curve in $\mathcal{T}$ (minimizing for curves disjoint from the s-axes). We first bound the lengths of such linking curves. A general comparison curve is prescribed by the sequence: a radial line segment in the t-coordinate, followed by an integer number of rotations about a t-coordinate circle and finally a radial line segment returning to the base point. From Corollary 4 the comparison curves can be prescribed with length uniformly bounded by a multiple of $|t'|$. It follows for $|s'|, |t'|$ small that the length-minimizing linking curves are all contained in a small neighborhood of the origin.

We consider length bounds involving the linking number. For a curve with linking number n and the minimal absolute value of the coordinate t on the curve $|t_0|$, then by Corollary 4 the length of the curve is at least a uniform multiple of $|n\,t_0^3|$. It follows that $|t_0|$ is bounded by $|t'/n|^{1/3}$; it further follows by considering only the t-radial component of the length that the linking curve has length at least a uniform multiple of $|t'| - |t'/n|^{1/3}$; the desired bound. A comparison curve with linking number one is the t-coordinate circle of radius $|t'|$; its length is bounded by a multiple of $|t'|^3$. We draw a simple conclusion: there is a length-minimizing linking curve γ of minimal length (presumably with linking number ± 1).

We bisect γ. Let p denote (s',t') and q denote the γ-midpoint. The length of γ is $O(|t'|^3)$; from Corollary 4 the length of a curve connecting p to the s-axes is at least a multiple of $|t'|$. With the length bounds and the fact that γ is a solution of the geodesic differential equation it now follows that the segments of γ connecting p to q are length-minimizing for the neighborhood of the origin. The neighborhood is not uniquely geodesic. The proof is complete.

6. Applications

We are interested in understanding the flat subspaces of $\overline{\mathcal{T}}$. Our purpose is to understand the flat geodesic simplices and in particular the flat geodesic triangles. Consider a geodesic triangle with distinct vertices o, p and q. Parameterize proportional to arc-length the sides $\widehat{op}$ and $\widehat{oq}$ by geodesics $\gamma(t)$ and $\gamma'(t)$, $t \in [0,1]$ with $\gamma(0) = \gamma'(0) = o$. The distance function $d(\gamma(t),\gamma'(t))$ is an important measure of the triangle. We also require a numerical invariant for noded Riemann surfaces. Let $\nu(R)$ be the number of components of $R - \{nodes\}$ that are not thrice-punctured spheres. The maximal dimension of a flat subspace of the stratum corresponding to R is given by $\nu(R)$. We use the description of flat subspaces to give a different proof of a Brock-Farb result: the WP metric is in general not Gromov-hyperbolic, [**9**]. Recall that a metric space (M,d) is *Gromov-hyperbolic* provided there exists a positive number δ such that for each geodesic triangle the δ-neighborhood of a pair of sides contains the third side, [**16**].

PROPOSITION 16. *On the augmented Teichmüller space $\overline{\mathcal{T}}$ of genus g, n punctured surfaces*

1. *For geodesics γ, γ' as above and an interior parameter value, consider that the values of the distance function $d(\gamma(t),\gamma'(t))$ and its supporting linear*

function coincide. The interiors of the geodesics γ, γ' then lie on a submanifold of $\overline{\mathcal{T}} - \mathcal{T}$ given as the Cartesian product of geodesics from component Teichmüller spaces.

2. *For a stratum corresponding to a noded Riemann surface R, the maximal dimension of a locally Euclidean isometric submanifold is $\nu(R)$. The maximal value of ν is $g - 1 + \lfloor \frac{g+n}{2} \rfloor$, which is achieved for an arrangement with g once-punctured tori and $\lfloor \frac{g+n}{2} \rfloor - 1$ four-punctured spheres.*
3. *For $3g-3+n \geq 3$ the Teichmüller space with the WP metric is not Gromov-hyperbolic.*

Proof. First, a convex function is necessarily linear if it shares a common interior value with the supporting linear function. From Bridson-Haefliger [**6**, Chap. I.1, Defn. 1.10 and Chap. II.3, Prop. 3.1] provided $d(\gamma(t), \gamma'(t))$ is linear then the comparison angles formed by the point triples $(\gamma(t), o, \gamma'(t))$ all coincide. The *flat triangle lemma* of A. D. Alexandrov can now be applied [**6**, Chap. II.2, Prop. 2.9]. The convex hull of o, p and q in $\overline{\mathcal{T}}$ is consequently isometric to the convex hull of a Euclidean triangle with the corresponding side lengths. An isometry is prescribed. There is an associated variation of geodesics $\beta(t,s)$, parameterized proportional to arc-length, such that $\beta(t,0) = \gamma(t)$, $\beta(t,1) = \gamma'(t)$, $\beta(0,s) = o$ and $\beta(1,s)$ lies on $\widehat{pq}$ with $d(p, \beta(1,s)) = sd(p,q)$. By Theorem 13 it follows for interior parameter values that $\beta(t,s)$ lies in a single stratum; by Proposition 7 it follows for interior parameter values that $\beta(t,s)$ is smooth. The stratum is a product of Teichmüller spaces. We may apply the techniques of Riemannian geometry, [**10**]. Since the triangle is flat the contribution to (5) from the term $\langle R(V,T)T, V\rangle$ is zero. For a product of negatively-curved metrics $\langle R(V,T)T, V\rangle$ vanishes only if the variation fields V and T everywhere have collinear projections to the tangent spaces of the factors [**33**, Chap. 3, Lemmas 39, 58]. Since the projections are collinear the triangle also projects to a geodesic segment in each component Teichmüller space. The desired first conclusion.

Second, as already indicated for a Riemannian product of negatively-curved metrics the maximal dimension of a flat subspace is the number of factors. The dimension of a maximal flat is given by $\nu(R)$ since a punctured Riemann surface has a positive dimensional Teichmüller space provided the surface is not the thrice-punctured sphere. Once-punctured tori have Euler characteristic -1 and $\dim \mathcal{T} > 0$; the general surface with $\dim \mathcal{T} > 0$ has Euler characteristic strictly less than -1. The maximal statement follows. The desired second conclusion.

Third, for the Euclidean plane, a positive number δ, and a non degenerate triangle, a large-scaling provides a triangle δ with a δ-neighborhoold of a pair of sides omitting an open segment on the third side. A stratum corresponding to a noded surface R with $\nu(R) \geq 2$ contains triangles isometric to Δ. For such a triangle a triple of points in $\mathcal{T}$, one close to each vertex, prescribes a triangle with measurements close to those of Δ. Independent of the length of an edge, the geodesic triangle in $\mathcal{T}$ is uniformly close to the triangle Δ with distance estimated only by the distance separating corresponding vertices. For a suitable triple, a δ-neighborhood of two joining sides omits an open segment on the third joining side. A stratum with $\nu(R) \geq 2$ exists provided $\dim \mathcal{T} \geq 3$. The proof is complete.

The maximal simplices in $C(F)$ serve an important role for the geometry of $\overline{\mathcal{T}}$. Since thrice-punctured spheres are conformally rigid, a $3g-4$-simplex σ in $C(F)$ corresponds by Λ^{-1} to a unique marked maximally noded Riemann surface R_σ in

$\overline{\mathcal{T}}$. The *Mod* stabilizer of a maximally noded Riemann surface R_σ is an extension of a finite group by a rank $3g-3$ Abelian group, the mapping classes of products of Dehn twists about the elements of σ. The maximally noded Riemann surfaces serve the role of the *maximal rank cusps* for the moduli space. Brock studied the finite length WP geodesics from a point of $\mathcal{T}$ to the marked noded Riemann surfaces, [**7**]. The geodesics from a point can be extended to include their endpoints in $\overline{\mathcal{T}}$. As a consequence of the $CAT(0)$ geometry the initial unit tangents for the family of geodesics from a point to a stratum provide for a Lipschitz map from the stratum to the unit tangent sphere. Accordingly the image of $\overline{\mathcal{T}} - \mathcal{T}$ in each unit tangent sphere has measure zero and thus the infinite length geodesic rays have tangents dense in each tangent sphere. Brock's method for approximating infinite length rays by finite length rays now provides the following.

THEOREM 17 ([**7**]). *The geodesic rays from a point of $\mathcal{T}$ to the maximally noded Riemann surfaces have initial tangents dense in the tangent space.*

The following is a new consequence of the result.

COROLLARY 18. *The geodesics connecting maximally noded Riemann surfaces have tangents dense in the tangent bundle of $\mathcal{T}$.*

Proof. We consider unit-speed geodesics. Given a unit-tangent v at a point p of $\mathcal{T}$ and a positive number ϵ, we proceed to determine an approximating geodesic. By the above Theorem let γ_- be a unit-speed geodesic connecting a point q, representing a maximally noded Riemann surface to p, with the final tangent w of γ_- within ϵ of v. For a small positive number δ similarly let γ_+ be a unit-speed geodesic connecting p to a point r, representing a maximally noded Riemann surface, with the initial tangent of γ_+ within δ of w. The geodesics γ_- and γ_+ form a vertex at p with angle in the interval $[\pi - \delta, \pi]$. The three points p, q and r determine a geodesic triangle $\Delta(p,q,r)$ in $\overline{\mathcal{T}}$ for which there is a comparison triangle $\Delta(\bar{p},\bar{q},\bar{r})$ in the Euclidean plane. By [**6**, Chap. II.4, Lem. 4.11] the vertex angles for $\Delta(p,q,r)$ are bounded by the corresponding vertex angles for the Euclidean triangle $\Delta(\bar{p},\bar{q},\bar{r})$. In particular the vertex angle at $\bar{p}$ is also in the interval $[\pi - \delta, \pi]$. It follows for $\Delta(\bar{p},\bar{q},\bar{r})$ that the angle at $\bar{q}$ is at most δ. For $\delta < \pi/2$ it follows that $d(\bar{q},\bar{r}) > d(\bar{q},\bar{p})$ and there is a point $\bar{s}$ with $d(\bar{q},\bar{s}) = d(\bar{q},\bar{p})$, $\bar{s}$ on the geodesic segment $\widehat{\bar{q}\bar{r}}$. By trigonometry $d(\bar{s},\bar{p}) = 2d(\bar{p},\bar{q})\sin\delta/2$ and thus the comparison point s on the geodesic segment $\widehat{qr}$ satisfies $d(s,p) \le 2d(p,q)\sin\delta/2$. Similarly the midpoints s_{mid} of $\widehat{qs}$ and p_{mid} of $\widehat{qp}$ satisfy $2d(s_{mid},p_{mid}) \le d(s,p)$. To summarize the considerations, the geodesic $\widehat{qr}$ contains a point s that is within distance $2d(p,q)\sin\delta/2$ of p, and the midpoints s_{mid} of $\widehat{qs}$ and p_{mid} of $\widehat{qp}$ are within distance $d(p,q)\sin\delta/2$.

¿From Proposition 7 the geodesic segments $\widehat{s_{mid}s}$ and $\widehat{p_{mid}p}$ are sufficiently close in the C^1-topology for δ sufficiently small. We now choose r' and γ'_+ to provide that the tangent at s' is within ϵ of the final tangent of γ_-, which in turn was chosen to be within ϵ of v. The proof is complete.

The following is an immediate consequence of the above result.

COROLLARY 19. *$\overline{\mathcal{T}}$ is the closed convex hull of the subset of marked maximally noded Riemann surfaces.*

We combine the above and follow the outline of the Masur-Wolf approach to give an immediate proof of the Masur-Wolf theorem, [**31**, Theorem A]. The classification of simplicial automorphisms of the curve complex $C(F)$ by N. Ivanov [**22**], M. Korkmaz [**27**], and F. Luo [**28**] is an essential consideration.

THEOREM 20. *For $3g - 3n > 1$ and $(g, n) \neq (1, 2)$, every WP isometry of $\mathcal{T}$ is induced by an element of the extended mapping class group.*

Proof. An isometry of $\mathcal{T}$ extends to an isometry of the completion $\overline{\mathcal{T}}$. By Theorem 13 an isometry of $\overline{\mathcal{T}}$ necessarily preserves the strata structure and the incidence relations. It follows that an isometry induces a simplicial automorphism of $C(F)$. From the results of Ivanov [**22**], Korkmaz [**27**] and Luo [**28**], every simplicial automorphism coincides with the induced automorphism of an extended mapping class. In particular for a WP isometry of $\overline{\mathcal{T}}$, there is an extended mapping class such that the two mappings coincide on the subset of maximally noded Riemann surfaces. The conclusion now follows from Corollary 19. The proof is complete.

A complete, convex subset $\mathcal{C}$ of a $CAT(0)$ space is the base for an *orthogonal projection*, [**6**, Chap. II.2, Prop. 2.4]. A fibre of the projection is the unique geodesic realizing the distance between its points and the base. The projection is a retraction that does not increase distance. The distance $d_{\mathcal{C}}$ to $\mathcal{C}$ is a convex function satisfying $|d_{\mathcal{C}}(p) - d_{\mathcal{C}}(q)| \leq d(p, q)$, [**6**, Chap. II.2, Prop. 2.5]. Examples of complete, convex sets $\mathcal{C}$ are: points, complete geodesics, and fixed-point sets of isometry groups. Furthermore in the case of $\overline{\mathcal{T}}$ with Theorem 13 the closure of each individual stratum is the base of a projection (local projections are prescribed on $\overline{\mathcal{M}}$). On $\overline{\mathcal{T}}$ a tubular neighborhood of an stratum is fibered by the projection-geodesics. For the local understanding of the distance to the stratum we now consider a refinement of the prescription for a product cuspidal metric. In particular by Corollary 4 the WP metric has an expansion with an order-three approximation about the x-axes

$$ds^2 = (d\mu^2 + dr^2 + r^6 d\theta^2)(1 + O(\|r\|^3)).$$

COROLLARY 21. *For a stratum σ defined by vanishing of the geodesic-length sum $\ell = \ell_1 + \cdots + \ell_n$, the distance to the stratum is given locally as $d(p, \sigma) = (2\pi\, \ell)^{1/2} + O(\ell^2)$.*

Proof. We begin by considering distances on $\overline{\mathcal{M}}$. For a prescribed stratum and point choose a local-manifold cover $\widehat{\mathcal{U}}$ with the point corresponding to the origin and (image) stratum σ corresponding to the x-axes for the normal form of ds^2. The distance to the stratum $d(p, \sigma)$ is estimated from above by considering the radial line from a point p to the origin. The bound is $\|p_y\| + O(\|p_y\|^4)$ in terms of the y-projection of p. We next consider a lower bound for $d(p, \sigma)$. A curve in $\widehat{\mathcal{U}}$ connecting p to σ can be isometrically lifted to $\overline{\mathcal{T}}$; it suffices to examine curves that are length-minimizing on $\overline{\mathcal{T}}$. From [**6**, Chap. II.2, Prop. 2.4] for p close to the origin there is a geodesic $\gamma \subset \widehat{\mathcal{U}}$ connecting p to $q \in \sigma$ and γ provides the length-minimizing curve connecting to σ for each point of γ. For $p' \in \gamma$, as noted, $d(p', \sigma)$ is bounded above by the Euclidean norm of p'_y. Since ds^2 is likewise bounded below in terms of dr^2 it follows that $d(p'; \sigma)$ is actually comparable to $\|p'_y\|$ and consequently that the maximum Euclidean y-height of the γ-segment $\widehat{p'q}$ is likewise bounded. It now follows overall that

$$d(p', \sigma) = \int_{\widehat{p'q}} (d\mu^2 + dr^2 + r^6 d\theta^2)^{1/2} + O(\|p'\|^4).$$

The explicit integral is minorized by choosing the radial line in the y-coordinate; the resulting lower bound is $||p'_y||$. Thus the distance to the stratum on $\overline{\mathcal{M}}$ and on $\overline{\mathcal{T}}$ is $d(p,\sigma) = ||p_y|| + O(||p_y||^4)$. In [**44**, Example 4.3] a relationship is provided for the geodesic-length functions and the present hyperbolic metric plumbing coordinates; it is shown that

$$\ell_j = \frac{2\pi^2}{-\log|t_j|} + O\Big(\frac{1}{(\log^2|t_j|)}\sum_{k=1}^{n}\frac{1}{(\log^2|t_k|)}\Big).$$

We may rearrange terms and substitute the relation $\varrho_j^2 = \frac{4\pi^3}{-\log|t_j|}$ to obtain the expansion

$$\varrho_j^2 = 2\pi\ell_j(1 + O(\ell_j\sum_k \ell_k^2)).$$

The distance expansion now follows from Corollary 4. The proof is complete.

The WP gradients of the geodesic-length functions also have general expansions, [**45**, II: Sec. 2.2, Lemmas 2.3 and 2.4]. We have for ϵ positive and geodesic-length functions $\ell_\alpha \le \ell_\beta \le \epsilon$ that

$$\langle \operatorname{grad}\ell_\alpha, \operatorname{grad}\ell_\alpha\rangle = \frac{2}{\pi}\ell_\alpha \,+\, O(\ell_\alpha^3)$$
$$\langle \operatorname{grad}\ell_\alpha, \operatorname{grad}\ell_\beta\rangle = O(\ell_\alpha\ell_\beta^2)$$

with constants independent of the surface, or in particular for $\lambda_* = (2\pi\ell_*)^{1/2}$ that

$$\langle \operatorname{grad}\lambda_\alpha, \operatorname{grad}\lambda_\alpha\rangle = 1 \,+\, O(\lambda_\alpha^4)$$
$$\langle \operatorname{grad}\lambda_\alpha, \operatorname{grad}\lambda_\beta\rangle = O(\lambda_\alpha\lambda_\beta^3).$$

We consider applications of the gradient bounds. We now consider simplices σ and τ with free homotopy classes α in σ and β in τ with (all) representatives intersecting. The simplice σ and τ do not precede a common simplex. There is also a positive lower bound for the corresponding geodesic-length sum $\ell_\alpha + \ell_\beta$. In particular from the *collar result* there is a positive constant $\ell_0 < 2$ such that about a geodesic α with $\ell_\alpha \le \ell_0$, there is an embedded collar of width $2\log 2/\ell_\alpha$: in which case ℓ_β is at least the width, [**34**]. Consider a WP length-minimizing curve γ connecting the strata $\mathcal{S}(\sigma)$ and $\mathcal{S}(\tau)$. On the curve γ the geodesic-length functions ℓ_α and ℓ_β each vanish (ℓ_α on $\mathcal{S}(\sigma)$; ℓ_β on $\mathcal{S}(\tau)$) and are each unbounded. The following is now the consequence of the universal bounds for the gradients.

COROLLARY 22. *There is a positive constant δ such that null strata $\mathcal{S}(\sigma)$ and $\mathcal{S}(\tau)$ either have intersecting closures or $d(\mathcal{S}(\sigma), \mathcal{S}(\tau)) \ge \delta$.*

We also recognize from the relations for a stratum σ, defined by the vanishing of the geodesic-length sum $\ell = \ell_1 + \cdots + \ell_n$, that the vector fields $\{\operatorname{grad}\lambda_1, \dots, \operatorname{grad}\lambda_n\}$ play the role of *normal Fermi fields* [**17**, Sec. 2.3]. Recall in particular for a tubular neighborhood of a submanifold in a Riemannian manifold, Fermi coordinates are the relative analog of *normal* coordinates about a point. First consider a neighborhood $\mathcal{N}$ of the 0-section of the normal subbundle of the tangent bundle of the submanifold. With the restriction of the exponential map $\mathcal{N}$ is identified with a tubular neighborhood of the submanifold. An orthonormal frame for the normal bundle provides Fermi coordinates for the fibres of $\mathcal{N}$, and also for a tubular neighborhood of the submanifold upon composition with the inverse of the exponential map. The unit-speed geodesics normal to the submanifold are given in terms of the Fermi coordinates as the unit-speed linear rays from the 0-section. The tangent fields of the unit-speed geodesics normal to the submanifold are constant sums of the Fermi coordinate tangent fields (the *normal Fermi fields*).

We are ready to present the analogy. For constant sums of the vector fields $\operatorname{grad}\lambda_*$, the WP distance between endpoints of integral curves nearly equals the integral curve length. Consider for a positive vector $c = (c_1, \ldots, c_n)$ the integral curves of $\mathfrak{v} = -\sum_j c_j \operatorname{grad}\lambda_j$. The time-one integral curve of $\mathfrak{v}$ with initial point $2\pi(\ell_1, \ldots, \ell_n) = (c_1^2, \ldots, c_n^2)$ has terminal point at distance $O(\|c\|^4)$ to σ. Further from Corollary 21 the point $2\pi(\ell_1, \ldots, \ell_n) = (c_1^2, \ldots, c_n^2)$ is at WP distance $\|c\| + O(\|c\|^4)$ from σ. From the gradient relations above we have that $\|\mathfrak{v}\|_{WP} = \|c\| + O(\|c\|^4)$ and that the time-one integral curve has the same WP length. For $\|c\|$ small, the time-one integral curves of $\mathfrak{v}$ have endpoints at distance nearly equal to the curve length. The integral curves approximate WP geodesics. The integral curves of $\mathfrak{n} = \sum_j \ell^{-1/2}\lambda_j \operatorname{grad}\lambda_j$ also have length nearly equal to the distance between endpoints. The time $(2\pi\ell)^{1/2}$ integral curve of $\mathfrak{n}$ connects $(\ell_1, \ldots, \ell_n)$ and σ; for ℓ small $\mathfrak{n}$ is approximately the WP unit normal field to σ. Also by Corollary 21 $\mathfrak{n}$ approximates $\operatorname{grad} d_{WP}(\cdot, \sigma)$.

7. The structure of geodesic limits

We investigate sequences of geodesics. $\overline{\mathcal{T}}$ is a complete metric space with a compact quotient $\overline{\mathcal{M}}$. We anticipate that the compactness is manifested in the structure of the space of geodesics for $\overline{\mathcal{T}}$. We find that geometric limits of geodesics are described by ***polygonal paths*** and products of Dehn twists. Specifically for a sequence of bounded length geodesics there is a subsequence of Mod-translates that converges geometrically (sequences of products of Dehn twists are applied to subsegments) to a polygonal path, a piecewise geodesic curve connecting strata. We consider an application of the result and show that each fixed-point free element of the mapping class group Mod has a geodesic axis in $\overline{\mathcal{T}}$; the axis is unique and lies in $\mathcal{T}$ when the element is irreducible. Furthermore irreducible elements have either coinciding or divergent axes. The present results provide a different approach for the considerations of G. Daskalopoulos and R. Wentworth [**13**].

Sequences of geodesics can have special behavior for product cuspidal metrics. We present an example. Consider the half-plane $\mathbb{R}_{\geq 0} \times \mathbb{R}$ with coordinates (r, θ) and the identification space $\mathbb{R}_{\geq 0} \times \mathbb{R}/\{(0,\theta) \sim (0,\theta')\}$ with *metric* $dr^2 + r^6 d\theta^2$. Denote the special point $\{(0,\theta) \sim (0,\theta')\}$ by O. For the isometry $T : (r,\theta) \to (r, \theta + 1)$ consider the unit-speed geodesics γ_n connecting (r_0, θ_0) and $T^n(r_1, \theta_1)$. For n large the length of γ_n is nearly $r_0 + r_1$; we can provide that the γ_n are essentially parameterized on the interval $[0, r_0+r_1]$. By elementary considerations of differential equations, on the parameter interval $[0, r_0]$ the sequence $\{\gamma_n\}$ converges to the $\theta = \theta_0$ line segment $\widehat{(r_0,\theta_0)O}$. On the parameter interval $[r_0, r_0 + r_1]$ the sequence $\{T^{-n}\gamma_n\}$ converges to the $\theta = \theta_1$ line segment $\widehat{O(r_1,\theta_1)}$. In effect the geodesic sequence $\{\gamma_n\}$ is described by the polygonal path $\widehat{(r_0,\theta_0)O} \cup \widehat{O(r_1,\theta_1)}$ and the sequence of transformations $\{T^n\}$. Furthermore the curve $\widehat{(r_0,\theta_0)O} \cup T^n\widehat{O(r_1,\theta_1)}$ is continuous and has distance in the sense of parameterized unit-speed curves to γ_n that tends to zero as n tends to infinity.

We consider the local description of the map from $\overline{\mathcal{T}}$ to $\overline{\mathcal{M}}$. Associated to a k-simplex σ is the rank $k+1$ Abelian group Mod_σ of mapping classes of products of Dehn twists about the elements of σ. Mod_σ stabilizes the σ-null stratum $\mathcal{S}(\sigma)$. For a point $p \in \mathcal{S}(\sigma)$ the stabilizer $Mod(p) \subset Mod$ is a group extension of a finite group $G(p)$ by Mod_σ. Furthermore for a point p we can prescribe a suitable basis $\{\mathcal{U}\}$ of $Mod(p)$ invariant neighborhoods. Neighborhoods of the projection of p to $\overline{\mathcal{M}}$ are given as $\mathcal{U}/Mod(p)$. Furthermore each quotient $\mathcal{U}/Mod_\sigma$ is a local manifold cover. We can further prescribe that the quotients $\mathcal{U}/Mod_\sigma$ are relatively compact in a fixed quotient $\mathcal{U}'/Mod_\sigma$; the quotients $\partial\mathcal{U}/Mod_\sigma$ are accordingly compact.

Since $\overline{\mathcal{M}}$ is compact, given a sequence of points of $\overline{\mathcal{T}}$ there exists a subsequence and associated elements of Mod such that the sequence of image points converges. Accordingly we consider sequences of unit-speed parameterized geodesics with initial points converging.

PROPOSITION 23. *Consider a sequence of unit-speed geodesics $\{\gamma'_n\}$ with initial points converging to p_0, lengths converging to a positive value L' and parameter intervals converging to $[t', t'']$ with $L' = t'' - t'$. There exists an associated partition $t' = t_0 < t_1 < \cdots < t_k = t''$ of the interval; simplices $\sigma_0 = \Lambda(p_0), \sigma_1, \dots, \sigma_k$; and points $p_1 \in \mathcal{S}(\sigma_1), \dots, p_k \in \mathcal{S}(\sigma_k)$ on the null strata.*

The data satisfies $L(\widehat{p_j p_{j+1}}) = t_{j+1} - t_j$ for $j = 0, \dots, k-1$ and for the stratum with label $\tau_j = \Lambda(p_j) \cap \Lambda(p_{j+1})$ then: τ_0 strictly precedes σ_1 if $k > 1$; τ_{k-1} strictly precedes σ_{k-1} if $k > 1$; τ_j strictly precedes σ_j and σ_{j+1} for $j = 1, \dots, k-2$. The concatenation of geodesic segments $\widehat{p_0p_1} \cup \widehat{p_1p_2} \cup \cdots \cup \widehat{p_{k-1}p_k}$ is the unique length-minimizing curve connecting p_0 to p_k and intersecting in order the closures of the strata $\mathcal{S}(\sigma_1), \dots, \mathcal{S}(\sigma_{k-1})$.

There is a subsequence $\{\gamma_n\}$ of the geodesics and sequences of products of Dehn twists $T_{(j,n)} \in Mod_{\sigma_j - \tau_j}$, $j = 0, \dots, k-1$, such that on the parameter interval $[t_j, t_{j+1}]$ the geodesic segments $T_{(j,n)} \circ \cdots \circ T_{(0,n)}\gamma_n$ converge to $\widehat{p_jp_{j+1}}$ in the sense of parameterized unit-speed curves. Furthermore the distance between the parameterized unit-speed curves γ_n and $\widehat{p_0 p_{(k,n)}}$, $p_{(k,n)} = (T_{(k-1,n)} \circ \cdots \circ T_{(0,n)})^{-1} p_k$, tends to zero for n tending to infinity. The sequence of transformations $\{T_{(0,n)}\}$ is either trivial or unbounded. The sequences of transformations $\{T_{(j,n)}\}$, $j = 1, \dots, k-1$ are unbounded.

Proof. The main argument is to provide the two steps for determining the individual geodesic segments $\widehat{p_jp_{j+1}}$. The overall argument is then a finite induction. For the first step choose a neighborhood $\mathcal{U}$ of p_0 with $\partial\mathcal{U}/Mod_{\sigma_0}$ compact. For each geodesic γ'_n let q_n be the first point of intersection with $\partial\mathcal{U}$. Either a subsequence of the points q_n converges to a point q' or we select elements $T_{(0,n)} \in Mod_{\sigma_0}$ such that the images $T_{(0,n)}q_n$ lie in a relatively compact fundamental domain for the action of Mod_{σ_0} on $\partial\mathcal{U}$. For the situation of selecting elements $T_{(0,n)}$ there is a subsequence $\{T_{(0,n)}q_n\}$ convergent to a point q' and the sequence $\{T_{(0,n)}\}$ is unbounded. Now the group Mod_{σ_0} fixes p_0 and a sequence of points converges to q'. The group Mod_{σ_0} is a direct product with a factor $Mod_{\sigma_0-\tau_0}$ for $\tau_0 = \Lambda(q')$; for $\{T_{(0,n)}q_n\}$ converging to q' it is a basic feature of the $\overline{\mathcal{T}}$ topology that the $T_{(0,n)}$ can be replaced with their $Mod_{\sigma_0-\tau_0}$ factors and the resulting sequence also converges. Finally since geodesics in a $CAT(0)$ space depend continuously on endpoints [**6**, Chap. II.1, Prop. 1.4], the appropriate geodesic segments from p_0 to q_n or to $T_{(0,n)}q_n$ converge to $\widehat{pq'}$, as claimed.

The second step is to show for a subsequence of the geodesics that maximal initial segments converge to segments of the prolongation of $\widehat{pq'}$ in the stratum τ_0. We now write $T_{(0,n)}\gamma'_n$ whether the transformations are trivial or not. We preliminarily note that the subsequence necessarily converges on a closed interval. In particular for a subsequence converging uniformly to a segment $\widehat{p_0q''} - \{q''\}$ and ϵ small, consider the point on $\widehat{p_0q''}$ distance ϵ before q''; a corresponding sequence of points, one on each $T_{(0,n)}\gamma'_n$, is determined. On each $T_{(0,n)}\gamma'_n$ consider the point ϵ further along than the referenced point; the resulting points are at distance 3ϵ from q'' for n large. The interval of convergence is indeed closed. There are now three possibilities for the interval: i) a subsequence $\{T_{(0,n)}\gamma'_n\}$ converges on the entire parameter interval $[t', t'']$ to a segment of the forward prolongation of $\widehat{p_0q'}$ (and the overall convergence argument is complete), ii) a subsequence $\{T_{(0,n)}\gamma'_n\}$ converges on $[t', t_1]$, $t_1 < t''$, to $\widehat{p_0p_1} - \{p_1\}$ and $\Lambda(p_1)$ properly succeeds τ_0, or iii) a subsequence $\{T_{(0,n)}\gamma'_n\}$ converges on $[t', t_*]$, $t_* < t''$, to $\widehat{p_0q''}$, with $\widehat{p_0q''}$ having a nontrivial forward prolongation in the stratum τ. We examine case iii). We examine the behavior of the subsequence $\{T_{(0,n)}\gamma'_n\}$ in a neighborhood of q''. We can again apply the argument from the beginning of the proof to determine elements $S_n \in Mod_{\tau_0}$ such that a subsequence of $\{S_n \circ T_{(0,n)}\gamma'_n\}$ converges in a neighborhood of q''. The limit is length-minimizing. From Proposition 11 the subsequence $\{S_n \circ T_{(0,n)}\gamma'_n\}$ converges to the prolongation of $\widehat{p_0q''}$; from the above observation concerning the $\overline{\mathcal{T}}$ topology and the Mod_{τ_0} action the subsequence $\{T_{(0,n)}\gamma'_n\}$ also converges to the prolongation. We now summarize the convergence considerations: for a maximal parameter interval of convergence of a subsequence $\{T_{(0,n)}\gamma'_n\}$ either: i) the parameter interval is $[t', t'']$, or ii) the interval is $[t', t_1]$, $t_1 < t''$, and the limit is $\widehat{p_0p_1}$ with $\Lambda(p_1)$ strictly succeeding $\Lambda(p_0) \cap \Lambda(p_1)$.

We now proceed and apply the considerations of the above two paragraphs to the subsequence $\{T_{(0,n)}\gamma'_n\}$ considered on the interval $[t_1, t'']$. The initial points converge to p_1. Elements $T_{(1,n)} \in Mod_{\sigma_1 - \tau_1}$ are determined such that a further subsequence $\{T_{(1,n)} \circ T_{(0,n)}\gamma'_n\}$ converges as in case i) or case ii). A stratum τ_1 is prescribed. The simplex σ_1 properly succeeds τ_1 for otherwise (from the observation concerning the $\overline{\mathcal{T}}$ topology and the Mod_* action) elements $T_{(1,n)}$ are not required and the subsequence $\{T_{(0,n)}\gamma'_n\}$ on $[t', t_1 + \epsilon)$ converges to a curve that is not length-minimizing by Proposition 11. We now note that Mod preserves the strata structure of $\overline{\mathcal{T}}$. Since the entire considerations including the initial-point convergence can be applied to the sequence $\{\gamma'_n\}$ *starting* from an arbitrary value t''' and proceeding in the negative t-direction, we observe that consequently a finite partition $t' = t_0 < t_1 < \cdots < t_k = t''$ is determined. Points $p_0, \ldots, p_k$; strata $\sigma_0, \ldots, \sigma_k, \tau_0, \ldots, \tau_{k-1}$; and sequences $\{\gamma_n\}$, $\{T_{(0,n)}\}, \ldots, \{T_{(k-1,n)}\}$ are determined. The desired properties are provided in the above construction with only two remaining matters: the sequences $\{T_{(j,n)}\}$, $j = 1, \ldots, k-1$, are unbounded and the length-minimizing property of the concatenation of the points $p_0, \ldots, p_k$.

We consider the length property first. A candidate length-minimizing curve is given as a concatenation $\mathcal{C} = \widehat{p_0q_1} \cup \widehat{q_1q_2} \cup \cdots \cup \widehat{q_{k-1}p_k}$ with $p_0 \in \mathcal{S}(\sigma_0)$ and $q_j \in \mathcal{S}(\sigma_j)$ for $j = 1, \ldots, k-1$. Since the group $Mod_{\sigma_j - \tau_j}$ stabilizes $\overline{\mathcal{S}(\sigma_j)}$ it follows that the concatenation $\mathcal{C}_T = T^{-1}_{(0,n)}\widehat{p_0q_1} \cup T^{-1}_{(0,n)} \circ T^{-1}_{(1,n)}\widehat{q_1q_2} \cup \cdots \cup T^{-1}_{(0,n)} \circ \cdots \circ T^{-1}_{(k-1,n)}\widehat{q_{k-1}p_k}$ is a continuous curve connecting p_0 to $p_{(k,n)} = (T_{(k-1,n)} \circ \cdots \circ T_{(0,n)})^{-1}p_k$. From the overall construction $d(\gamma_n(t''), p_{(k,n)})$ tends to zero; from [**6**, Chap. II.1, Prop.

1.4] the distance between γ_n and $\widehat{p_0p_{(k,n)}}$ is consequently small for n large, as claimed. The three curves γ_n, $\mathcal{C}_T$ and $\widehat{p_0p_{(k,n)}}$ each approximately connect p_0 and $p_{(k,n)}$. Since $L(\mathcal{C}) = L(\mathcal{C}_T)$, it follows that $L(\mathcal{C}) \geq \lim_n L(\gamma_n) = L'$. Thus each suitable concatenation has length at least L' with the minimum achieved for the arrangement of points $\{p_0, p_1, \ldots, p_k\}$. It remains for $k > 1$ to establish uniqueness. Consider geodesics $\alpha_j(s)$, parameterized on the unit-interval, connecting p_j to q_j, $j = 1, \ldots, k-1$. The concatenation $\widehat{p_0\alpha_1(s)} \cup \widehat{\alpha_1(s)\alpha_2(s)} \cup \cdots \cup \widehat{\alpha_{k-1}(s)p_k}$ satisfies the strata hypothesis and by Theorem 14 has length a convex function of the parameter s. Now for a $CAT(0)$ space either the distance from a point to a geodesic is a strictly convex function, or the point lies on the prolongation of the geodesic, [**6**, Chap. II.1, Defn. 1.1]. The points $\{p_0, p_1, q_1\}$ do not lie on a common geodesic since $\tau_0 = \Lambda(p_0) \cap \Lambda(p_1)$ strictly precedes σ_1 and $\sigma_1 \subset \Lambda(p_1) \cap \Lambda(q_1)$. In consequence either $L(\widehat{p_0\alpha_1(s)})$ is a strictly convex function or $\alpha_1(s)$ is constant. It follows that the length of a non constant family is a strictly convex function. The minimal-length concatenation is unique.

The final matter is the unboundedness of the sequences $\{T_{(j,n)}\}$. The elements $T_{(j,n)}$ are selected to provide that the initial curve segments lie in relatively compact sets. We can prescribe that each sequence $\{T_{(j,n)}\}$ is either trivial or unbounded. In a neighborhood of the parameter value t_j the concatenation $\mathcal{C}_j = \widehat{p_{j-1}p_j} \cup T^{-1}_{(j,n)}\widehat{p_jp_{j+1}}$ is the approximation to $T_{(j-1,n)} \circ \cdots \circ T_{(0,n)}\gamma_n$. Since γ_n is length-minimizing, it follows for n large that the concatenation $\mathcal{C}_j$ is arbitrarily close to length-minimizing. If $\{T_{(j,n)}\}$ were trivial then $\widehat{p_{j-1}p_j} \cup \widehat{p_jp_{j+1}}$ would be length-minimizing in contradiction of Theorem 13, since $p_j \in \mathcal{S}(\sigma_j)$ and σ_j strictly succeeds the stratum of one of the connecting geodesic segments. The proof is complete.

We now introduce terminology for the data for a convergent sequence of geodesics. Consider as above a convergent sequence $\{\gamma_n\}$ with data $\{p_j\}$ and $\{T_{(j,n)}\}$.

DEFINITION 24. *For a convergent sequence of geodesics $\{\gamma_n\}$ the vertices are the points p_j, $j = 0, \ldots, k$; the vertex concatenation is $\widehat{p_0p_1} \cup \widehat{p_1p_2} \cup \cdots \cup \widehat{p_{k-1}p_k}$ and an approximating concatenation is $T^{-1}_{(0,n)}\widehat{p_0p_1} \cup T^{-1}_{(0,n)} \circ T^{-1}_{(1,n)}\widehat{p_1p_2} \cup \cdots \cup T^{-1}_{(0,n)} \circ \cdots \circ T^{-1}_{(k-1,n)}\widehat{p_{k-1}p_k}$.*

We are ready to consider the matter of existence of axes for elements of Mod. We present a different approach towards certain results of G. Daskalopoulos and R. Wentworth [**13**]. They show that each irreducible (pseudo Anosov) mapping class has a unique invariant axis and that non commuting irreducible mapping classes have divergent axes. To provide a context we first recall the Thurston-Nielsen classification of mapping classes [**1**, Exposés 9, 11]. A mapping class is *irreducible* provided no power fixes the free homotopy class of a simple closed curve. A mapping class is precisely one of: periodic, irreducible or reducible, [**1**]. Reducible classes are first analyzed in terms of mappings of proper subsurfaces. For a reducible mapping class $[h]$ an invariant is $\sigma_{[h]}$ the maximal simplex fixed by a power. A general invariant of a transformation S is its translation length: $\inf_p d(p, Sp)$.

THEOREM 25. *A mapping class S acting on $\overline{\mathcal{T}}$ either has fixed-points or positive translation length realized on a closed, convex set $\mathcal{A}_S$, isometric to a metric space product $\mathbb{R} \times Y$. In the latter case the isometry S acts on $\mathbb{R} \times Y$ as the product of a translation of $\mathbb{R}$ and id_Y. For S irreducible the translation length is positive* *and $\mathcal{A}_S$ is a geodesic in $\mathcal{T}$. For S reducible the null stratum $\mathcal{S}(\sigma_{[h]})$ is a product of Teichmüller spaces $\prod \mathcal{T}' \times \prod \mathcal{T}''$ with a power S^m fixing the factors, acting by a product of: irreducible elements S' on $\mathcal{T}'$ with axis $\gamma_{S'}$ and the identity on each $\mathcal{T}''$; $\mathcal{A}_S \subset \prod \gamma_{S'} \times \prod \overline{\mathcal{T}}''$.*

Proof. The first matter is to establish that the translation length is realized. We consider a sequence of geodesics $\{\gamma_n'\}$ parameterized on $[a,b]$ with S connecting endpoints, $S(\gamma_n'(a)) = \gamma_n'(b)$, and $\lim_n L(\gamma_n')$ the translation length. Apply elements of Mod and according to Proposition 23 select a convergent subsequence $\{\gamma_n\}$ with vertices $\{p_0, \dots, p_k\}$; each γ_n has endpoints connected by a conjugate of S. For the special situation of translation length zero then $a = b$ and the vertex concatenation is the singleton $\{p_0\}$. The main matter is to determine the distance between the Mod orbit of p_0 and that of p_k. For an ϵ approximating concatenation to γ_n and a conjugate Q we have $Q(\gamma_n(a)) = \gamma_n(b)$ with $\gamma_n(a)$ within ϵ of p_0 and $\gamma_n(b)$ within ϵ of the prescribed endpoint $p_{(k,n)}$; it follows that $d(Q(p_0), p_{(k,n)}) < 2\epsilon$. It follows that the distance between the Mod orbit of p_0 and that of p_k is zero. The distance is also given by considering Mod translates of p_k in a neighborhood of p_0. From the preliminary discussion there is a positive lower bound for the distance between the points of the p_k orbit. It now follows that for a suitable ϵ and n' above, the distance inequality implies that $Q(p_0) = p_{(k,n')}$. For the value n' the approximating concatenation connects p_0 and $p_{(k,n')}$ and has length $\lim_n L(\gamma_n)$, the minimal translation length. It follows that there is only one geodesic segment in the concatenation and that Q realizes its translation length for the segment $\widehat{p_0 Q(p_0)}$. S realizes its translation length for an image of the segment.

By general considerations for positive translation length S realizes its translation length on axes in $\overline{\mathcal{T}}$, [**6**, Chap. II.6, Defn 6.3, Thrm. 6.8]. An axis is isometric to $\mathbb{R}$ and may not be unique; we consider specifics. S stabilizes each axis and thus stabilizes the stratum of an axis. Since in fact an irreducible mapping class only stabilizes the single stratum $\mathcal{T}$, it follows that an irreducible class is fixed-point free with axes in $\mathcal{T}$. Since axes are parallel and the distance between geodesics in a Teichmüller space is a strictly convex function, it follows that an irreducible axis is unique in $\overline{\mathcal{T}}$. Now for a reducible mapping class S a power P fixes a simplex if and only if the power fixes the corresponding null stratum. For the maximal simplex σ_S the geodesic-length sum $\mathcal{L}_S = \sum_{\alpha \in \sigma_S} \ell_\alpha$ restricts on each geodesic of $\overline{\mathcal{T}}$ to either the zero-function or a strictly convex function. Since $\mathcal{L}_S$ is P-invariant on an axis of P the restriction is the zero-function. Thus the axes of P are contained in the null stratum $\mathcal{S}(\sigma_S)$. On considering a power of P we can further arrange that P^m stabilizes the components $F - \cup_{\alpha\in\sigma_S}\{\alpha\}$ of the reference surface, and by [**1**, Exposé 11, Thrm 4.2] that on each component of $F - \cup_{\alpha\in\sigma_S}\{\alpha\}$ the restriction of P^m is either the trivial or an irreducible mapping class. For positive translation length at least one factor is an irreducible mapping class since by [**6**, Chap. II.6, Thrm. 6.8(2)] the translation length of the power is also positive. It follows that P^m realizes its translation length on a product $\mathcal{A}_{P^m} = \prod \gamma' \times \prod \overline{\mathcal{T}}''$ contained in the closure of $\mathcal{S}(\sigma_S)$. By [**6**, Chap. II.6, Thrm. 6.8(4)] $\mathcal{A}_{P^m}$ is isometric to a product $\mathbb{R} \times Y$ with S stabilizing the product and acting thereon by the product of a translation and a periodic element. $\mathcal{A}_{P^m}$ is itself a complete $CAT(0)$ space. It follows that Y is a complete $CAT(0)$ space and by [**6**, Chap. II.2, Cor. 2.8] that a periodic element has a non-empty closed convex fixed-point set. The proof is complete.

We recall notions regarding the behavior of geodesic rays. Unit-speed rays $\gamma(t)$, $\gamma'(t)$ are *asymptotic* provided the limit $\lim_{t\to\infty} d(\gamma(t),\gamma'(t))$ is zero and are *divergent* provided the limit is infinity. For values t, τ the points $\gamma(0)$, $\gamma'(0)$, $\gamma'(\tau)$ and $\gamma(t)$ determine a quadrilateral; on comparing side lengths we have that $|\tau - t| \leq d(\gamma(0),\gamma'(0)) + d(\gamma(t),\gamma'(\tau))$. Since also $d(\gamma(t),\gamma'(t)) \leq d(\gamma(t),\gamma'(\tau)) + |\tau - t|$ it follows on substituting for $|\tau - t|$ that $d(\gamma(t),\gamma'(t)) \leq 2d(\gamma(t),\gamma'(\tau)) + d(\gamma(0),\gamma'(0))$. In particular for divergent rays the distance from a point on one ray to the other ray tends to infinity with the point.

COROLLARY 26. *A geodesic ray in $\mathcal{T}$ and the axes of an irreducible mapping class are either asymptotic or divergent. Two irreducible mapping classes have axes that coincide or are divergent.*

Proof. We first consider a ray γ and an irreducible element S with axis γ_S. By Proposition 7 for unit-speed parameterizations the distance between corresponding points of γ and γ_S is a convex function. In particular the distance has a limit $L_\infty = \lim_{t\to\infty} d(\gamma(t),\gamma'(t))$. We will use that the WP geometry along γ_S is periodic to show that: either L_∞ is zero or there is a positive lower bound for the convexity of the distance and thus L_∞ is infinite.

We revisit formula (5) for the one-parameter variation $\beta(t,s)$ (t is now the parameter for the geodesics *connecting* γ_S to γ.) The integrand of (5) is bounded below by the contribution of the curvature term $-\langle R(V,T)T,V\rangle$, which in turn is non-negative. The curve $\beta(t,s')$, $t_0 \leq t \leq t_1$, is a geodesic with initial point on γ_S and length at least L_∞. T is the tangent field of $\beta(t,s')$ and V is a Jacobi field along $\beta(t,s')$ with initial vector having unit length. Now choose t', $t_0 < t' \leq L_\infty$ such that the neighborhood of radius $t' - t_0$ about a point of γ_S is relatively compact in $\mathcal{T}$. For the parameter range $t_0 \leq t \leq t'$ the geodesic segments $\beta(t,s')$, the tangent fields T, and the Jacobi fields V are all modulo the action of S supported on a compact set: the closure of the neighborhood of a fundamental segment of γ_S. Since the WP curvature is strictly negative on the compact set, we obtain a positive lower bound for the evaluations, the desired convexity bound for the distance function. It follows that L_∞ is either zero or infinite.

Consider next irreducible elements S, resp. Q, with translation lengths L_S, resp. L_Q, and axis γ_S, resp. γ_Q; assume the axes are asymptotic in the forward direction. Choose a reference point p' on γ_S. Given ϵ positive, choose positive integers n,m such that $|nL_S - mL_Q| < \epsilon$. Further choose a positive integer k_0 such that for $k \geq k_0$ the point $p = S^k p'$ on γ_S and the corresponding point q on γ_Q are at distance at most ϵ. We have then that $d(S^n p, Q^m q) < 2\epsilon$; we thus have that $d(Q^{-m}S^n p, p) < 3\epsilon$. Since there is a positive lower bound for the distance between distinct points of the Mod orbit of p', it follows for ϵ small that the transformation $Q^{-m}S^n$ fixes the sequence of points $S^k p'$, $k \geq k_0$. It follows that Q^m stabilizes γ_S. Since the axes are asymptotic, for p far along γ_S the displacement $d(Q^m p, p)$ is close to $d(Q^m q, q)$; Q^m realizes its translation length on γ_S; by the Theorem the axes coincide. The proof is complete.

References

[1] *Travaux de Thurston sur les surfaces.* Société Mathématique de France, Paris, 1991. Séminaire Orsay, Reprint of *Travaux de Thurston sur les surfaces*, Soc. Math. France, Paris, 1979 [MR 82m:57003], Astérisque No. 66-67 (1991).

[2] William Abikoff. Degenerating families of Riemann surfaces. *Ann. of Math. (2)*, 105(1):29–44, 1977.

[3] William Abikoff. *The real analytic theory of Teichmüller space.* Springer, Berlin, 1980.

[4] Lars V. Ahlfors. Some remarks on Teichmüller's space of Riemann surfaces. *Ann. of Math. (2)*, 74:171–191, 1961.

[5] Lipman Bers. Spaces of degenerating Riemann surfaces. In *Discontinuous groups and Riemann surfaces (Proc. Conf., Univ. Maryland, College Park, Md., 1973)*, pages 43–55. Ann. of Math. Studies, No. 79. Princeton Univ. Press, Princeton, N.J., 1974.

[6] Martin R. Bridson and André Haefliger. *Metric spaces of non-positive curvature.* Springer-Verlag, Berlin, 1999.

[7] Jeffrey F. Brock. The Weil-Petersson visual sphere. preprint, 2002.

[8] Jeffrey F. Brock. The Weil-Petersson metric and volumes of 3-dimensional hyperbolic convex cores. *J. Amer. Math. Soc.*, 16(3):495–535 (electronic), 2003.

[9] Jeffrey F. Brock and Benson Farb. Curvature and rank of Teichmüller space. preprint, 2001.

[10] Jeff Cheeger and David G. Ebin. *Comparison theorems in Riemannian geometry.* North-Holland Publishing Co., Amsterdam, 1975. North-Holland Mathematical Library, Vol. 9.

[11] Kevin Corlette. Archimedean superrigidity and hyperbolic geometry. *Ann. of Math. (2)*, 135(1):165–182, 1992.

[12] Georgios Daskalopoulos, Ludmil Katzarkov, and Richard Wentworth. Harmonic maps to Teichmüller space. *Math. Res. Lett.*, 7(1):133–146, 2000.

[13] Georgios Daskalopoulos and Richard Wentworth. Classification of Weil-Petersson Isometries. preprint, 2002.

[14] Benson Farb, Alexander Lubotzky, and Yair Minsky. Rank-1 phenomena for mapping class groups. *Duke Math. J.*, 106(3):581–597, 2001.

[15] Benson Farb and Howard Masur. Superrigidity and mapping class groups. *Topology*, 37(6):1169–1176, 1998.

[16] Étienne Ghys and Pierre de la Harpe. Le bord d'un arbre. In *Sur les groupes hyperboliques d'après Mikhael Gromov (Bern, 1988)*, volume 83 of *Progr. Math.*, pages 103–116. Birkhäuser Boston, Boston, MA, 1990.

[17] Alfred Gray. *Tubes.* Addison-Wesley Publishing Company Advanced Book Program, Redwood City, CA, 1990.

[18] Mikhail Gromov and Richard Schoen. Harmonic maps into singular spaces and p-adic superrigidity for lattices in groups of rank one. *Inst. Hautes Études Sci. Publ. Math.*, (76):165–246, 1992.

[19] Zheng Huang. Asymptotic Flatness of the Weil-Petersson Metric on Teichmüller Space. preprint, 2002.

[20] Y. Imayoshi and M. Taniguchi. *An introduction to Teichmüller spaces.* Springer-Verlag, Tokyo, 1992. Translated and revised from the Japanese by the authors.

[21] Nikolai V. Ivanov. The rank of Teichmüller modular groups. *Mat. Zametki*, 44(5):636–644, 701, 1988.

[22] Nikolai V. Ivanov. Automorphisms of complexes of curves and of Teichmüller spaces. In *Progress in knot theory and related topics*, volume 56 of *Travaux en Cours*, pages 113–120. Hermann, Paris, 1997.

[23] Nikolai V. Ivanov. Mapping class groups. In *Handbook of geometric topology*, pages 523–633. North-Holland, Amsterdam, 2002.

[24] Shoshichi Kobayashi and Katsumi Nomizu. *Foundations of differential geometry. Vol. II.* Interscience Publishers John Wiley & Sons, Inc., New York-London-Sydney, 1969.

[25] Kunihiko Kodaira. *Complex manifolds and deformation of complex structures.* Springer-Verlag, New York, 1986. Translated from the Japanese by Kazuo Akao, With an appendix by Daisuke Fujiwara.

[26] Nicholas J. Korevaar and Richard M. Schoen. Sobolev spaces and harmonic maps for metric space targets. *Comm. Anal. Geom.*, 1(3-4):561–659, 1993.

[27] Mustafa Korkmaz. Automorphisms of complexes of curves on punctured spheres and on punctured tori. *Topology Appl.*, 95(2):85–111, 1999.

[28] Feng Luo. Automorphisms of the complex of curves. *Topology*, 39(2):283–298, 2000.

[29] Bernard Maskit. Matrices for Fenchel-Nielsen coordinates. *Ann. Acad. Sci. Fenn. Math.*, 26(2):267–304, 2001.

[30] Howard Masur. Extension of the Weil-Petersson metric to the boundary of Teichmuller space. *Duke Math. J.*, 43(3):623–635, 1976.

[31] Howard Masur and Michael Wolf. The Weil-Petersson isometry group. *Geom. Dedicata*, 93:177–190, 2002.

[32] Curtis T. McMullen. The moduli space of Riemann surfaces is Kähler hyperbolic. *Ann. of Math. (2)*, 151(1):327–357, 2000.

[33] Barrett O'Neill. *Semi-Riemannian geometry*. Academic Press Inc. [Harcourt Brace Jovanovich Publishers], New York, 1983. With applications to relativity.

[34] Burton Randol. Cylinders in Riemann surfaces. *Comment. Math. Helv.*, 54(1):1–5, 1979.

[35] H. L. Royden. Intrinsic metrics on Teichmüller space. In *Proceedings of the International Congress of Mathematicians (Vancouver, B. C., 1974), Vol. 2*, pages 217–221. Canad. Math. Congress, Montreal, Que., 1975.

[36] A. J. Tromba. On a natural algebraic affine connection on the space of almost complex structures and the curvature of Teichmüller space with respect to its Weil-Petersson metric. *Manuscripta Math.*, 56(4):475–497, 1986.

[37] Michael Wolf. Infinite energy harmonic maps and degeneration of hyperbolic surfaces in moduli space. *J. Differential Geom.*, 33(2):487–539, 1991.

[38] Michael Wolf and Scott Wolpert. Real analytic structures on the moduli space of curves. *Amer. J. Math.*, 114(5):1079–1102, 1992.

[39] Scott Wolpert. Noncompleteness of the Weil-Petersson metric for Teichmüller space. *Pacific J. Math.*, 61(2):573–577, 1975.

[40] Scott Wolpert. The Fenchel-Nielsen deformation. *Ann. of Math. (2)*, 115(3):501–528, 1982.

[41] Scott A. Wolpert. Chern forms and the Riemann tensor for the moduli space of curves. *Invent. Math.*, 85(1):119–145, 1986.

[42] Scott A. Wolpert. Geodesic length functions and the Nielsen problem. *J. Differential Geom.*, 25(2):275–296, 1987.

[43] Scott A. Wolpert. Cut-and-paste deformations of Riemann surfaces. *Ann. Acad. Sci. Fenn. Ser. A I Math.*, 13(3):401–413, 1988.

[44] Scott A. Wolpert. The hyperbolic metric and the geometry of the universal curve. *J. Differential Geom.*, 31(2):417–472, 1990.

[45] Scott A. Wolpert. Spectral limits for hyperbolic surfaces. I, II. *Invent. Math.*, 108(1):67–89, 91–129, 1992.

[46] Sumio Yamada. Oral communication.

[47] Sumio Yamada. Weil-Petersson Completion of Teichmüller Spaces and Mapping Class Group Actions. preprint, 2001.

Made in the USA
Monee, IL
08 November 2026

68055341R00221